STATS
In Your World

DAVID E. BOCK
Ithaca High School (retired)

THOMAS J. MARIANO
Greece Central School District

With Contributions by

PAUL F. VELLEMAN
Cornell University

RICHARD D. DE VEAUX
Williams College

Addison Wesley

Boston   Columbus   Indianapolis   New York   San Francisco   Upper Saddle River
Amsterdam   Cape Town   Dubai   London   Madrid   Milan   Munich   Paris
Montreal   Toronto   Delhi   Mexico City   Sao Paulo   Sydney
Hong Kong   Seoul   Singapore   Taipei   Tokyo

Editor in Chief	Deirdre Lynch
Acquisitions Editor	Christopher Cummings
Senior Editor, AP and Electives	Andrea Sheehan
Associate Editor	Christine Lepre
Senior Content Ediror	Joanne Dill
Editorial Assistants	Dana Bettez and Sonia Ashraf
Senior Managing Editor	Karen Wernholm
Associate Managing Editor	Tamela Ambush
Senior Production Project Manager	Sheila Spinney
Senior Cover Designer	Barbara T. Atkinson
Digital Assets Manager	Marianne Groth
Senior Media Producer	Christine Stavrou
Software Development	Bob Carroll (MathXL) and Marty Wright (TestGen)
Executive Marketing Manager	Becky Anderson
Marketing Coordinator	Kathleen DeChavez
Senior Author Support/ Technology Specialist	Joe Vetere
Senior Prepress Supervisor	Caroline Fell
Manufacturing Manager	Evelyn Beaton
Senior Manufacturing Buyer	Carol Melville
Senior Media Buyer	Ginny Michaud
Production Coordination, Composition, and Illustrations	PreMediaGlobal
Interior Design	The Davis Group Inc
Cover Image	Cloud Formation in World Map Over Country Paved Road ©iStockphoto; City Skyline of Chicago, IL, USA ©Shutterstock

For permission to use copyrighted material, grateful acknowledgment has been made to the copyright holders listed in Appendix D, which is hereby made part of this copyright page.

Many of the designations used by manufacturers and sellers to distinguish their products are claimed as trademarks. Where those designations appear in this book, and Addison-Wesley was aware of a trademark claim, the designations have been printed in initial caps or all caps. TI-Nspire and the TI-Nspire logo are trademarks of Texas Instruments, Inc.

8-V082-14

Library of Congress Cataloging-in-Publication Data
Bock, David E.
   Stats in your World / Dave Bock, Thomas Mariano ; with contributions
by Richard De Veaux and Paul Velleman.—1st ed.
     p. cm.
   Includes index.
   ISBN 978-0-13-138489-7
   1. Mathematical statistics—Textbooks.   I. Mariano, Thomas J.   II. Title.
QA276.12.B6277 2012
519.5—dc22

2010020086

To Greg and Becca, great fun as kids and great friends as adults,
and especially to my wife and best friend, Joanna, for her
understanding, encouragement, and love
—Dave

To my beautiful wife Stacy, whose love and encouragement defies logic, to Benjamin,
our artist/dreamer who expands our world, to my hero, Seth, a constant source of
inspiration and pride, and to Abigail, the princess of my heart.
—Tom

To my sons, David and Zev, from whom I've learned so much,
and to my wife, Sue, for taking a chance on me.
—Paul

To Sylvia, who has helped me in more ways than she'll ever know,
and to Nicholas, Scyrine, Frederick, and Alexandra,
who make me so proud in everything that they are and do.
—Dick

Meet the Authors

David E. Bock taught mathematics at Ithaca High School for 35 years. He has taught Statistics at Ithaca High School, Tompkins-Cortland Community College, Ithaca College, and Cornell University. Dave has won numerous teaching awards, including the MAA's Edyth May Sliffe Award for Distinguished High School Mathematics Teaching (twice), Cornell University's Outstanding Educator Award (three times), and has been a finalist for New York State Teacher of the Year. Dave holds degrees from the University at Albany in Mathematics (B.A.) and Statistics/Education (M.S.).

Dave has been a Reader and Table Leader for the AP Statistics exam and recently served as K–12 Education and Outreach Coordinator and a senior lecturer for the Mathematics Department at Cornell University. He currently serves as a Statistics consultant to the College Board, leads workshops and institutes for AP Statistics teachers, teaches online Statistics cources, and mentors inner-city teachers and their students for the National Maths + Science Initiative. His understanding of how students learn guides much of this book's apporach.

Dave relaxes by biking, hiking, and attempting to play golf. He and his wife have enjoyed many days camping across Canada and through the Rockies. They have a son, a daughter, and four grandchildren.

Thomas J. Mariano has 15 years experience teaching middle and high school mathematics and computer science and eight years school administration experience in private and public schools in New York State and Puerto Rico. He also taught Statistics at Ithaca College. He is currently the Director of Mathematics and Science for the Greece Central School District in Rochester, New York where he oversees the implementation, professional development, and evaluation of the mathematics and science programs of 13 elementary schools and seven secondary schools. In this role, he works closely with teachers to develop and refine effective student-focused mathematics lessons that promote high expectations for all students, especially those with learning challenges. His most recent initiative is developing a cadre of classroom-embedded teacher coaches to increase the level of mathematical rigor in special education classroom settings. As a consultant with the National Science Foundation-funded COMPASS center at Ithaca College, Tom has trained teachers in reform- and inquiry-based elementary, middle, and high school mathematics programs and has presented at state and regional NCTM conferences.

Tom has a B.A. in Mathematics from Houghton College, and an M.S. in Curriculum and Instruction from SUNY Oswego. He holds School District and Building Administration Certificates and is currently an Ed D. student in Teaching and Curriculum at the Warner School of the University of Rochester. Tom enjoys teaching adults and children at his church, long walks with his wife, and trying to keep up with his three school-age children.

Paul F. Velleman has an international reputation for innovative Statistics education. He is the author and designer of the multimedia statistics CD-ROM *ActivStats*, for which he was awarded the EDUCOM Medal for innovative uses of computers in teaching statistics, and the ICTCM Award for Innovation in Using Technology in College Mathematics. He also developed the award-winning statistics program, Data Desk, and the Internet site Data And Story Library (DASL) (http://dasl.datadesk.com), which provides data sets for teaching Statistics. Paul's understanding of using and teaching with technology informs much of this book's approach.

Paul has taught Statistics at Cornell University since 1975. He holds an A.B. from Dartmouth College in Mathematics and Social Science, and M.S. and Ph.D. degrees in Statistics from Princeton University, where he studied with John Tukey. His research often deals with statistical graphics and data analysis methods. Paul coauthored (with David Hoaglin) *ABCs of Exploratory Data Analysis*. Paul is a Fellow of the American Statistical Association and of the American Association for the Advancement of Science.

Out of class, Paul sings baritone in a barbershop quartet. He is the father of two boys.

Richard D. De Veaux is an internationally known educator and consultant. He has taught at the Wharton School and the Princeton University School of Engineering, where he won a "Lifetime Award for Dedication and Excellence in Teaching." Since 1994, he has been Professor of Statistics at Williams College. Dick has won both the Wilcoxon and Shewell awards from the American Society for Quality. He is a fellow of the American Statistical Association. Dick is also well known in industry, where for the past 20 years he has consulted for such companies as Hewlett-Packard, Alcoa, DuPont, Pillsbury, General Electric, and Chemical Bank. He has also sometimes been called the "Official Statistician for the Grateful Dead." His real-world experiences and anecdotes illustrate many of this book's chapters.

Dick holds degrees from Princeton University in Civil Engineering (B.S.E.) and Mathematics (A.B.) and from Stanford University in Dance Education (M.A.) and Statistics (Ph.D.), where he studied with Persi Diaconis. His research focuses on the analysis of large data sets and data mining in science and industry.

In his spare time he is an avid cyclist and swimmer. He also is the founder and bass for the "Diminished Faculty," an a cappella Doo-Wop quartet at Williams College. Dick is the father of four children.

Contents

v

Preface

About This Book

We've created *Stats In Your World* out of a sense of civic duty. Seriously. We think there's no course that better prepares people for life than Statistics. Everyone is bombarded daily by charts and graphs, by data, by polls, by results of studies, and by assertions and claims made by people wanting to sell us things or convince us of something. The ability to sort out what's dubious (or even pure nonsense) from important and meaningful insights not only enlightens people, it allows them to make good decisions as consumers, as parents, and as citizens. Sure, Statistics is a math course, but it's also a course in critical thinking and civics that will prepare our students for greater success in this age of information.

This text presents Statistics as a key tool for thinking about the world. By leading with real-world examples, clear graphics, and practical data analysis, we get students "doing Statistics" quickly and "thinking statistically" right from the start. The questions that motivate our hundreds of examples and exercises highlight the wide applicability of Statistics, teach the methods and procedures of proper data analysis, and—most importantly—emphasize thinking about what the results mean.

We have modeled this textbook on *Stats: Modeling the World*, the highly successful AP* Statistics book that's winning praise from both students and teachers. We've been thrilled by their feedback. Teachers are delighted that students can and do learn from reading the text, and students write to tell us that they find the book easy and (to their amazement) even enjoyable to read. With a new audience now in mind, we've worked hard to enhance readability even further, to offer greater support for the underlying math skills, to eliminate some of the more advanced topics while maintaining those that develop broad-based statistical literacy, to develop scaffolded materials that enable success for all students, and to clearly tie all the skills and concepts to the world we live in. In this course and with this textbook, you can say goodbye to that oft-uttered math class question, "When does anyone ever use this stuff?"!

What Stats Course?

Given the previously-outlined benefits, were it up to us, we'd make taking a Statistics course a requirement for high school graduation. But what should that course look like? Well, that depends on several factors, among them the math backgrounds of the students and whether it's to be a one-semester or full-year course. By selecting appropriate topics and pacing, teachers can use this textbook for a wide variety of courses. If your course includes...

- *Basic data analysis*—choose Chapters 1–8 to cover graphical displays (pie charts, bar graphs, stem-and-leaf plots, boxplots, histograms, and scatterplots), summary statistics (mean, median, range, standard deviation, correlation), lines of best fit, and use of the Normal curve;

- *Curve fitting*—add Chapter 9 to look at exponential and power models;

- *Data collection*—include Chapter 10 (sampling, polls, and surveys), Chapter 11 (observational studies and experiments), and Chapter 12 (simulations);

- *Probability*—choose Chapter 13 (sample spaces and counting strategies) and Chapter 14 (probability basics), perhaps enriching your course with Chapter 15 (probability rules, independence, Venn and tree diagrams) and Chapter 16 (expected value and binomial probability);

- *Inference Concepts*—explore inferences about a population proportion with Chapter 17 (confidence intervals) and Chapter 18 (hypothesis tests);

- *More advanced inference*—add Chapter 19 (inference techniques for a population mean), Chapter 20 (inferences for the difference of two proportions or means), and/or Chapter 21 (χ^2 procedures).

We've been guided in the choice and order of topics by several fundamental principles. First, we have tried to ensure that each new topic fits into the growing structure of understanding that we hope students will build. We have worked to provide scaffolded materials to help each class, in its own way, follow the guidelines of the GAISE (Guidelines for Assessment and Instruction in Statistics Education) project sponsored by the American Statistical Association. That report urges that Statistics education should:

1. emphasize Statistical literacy and develop Statistical thinking,
2. use real data,
3. stress conceptual understanding rather than mere knowledge of procedures,
4. foster active learning,
5. use technology for developing concepts and analyzing data, and
6. make assessment a part of the learning process.

Our Goal: Read This Book!

The best text in the world is of little value if students don't read it. Here are some of the ways we have tried to encourage students to read *Stats In Your World*:

- *Readability.* You'll see immediately that this book doesn't read like other math texts. The style, both colloquial and informative, engages students to actually read the book to see what it says.

- *Humor.* We know that humor is a great way to promote interest and learning. You'll find quips and wry comments sprinkled through the narrative, margin notes, and footnotes, as well as a number of cartoons.

- *Informality.* Don't let our informal diction fool you. It doesn't mean that the subject matter is covered lightly or sloppily. You'll find we've been quite precise and that the narrative offers sound explanations and justifications.

- *Consistency*. We've worked hard to avoid the "do what we say, not what we do" trap. From the very start, we teach the importance of plotting data and checking assumptions and conditions to ensure a statistical method will produce reliable results. We have been careful to model that behavior throughout the book.

- ***The benefit of reading.*** Most high school students have not yet learned the value of reading a textbook—and especially not a math book! We've worked hard to make the text engaging, and to weave important concepts, definitions, and sample solutions into a narrative that will reward students who read the book.

Features

We have written *Stats In Your World* with several themes and pedagogical features in mind.

▶ ***Think, Show, Tell.*** We repeat the mantra of *Think*, *Show*, and *Tell* in every chapter, providing a consistent framework for how to do Statistics. *Think* emphasizes the importance of clearly understanding a Statistics question (What do we know? What do we hope to learn? What procedure should we use? Are the assumptions and conditions satisfied?). The *Show* step contains the mechanics of calculating results and conveys the message that number-crunching is only one part of the process. *Tell* reminds students that interpreting the findings is where the action is, reporting what we have learned about the world.

▶ ***Do The Math.*** We often offer explanations and some practice exercises to allow students to review and master important arithmetic or algebra skills they'll need in order to perform a statistical procedure successfully. Answers are at the end of the chapter's exercise sets so students can easily check themselves.

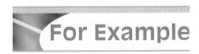

▶ ***For Example.*** In every chapter, you'll find worked examples that illustrate how to apply each new concept and methods. As we move through the chapter, each *For Example* further develops a statistical analysis, picking up the story and moving it forward as students learn to do each new step.

▶ ***Just Checking.*** After each new skill or concept has been presented, we ask students to pause and think about what they've just learned. These questions are designed to allow a quick check that they understand the material. These answers also appear at the end of the chapter's exercise sets.

▶ ***Step-by-Step Worked Examples.*** When students have mastered the individual steps, we present the big picture: a complete statistical analysis following our *Think-Show-Tell* rubric. These *Step-by-Step* examples ask students real-world questions, then guide them through the process by developing answers in a unique and pedagogically effective two-column format: a general explanation of what to do on the left paralleling each worked-out step on the right. The result is better understanding of statistical concepts and procedures, not just number crunching.

▶ ***TI Tips.*** We emphasize sound understanding of formulas and methods, but we want students to use technology for actual calculations. Easy-to-read *TI Tips* in the chapters show students how to use TI-83/84 Plus® statistics functions. (Help using a TI-89 or TI-Nspire appears in Appendix B.) We do remind students that calculators are just for *Show*—the technology cannot *Think* about what to do nor *Tell* what it all means.

▶ ***What Can Go Wrong?*** Each chapter contains innovative *What Can Go Wrong?* alerts highlighting the most common errors people make and the misconceptions they have about Statistics. Our goals are to help students avoid these pitfalls, and to arm them with the tools to detect statistical errors and to debunk misuses of statistics, whether intentional or not.

▶ *In Your World.*   Each chapter closes with a recent news story or magazine article that applies the new Statistics knowledge in an interesting context, amplifying the book's central message that Statistics is about the real world.

WHAT HAVE WE LEARNED?

▶ *What Have We Learned?*   These chapter-ending summaries are great study guides providing complete overviews that highlight the new concepts, define the new terms, and list the skills that the student should have acquired in the chapter.

EXERCISES

▶ *Exercises.*   We've created exercises at three levels of complexity. The **A** exercises offer straightforward practice of individual concepts or skills. The **B** exercises ask students to combine several steps in a more complete analysis. Teachers who are looking for more challenging exercises that enrich student understanding will find them in the **C** sections. Throughout, examples are paired so that each odd-numbered exercise (with an answer in the back of the book) is followed by an even-numbered exercise about the same concept. Almost all exercises are based on real-world studies and data sets, the data often included in the exercise or available on the CD (in the Teacher's Edition) or at the book's website.

NOTATION ALERT

▶ *Notation Alerts.*   Throughout this book we emphasize the importance of clear communication, and proper notation is part of the vocabulary of Statistics. We've found that it helps students when we call attention to the letters and symbols statisticians use to mean very specific things.

ON THE COMPUTER

▶ *On the Computer.*   In the real world Statistics is implemented on computers, so at the end of each chapter we show an annotated example of software output, usually based on StatCrunch®, a great online stats package that nicely complements this course.

TI-nspire

▶ *TI-Nspire Activities.*   We've created many demonstrations and investigations for TI-Nspire handhelds to enhance each chapter. They're on the book's web site www.pearsonhighered.com/bock.

Mathematics, Technology, and Data

This book is not concerned with proving theorems about Statistics, nor do we present concocted data sets or ask students to do tedious calculations by hand. To experience the power of Statistics it's best to explore real data sets using modern technology.

▶ *Technology.*   We assume that you will use some form of technology in your Statistics course. That could be a calculator, a spreadsheet, or a statistics package. In each chapter we present easy-to-understand instructions for using a TI-83/84, and we explain typical computer output. In Appendix B, we offer general guidance (organized by chapter) to help students use StatCrunch, Excel®, a TI-89 calculator, or a TI-Nspire.

▶ *Data.*   Because we use technology for computing, we don't limit ourselves to small, artificial data sets. In addition to including some small data sets, we have built our examples and exercises on real data. Many of these data are included on the CD (in the Teacher's Edition) as well as on the book's website www.pearsonhighered.com/bock.

Supplements

Student Supplements

The following supplements are available for purchase:

Statistics Study Card is a resource for students containing important formulas, definitions, and tables that correspond precisely to the De Veaux/Velleman/Bock Statistics series. This card can work as a reference for completing homework assignments or as an aid in studying.

Graphing Calculator Tutorial for Statistics will guide students through the keystrokes needed to most efficiently use their graphing calculator. Although based on the TI-84 Plus Silver Edition, operating system 2.30, the keystrokes for this calculator are identical to those on the TI-84 Plus, and very similar to the TI-83 and TI-83 Plus. This tutorial should be helpful to students using any of these calculators, though there may be differences in some lessons. The tutorial is organized by topic.

Teacher Supplements

Some of the teacher supplements and resources available for this text are available electronically for download on the Instructor Resource Center (IRC) to qualified adopters. Upon adoption or to preview the online resources, please go to www.PearsonSchool .com/Access_Request and select "Online Teacher Supplements." You will be required to complete a one-time registration subject to verification. Upon verification of educator status, access information and instructions will be sent via email.

Teacher's Edition contains answers to all exercises. Packaged with the Teacher's Edition is the Teacher's Resource CD. The Teacher's Resource CD includes the files for the Teachers' Solutions Manual, a Resource Guide with Worksheets, and a Test Bank.

Teacher's Resource Guide with Worksheets by Thomas J. Mariano, contains chapter-by-chapter comments on the major concepts, tips on presenting topics (and what to avoid), teaching examples, suggested assignments, Web links and lists of other resources. Also provided are individual and group performance tasks to give students realistic settings in which to apply and demonstrate their new skills. Structured classroom activities and worksheets are included to scaffold student learning. Finally, end of chapter review sheets are provided at three levels to meet the needs of a broad range of students. This is an indispensable guide to help teachers prepare for class. The Resource Guide with Worksheets is on the Teacher's Resource CD and available for download.

Teacher's Solutions Manual, by William B. Craine III, contains detailed solutions to all of the exercises. The Solutions Manual is on the Teacher's Resource CD and available for download.

Test Bank (Download Only), by Thomas J. Mariano, contains one to two quizzes per chapter and a chapter test, each at three levels to work in any classroom. The Test Bank is on the Teacher's Resource CD and available for download.

TestGen® enables teachers to build, edit, print, and administer tests using a computerized bank of questions developed to cover all the objectives of the text. TestGen is algorithmically-based, allowing teachers to create multiple but equivalent versions of the same question or test with the click of a button. Teachers can also modify test bank questions or add new questions. Tests can be printed or administered online. TestGen is also available for download.

Technology Resources

Teacher's Resource CD, packaged with every new Teacher's Edition, includes the Teacher's Solutions Manual, Teacher's Resource Guide with Worksheets, Test Bank, Statistics Study Card, and Graphing Calculator Tutorial for Statistics. Data for exercises marked are also available and at www.pearsonhighered.com/bock formatted for StatCrunch, Excel, and the TI calculators, and as text files suitable for these and virtually any other statistics software.

MathXL® for School is a powerful online homework, tutorial, and assessment system that accompanies Pearson textbooks in Math and Statistics. With MathXL for School, teachers can create, edit, and assign online homework and tests using algorithmically-generated exercises correlated at the objective level to the textbook. They can also create and assign their own online exercises and import TestGen tests for added flexibility. All student work is tracked in MathXL for School's online gradebook. Students can take chapter tests in MathXL for School and receive a personalized study plan based on their test results. The study plan diagnoses weaknesses and links students directly to tutorial exercises for the objectives they need to study and retest. Students can also access interactive learning aids directly from selected exercises. For more information options for purchasing MathXL for School visit our website at www.MathXLforSchool.com, or contact your Pearson sales representative.

StatCrunch is web-based statistical software that allows users to perform complex analyses, share data sets, and generate compelling reports of their data. Users can upload their own data to StatCrunch or search the library of over twelve thousand publicly shared data sets, covering almost any topic of interest.

Interactive graphical outputs help users understand statistical concepts, and are available for export to enrich reports with visual representations of data. Additional features include:

- A full range of numerical and graphical methods that allow users to analyze and gain insights from any data set.
- Reporting options that help users create a wide variety of visually-appealing representations of their data.
- An online survey tool that allows users to quickly build and administer surveys via a web form.

StatCrunch requires an access code, which is available for purchase. For more information, visit our website at www.statcrunch.com or contact your Pearson representative.

Flexible Content Solutions Pearson offers a 100% digital solution for many of our products, please contact your Pearson representative for more information.

Companion website (www.pearsonhighered.com/bock) provides additional resources for teachers and student.

Acknowledgments

Many people have contributed to the development of this book. Great writing and keen insights from Dick De Veaux and Paul Velleman helped fuel the success of Stats: Modeling the World and are inherited here. This edition would have never seen the light of day without the assistance of the incredible team at Pearson. Our editor in chief, Deirdre Lynch, was central to the genesis, development, and realization of the book from day one. Chris Cummings, acquisitions editor, provided much needed support. Joanne Dill, senior content editor, kept us on task as much as humanly possible. Sheila Spinney, senior production project manager, kept the cogs from getting into the wheels where they often wanted to wander. Christina Lepre, assistant editor, Dana Bettez, editorial assistant, and Kathleen DeChavez, marketing assistant, were essential in managing all of the behind-the-scenes work that needed to be done. Christine Stavrou, senior media producer, put together a top-notch media package for this book. Barbara T. Atkinson, senior designer, and Geri Davis are responsible for the wonderful way the book looks. Carol Melville, senior manufacturing buyer, and Ginny Michaud, senior media buyer, worked miracles to get this book and CD in your hands, and Greg Tobin, publisher, was supportive and good-humored throughout all aspects of the project. Special thanks go out to PreMediaGlobal, the compositor, for the wonderful work they did on this book, and in particular to Laura Hakala, senior project manager, for her close attention to detail. We'd also like to thank our accuracy checkers whose monumental task was to make sure we said what we thought we were saying. They are Douglas Cashing, St. Bonaventure University; Rich Crecelius, Everglades High School; Jared Derksen, Rancho Cucamonga High School; John Diehl, Hinsdale Central High School; Nathan Kidwell, Holly High School; Stan Seltzer, Ithaca College.

We extend our sincere thanks for the suggestions and contributions made by the following reviewers:

John Arko,
Glenbrook South High School, IL

Scott Armstrong,
Andover High School, MA

Kathleen Arthur,
Shaker High School, NY

Allen Back,
Cornell University, New York

Kathleen Barnes,
Oxford High School, MS

Tom Beatini,
Glen Rock High School, NJ

Beverly Beemer,
Ruben S. Ayala High School, CA

Judy Bevington,
Santa Maria High School, CA

Susan Blackwell,
First Flight High School, NC

John Bowman,
Hinsdale Central High School, IL

Gwendolyn Bright,
Wilbur Cross High School, CT

Gail Brooks,
McLennan Community College, TX

Walter Brown,
Brackenridge High School, TX

Darin Clifft,
Memphis University School, TN

Lawrence Cohan,
Hyde Park Academy, IL

Bill Craine,
Lansing High School, NY and Ithaca High School, NY

Sybil Coley,
Woodward Academy, CA

Rich Crecelius,
Everglades High School, FL

Kevin Crowther,
Lake Orion High School, MI

Caroline DiTullio,
Summit High School, NJ

Jared Derksen,
Rancho Cucamonga High School, CA

Sam Erickson,
North High School, Wisconsin

Laura Estersohn,
Scarsdale High School, NY

Elissa Farmer,
Garfield High School, WA

Laura Favata,
Niskayuna High School, NY

Katie Fay,
Williston Northampton School, MA

David Ferris,
Noblesville High School, IN

Matt Freeman,
Madeira High School, OH

Linda Gann,
Sandra Day O'Connor High School, TX

Robert Gerver,
North Shore High School, NY

Bill Gillam,
Blythewood High School, SC

Randall Groth,
Illinois State University, IL

Donnie Hallstone,
Green River Community College, WA

Howard W. Hand,
St. Marks School of Texas, TX

Bill Hayes,
Foothill High School, CA

Miles Hercamp,
New Palestine High School, IN

Michelle Hipke,
Glen Burnie Senior High School, MD

Carol Huss,
Independence High School, NC

Karen Hynes,
East Lake High School, FL

Sam Jovell,
Niskayuna High School, NY

Peter Kaczmar,
Lower Merion High School, PA

John Kotmel,
Lansing High School, NY

Lee E. Kucera,
Capistrano Valley High School, CA

Beth Lazerick,
St. Andrews School, FL

Michael Legacy,
Greenhill School, TX

Guillermo Leon,
Coral Reef High School, Florida

John Lieb,
The Roxbury Latin School, MA

Martha Lowther,
The Tatnall School, Delaware

John Maceli,
Ithaca College, NY

Jim Miller,
Alta High School, UT

Timothy E. Mitchell,
King Philip Regional High School, MA

Maxine Nesbitt,
Carmel High School, IN

Cindy Percival,
Roosevelt High School, IA

Elizabeth Ann Przybysz,
Dr. Phillips High School, FL

Diana Podhrasky,
Hillcrest High School, TX

Rochelle Robert,
Nassau Community College, NY

Karl Ronning,
Davis Senior High School, California

Bruce Saathoff,
Centennial High School, CA

Donald Saunders,
King George High School, VA

Eric Sever,
Mill Creek High School, GA

Agatha Shaw,
Valencia Community College, FL

Murray Siegel,
Sam Houston State Universtity, TX

Chris Sollars,
Alamo Heights High School, TX

John Souris,
Oakville High School, IL

Darren Starnes,
The Webb Schools, CA

Josh Tabor,
Canyon del Oro High School, AZ

Katie Thayer,
Heritage High School, VA

Index of Applications

Surveys and Opinion Polls

Technology

Transportation

Exploring and Understanding Data

Stats Starts Here[1]

S tatistics gets no respect. People say things like "You can prove anything with Statistics." People will write off a claim based on data as "just a statistical trick." And Statistics courses don't have the reputation of being fun.

But Statistics *is* fun. That's probably not what you heard on the street, but it's true. Statistics is about how to think clearly with data. A little practice thinking statistically is all it takes to start seeing the world more clearly and accurately.

So, What Is (Are?) Statistics?

It seems every time we turn around, someone is collecting data on us, from every purchase we make in the grocery store, to every click of our mouse as we surf the Web. The United Parcel Service (UPS) tracks every package it ships from one place to another around the world and stores these records in a giant database. What can anyone hope to do with all these data?

---

[1]This chapter might have been called "Introduction," but nobody reads the introduction, and we wanted you to read this. We feel safe admitting this here, in the footnote, because nobody reads footnotes either.

Statistics plays a role in making sense of the complex world in which we live. Statisticians judge the risk of genetically engineered foods or of a new drug being considered by the Food and Drug Administration (FDA). They predict the number of new cases of AIDS for each region of the country or the number of customers likely to respond to a sale at the mall. And statisticians help scientists and social scientists understand how unemployment is related to environmental controls, whether attending preschool affects later performance of children, and whether vitamin C really prevents illness. Whenever there are data and a need for understanding the world, you need Statistics.

This book will help you think clearly about the questions under investigation, find out what the data are saying, and tell clearly what it all means.

> Q: What is Statistics?
> A: Statistics is a way of reasoning, along with a collection of tools and methods, designed to help us understand the world.
> Q: What are statistics?
> A: Statistics (plural) are particular calculations made from data.
> Q: So what is data?
> A: You mean, "what *are* data?" Data is the plural form. The singular is datum.
> Q: OK, OK, so what are data?
> A: Data are values along with their context.

> The ads say, "Don't drink and drive; you don't want to be a statistic." But you can't be a statistic.
>   We say: "Don't be a datum."

FRAZZ reprinted by permission of United Feature Syndicate, Inc.

Statistics in a Word

> Statistics is about variation.
>   Data vary because we don't see everything and because even what we do see and measure, we measure imperfectly.
>   So, in a very basic way, Statistics is about the real, imperfect world in which we live.

It can be fun, and sometimes useful, to summarize a discipline in only a few words. So,

Economics is about . . . *Money (and why it is good)*.

Psychology: *Why we think what we think (we think)*.

Biology: *Life*.

Anthropology: *Who?*

History: *What, where, and when?*

Philosophy: *Why?*

Engineering: *How?*

Accounting: *How much?*

In such a caricature, Statistics is about . . . **Variation.**

Data vary. People are different. We can't see everything, let alone measure it all. And even what we do measure, we measure imperfectly. So the data we look at and base decisions on provide, at best, an imperfect picture of the world. How to make sense of it is what Statistics is all about.

So, How Will This Book Help?

A fair question. Most likely, this book will not turn out to be quite what you expected.

What's different?

Close your eyes and open the book to a page at random. Is there a graph or table on that page? Do that again, say, 10 times.

We'll bet you saw data displayed in many different ways. We can better understand data by making pictures.

You looked at only a few randomly selected pages to get an impression of the entire book. We'll see soon that doing so was sound Statistics practice and reasoning.

Next, pick a chapter and read the first two sentences. (Go ahead; we'll wait.)

We'll bet you didn't see anything about Statistics. Why? Because the best way to understand Statistics is to see it at work. In this book, chapters usually start by presenting a story and posing questions. That's when Statistics really gets down to work.

There are three simple steps to doing Statistics right: *think, show,* and *tell:*

Think first. Know where you're headed and why. It will save you a lot of work.

Show is what most folks think Statistics is about. The *mechanics* of calculating statistics and making displays is important, but not the most important part of Statistics.

Tell what you've learned. Until you've explained your results so that someone else can understand your conclusions, the job is not done.

Of course, doing Statistics often rests on some important math skills. As these come up we'll offer you a few **Do the Math** review and practice problems, with answers at the end of the chapter's exercises so you can check yourself quickly.

The best way to learn new statistical skills is to take them out for a spin. At important points in each chapter, we've put a section called **Just Checking**. There you'll see a few short questions you can answer— a quick way to check to see if you've understood the basic ideas. You'll find these answers at the end of the chapter's exercises, too. In **For Example** boxes you'll see brief ways to apply new ideas and methods as you learn them. The ultimate goal is to put it all together **Step-by-Step.** These fully worked solutions come side by side with explanations, showing you the way statisticians attack and solve problems. You'll see how to *Think* about the problem, what to *Show,* and how to *Tell* what it all means.

> **Time out.** From time to time, we'll take time out to discuss an interesting or important side issue. We indicate these by setting them apart like this.[2]

TI Tips   Do statistics on your calculator!

> How do I use
> this thing?

Although we'll show you all the formulas you need to understand the calculations, you will most often use a calculator or computer to do the mechanics of a statistics problem. Your graphing calculator has a specialized program called a "statistics package." Each chapter's **TI Tips** teach you how to use it (and duck most of the messy calculations).

---

[2]Or in a footnote.

"Get your facts first, and then you can distort them as much as you please. (Facts are stubborn, but statistics are more pliable.)"

—Mark Twain

WHAT HAVE WE LEARNED?

ON THE COMPUTER

You'll find all sorts of stuff in margin notes, such as stories and quotations. For example:

"Computers are useless. They can only give you answers."

—Pablo Picasso

While Picasso underestimated the value of good statistics software, he did know that creating a solution requires more than just *Showing* an answer—it means you have to *Think* and *Tell,* too!

If you are using TI-Nspire™ technology, these margin icons will alert you to demonstrations that can help you understand important ideas in the text. If you have the DVD that's available with this book, you'll find these demos there; if not, they're also available on the book's website http://www.pearsonhighered.com/bock.

One of the interesting challenges of Statistics is that, unlike in some math and science courses, there can be more than one right answer. That's why two statisticians can testify honestly on opposite sides of a court case. And it's why some people think that you can prove anything with statistics. But that's not true. People make mistakes using statistics, sometimes on purpose to mislead others. Most of the unintentional mistakes people make, though, can be avoided. We're not talking about arithmetic. More often, mistakes come from using a method in the wrong situation or misinterpreting the results. Each chapter has a section called **What Can Go Wrong?** to help you avoid some of the most common mistakes.

All around the country and throughout the world, people use Statistics every day. Each chapter of this book concludes by showing you an example of the important statistical ideas at work **... In Your World.**

At the end of each chapter, there's a brief summary of the important concepts in a section called **What Have We Learned?** That section includes a list of the **Terms** and a summary of the important **Skills** you've acquired in the chapter. You won't be able to learn the material from these summaries, but you can use them to check your knowledge of the important ideas in the chapter. If you have the skills, know the terms, and understand the concepts, you should be ready to use Statistics!

Most people actually do the messy work of Statistics **... On the Computer**, so in many chapters just before the exercises we'll show you what to look for in typical output. You can try it yourself using *StatCrunch*, available at http://www.pearsonhighered.com/bock.

Beware: No one can learn Statistics just by reading or listening. The only way to learn it is to do it. So, of course, at the end of each chapter (except this one) you'll find **Exercises** designed to help you learn to use the Statistics you've just read about.

Some exercises are marked with an orange Ⓣ. You'll find the data for these exercises on the DVD in the back of the book or on the book's website at http://www.pearsonhighered.com/bock.

You'll find answers to the odd-numbered exercises at the back of the book. But these are only "answers" and not complete "solutions." Huh? What's the difference? The answers are sketches of the complete solutions. If your calculations match the numerical parts of the "answer," you're on the right track. Your complete solution should follow the model of the Step-By-Step Examples: explain the context, show your reasoning and calculations, and state your conclusions. Don't fret too much if your numbers don't match the printed answers to every decimal place. Statistics is more about getting the reasoning correct—pay more attention to how you interpret a result than what the digit in the third decimal place was.

Onward!

It's only fair to warn you: You can't get there by just picking out the highlighted sentences and the summaries. This book is different. It's not about memorizing definitions and learning equations. It's deeper than that. And much more fun. But . . .

You have to read the book![3]

---

[3]So, turn the page.

2

What Are Data?

Many years ago, most stores in small towns knew their customers personally. If you walked into the hobby shop, the owner might tell you about a new bridge that had come in for your Lionel train set. The tailor knew your dad's size, and the hairdresser knew how your mom liked her hair. There are still some stores like that around today, but we're increasingly likely to shop at large stores, by phone, or on the Internet. Even so, when you phone an 800 number to buy new running shoes, customer service representatives may call you by your first name or ask about the socks you bought 6 weeks ago. Or the company may send an e-mail in October offering new head warmers for winter running. This company has millions of customers, and you were recognized without identifying yourself. How did the sales rep know who you are, where you live, and what you had bought?

The answer is data. Collecting data on their customers, transactions, and sales lets companies track their inventory and helps them predict what their customers prefer. These data can help them predict what their customers may buy in the future so they know how much of each item to stock. The store can use the data and what it learns from the data to improve customer service, mimicking the kind of personal attention a shopper had 50 years ago.

"Data is king at Amazon. Clickstream and purchase data are the crown jewels at Amazon. They help us build features to personalize the website experience."

—Ronny Kohavi,
Director of Data Mining and
Personalization, Amazon.com

Amazon.com opened for business in July 1995, billing itself as "Earth's Biggest Bookstore." By 1997, Amazon had a catalog of more than 2.5 million book titles and had sold books to more than 1.5 million customers in 150 countries. In 2007, the company's revenue reached $14.8 billion. Amazon has expanded into selling a wide selection of merchandise, from $400,000 necklaces[1] to yak cheese from Tibet to the largest book in the world.

Amazon is constantly monitoring and evolving its website to serve its customers better and maximize sales performance. To decide which changes to make to the site, the company experiments, collecting data and analyzing what works best. When you visit the Amazon website, you may encounter a different look or different suggestions and offers. Amazon statisticians want to know whether you'll follow the links offered, purchase the items suggested, or even spend a longer time browsing the site. As Ronny Kohavi, director of Data Mining and Personalization, said, "Data trumps intuition. Instead of using our intuition, we experiment on the live site and let our customers tell us what works for them."

But What *Are* Data?

We bet you thought you knew this instinctively. Think about it for a minute. What exactly *do* we mean by "data"?

Do data have to be numbers? The amount of your last purchase in dollars is numerical, but some data are names or other labels. The names in Amazon.com's database are data, but not numerical.

Sometimes, data can have values that look like numerical values but are just numerals serving as labels. This can be confusing. For example, the ASIN (Amazon Standard Item Number) of a book, like 0321599055, may have a numerical value, but it's really just another name for *Stats in Your World*.

THE W'S:
: WHO
: WHAT
: and in what units
: WHEN
: WHERE
: WHY
: HOW

Data values, no matter what kind, are useless without their context. Newspaper journalists know that the lead paragraph of a good story should establish the "Five W's": *Who, What, When, Where,* and (if possible) *Why.* Often we add *How* to the list as well. Answering these questions can provide the **context** for data values. The answers to the first two questions are essential. If you can't answer *Who* and *What,* you don't have **data,** and you don't have any useful information.

Data Tables

Here are some data Amazon might collect:

B0000010AA	10.99	Chris G.	902	15783947	15.98	Kansas	Illinois	Boston
Canada	Samuel P.	Orange County	N	B000068ZVQ	Bad Blood	Nashville	Katherine H.	N
Mammals	10783489	Ohio	N	Chicago	12837593	11.99	Massachusetts	16.99
312	Monique D.	10675489	413	B00000I5Y6	440	B000002BK9	Let Go	Y

---

[1]Please get credit card approval before purchasing online.

Try to guess what they represent. Why is that hard? Because these data have no *context*. If we don't know *Who* they're about or *What* they measure, these values are meaningless. We can make the meaning clear if we organize the values into a **data table** such as this one:

Purchase Order	Name	Ship to State/Country	Price	Area Code	Previous CD Purchase	Gift?	ASIN	Artist
10675489	Katharine H.	Ohio	10.99	440	Nashville	N	B00000I5Y6	Kansas
10783489	Samuel P.	Illinois	16.99	312	Orange County	Y	B000002BK9	Boston
12837593	Chris G.	Massachusetts	15.98	413	Bad Blood	N	B000068ZVQ	Chicago
15783947	Monique D.	Canada	11.99	902	Let Go	N	B000001OAA	Mammals

Now we can see that these are Amazon's records of four CD purchases. The column titles tell *What* has been recorded. The rows tell us *Who*. But be careful. Look at all the variables to see *Who* the data are about. Even if people are involved, they may not be the *Who* of the data. For example, the *Who* here are the purchases (not the people who made them).

Who

In general, the rows of a data table are individual **cases** about *Who*m (or about which—if they're not people) we record some information. What we call these cases depends on the situation. Individuals who answer a survey are referred to as *respondents*. In an experiment people are *subjects* or *participants*, but animals, plants, and inanimate subjects are often just called *experimental units*. In a database, the cases (rows) are sometimes called *records*. In the Amazon table, the cases are the records of individual CD orders.

Often, the cases are a **sample** selected from some larger **population** that we'd like to understand. Amazon certainly cares about its customers, but also wants to know how to attract all those other Internet users who may never have made a purchase from Amazon's site. To be able to generalize from the sample of cases to the larger population, we'll want the sample to be *representative* of that population—a kind of snapshot of the larger world.

For Example   Identifying the "Who"

In March 2007, *Consumer Reports* published an evaluation of large-screen, high-definition television sets (HDTVs). The magazine purchased and tested 98 different models from a variety of manufacturers.

Question: What are the population of interest, the sample, and the *Who* of this study?

The magazine is interested in the performance of all HDTVs currently being offered for sale. It tested a sample of 98 sets, the "Who" for these data. Each HDTV set represents all similar sets offered by that manufacturer.

Just Checking

The marketing company J.D. Power and Associates <http://www.jdpower.com/> surveys consumers to learn about buyer behavior and customer satisfaction for many products worldwide. For example, J.D. Power's Automotive Performance, Execution, and Layout (APEAL) studies examine what people like about their car's performance and design. The 2008 Japan Mini-Car APEAL Study looked at 38 different mini-car models from seven manufacturers. J.D. Power based this report on responses from 4255 people who had recently purchased new mini-cars.

1. Describe the population of interest.

2. What was the sample?

3. Identify the *Who* of this study.

(Check your answers on page 19.)

What and Why

The characteristics recorded about each individual are called **variables.** These are usually shown as the columns of a data table, and they should have a name that identifies *What* has been measured. Variables may seem simple, but to really understand your variables, you must *Think* about what you want to know.

Although area codes are numbers, do we use them that way? Is 610 twice 305? Of course it is, but is that the question? Why would we want to know whether Allentown, PA (area code 610), is twice Key West, FL (305)? Variables play different roles, and you can't tell a variable's role just by looking at it.

Some variables just tell us what group or category each individual belongs to. Are you male or female? Pierced or not? . . . When a variable names categories and answers questions about how cases fall into those categories, we call it a **categorical variable.**[2] What kinds of things can we learn about variables like these? A natural start is to *count* how many cases belong in each category. (Are you listening to music while reading this? We could count the number of students in the class who were and the number who weren't.) We'll look for ways to compare and contrast the sizes of such categories.

Other variables have measurement **units.** Units such as yen, cubits, carats, angstroms, nanoseconds, miles per hour, or degrees Celsius tell us the *scale* of measurement. The units tell us how much of something we have or how far apart two values are. Without units, the values of a measured variable have no meaning. It does little good to be promised a raise of 5000 a year if you don't know whether it will be paid in euros, dollars, yen, or Estonian krooni.

What kinds of things can we learn about measured variables? We can do a lot more than just counting categories. We can look for patterns and trends. (How much did you pay for your last movie ticket? What is the range of ticket prices available in your town? How has the price of a ticket changed over the past 20 years?) When a measured variable with units answers questions about the quantity of what is measured, we call it a **quantitative variable.**

By international agreement, the International System of Units links together all systems of weights and measures. There are seven base units from which all other physical units are derived:

- Distance   Meter
- Mass   Kilogram
- Time   Second
- Electric
  current   Ampere
- Temperature   °Kelvin
- Amount
  of substance   Mole
- Intensity
  of light   Candela

---

[2]You may also see it called a *qualitative variable.*

It's important to think about whether a variable is categorical or quantitative. Questions we ask a variable (the *Why* of our analysis) shape how we think about it and how we treat it.

Some variables can answer questions only about categories. If the values of a variable are words rather than numbers, it's a good bet that it is categorical. But some variables can answer both kinds of questions. Amazon could ask for your *Age* in years. That seems quantitative, and would be if the company wanted to know the average age of those customers who visit their site after 3 A.M. But suppose Amazon wants to decide which CD to offer you in a special deal—one by Lady Gaga, Dave Matthews, Carly Simon, or Mantovani—and needs to be sure to have adequate supplies on hand to meet the demand. Then thinking of your age in one of the categories—child, teen, adult, or senior— might be more useful. If it isn't clear whether a variable is categorical or quantitative, think about *Why* you are looking at it and what you want it to tell you.

A typical course evaluation survey asks, "How valuable do you think this course will be to you?": 1 = Worthless; 2 = Slightly; 3 = Middling; 4 = Reasonably; 5 = Invaluable. Is *Educational Value* categorical or quantitative? Once again, we'll look to the *Why*. A teacher might just count the number of students who gave each response for her course, treating *Educational Value* as a categorical variable. When she wants to see whether the course is improving, she might treat the responses as the *amount* of perceived value—in effect, treating the variable as quantitative. She'll have to imagine that it has "educational value units."

Beware: some variables are neither categorical nor quantitative. What's your student ID number? It is numerical, but is it a quantitative variable? No, it doesn't have units. Is it categorical? Sort of, but this category contains only *you*—not very interesting.[3]

You'll want to recognize when a variable is playing the role of an **identifier** so you won't be tempted to analyze it. If this year's average ID number happens to be higher than last year's, it doesn't mean that the students are better.

IDENTIFYING IDENTIFIERS

Identifier variables are crucial in this age of large data sets. They make it possible to combine data from different sources, to protect confidentiality, and to provide unique labels. The variables *UPS Tracking Number, Social Security Number,* and Amazon's *ASIN* are all examples of identifier variables.

For Example　Identifying "What" and "Why" of HDTVs.

Recap: A *Consumer Reports* article about 98 HDTVs lists each set's manufacturer, cost, screen size, type (LCD, plasma, or rear projection), and overall performance score (0–100).

Question: Are these variables categorical or quantitative? Include units where appropriate, and describe the "Why" of this investigation.

The "what" of this article includes the following variables:

- manufacturer (categorical);
- cost (in dollars, quantitative);
- screen size (in inches, quantitative);
- type (categorical);
- performance score (quantitative).

The "why": The magazine hopes to help consumers pick a good HDTV set.

---

[3]No offense intended!

Just Checking

In the Japan Mini-Car APEAL Study, J.D. Power investigated customer satisfaction by asking people about cars they had recently purchased.

  4. Describe the *Why* of this study.

  5. The *What* of the APEAL study included the variables listed here. Classify each as categorical or quantitative and suggest possible units where appropriate.

    a) Manufacturer
    b) Passengers
    c) Fuel efficiency
    d) Engine displacement (size)
    e) Type of sound system
    f) Satisfaction rating

(Check your answers on page 19.)

Counts Count

In Statistics, we use counts two different ways. When Amazon considers a special offer of free shipping to customers, it might first analyze how purchases are shipped. They'd probably start by counting the number of purchases shipped by ground transportation, by second-day air, and by overnight air.

Shipping Method	Number of Purchases
Ground	20,345
Second-day	7,890
Overnight	5,432

Counting the number of purchases in each category is a natural way to summarize the categorical variable *Shipping Method*.

So every time we see counts, does that mean the variable is categorical? Actually, no. We also use counts to measure the amounts of things. How many songs are on your digital music player? How many classes are you taking this semester? To measure these quantities, we'd naturally count. The variables (*Songs, Classes*) would be quantitative, and we'd consider the units to be "number of . . ." or, generically, just "counts" for short. Amazon might record the number of teenage customers visiting their site each month to track customer growth and forecast CD sales (the *Why*).

Month	Teens
January	123,456
February	234,567
March	345,678
April	456,789
May	. . .
. . .	. . .

Here *Teens* is a quantitative variable whose amount is measured by counting the number of customers (the units).

Where, When, and How

We must know *Who*, *What*, and *Why* to analyze data. Without knowing these three, we don't have enough information. Of course, we'd always like to know more. The more we know about the data, the more we'll understand about the world.

If possible, we'd like to know the **When** and **Where** of data as well. Values recorded in 1803 may mean something different than similar values recorded last year. Values measured in Tanzania may differ in meaning from similar measurements made in Mexico.

How the data are collected can make the difference between insight and nonsense. As we'll see later, data that come from a voluntary survey on the Internet are almost always worthless. One primary concern of Statistics is designing methods for collecting meaningful data. These include surveys, polls, experiments, and other kinds of studies.[4]

Throughout this book, whenever we introduce data, we'll provide a margin note listing the W's (and H) of the data. It's a habit we recommend. The first step of any data analysis is to know why you are examining the data (what you want to know), whom each row of your data table refers to, and what the variables (the columns of the table) record. These are the *Why*, the *Who*, and the *What*. Identifying them is a key part of the *Think* step of any analysis. Make sure you know all three before you proceed to *Show* or *Tell* anything about the data.

Just Checking

In the 2003 Tour de France, Lance Armstrong averaged 40.94 kilometers per hour (km/h) for the entire course, making it the fastest Tour de France in its 100-year history. In 2004, he made history again by winning the race for an unprecedented sixth time. In 2005, he became the only 7-time winner and once again set a new record for the fastest average speed. You can find data on all the Tour de France races on the DVD. Here are the first 3 and last 12 lines of the data set. Keep in mind that the entire data set has nearly 100 entries.

6. List as many of the W's as you can for this data set.

7. Classify each variable as categorical or quantitative; if quantitative, identify the units.

Year	Winner	Country of origin	Total time (h/min/s)	Avg. speed (km/h)	Stages	Total distance ridden (km)	Starting riders	Finishing riders
1903	Maurice Garin	France	94.33.00	25.3	6	2428	60	21
1904	Henri Cornet	France	96.05.00	24.3	6	2388	88	23
1905	Louis Trousselier	France	112.18.09	27.3	11	2975	60	24
⋮								
1999	Lance Armstrong	USA	91.32.16	40.30	20	3687	180	141
2000	Lance Armstrong	USA	92.33.08	39.56	21	3662	180	128
2001	Lance Armstrong	USA	86.17.28	40.02	20	3453	189	144
2002	Lance Armstrong	USA	82.05.12	39.93	20	3278	189	153
2003	Lance Armstrong	USA	83.41.12	40.94	20	3427	189	147
2004	Lance Armstrong	USA	83.36.02	40.53	20	3391	188	147

(Check your answers on page 19.)

(continued)

[4]Coming attractions: Part III of this book. We can tell you're excited.

Year	Winner	Country of origin	Total time (h/min/s)	Avg. speed (km/h)	Stages	Total distance ridden (km)	Starting riders	Finishing riders
2005	Lance Armstrong	USA	86.15.02	41.65	21	3608	189	155
2006	Óscar Periero	Spain	89.40.27	40.78	20	3657	176	139
2007	Alberto Contador	Spain	91.00.26	38.97	20	3547	189	141
2008	Carlos Sastre	Spain	87.52.52	40.50	21	3559	199	145
2009	Alberto Contador	Spain	85.48.35	40.31	21	3459	180	156
2010	Alberto Contador	Spain	91.58.48	39.60	20	3642	198	170

There's a world of data on the Internet. These days, one of the richest sources of data is the Internet. With a bit of practice, you can learn to find data on almost any subject. Many of the data sets we use in this book were found in this way. The Internet has both advantages and disadvantages as a source of data. Among the advantages are the fact that often you'll be able to find even more current data than those we present. The disadvantage is that not all Internet sources can be trusted to have reliable data. Beyond that, references to Internet addresses can "break" as sites evolve, move, and die.

Our solution to these challenges is to offer the best advice we can to help you search for the data, wherever they may be residing. We usually point you to a website. We'll sometimes suggest search terms and offer other guidance.

DATA **IN YOUR WORLD**

TV queen Oprah Reigns as Richest Woman Entertainer

NEW YORK, Jan 19, 2007 (Reuters Life!)—U.S. talk show host and media mogul Oprah Winfrey has earned an estimated $1.5 billion during her career, putting her at the top of a *Forbes* magazine list of the wealthiest women in entertainment.

Runner-up to Winfrey on *Forbes'* new ranking was author J.K. Rowling, creator of boy wizard Harry Potter, with career earnings of $1 billion, followed by lifestyle guru Martha Stewart with $638 million and pop singer Madonna with $325 million. *Forbes* said to make its top-20 list, posted at www.forbes.com/women-stars . . . , these female megastars needed a minimum net worth of $45 million amassed over the course of their careers.

Rank	Entertainer	Country	Earnings (in millions)	Source	Kids
1	Oprah Winfrey	USA	$1500	TV	0
2	J. K. Rowling	UK	$1000	Books	3
3	Martha Stewart	USA	$638	Media	1
4	Madonna	USA	$325	Music	3
5	Celine Dion	USA	$250	Music	1
6	Mariah Carey	USA	$225	Music	0
7	Janet Jackson	USA	$150	Music	0
8	Julia Roberts	USA	$140	Movies	2
9	Jennifer Lopez	USA	$110	Music/Movies	0
10	Jennifer Aniston	USA	$110	TV/Movies	0

WHAT CAN GO WRONG?

- **Don't label a variable as categorical or quantitative without thinking about the question you want it to answer.** The same variable can sometimes take on different roles.

- **Just because your variable's values are numbers, don't assume that it's quantitative.** Categories and identifiers are often given numerical labels. Don't let that fool you into thinking they have quantitative meaning. Look at the context.

- **Always be skeptical.** One reason to analyze data is to discover the truth. Even when you are told a context for the data, it may turn out that the truth is a bit (or even a lot) different. The context colors our interpretation of the data, so those who want to influence what you think may slant the context. A survey that seems to be about all students may in fact report just the opinions of those who visited a fan website. The question that respondents answered may have been posed in a way that influenced their responses.

TI Tips   Working with data

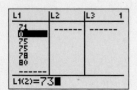

You'll need to be able to enter and edit data in your calculator. Here's how.

To enter data:
Hit the **STAT** button, and choose **EDIT** from the menu. You'll see a set of columns labeled **L1**, **L2**, and so on. Here is where you can enter, change, or delete a set of data.

Let's enter the heights (in inches) of the five starting players on a basketball team: 71, 75, 75, 76, and 80. Move the cursor to the space under **L1**, type in 71, and hit **ENTER** (or the down arrow). There's the first player. Now enter the data for the rest of the team.

To change a datum:
Suppose the 76″ player grew since last season; his height should be listed as 78″. Use the arrow keys to move the cursor onto the 76, then change the value and **ENTER** the correction.

To add more data:
We want to include the sixth man, 73″ tall. It would be easy to simply add this new datum to the end of the list. However, sometimes the order of the data matters, so let's place this datum in numerical order. Move the cursor to the desired position (atop the first 75). Hit **2ND INS**, then **ENTER** the 73 in the new space.

To delete a datum:
The 78″ player just quit the team. Move the cursor there. Hit **DEL**. Bye.

To clear the datalist:
Finished playing basketball? Move the cursor atop the L1. Hit **CLEAR**, then **ENTER** (or down arrow). You should now have a blank datalist, ready for you to enter your next set of values.

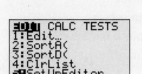

Lost a datalist?
Oops! Is L1 now missing entirely? Did you delete L1 by mistake, instead of just *clearing* it? Easy problem to fix: buy a new calculator. No? OK, then simply go to the **STAT EDIT** menu, and run **SetUpEditor** to recreate all the lists.

WHAT HAVE WE LEARNED?

We've learned that data are information in a context.

▶ The W's help nail down the context: *Who, What, Why, Where, When,* and *hoW.*

▶ We must know at least the *Who, What,* and *Why* to be able to say anything useful based on the data. The *Who* are the *cases.* The *What* are the *variables.* A variable gives information about each of the cases. The *Why* helps us decide which way to treat the variables.

We treat variables in two basic ways: as *categorical* or *quantitative.*

▶ Categorical variables identify a category for each case. Usually, we think about the counts of cases that fall into each category. (An exception is an identifier variable that just names each case.)

▶ Quantitative variables record measurements or amounts of something; they must have *units.*

▶ Sometimes we treat a variable as categorical or quantitative depending on what we want to learn from it, which means that some variables can't be pigeonholed as one type or the other. That's an early hint that in Statistics we can't always pin things down precisely.

Terms

Context	The context ideally tells *Who* was measured, *What* was measured, *How* the data were collected, *Where* the data were collected, and *When* and *Why* the study was performed.
Data	Systematically recorded information, whether numbers or labels, together with its context.
Data table	An arrangement of data in which each row represents a case and each column represents a variable.
Case	A case is an individual about whom or which we have data.
Population	All the cases we wish we knew about.
Sample	The cases we actually examine in seeking to understand the much larger population.
Variable	A variable holds information about the same characteristic for many cases.
Units	A quantity or amount adopted as a standard of measurement, such as dollars, hours, or grams.
Categorical variable	A variable that names categories (whether with words or numerals).
Quantitative variable	A variable in which the numbers act as numerical values. Quantitative variables always have units.

(continued)

Skills

▶ Be able to identify the *Who, What, When, Where, Why,* and *How* of data, or recognize when some of this information has not been provided.

▶ Be able to identify the cases and variables in any data set.

▶ Be able to identify the population from which a sample was chosen.

▶ Be able to classify a variable as categorical or quantitative, depending on its use.

▶ For any quantitative variable, be able to identify the units in which the variable has been measured (or note that they have not been provided).

▶ Be able to describe a variable in terms of its *Who, What, When, Where, Why,* and *How* (and be prepared to remark when that information is not provided).

DATA ON THE COMPUTER

Most often we find statistics on a computer using a program, or *package,* designed for that purpose. There are many different statistics packages, but they all *do* essentially the same things. If you understand what the computer needs to know to do what you want and what it needs to show you in return, you can figure out the specific details of most packages pretty easily.

For example, to get your data into a computer statistics package, you need to tell the computer:

- Where to find the data. This usually means directing the computer to a file stored on your computer's disk or to data on a database. Or it might just mean that you have copied the data from a spreadsheet program or Internet site and it is currently on your computer's clipboard. Usually, the data should be in the form of a data table. Most computer statistics packages prefer the *delimiter* that marks the division between elements of a data table to be a *tab* character and the delimiter that marks the end of a case to be a *return* character.
- Where to put the data. (Usually this is handled automatically.)
- What to call the variables. Some data tables have variable names as the first row of the data, and often statistics packages can take the variable names from the first row automatically.

93cars.dat
Link Embed Twitter Facebook

| StatCrunch | Data | Stat | Graphics | Help |

Row	Manufacturer	Model	Type	Min Price	Mid Price	Max Price	City MPG	Highway M
1	Acura	Integra	Small	12.9	15.9	18.8	25	
2	Acura	Legend	Midsize	29.2	33.9	38.7	18	
3	Audi	90	Compact	25.9	29.1	32.3	20	
4	Audi	100	Midsize	30.8	37.7	44.6	19	
5	BMW	535i	Midsize	23.7	30	36.2	22	
6	Buick	Century	Midsize	14.2	15.7	17.3	22	
7	Buick	LeSabre	Large	19.9	20.8	21.7	19	
8	Buick	Roadmaster	Large	22.6	23.7	24.9	16	
9	Buick	Riviera	Midsize	26.3	26.3	26.3	19	
10	Cadillac	DeVille	Large	33	34.7	36.3	16	
11	Cadillac	Seville	Midsize	37.5	40.1	42.7	16	

EXERCISES

A

1. **Voters.** A February 2007 Gallup Poll question asked, "In politics, as of today, do you consider yourself a Republican, a Democrat, or an Independent?" The possible responses were "Democrat", "Republican", "Independent", "Other", and "No Response". What kind of variable is the response?

2. **Mood.** A January 2007 Gallup Poll question asked, "In general, do you think things have gotten better or gotten worse in this country in the last five years?" Possible answers were "Better", "Worse", "No Change", "Don't Know", and "No Response". What kind of variable is the response?

3. **Medicine.** A pharmaceutical company conducts an experiment in which a subject takes 100 mg of a substance orally. The researchers measure how many minutes it takes for half of the substance to exit the bloodstream. What kind of variable is the company studying?

4. **Stress.** A medical researcher measures the increase in heart rate of patients under a stress test. What kind of variable is the researcher studying?

(Exercises 5–8) For each description of data, identify the population of interest and the sample.

5. **Weighing bears.** Because of the difficulty of weighing a bear in the woods, researchers caught and measured 54 bears, recording their weight, neck size, length, and sex. They hoped to find a way to estimate weight from the other, more easily determined quantities.

6. **Flowers.** In a study appearing in the journal *Science,* a research team reports that plants in southern England are flowering earlier in the spring. Records of the first flowering dates for 385 species over a period of 47 years show that flowering has advanced an average of 15 days per decade, an indication of climate warming, according to the authors.

7. **Fitness.** Are physically fit people less likely to die of cancer? An article in the May 2002 issue of *Medicine and Science in Sports and Exercise* reported results of a study that followed 25,892 men aged 30 to 87 for 10 years. The most physically fit men had a 55% lower risk of death from cancer than the least fit group.

8. **Movies.** Some motion pictures are profitable and others are not. Understandably, the movie industry would like to know what makes a movie successful. Data from 120 first-run movies released in 2005 suggest that longer movies actually make *less* profit.

(Exercises 9–12) For each description of data, identify Who and What were investigated.

9. **Honesty.** Coffee stations in offices often just ask users to leave money in a tray to pay for their coffee, but many people cheat. Researchers at Newcastle University alternately taped two posters over the coffee station. During one week, it was a picture of flowers; during the other, it was a pair of staring eyes. They found that the average contribution was significantly higher when the eyes poster was up than when the flowers were there. Apparently, the mere feeling of being watched—even by eyes that were not real—was enough to encourage people to behave more honestly. [*NY Times,* Dec. 10, 2006]

10. **Investments.** Some companies offer 401(k) retirement plans to employees, permitting them to shift part of their before-tax salaries into investments such as mutual funds. Employers typically match 50% of the employees' contribution up to about 6% of salary. One company, concerned with what it believed was a low employee participation rate in its 401(k) plan, sampled 30 other companies with similar plans and asked for their 401(k) participation rates.

11. **Fitness, again.** Look once more at the study described in Exercise 7.

12. **Molten iron.** The Cleveland Casting Plant is a large, highly automated producer of gray and nodular iron automotive castings for Ford Motor Company. The company is interested in keeping the pouring temperature of the molten iron (in degrees Fahrenheit) close to the specified value of 2550 degrees. Cleveland Casting measured the pouring temperature for 10 randomly selected crankshafts.

(Exercises 13–16) For each description of data, identify the variables and tell whether each should be treated as categorical or quantitative.

13. **Bicycle safety.** Ian Walker, a psychologist at the University of Bath, wondered whether drivers treat bicycle riders differently when they wear helmets. He rigged his bicycle with an ultrasonic sensor that could measure how close each car was that passed him. He then rode on alternating days with and without a helmet. Out of 2500 cars passing him, he found that when he wore his helmet, motorists passed 3.35 inches closer to him, on average, than when his head was bare. [*NY Times*, Dec. 10, 2006]

14. **More bears.** Look once more at the study described in Exercise 5.

15. **Arby's menu.** A listing posted by the Arby's restaurant chain gives, for each of the sandwiches it sells, the type of meat in the sandwich, the number of calories, and the serving size in ounces. The data might be used to assess the nutritional value of the different sandwiches.

16. **Schools.** The State Education Department requires local school districts to keep these records on all students: age, race or ethnicity, days absent, current grade level, standardized test scores in reading and mathematics, and any disabilities or special educational needs.

(*Exercises 17–20*) *For each description of data, identify Who and What were investigated and the population of interest. Name the variables and indicate whether each should be treated as categorical or quantitative, and, (if quantitative) the units in which it was measured (if known).*

17. Babies. Medical researchers at a large city hospital investigating the impact of prenatal care on newborn health collected data from 882 births during 1998–2000. They kept track of the mother's age, the number of weeks the pregnancy lasted, the type of birth (cesarean, induced, natural), the level of prenatal care the mother had (none, minimal, adequate), the birth weight and sex of the baby, and whether the baby exhibited health problems (none, minor, major).

18. Age and party. The Gallup Poll conducted a representative telephone survey of 1180 American voters during the first quarter of 2009. Among the reported results were the voter's region (Northeast, South, etc.), age, party affiliation, and whether or not the person had voted in the 2008 presidential election.

19. Refrigerators. In 2006, *Consumer Reports* published an article evaluating refrigerators. It listed 41 models, giving the brand, cost, size (cu ft), type (such as top freezer), estimated annual energy cost, an overall rating (good, excellent, etc.), and the repair history for that brand (percentage requiring repairs over the past 5 years).

20. Walking in circles. People who get lost in the desert, mountains, or woods often seem to wander in circles rather than walk in straight lines. To see whether people naturally walk in circles in the absence of visual clues, researcher Andrea Axtell tested 32 people on a football field. One at a time, they stood at the center of one goal line, were blindfolded, and then tried to walk to the other goal line. She recorded each individual's sex, height, handedness, the number of yards each was able to walk before going out of bounds, and whether each wandered off course to the left or the right. No one made it all the way to the far end of the field without crossing one of the sidelines. [*STATS* No. 39, Winter 2004]

(*Exercises 21–24*) *For each description of data, identify the W's, name the variables, specify for each variable whether its use indicates that it should be treated as categorical or quantitative, and, for any quantitative variable, identify the units in which it was measured (or note that they were not provided).*

21. Herbal medicine. Scientists at a major pharmaceutical firm conducted an experiment to study the effectiveness of an herbal compound to treat the common cold. They exposed each patient to a cold virus, then gave them either the herbal compound or a sugar solution known to have no effect on colds. Several days later they assessed each patient's condition, using a cold severity scale ranging from 0 to 5. They found no evidence of the benefits of the compound.

22. Vineyards. Business analysts hoping to provide information helpful to American grape growers compiled these data by surveying several vineyards: size (acres), number of years in existence, state, varieties of grapes grown, average case price, gross sales, and percent profit.

23. Streams. In performing field research for an ecology class, students at a college in upstate New York collect data on streams each year. They record a number of biological, chemical, and physical variables, including the stream name, the substrate of the stream (limestone, shale, or mixed), the acidity of the water (pH), the temperature (°C), and the BCI (a numerical measure of biological diversity).

24. Fuel economy. The Environmental Protection Agency (EPA) tracks fuel economy of automobiles based on information from the manufacturers (Ford, Toyota, etc.). Among the data the agency surveys are the manufacturer, vehicle type (car, SUV, etc.), weight, horsepower, and gas mileage (mpg) for city and highway driving.

C

(*Exercises 25–28*) *Write a report about each, including a complete description of the data.*

 25. Horse race 2010. The Kentucky Derby is a horse race that has been run every year since 1875 at Churchill Downs, Louisville, Kentucky. The race started as a 1.5-mile race, but in 1896, it was shortened to 1.25 miles because experts felt that 3-year-old horses shouldn't run such a long race that early in the season. (It has been run in May every year but one—1901—when it took place on April 29.) Here are the data for the first four and several recent races.

Date	Winner	Margin (lengths)	Jockey	Winner's Payoff ($)	Duration (min:sec)	Track Condition
May 17, 1875	Aristides	2	O. Lewis	2850	2:37.75	Fast
May 15, 1876	Vagrant	2	B. Swim	2950	2:38.25	Fast
May 22, 1877	Baden-Baden	2	W. Walker	3300	2:38.00	Fast
May 21, 1878	Day Star	1	J. Carter	4050	2:37.25	Dusty
. . . . . .						
May 6, 2006	Barbaro	6 1/2	E. Prado	2000000	2:01.36	Fast
May 5, 2007	Street Sense	2 1/4	C. Borel	2210000	2:02.17	Fast
May 3, 2008	Big Brown	4 3/4	K. Desormeaux	2000000	2:01.82	Fast
May 2, 2009	Mine That Bird	6 3/4	C. Borel	1417000	2:02.66	Sloppy
May 1, 2010	Super Saver	2 1/2	C. Borel	1425200	2:04.45	Sloppy

26. Indy 2010. The 2.5-mile Indianapolis Motor Speedway has been the home to a race on Memorial Day nearly every year since 1911. Even during the first race, there were controversies. Ralph Mulford was given the checkered flag first but took three extra laps just to make sure he'd completed 500 miles. When he finished, another driver, Ray Harroun, was being presented with the winner's trophy, and Mulford's protests were ignored. Harroun averaged 74.6 mph for the 500 miles. In 2010, the winner, Dario Franchitti, averaged 161.623 mph.

Here are the data for the first five races and five recent Indianapolis 500 races. Included also are the pole winners (the winners of the trial races, when each driver drives alone to determine the position on race day).

Year	Winner	Pole Position	Average Speed (mph)	Pole Winner	Average Pole Speed (mph)
1911	Ray Harroun	28	74.602	Lewis Strang	.
1912	Joe Dawson	7	78.719	Gil Anderson	.
1913	Jules Goux	7	75.933	Caleb Bragg	.
1914	René Thomas	15	82.474	Jean Chassagne	.
1915	Ralph DePalma	2	89.840	Howard Wilcox	98.580
. . .					
2006	Sam Hornish Jr.	1	157.085	Sam Hornish Jr.	228.985
2007	Dario Franchitti	3	151.744	Hélio Castroneves	225.817
2008	Scott Dixon	1	143.567	Scott Dixon	221.514
2009	Hélio Castroneves	1	150.318	Hélio Castroneves	222.864
2010	Dario Franchitti	3	161.623	Hélio Castroneves	227.970

27. The news. Find a newspaper or magazine article in which some data are reported. Write a complete description of the data discussed in the article. Include a copy of the article with your report.

28. The Internet. Find an Internet source that reports on a study and describes the data. Print out the report and write a complete description of the data.

Answers

Just Checking

1. All mini-cars

2. 4225 recently purchased mini-cars

3. The mini-cars purchased by the 4225 owners. (The owners are not the *Who* because we are interested in making generalizations about the cars.)

4. To find out what buyers like about their cars (in order to improve future cars?)

5. a) Categorical
   b) Quantitative (though could be considered categorical—e.g., "5-passenger cars")
   c) Quantitative, miles per gallon
   d) Quantitative, probably liters
   e) Categorical
   f) Categorical ("Very satisfied", "Satisfied", etc.) or quantitative (a 1–5 scale?)

6. Who—Tour de France races; What—year, winner, country of origin, total time, average speed, stages, total distance ridden, starting riders, finishing riders; How—official statistics at race; Where—France (for the most part); When—1903 to 2009; Why—not specified (To see progress in speeds of cycling racing?)

7.

Variable	Type	Units
Year	Quantitative or Categorical	Years
Winner	Categorical	
Country of Origin	Categorical	
Total Time	Quantitative	Hours/minutes/seconds
Average Speed	Quantitative	Kilometers per hour
Stages	Quantitative	Counts (stages)
Total Distance	Quantitative	Kilometers
Starting Riders	Quantitative	Counts (riders)
Finishing Riders	Quantitative	Counts (riders)

Stories Categorical Data Tell

hat happened on the *Titanic* at 11:40 on the night of April 14, 1912, is well known. Frederick Fleet's cry of "Iceberg, right ahead" was the beginning of a nightmare that has become legend. By 2:15 a.m., the "unsink-able" *Titanic* had sunk, leaving more than 1500 people to meet their icy fates.

At the top of the next page are some data about the passengers and crew aboard the *Titanic*. Each case (row) of the data table represents a person on board the ship. The variables are the person's *Survival* status (Dead or Alive), *Age* (Adult or Child), *Sex* (Male or Female), and ticket *Class* (First, Second, Third, or Crew).

The problem with a data table like this—and in fact with all data tables—is that you can't *see* what's going on. And seeing is just what we want to do. We need ways to show the data so that we can see patterns, relationships, trends, and exceptions.

WHO	People on the *Titanic*
WHAT	Survival status, age, sex, ticket class
WHEN	April 14, 1912
WHERE	North Atlantic
HOW	A variety of sources and Internet sites
WHY	Historical interest

Survival	Age	Sex	Class
Dead	Adult	Male	Third
Dead	Adult	Male	Crew
Dead	Adult	Male	Third
Dead	Adult	Male	Crew
Dead	Adult	Male	Crew
Dead	Adult	Male	Crew
Alive	Adult	Female	First
Dead	Adult	Male	Third
Dead	Adult	Male	Crew

TABLE 3.1 Part of a data table showing four variables for nine people aboard the *Titanic*.

The Three Rules of Data Analysis

So, what should we do with data like these? There are three things you should always do first with data:

1. **Make a picture.** A display of your data will reveal things you're not likely to see in a table of numbers and will help you to *Think* clearly about the patterns and relationships that may be hiding in your data.
2. **Make a picture.** A well-designed display will *Show* the important features and patterns in your data. It could also show you things you did not expect to see: extraordinary (possibly wrong) data values or unexpected patterns.
3. **Make a picture.** The best way to *Tell* others about your data is with a well-chosen picture.

These are the three rules of data analysis. There are pictures of data throughout the book, and new kinds keep showing up. These days, technology makes drawing pictures of data easy, so there is no reason not to follow the three rules.

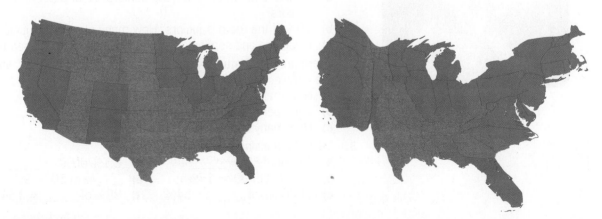

FIGURE 3.1 A Picture to Tell a Story. In November 2008, Barack Obama was elected the 44th president of the United States. News reports commonly showed the election results with maps like the one on the left, coloring states won by Obama blue and those won by his opponent John McCain red. Even though McCain lost, doesn't it look like there's more red than blue? That's because some of the larger states like Montana and Wyoming have far fewer voters than some of the smaller states like Maryland and Connecticut. The strange-looking map on the right cleverly distorts the states to resize them proportional to their populations. By sacrificing an accurate display of the land areas, we get a better impression of the votes cast, giving us a clear picture of Obama's historic victory.

(http://www-personal.umich.edu/~mejn/election/2008/)

Frequency Tables: Making Piles

Class	Count
First	325
Second	285
Third	706
Crew	885

TABLE 3.2 A frequency table of the *Titanic* passengers.

Class	%
First	14.77
Second	12.95
Third	32.08
Crew	40.21

TABLE 3.3 A relative frequency table for the same data.

To make a picture of data, the first thing we have to do is to make piles. Making piles is the beginning of understanding data. We pile together things that seem to go together. For categorical data, piling is easy. We just count the number of cases corresponding to each category and pile them up.

One way to put all 2201 people on the *Titanic* into piles is by ticket *Class*, counting up how many had each kind of ticket. We can organize these counts into a **frequency table,** which records the totals and the category names.

Even when we have thousands of cases, a variable like ticket *Class,* with only a few categories, has a frequency table that's easy to read. We use the names of the categories to label each row in the frequency table. For ticket *Class,* these are "First," "Second," "Third," and "Crew."

Counts are useful, but usually we want to know the fraction or **proportion** of the data in each category, so we divide the counts by the total number of cases. Usually we multiply by 100 to express these proportions as **percentages**. A **relative frequency table** displays these *percentages*. Both types of tables show the **distribution** of a categorical variable by naming the possible categories and telling how frequently each occurs.

Do The Math

By now you've probably figured out that in Statistics we'll be using percentages a lot. If you need some practice, try working on these questions.

1. Last season there were 26 students on the Springfield High School varsity football team, all freshmen, sophomores, juniors, or seniors.
   a) 18 students on the team are seniors. What percent of the team are seniors?
   b) 19% of the team is juniors. How many players are juniors?
   c) Freshman and sophomores together make up what percent of the team?

2. There are 1800 students enrolled at Springfield College, and 65% of them are female.
   a) How many students are female?
   b) What percent of the student body is male?
   c) How many male students are there?

3. Fill in the missing values:
   a) 17 out of 85 = _____%
   b) _____ out of 90 = 17%
   c) 1026 out of _____ = 54%
   d) 8.1% of 90 = _____
   e) _____% of 50 = 90
   f) 88% of _____ = 154

(Check your answers on page 43.)

The Area Principle

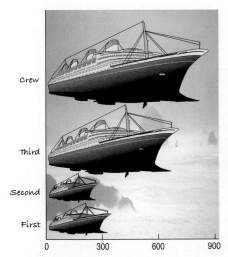

FIGURE 3.2 How many people were in each class on the Titanic? From this display, it looks as though the service must have been great, since most aboard were crew members. Although the length of each ship here corresponds to the correct number, the impression is all wrong. In fact, only about 40% were crew.

Now that we have the frequency table, we're ready to follow the three rules of data analysis and make a picture of the data. But a bad picture can distort our understanding rather than help it. Here's a graph of the *Titanic* data. What impression do you get about who was aboard the ship?

It sure looks like most of the people on the *Titanic* were crew members, with a few passengers along for the ride. That doesn't seem right. What's wrong? The lengths of the ships *do* match the totals in the table. (You can check the scale at the bottom.) However, experience and psychological tests show that our eyes tend to be more impressed by the *area* of each ship image. There were about 3 times as many crew as second-class passengers, but making the ship 3 times longer gives it 9 times the area. That creates a misleading impression.

The best data displays observe a fundamental principle of graphing data called the **area principle**. The area principle says that the area occupied by a part of the graph should correspond to the size of the value it represents.

Look back at the red state-blue state maps of the 2008 election. Basing the areas of states on voters gives us a much clearer picture of the Obama victory. Violations of the area principle are a common way to lie (or, since most mistakes are unintentional, we should say err) with Statistics.

Bar Charts

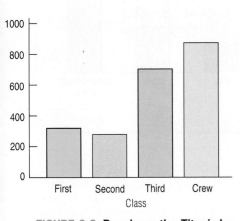

FIGURE 3.3 People on the *Titanic* by Ticket Class. With the area principle satisfied, we can see the true distribution more clearly.

Here's a chart that obeys the area principle. It's not as visually entertaining as the ships, but it does give an *accurate* visual impression of the distribution. The height of each bar shows the count for its category. The bars are the same width, so their heights make their areas proportional to the counts in each class. Now it's easy to see that the majority of people on board were *not* crew, as the ships picture led us to believe. We can also see that there were about 3 times as many crew as second-class passengers. And there were more than twice as many third-class passengers as either first- or second-class passengers, something you may have missed in the frequency table. Bar charts make these kinds of comparisons easy and natural.

A **bar chart** displays the distribution of a categorical variable, showing the counts for each category next to each other for easy comparison. Bar charts should have small spaces between the bars to indicate that these are separate bars that could be arranged in any order.

Usually they stick up like this  but sometimes they run

sideways like this .

For some reason, some computer programs give the name "bar chart" to any graph that uses bars. And others use different names according to whether the bars are horizontal or vertical. Don't be misled. "Bar chart" is the term for a *display of counts of a categorical variable* with bars.

If we really want to draw attention to the relative *proportion* of passengers falling into each of these classes, we could replace the counts with percentages and use a **relative frequency bar chart.**

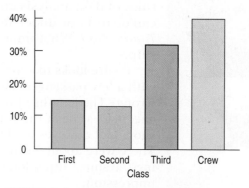

FIGURE 3.4 The relative frequency bar chart looks the same as the bar chart (Figure 3.3) but shows the proportion (percentages) of people in each category rather than the counts.

Just Checking

A Statistics class collected data on *Sex* and *Eye Color* for the students in the course. Here's a bar chart showing the distribution of *Eye Color* for the females in the class.

1. What percent of the females have each eye color?

2. Write a sentence describing what the middle bar tells you. (Be sure that your description is clear and pays attention to the W's.)

(Check your answers on page 43.)

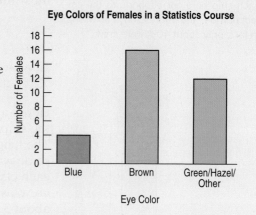

Eye Colors of Females in a Statistics Course

Pie Charts

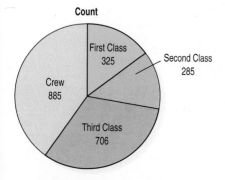

FIGURE 3.5 Number of *Titanic* passengers in each class.

Another common display that shows how a whole group breaks into several categories is a pie chart. **Pie charts** show the whole group of cases as a circle. They slice the circle into pieces whose sizes are proportional to the fraction of the whole in each category.

Pie charts give a quick impression of how a whole group is partitioned into smaller groups. Because we're used to cutting up pies into 2, 4, or 8 pieces, pie charts are good for seeing relative frequencies near 1/2, 1/4, or 1/8. For example, you may be able to tell that the pink slice, representing the second-class passengers, is very close to 1/8 of the total. It's harder to see that there were about twice as many third-class as first-class passengers. Which category had the most passengers? Were there more crew or more third-class passengers? Comparisons such as these are easier in a bar chart.

> **Think before you draw.** Our first rule of data analysis is *Make a picture*. But what kind of picture? We don't have a lot of options—yet. There's more to Statistics than pie charts and bar charts, and knowing when to use each type of graph is a critical first step in data analysis. That decision depends in part on what type of data we have.
>
> It's important to check that the data are appropriate for whatever method of analysis you choose. Before you make a bar chart or a pie chart, always check the **Categorical Data Condition:** The data are counts or percentages of individuals in categories.
>
> If you want to make a relative frequency bar chart or a pie chart, you'll need to also make sure that the categories don't overlap so that no individual is counted twice. If the categories do overlap, you can still make a bar chart, but the percentages won't add up to 100%. For the *Titanic* data, either kind of display is appropriate because the categories don't overlap.
>
> Throughout this course, you'll see that doing Statistics right means selecting the proper methods. That means you have to *Think* about the situation at hand. An important first step, then, is to check that the type of analysis you plan is appropriate. The Categorical Data Condition is just the first of many such checks.

Contingency Tables: Children and First-Class Ticket Holders First?

We know how many tickets of each class were sold on the *Titanic*, and we know that only about 32% of all those aboard the *Titanic* survived. We've looked at each variable by itself. It's more interesting to ask if the variables are related. Were some classes of passengers more likely to make it into a lifeboat? To answer this question, we need to look at the two categorical variables *Class* and *Survival* together.

To look at two categorical variables together, we often arrange the counts in a two-way table, called a **contingency table**. Here's a contingency table of those aboard the *Titanic*, showing both the class of ticket and whether the person survived.

		Class				
		First	Second	Third	Crew	Total
Survival	**Alive**	203	118	178	212	**711**
	Dead	122	167	528	673	**1490**
	Total	**325**	**285**	**706**	**885**	**2201**

TABLE 3.4 **Contingency table of ticket *Class* and *Survival*.** The bottom line of "Totals" is the same as the first frequency table.

The margins of the table, both on the right and at the bottom, give totals. The bottom line of the table is just the frequency distribution of ticket *Class*. The right column of the table is the frequency distribution of the variable *Survival*. When shown like this, in the margins of a contingency table, the frequency distribution of one of the variables is called its **marginal distribution**.

For Example Finding Marginal Distributions

In January 2007, a Gallup poll asked 1008 Americans age 18 and over whether they planned to watch the upcoming Super Bowl. The pollster also asked those who planned to watch whether they were looking forward more to seeing the football game or the commercials. The results are summarized in the table:

Question: What's the marginal distribution of the responses?

		Sex		
		Male	**Female**	**Total**
Response	**Game**	279	200	**479**
	Commercials	81	156	**237**
	Won't watch	132	160	**292**
	Total	**492**	**516**	**1008**

To determine the percentages for the three responses, divide the count for each response by the total number of people polled:

$$\frac{479}{1008} = 47.5\% \quad \frac{237}{1008} = 23.5\% \quad \frac{292}{1008} = 29.0\%$$

According to the poll, 47.5% of American adults were looking forward to watching the Super Bowl game, 23.5% were looking forward to watching the commercials, and 29% didn't plan to watch at all.

Just Checking

Here's a contingency table summarizing the Statistics class's *Sex* and *Eye Color* data.

		Eye Color			
		Blue	**Brown**	**Green/Hazel/Other**	**Total**
Sex	**Males**	8	20	6	?
	Females	4	16	12	?
	Total	**12**	**36**	**18**	?

3. How many male and females are in this class? (That's the marginal distribution of *Sex*.)

4. What's the marginal distribution of *Eye Color*?

5. Write a sentence that uses percents to describe the distribution of *Eye Color* among these students.

(Check your answers on page 43.)

Percents of What?

Now let's think some more about the story the *Titanic* table tells. If you look down the column for second-class passengers to the first **cell**, you can see that 118 second-class passengers survived. It's helpful to describe this number as a percentage—but as a percentage of what?

	–	**Second**	–	–	**Total**
Alive	–	118	–	–	**711**
–	–	–	–	–	–
Total	–	**285**	–	–	**2201**

- The total number of passengers? (118 is 5.4% of the total: 2201.)
- The number of second-class passengers? (118 is 41.4% of the 285 second-class passengers.)
- The number of survivors? (118 is 16.6% of the 711 survivors.)

All of these are possibilities, and all are potentially useful or interesting. To decide which percentage to use, you'll need to *Think* about what question you hope to answer.

The English language can be tricky when we talk about percentages. If you wonder "What percent *of the survivors* were in second class?" it's pretty clear that we're interested only in survivors. It's as if we're restricting the *Who* in the question to the survivors, so we should look at the number of second-class passengers among all the survivors—in other words, 118 of 711, or 16.6%.

But if you ask "What percent were second-class passengers who survived?" that's a different question. Be careful; here, the *Who* is everyone on board, so the answer is 118 of 2201, or 5.4%.

And if you want to know "What percent of the second-class passengers survived?" you have a third question. Now the *Who* is the second-class passengers, so the answer is 118 of 285, or 41.4%.

Whenever you want to know something about the variables in a contingency table, always be sure to ask "percent of what?" That will help you to know the *Who* and find the appropriate percentage. And this will also help you *Tell* what you have discovered about your data:

- Of all the people who sailed on the *Titanic*, 5.4% were second-class passengers who survived.
- 16.6% of the people who survived the sinking of the *Titanic* were ticketed in second class.
- When the *Titanic* sank, 41.4% of the second-class passengers survived.

A bell-shaped artifact from the *Titanic*.

Just Checking

Here again is the contingency table of *Sex* and *Eye Color* for that Stats class.

		Eye Color			
		Blue	**Brown**	**Green/Hazel/Other**	**Total**
Sex	**Males**	8	20	6	**34**
	Females	4	16	12	**32**
	Total	**12**	**36**	**18**	**66**

Write a sentence that answers each of these questions:

6. What percent of the blue-eyed students are males?

7. What percent of the class are blue-eyed males?

8. What percent of the male students have blue eyes?

9. What percent of the students who do *not* have blue eyes are females?

(Check your answers on pages 43–44.)

Conditional Distributions

To find out whether the chance of surviving the *Titanic* sinking *depended* on ticket class, we can look at the question in two ways. First, we could ask how the distribution of ticket *Class* changes between survivors and non-survivors. To do that, we look at the *row percentages*:

		Class				
		First	**Second**	**Third**	**Crew**	**Total**
Survival	**Alive**	203 28.6%	118 16.6%	178 25.0%	212 29.8%	711 **100%**
	Dead	122 8.2%	167 11.2%	528 35.4%	673 45.2%	1490 **100%**

TABLE 3.5 **The conditional distribution of ticket *Class* for each value of *Survival* (Alive and Dead).**

By focusing on each row separately, we see the distribution of class under the *condition* of surviving or not. First we restrict the *Who* to survivors and find percents for each ticket class. Then we refocus the *Who* on the nonsurvivors and find percentages again. The distributions we create this way are called **conditional distributions,** because they show the distribution of one variable for just those cases that satisfy a condition on another variable.

The sum of the percentages in each row is 100%, split up by ticket class. Now we can make separate pie charts for each row: survivors and nonsurvivors.

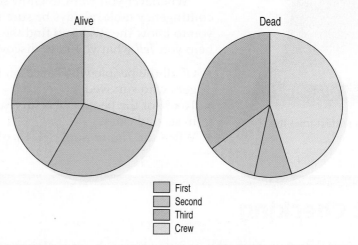

Alive   Dead

- First
- Second
- Third
- Crew

FIGURE 3.6 Pie charts of the conditional distributions of ticket *Class* for the survivors and nonsurvivors, separately. We're primarily concerned with percentages here, so pie charts are a reasonable choice. Which ticket class would you want to be in?

This picture makes the advantage of being a first-class passenger pretty obvious. (And too bad about the crew!)

For Example   Finding Conditional Distributions

Recap: The table shows results of a poll asking adults whether they were looking forward to the Super Bowl game, looking forward to the commercials, or didn't plan to watch.

Question: What does the conditional distribution of *Sex* tell us about viewers' interest in the commercials?

Look at the group of people who responded "Commercials" and determine what percent of them were male and female:

$$\frac{81}{237} = 34.2\% \qquad \frac{156}{237} = 65.8\%$$

		Sex		
		Male	**Female**	**Total**
	Game	279	200	**479**
Response	**Commercials**	81	156	**237**
	Won't watch	132	160	**292**
	Total	**492**	**516**	**1008**

Women make up a sizable majority of the adult Americans who look forward to seeing Super Bowl commercials more than the game itself. Nearly 66% of people who were more interested in the commercials were women, and only 34% were men.

But we can also turn the question around. We can look at the distribution of *Survival* for each ticket *Class*. To do this, we restrict the *Who* to one *column* at a time. Now the percentages in each column add to 100%. These conditional distributions show us whether the chance of surviving was roughly the same for each of the four ticket classes.

		First	Second	Third	Crew	Total
				Class		
Alive	Count % of Column	203 62.5%	118 41.4%	178 25.2%	212 24.0%	**711** **32.3%**
Dead	Count % of Column	122 37.5%	167 58.6%	528 74.8%	673 76.0%	**1490** **67.7%**
Total	Count	**325** **100%**	**285** **100%**	**706** **100%**	**885** **100%**	**2201** **100%**

(Survival is the row label spanning Alive/Dead/Total.)

TABLE 3.6 A contingency table of *Class* by *Survival*. Each column shows the conditional distribution of *Survival* for a given ticket *Class*.

Just Checking

One more time! Here's that contingency table of *Sex* and *Eye Color* for students in a Statistics class.

		Blue	Brown	Green/Hazel/Other	Total
			Eye Color		
Males		8	20	6	**34**
Females		4	16	12	**32**
Total		**12**	**36**	**18**	**66**

(Sex is the row label spanning Males/Females/Total.)

Write a sentence that answers each of these questions:

10. What's the conditional distribution of *Sex* for the brown-eyed students?

11. What's the conditional distribution of *Eye Color* for the males?

(Check your answers on page 44.)

Associations vs. Independence

Look back at the last table we made for the *Titanic* data. See how the survival percentages change across each row? It sure looks like ticket class mattered in whether a passenger lived through the disaster. To make it more vivid, follow the rules: Make a picture. Here's a side-by-side bar chart showing percentages of people surviving and not surviving for each ticket *Class*:

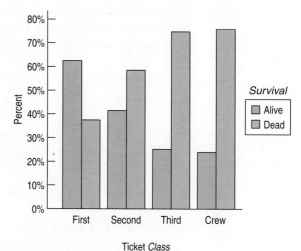

FIGURE 3.7 **Side-by-side bar chart.** Showing the conditional distribution of *Survival* for each category of ticket *Class*. Pie charts would have only two categories in each of four pies, so this works better.

We can make the picture even simpler by showing only one set of bars, the death rates for each ticket *Class*:

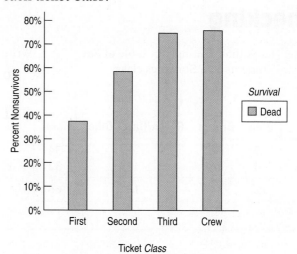

FIGURE 3.8 **Bar chart showing just nonsurvivor percentages for each ticket *Class*.**

Now it's easy to compare the risks. Among first-class passengers, 37.5% perished, compared to 58.6% for second-class ticket holders, 74.8% for those in third class, and 76.0% for crew members.

If the risk had been about the same for all ticket classes, we'd say that survival was *independent* of class. But it's not. It appears survival may have depended on ticket class. The death rates for third-class passengers and crew members were about twice as high as for first-class passengers.

It is interesting to know that there's an **association** between *Class* and *Survival*. That's an important part of the *Titanic* story. Variables can be *associated* in many ways and to different degrees. The best way to tell whether two variables are associated is to ask whether they are *not*.[1] In a contingency

---

[1]This kind of "backwards" reasoning shows up surprisingly often in science—and in Statistics. We'll see it again.

table, when the distribution of *one* variable is almost the same for all categories of another, we say that the variables are **independent**. That tells us there's no association between these variables.

For Example   Looking for Associations Between Variables

Recap: The table shows results of a poll asking adults whether they were looking forward to the Super Bowl game, looking forward to the commercials, or didn't plan to watch.

Question: Does it seem that there's an association between interest in Super Bowl TV coverage and a person's sex?

		Sex		
		Male	**Female**	**Total**
Response	**Game**	279	200	**479**
	Commercials	81	156	**237**
	Won't watch	132	160	**292**
	Total	**492**	**516**	**1008**

First find the distribution of the three responses for the men (the column percentages):

$$\frac{279}{492} = 56.7\% \qquad \frac{81}{492} = 16.5\% \qquad \frac{132}{492} = 26.8\%$$

Then do the same for the women who were polled, and display the two distributions with a side-by-side bar chart:

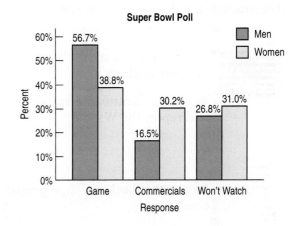

Based on this poll it appears that women were only slightly less interested than men in watching the Super Bowl telecast: 31% of the women said they didn't plan to watch, compared to just under 27% of men. Among those who planned to watch, however, there appears to be an association between the viewer's sex and what the viewer is most looking forward to. While more women are interested in the game (39%) than the commercials (30%), the margin among men is much wider: 57% of men said they were looking forward to seeing the game, compared to only 16.5% who cited the commercials.

Just Checking

Take one last look at our Statistics class data on *Sex* and *Eye Color*:

		Eye Color			
		Blue	**Brown**	**Green/Hazel/Other**	**Total**
Sex	**Males**	8	20	6	**34**
	Females	4	16	12	**32**
	Total	**12**	**36**	**18**	**66**

12. Compare the overall percent of students who are female to the percents of females with each eye color.

13. Does it seem that *Eye Color* and *Sex* are independent? Explain.

(Check your answers on page 44.)

STEP–BY–STEP EXAMPLE   Examining Contingency Tables

Medical researchers followed 6272 Swedish men for 30 years to see if there was any association between the amount of fish in their diet and prostate cancer ("Fatty Fish Consumption and Risk of Prostate Cancer," *Lancet,* June 2001). Their results are summarized in this table:

We asked for a picture of a man eating fish. This is what we got.

Fish Consumption	Prostate Cancer	
	No	Yes
Never/seldom	110	14
Small part of diet	2420	201
Moderate part	2769	209
Large part	507	42

TABLE 3.7

Question: Is there an association between fish consumption and prostate cancer?

Plan Be sure to state what the problem is about.

Variables Identify the variables and report the W's.

Be sure to check the appropriate condition.

I want to know if there is an association between fish consumption and prostate cancer.

The individuals are 6272 Swedish men followed by medical researchers for 30 years. The variables record their fish consumption and whether or not they were diagnosed with prostate cancer.

✔ **Categorical Data Condition:** I have counts for categories of both fish consumption and cancer diagnosis. It's okay to draw pie charts or bar charts.

Mechanics It's a good idea to check the marginal distributions first before looking at the two variables together.

Fish Consumption	Prostate Cancer		
	No	Yes	Total
Never/seldom	110	14	124 (2.0%)
Small part of diet	2420	201	2621 (41.8%)
Moderate part	2769	209	2978 (47.5%)
Large part	507	42	549 (8.8%)
Total	5806 (92.6%)	466 (7.4%)	6272 (100%)

Two categories of the diet are quite small, with only 2.0% Never/Seldom eating fish and 8.8% in the "Large part" category. Overall, 7.4% of the men in this study had prostate cancer.

Then, make appropriate displays to see whether there is a difference in the relative proportions. These pie charts compare fish consumption for men who have prostate cancer to fish consumption for men who don't.

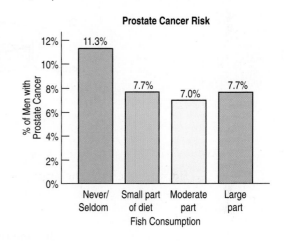

It's hard to see much difference in the pie charts. So, I made a display of the row percentages showing the risk of prostate cancer for each group:

Both pie charts and bar charts can be used to compare conditional distributions. Here we compare prostate cancer rates based on differences in fish consumption.

Conclusion   Interpret the patterns in the table and displays in context. If you can, discuss possible real-world consequences. Be careful not to overstate what you see. The results may not generalize to other situations.

Overall, there is a 7.4% rate of prostate cancer among men in this study. Most of the men (89.3%) ate fish either as a moderate or small part of their diet. It looks like this group of men had a somewhat lower cancer rate than those who never/seldom ate fish.

However, only 124 of these 6272 men fell into this category, and only 14 of them developed prostate cancer. More study would probably be needed before we would recommend that men change their diets.[2]

(continued)

---

[2]The original study actually used pairs of twins, which enabled the researchers to discern that the risk of cancer for those who never ate fish actually *was* substantially greater. Using pairs is a special way of gathering data. We'll discuss such study design issues and how to analyze the data in the later chapters.

This study is an example of looking at a sample of data to learn something about a larger population. We care about more than these particular 6272 Swedish men. We hope that learning about their experiences will tell us something about the value of eating fish in general. That raises the interesting question of what population we think this sample might represent. Do we hope to learn about all Swedish men? About all men? About the value of eating fish for all adult humans?[3] Often, it can be hard to decide just which population our findings may tell us about, but that also is how researchers decide what to look into in future studies.

🚫 WHAT CAN GO WRONG?

- **Don't violate the area principle.** This is probably the most common mistake in a graphical display. It is often made in the cause of artistic presentation. Here, for example, are two displays of the pie chart of the *Titanic* passengers by class:

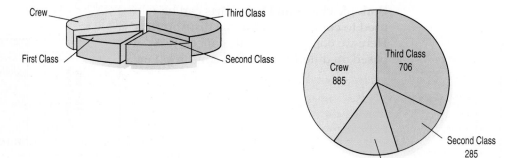

The one on the left looks pretty, doesn't it? But showing the pie on a slant violates the area principle and makes it much more difficult to compare fractions of the whole made up of each class—the principal feature that a pie chart ought to show.

- **Keep it honest.** Here's a pie chart that displays data on the percentage of high school students who engage in specified dangerous behaviors as reported by the Centers for Disease Control and Prevention. What's wrong with this plot?

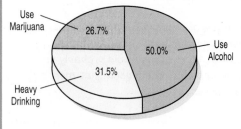

Try adding up the percentages. Or look at the 50% slice. Does it look right? Then think: What are these percentages of? Is there a "whole" that has been sliced up? In a pie chart, the proportions shown by each slice of the pie must add up to 100% and each individual must fall into only one category. But shouldn't the heavy drinkers be included in the "Use Alcohol" crowd? Of course, showing the pie on a slant makes it even harder to detect the error.

---

[3]Probably not, since we're looking only at prostate cancer risk; women don't have prostates.

Here's another. This bar chart shows the number of airline passengers searched in security screening, by year:

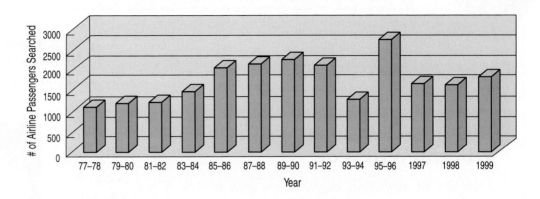

Looks like things didn't change much in the final years of the 20th century—until you read the bar labels and see that the last three bars represent single years while all the others are for *pairs* of years. Of course, the false depth makes it harder to see the problem.

- **Don't confuse similar-sounding percentages.** These percentages sound similar but are different:

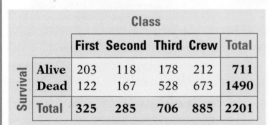

	First	Second	Third	Crew	Total
Alive	203	118	178	212	**711**
Dead	122	167	528	673	**1490**
Total	**325**	**285**	**706**	**885**	**2201**

- The percentage of the passengers who were both in first class and survived: This would be 203/2201, or 9.2%.
- The percentage of the first-class passengers who survived: This is 203/325, or 62.5%.
- The percentage of the survivors who were in first class: This is 203/711, or 28.6%.

  In each instance, pay attention to the *Who* implicitly defined by the phrase. Often there is a restriction to a smaller group (all aboard the *Titanic*, those in first class, and those who survived, respectively) before a percentage is found. Be sure to make these differences clear when you Tell your results.

- **Don't forget to look at the variables separately, too.** When you make a contingency table or display a conditional distribution, be sure you also examine the marginal distributions. It's important to know how many cases are in each category.

- **Be sure to use enough individuals.** When you consider percentages, take care that they are based on a large enough number of individuals. Take care not to make a report such as this one:

  We found that 66.67% of the rats improved their performance with training. The other rat died.

- **Don't overstate your case.** Independence is an important concept, but it is rare for two variables to be *entirely* independent. We can't conclude that one variable has no effect whatsoever on another. Usually, all we know is that little effect was observed in our study. Other studies of other groups under other circumstances could find different results.

CATEGORICAL DATA **IN YOUR WORLD**

In 2008, Senator Barack Obama was the Democratic candidate for U.S. President and Alaska Governor Sarah Palin was the Republican candidate for Vice President. Before this election, neither a woman nor an African-American had ever served in these offices. It is no surprise then that the interest in political polls was quite high in the days leading up to the election.

Poll: 7 in 10 Americans Say Obama Will Win

Even McCain supporters slightly more likely to say Obama, rather than McCain, will win

by Frank Newport

PRINCETON, NJ—By a 71% to 23% margin, Americans expect that Barack Obama will be elected president in next Tuesday's election.

The current Gallup Poll data, from Oct. 23–26, suggest that McCain's own supporters [doubt] that he will be able to pull off a victory in the election. By a 49% to 46% margin, McCain voters say Obama will win. On the other hand, perhaps not surprisingly, Obama voters overwhelmingly believe that their candidate will win, by a 94% to 2% margin.

WHO	1010 adults, age 18 and older
WHAT	Opinions of who will win election.
WHERE/ WHEN/ HOW	Telephone interviews conducted on October 23–26, 2008
WHY	To determine if voters think their chosen candidate will win.

Regardless of whom you support, and trying to be as objective as possible, who do you think will win the presidential election in November?

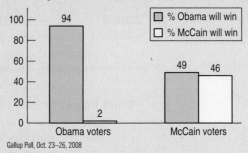

Gallup Poll, Oct. 23–26, 2008

http://www.gallup.com/poll/111559/Seven-Americans-Believe-Obama-Will-Win-Election.aspx

WHAT HAVE WE LEARNED?

We've learned that we can summarize categorical data by counting the number of cases in each category, sometimes expressing the resulting distribution as percents. We can display the distribution in a bar chart or a pie chart. When we want to see how two categorical variables are related, we put the counts (and/or percentages) in a two-way table called a contingency table.

▶ We look at the marginal distribution of each variable (found in the margins of the table).

▶ We also look at the conditional distribution of a variable within each category of the other variable.

▶ We can display these conditional and marginal distributions by using bar charts or pie charts.

▶ If the conditional distributions of one variable are (roughly) the same for every category of the other, the variables are independent. If not, we say there's an association.

Terms

Frequency table (Relative frequency table)	A frequency table lists the categories in a categorical variable and gives the count (or percentage) of observations for each category.
Distribution	The distribution of a variable gives ▶ the possible values of the variable and ▶ the relative frequency of each value.
Area principle	In a statistical display, each data value should be represented by the same amount of area.
Bar chart (Relative frequency bar chart)	Bar charts show a bar whose area represents the count (or percentage) of observations for each category of a categorical variable.
Pie chart	Pie charts show how a "whole" divides into categories by showing a wedge of a circle whose area corresponds to the proportion in each category.
Categorical data condition	The methods in this chapter are appropriate for displaying and describing categorical data. Be careful not to use them with quantitative data.
Contingency table	A contingency table displays counts and, sometimes, percentages of individuals falling into categories on two or more variables. The table categorizes the individuals on all variables at once to reveal possible patterns in one variable that may depend on the category of the other.
Marginal distribution	In a contingency table, the distribution of either variable alone is called the marginal distribution. The counts or percentages are the totals found in the margins (last row or column) of the table.
Conditional distribution	A conditional distribution restricts the *Who* to consider only a smaller group of individuals.
Independence	Variables are said to be independent if the conditional distribution of one variable is roughly the same for each category of the other.
Association	When we see evidence that one variable depends on another, we say there's an association between them.

Skills

▶ Be able to recognize when a variable is categorical and choose an appropriate display for it.

▶ Understand how to examine the association between categorical variables by comparing conditional and marginal percentages.

▶ Be able to summarize the distribution of a categorical variable with a frequency table.

▶ Be able to display the distribution of a categorical variable with a bar chart or pie chart.

▶ Know how to make and interpret a contingency table.

▶ Know how to make and interpret displays of the conditional distributions of one variable for two or more groups.

▶ Be able to describe the distribution of a categorical variable in terms of its possible values and relative frequencies.

▶ Be able to describe and discuss patterns and associations found in a contingency table and displays of conditional distributions.

DISPLAYING CATEGORICAL DATA ON THE COMPUTER

Although every package makes a slightly different bar chart, they all have similar features:

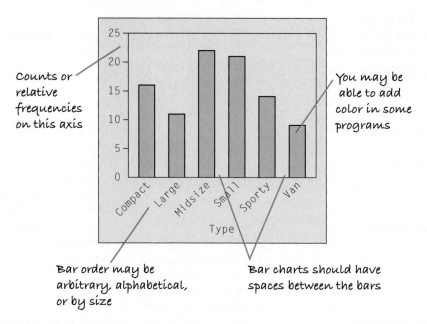

Counts or relative frequencies on this axis

You may be able to add color in some programs

Bar order may be arbitrary, alphabetical, or by size

Bar charts should have spaces between the bars

Sometimes the count or a percentage is printed above or on top of each bar to give some additional information. A pie chart may display those counts and/or percentages in each sector of the pie, or in a separate key as seen here:

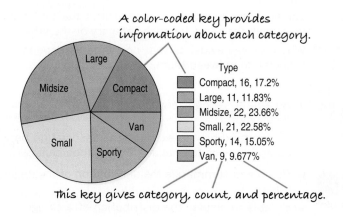

A color-coded key provides information about each category.

Type
Compact, 16, 17.2%
Large, 11, 11.83%
Midsize, 22, 23.66%
Small, 21, 22.58%
Sporty, 14, 15.05%
Van, 9, 9.677%

This key gives category, count, and percentage.

EXERCISES

A

1. **Movie genres.** The pie chart summarizes the genres of 120 first-run movies released in 2005.
   a) Is this an appropriate display for the genres? Why/why not?
   b) Which genre was least common?

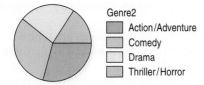

2. **Movie ratings.** The pie chart shows the ratings assigned to 120 first-run movies released in 2005.
   a) Is this an appropriate display for these data? Explain.
   b) Which was the most common rating?

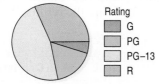

3. **Genres again.** Here is a bar chart summarizing the 2005 movie genres, as seen in the pie chart in Exercise 1.
   a) Which genre was most common?
   b) Is it easier to see that in the pie chart or the bar chart? Explain.

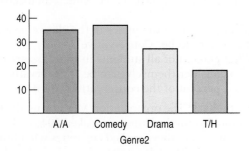

4. **Ratings again.** Here is a bar chart summarizing the 2005 movie ratings, as seen in the pie chart in Exercise 2.
   a) Which was the least common rating?
   b) An editorial claimed that there's been a growth in PG-13 rated films that, according to the writer, "have too much sex and violence," at the expense of G-rated films that offer "good, clean fun." The writer offered the bar chart below as evidence to support his claim. Does the bar chart support his claim? Explain.

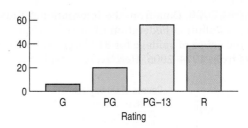

5. **Magnet schools.** An article in the Winter 2003 issue of *Chance* magazine reported on the Houston Independent School District's magnet schools programs. Of the 1755 qualified applicants, 931 were accepted, 298 were wait-listed, and 526 were turned away for lack of space. Find the relative frequency distribution of the decisions made, and write a sentence describing it.

6. **Magnet schools again.** The *Chance* article about the Houston magnet schools program described in Exercise 5 also indicated that 517 applicants were black or Hispanic, 292 Asian, and 946 white. Summarize the relative frequency distribution of ethnicity with a sentence or two (in the proper context, of course).

7. **Causes of death.** The Centers for Disease Control and Prevention (www.cdc.gov) lists causes of death in the United States:

Cause of Death	Percent
Heart disease	27.2
Cancer	23.1
Circulatory diseases and stroke	6.3
Respiratory diseases	5.1
Accidents	4.7

   a) Is it reasonable to conclude that heart or respiratory diseases were the cause of approximately 33% of U.S. deaths?
   b) What percent of deaths were from causes not listed here?

8. **Plane crashes.** An investigation compiled information about recent nonmilitary plane crashes (www.planecrashinfo.com). The causes, to the extent that they could be determined, are summarized in the table.

Cause	Percent
Pilot error	40
Other human error	5
Weather	6
Mechanical failure	14
Sabotage	6

   a) Is it reasonable to conclude that the weather or mechanical failures caused only about 20% of recent plane crashes?
   b) In what percent of crashes were the causes not determined?

9. **Causes of death again.** Create an appropriate display for the data described in Exercise 7.

10. **Plane crashes again.** Create an appropriate display for the data described in Exercise 8.

11. **Oil spills 2006.** Data from the International Tanker Owners Pollution Federation Limited (www.itopf.com) give the cause of spillage for 312 large oil tanker accidents from 1974–2006. Here's a bar graph:

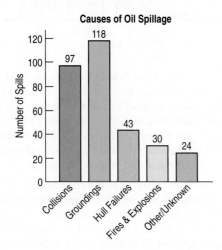

a) Would it be appropriate to make a pie chart for these data? Explain.
b) Summarize the findings of the poll in a few sentences that might appear in a newspaper article.

12. **Auditing reform.** In the wake of some corporate financing scandals, the Gallup Organization asked 1001 American adults what kind of changes, if any, are needed in the way major corporations are audited. Here's a display of the results.

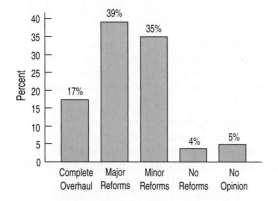

a) Would it be appropriate to make a pie chart for these data? Explain.
b) Summarize the findings of the poll in a few sentences that might appear in a newspaper article.

13. **Seniors.** Prior to graduation, a high school class was surveyed about its plans. The following table displays the results for white and minority students (the

"Minority" group included African-American, Asian, Hispanic, and Native American students):

	Seniors		
	White	**Minority**	**Total**
4-year college	198	44	**242**
2-year college	36	6	**42**
Military	4	1	**5**
Employment	14	3	**17**
Other	16	3	**19**
Total	**268**	**57**	**325**

(The row labels 4-year college, 2-year college, Military, Employment, Other are grouped under "Plans".)

a) What percent of the seniors are white?
b) What percent of the seniors are planning to attend a 2-year college?
c) What percent of the seniors are white and planning to attend a 2-year college?
d) What percent of the white seniors are planning to attend a 2-year college?
e) What percent of the seniors planning to attend a 2-year college are white?

14. **Politics.** Students in an Intro Stats course were asked to describe their politics as "Liberal," "Moderate," or "Conservative." Here are the results:

	Politics			
	L	**M**	**C**	**Total**
Female	35	36	6	**77**
Male	50	44	21	**115**
Total	**85**	**80**	**27**	**192**

(The row labels Female and Male are grouped under "Sex".)

a) What percent of the class is male?
b) What percent of the class considers themselves to be "Conservative"?
c) What percent of the males in the class consider themselves to be "Conservative"?
d) What percent of all students in the class are males who consider themselves to be "Conservative"?
e) What percent of the conservatives are male?

15. **Teen smokers.** The organization Monitoring the Future (www.monitoringthefuture.org) asked 2048 eighth graders who said they smoked cigarettes what brands they preferred. The table below shows brand preferences for two regions of the country. Is there an association between brand preference and region? Explain.

Brand preference	South	West
Marlboro	58.4%	58.0%
Newport	22.5%	10.1%
Camel	3.3%	9.5%
Other (over 20 brands)	9.1%	9.5%
No usual brand	6.7%	12.9%

16. Movies by genre and rating. Here's a table that classifies movies released in 2005 by genre and MPAA rating:

	G	PG	PG-13	R	Total
Action/Adventure	66.7	25	30.4	23.7	**29.2**
Comedy	33.3	60.0	35.7	10.5	**31.7**
Drama	0	15.0	14.3	44.7	**23.3**
Thriller/Horror	0	0	19.6	21.1	**15.8**
Total	100%	100%	100%	100%	100%

Is there an association between types of movies and their ratings? Explain.

17. Global warming. The Pew Research Center for the People and the Press (http://people-press.org) has asked a representative sample of U.S. adults about global warming, repeating the question over time. In January 2007, the responses reflected an increased belief that global warming is real and due to human activity. Here's a display of the percentages of respondents choosing each of the major alternatives offered:

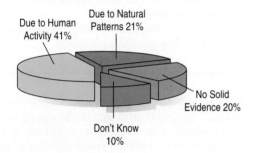

List the errors in this display.

18. Teens and technology. The Gallup organization surveyed 744 teenagers, asking them what technologies they use every day. Here's a graph summarizing the responses.
a) How is this graph misleading?
b) Explain why we could not use a pie chart for these data.

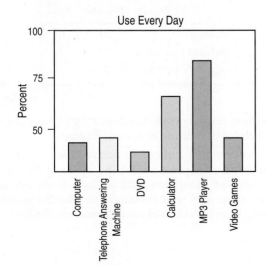

19. Cars. A survey of autos parked in student and staff lots at a large university classified the brands by country of origin, as seen in the table.

		Driver	
		Student	**Staff**
Origin	**American**	107	105
	European	33	12
	Asian	55	47

a) What percent of all the cars surveyed were American?
b) What percent of the American cars were owned by students?
c) What percent of the students owned American cars?
d) What is the marginal distribution of the drivers?
e) What is the conditional distribution of driver for American cars?

20. Politics revisited. Here again is the table summarizing the political views of males and females in an Intro Stats class.

		Politics			
		L	**M**	**C**	**Total**
Sex	**Female**	35	36	6	**77**
	Male	50	44	21	**115**
	Total	85	80	27	**192**

a) What percent of the class called themselves "Liberal"?
b) What percent of the females were liberals?
c) What percent of the liberals were females?
d) What's the marginal distribution of *Sex*?
e) Write a sentence describing the conditional distribution of *Sex* among the liberals.

21. Cars again. Look again at the table about the drivers and origins of cars in Exercise 19.
a) What's the marginal distribution of *Origin*?
b) What's the conditional distribution of *Origin* for the students' cars?
c) What's the conditional distribution of *Origin* for the cars driven by staff?
d) Do you think the origin of the car is independent of the type of driver? Explain.

22. Politics again. Have another look at the table showing Stats students' political views in Exercise 20.
a) What's the marginal distribution of *Politics*?
b) What's the conditional distribution of *Politics* among the males?

c) What's the conditional distribution of *Politics* among the females?

d) Do you think a student's political outlook is independent of the student's sex? Explain.

23. Twins. In 2000, the *Journal of the American Medical Association (JAMA)* published a study that examined pregnancies that resulted in the birth of twins. Births were classified as preterm with intervention (induced labor or cesarean), preterm without procedures, or term/post-term. Researchers also classified the pregnancies by the level of prenatal medical care the mother received (inadequate, adequate, or intensive). The data, from the years 1995–1997, are summarized in the table below. Figures are in thousands of births. (*JAMA* 284 [2000]:335–341)

Twin Births 1995–1997 (In Thousands)

	Preterm (induced or cesarean)	Preterm (without procedures)	Term or post-term	Total
Intensive	18	15	28	**61**
Adequate	46	43	65	**154**
Inadequate	12	13	38	**63**
Total	**76**	**71**	**131**	**278**

Level of Prenatal Care

a) What percent of these mothers received inadequate medical care during their pregnancies?

b) What percent of all twin births were preterm?

c) Among the mothers who received inadequate medical care, what percent of the twin births were preterm?

d) Write a sentence describing the marginal distribution of *Care*.

e) Write a sentence describing the conditional distribution of *Care* among mothers who had induced or cesarean preterm births.

24. Blood pressure. A company held a blood pressure screening clinic for its employees. The results are summarized in the table below by age group and blood pressure level:

		Age	
	Under 30	**30–49**	**Over 50**
Low	27	37	31
Normal	48	91	93
High	23	51	73

Blood Pressure

a) What percent of the employees had high blood pressure?

b) What percent of those employees over 50 had high blood pressure?

c) What percent of the employees with high blood pressure were over 50?

d) Write a sentence describing the marginal distribution of *Blood Pressure* for the employees.

e) Write a sentence describing the conditional distribution of *Blood Pressure* among the employees who were over 50.

25. Twins again. Look again at the table in Exercise 23, summarizing data about the level of prenatal care received by mothers of twins and the type of birth they had.

a) Find the marginal distribution of the types of *Birth*.

b) Find the conditional distribution of the types of *Birth* among those women who received inadequate prenatal care.

c) Do you think there's an association between type of *Birth* and level of *Care*? Explain.

d) Does this mean that inadequate prenatal care may prevent premature births? Explain.

26. Blood pressure revisited. Look again at the table in Exercise 24, summarizing data for the ages and blood pressure of a company's employees.

a) Find the marginal distribution of *Age* for the employees.

b) Find the conditional distribution of *Age* among the employees with high blood pressure.

c) Do you think there's an association between *Age* and *Blood Pressure*? Explain.

d) Does this prove that people's blood pressure increases as they get older? Explain.

C

27. Anorexia. Hearing anecdotal reports that some patients undergoing treatment for the eating disorder anorexia seemed to be responding positively to the antidepressant Prozac, medical researchers conducted an experiment to investigate. They found 93 women being treated for anorexia who volunteered to participate. For one year, 49 randomly selected patients were treated with Prozac and the other 44 were given an inert substance called a placebo. At the end of the year, patients were diagnosed as healthy or relapsed, as summarized in the table:

	Prozac	Placebo	Total
Healthy	35	32	67
Relapse	14	12	26
Total	49	44	93

Do these results provide evidence that Prozac might be helpful in treating anorexia? Explain.

28. Antidepressants and bone fractures. For a period of five years, physicians at McGill University Health Center followed more than 5000 adults over the age of 50. The researchers were investigating whether people taking a certain class of antidepressants (SSRIs) might be at greater risk of bone fractures. Their observations are summarized in the table:

	Taking SSRI	No SSRI	Total
Experienced fractures	14	244	258
No fractures	123	4627	4750
Total	137	4871	5008

Do these results suggest there's an association between taking SSRI antidepressants and experiencing bone fractures? Explain.

29. Magnet schools revisited. The *Chance* magazine article described in Exercise 5 further examined the impact of an applicant's ethnicity on the likelihood of admission to the Houston Independent School District's magnet schools programs. Those data are summarized in the following table:

		Admission Decision		
		Accepted	Wait-listed	Turned away	Total
Ethnicity	**Black/Hispanic**	485	0	32	517
	Asian	110	49	133	292
	White	336	251	359	946
	Total	931	300	524	1755

Does it appear that the admissions decisions are made independent of the applicant's ethnicity? Explain.

30. Tattoos. A study by the University of Texas Southwestern Medical Center examined 626 people to see if an increased risk of contracting hepatitis C was associated with having a tattoo. If the subject had a tattoo, researchers asked whether it had been done in a commercial tattoo parlor or elsewhere. Write a brief description of the association between tattooing and hepatitis C, including an appropriate graphical display.

	Tattoo done in commercial parlor	Tattoo done elsewhere	No tattoo
Has hepatitis C	17	8	18
No hepatitis C	35	53	495

31. Graphs in the news. Find a bar graph of categorical data from a newspaper, a magazine, or the Internet.
a) Is the graph clearly labeled?
b) Does it violate the area principle?
c) Does the accompanying article tell the W's of the variable?
d) Do you think the article correctly interprets the data? Explain.

32. Graphs in the news II. Find a pie chart of categorical data from a newspaper, a magazine, or the Internet.
a) Is the graph clearly labeled?
b) Does it violate the area principle?
c) Does the accompanying article tell the W's of the variable?
d) Do you think the article correctly interprets the data? Explain.

33. Tables in the news. Find a frequency table of categorical data from a newspaper, a magazine, or the Internet.
a) Is it clearly labeled?
b) Does it display percentages or counts?
c) Does the accompanying article tell the W's of the variable?
d) Do you think the article correctly interprets the data? Explain.

34. Tables in the news II. Find a contingency table of categorical data from a newspaper, a magazine, or the Internet.
a) Is it clearly labeled?
b) Does it display percentages or counts?
c) Does the accompanying article tell the W's of the variables?
d) Do you think the article correctly interprets the data? Explain.

Answers

Do The Math

1. a) 69%   b) 5   c) 12%

2. a) 1170   b) 35%   c) 630

3. a) 20%   b) 15.3   c) 1900
d) 7.29   e) 180%   f) 175

Just Checking

1. $\frac{4}{32} = 12.5\%$ Blue, $\frac{16}{32} = 50\%$ Brown,

$\frac{12}{32} = 37.5\%$ Green/Hazel/Other

2. In this Statistics class there are 16 females with brown eyes.

3. 34 males, 32 females

4. $\frac{12}{66} \approx 18.2\%$ Blue, $\frac{36}{66} \approx 54.5\%$ Brown,

$\frac{18}{66} \approx 27.3\%$ Green/Hazel/Other

5. In this Statistics class most students (nearly 55%) had brown eyes, compared to only 18.2% with blue eyes and 27.3% whose eyes were green, hazel, or some other color.

6. 66.7% of the students with blue eyes are male. (8/12)

7. 12.1% of these Statistics students are males with blue eyes. (8/66)

8. Among the male students in this class, 23.5% have blue eyes. (8/34)

9. 51.9% of the non-blue-eyed students are female. (28/54)

10. The brown-eyed students are 55.6% male and 44.4% female.

11. Of the males, 23.5% have blue eyes, 58.8% brown eyes, and the remaining 17.6% have green, hazel, or some other color eyes.

12. Overall: 48.5% female. However, 33.3% of students with blue eyes are female, as are 44.4% of the brown-eyed students and 66.7% of the others.

13. Blue-eyed students appear to be less likely to be female, and those with green, hazel, or other color eyes more likely. It appears *Eye Color* and *Sex* may not be independent. (But the number of students studied is small.)

Exploring Quantitative Data

T sunamis are potentially destructive waves, most often caused by earthquakes beneath the sea. The tsunami of December 26, 2004, with epicenter off the west coast of Sumatra, was caused by an earthquake of magnitude 9.0 on the Richter scale. It killed an estimated 297,248 people, making it the most disastrous tsunami on record. But was the earthquake that caused it truly extraordinary, or did it just happen at an unlucky place and time? The U.S. National Geophysical Data Center[1] has information on more than 2400 tsunamis dating back to 2000 B.C.E., and we have estimates of the magnitude of the underlying earthquake for 1240 of them. What can we learn from these data?

[1]www.ngdc.noaa.gov

Histograms

Let's start with a picture. For categorical variables, it is easy to draw the distribution because each category is a natural "pile." But for quantitative variables, there's no obvious way to choose piles. So, usually, we slice up all the possible values into equal-width bins. We then count the number of cases that fall into each bin. The bins, together with these counts, give the **distribution** of the quantitative variable and provide the building blocks for a picture by representing each count as a bar at each bin value. A **histogram** displays the distribution at a glance.

For example, here are the *Magnitudes* (on the Richter scale) of the 1240 earthquakes in the NGDC data:

WHO	1240 earthquakes known to have caused tsunamis for which we have data or good estimates
WHAT	Magnitude (Richter scale[2]), depth (m), date, location, and other variables
WHEN	From 2000 B.C.E. to the present
WHERE	All over the earth

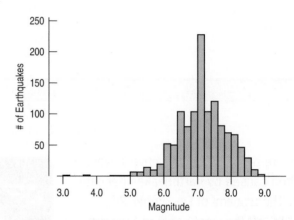

FIGURE 4.1 **A histogram of earthquake magnitudes** shows the number of earthquakes with magnitudes (in Richter scale units) in each bin.

One surprising feature of the earthquake magnitudes is the spike around magnitude 7.0. Only one other bin holds even half that many earthquakes. These values include historical data for which the magnitudes were estimated by experts and not measured by modern seismographs. Perhaps the experts thought 7 was a typical and reasonable value for a tsunami-causing earthquake when they lacked detailed information. That would explain the overabundance of magnitudes right at 7.0 rather than spread out near that value.

Like a bar chart, a histogram plots the bin counts as the heights of bars. In this histogram of earthquake magnitudes, each bin has a width of 0.2, so, for example, the height of the tallest bar says that there were about 230 earthquakes with magnitudes between 7.0 and 7.2.

Does the distribution of earthquake magnitudes look as you expected? It is often a good idea to *imagine* what a distribution might look like before you make a display. That way you'll be less likely to be fooled by errors in the data or when you accidentally graph the wrong variable.

From the histogram, we can see that these earthquakes typically have magnitudes around 7. Most are between 5.5 and 8.5, and some are as small as 3 and as big as 9. Now we can answer the question about the Sumatra tsunami. With a value of 9.0 it's clear that the earthquake that caused it was an extraordinarily powerful earthquake—one of the largest on record.[3]

The bar charts of categorical variables we saw in Chapter 3 had spaces between the bars to separate the counts of different categories. But in a histogram, the bins slice up *all the values* of the quantitative variable, so any spaces in a histogram are actual **gaps** in the data, indicating a region where there are no values.

---

[2]Technically, Richter scale values are in units of log dyne-cm. But the Richter scale is so common now that usually the units are assumed. The U.S. Geological Survey gives the background details of Richter scale measurements on its website www.usgs.gov/.
[3]Some experts now estimate the magnitude at between 9.1 and 9.3.

Sometimes we make a **relative frequency histogram**, showing the *percentage* of the total number of cases falling in each bin. Of course, the shape of the histogram is exactly the same; only the vertical scale is different.

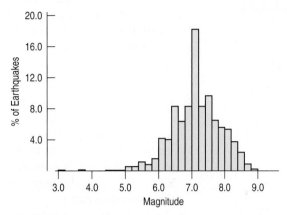

FIGURE 4.2 A relative frequency histogram looks just like a frequency histogram except for the labels on the y-axis, which now show the percentage of earthquakes in each bin.

TI Tips — Making a histogram

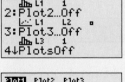

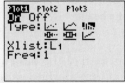

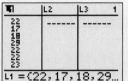

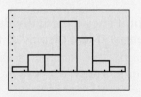

Your calculator can create histograms. First you need some data. For an agility test, fourth-grade children jump from side to side across a set of parallel lines, counting the number of lines they clear in 30 seconds. Here are their scores:

22, 17, 18, 29, 22, 22, 23, 24, 23, 17, 21, 25, 20

12, 19, 28, 24, 22, 21, 25, 26, 25, 16, 27, 22

Enter these data into L1.

Now set up the calculator's plot:

- Go to **2nd STATPLOT**, choose **Plot1**, then **ENTER**.
- In the **Plot1** screen choose **On**, select the little histogram icon, then specify **Xlist:L1** and **Freq:1**.
- Be sure to turn off any other graphs the calculator may be set up for. Just hit the **Y=** button, and deactivate any functions seen there.

All set? To create your preliminary plot go to **ZOOM**, select **9:ZoomStat**, and then **ENTER**.

You now see the calculator's initial attempt to create a histogram of these data. Not bad. We can see that the distribution is roughly symmetric. But it's hard to tell exactly what this histogram shows, right? Let's fix it up a bit.

- Under **WINDOW**, let's reset the bins to convenient, sensible values. Try **Xmin=12**, **Xmax=30**, and **Xscl=2**. That specifies the range of values along the *x*-axis and makes each bar span two lines.
- Hit **GRAPH** (*not* **ZoomStat**—this time we want control of the scale!).

There. We still see rough symmetry, but also see that one of the scores was much lower than the others. Note that you can now find out exactly what the bars indicate by activating **TRACE** and then moving across the histogram using the arrow keys. For each bar the calculator will indicate the interval of values

(continued)

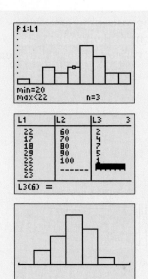

and the number of data values in that bin. We see that 3 kids had agility scores of 20 or 21.

Play around with the WINDOW settings. A different Ymax will make the bars appear shorter or taller. What happens if you set the bar width (Xscl) smaller? Or larger? You don't want to lump lots of values into just a few bins or make so many bins that the overall shape of the histogram is not clear. Choosing the best bar width takes practice.

Finally, suppose the data are given as a frequency table. Consider a set of test scores, with two grades in the 60s, four in the 70s, seven in the 80s, five in the 90s, and one 100. Enter the group cutoffs 60, 70, 80, 90, 100 in L2 and the corresponding frequencies 2, 4, 7, 5, 1 in L3. When you set up the histogram STATPLOT, specify Xlist:L2 and Freq:L3. Can you specify the WINDOW settings to make this histogram look the way you want it? (By the way, if you get a DIM MISMATCH error, it means you can't count. Look at L2 and L3; you'll see the two lists don't have the same number of entries. Fix the problem by correcting the data you entered.)

Stem-and-Leaf Displays

Histograms are easy to understand, but they don't show the data values themselves. Here's a histogram of the pulse rates of 24 women, taken by a researcher at a health clinic:

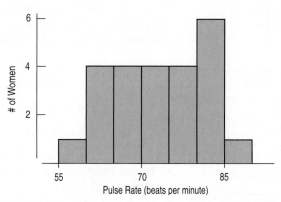

FIGURE 4.3 **The pulse rates of 24 women at a health clinic**

The story seems pretty clear. We can see the entire span of the data and can easily see what a typical pulse rate might be. But is that all there is to these data?

A stem-and-leaf display is like a histogram, but it shows the individual values. It's also easier to make by hand. Here's a stem-and-leaf display of the same data:

The stem-and-leaf display was devised by John W. Tukey, one of the greatest statisticians of the 20th century. It is called a "Stemplot" in some texts and computer programs, but we prefer Tukey's original name for it.

```
8 | 8
8 | 000044
7 | 6666
7 | 2222
6 | 8888
6 | 0444
5 | 6
```
Pulse Rate
(8|8 means 88 beats/min)

Turn the stem-and-leaf on its side (or turn your head to the right) and squint at it. It should look roughly like the histogram of the same data. Does it? (Well, it's backwards because now the higher values are on the left, but other than that, it has the same shape.[4])

See the line at the top that says 8 | 8? It stands for a pulse of 88 beats per minute (bpm). We've taken the tens place of the number and made that the "stem." Then we sliced off the ones place and made it a "leaf." The next line down is 8 | 000044. That shows that there were four pulse rates of 80 and two of 84 bpm.

Stem-and-leaf displays are especially useful when you make them by hand for batches of fewer than a few hundred data values. They are a quick way to display—and even to record—numbers. Because the leaves show the individual values, we can sometimes see even more in the data than the distribution's shape. Take another look at all the leaves of the pulse data. See anything unusual? At a glance you can see that they are all even. With a bit more thought you can see that they are all multiples of 4—something you couldn't possibly see from a histogram. How do you think the nurse took these pulses? Counting beats for a full minute or counting for only 15 seconds and multiplying by 4?

How do stem-and-leaf displays work? Stem-and-leaf displays work like histograms, but they show more information. They use part of the number itself (called the stem) to name the bins. To make the "bars," they use the next digit of the number. For example, if we had a test score of 83, we could write it 8 | 3, where 8 serves as the stem and 3 as the leaf. Then, to display the scores 83, 76, and 88 together, we would write

```
8 | 38
7 | 6
```

For the pulse data, we have

```
8 | 0000448
7 | 22226666
6 | 04448888
5 | 6
```
Pulse Rate
(5 | 6 means 56 beats/min)

This display is OK, but a little crowded. A histogram might split each line into two bars. With a stem-and-leaf, we can do the same by putting the leaves 0–4 on one line and 5–9 on another, as we saw above:

```
8 | 8
8 | 000044
7 | 6666
7 | 2222
6 | 8888
6 | 0444
5 | 6
```
Pulse Rate
(8 | 8 means 88 beats/min)

(*continued*)

---

[4]You could make the stem-and-leaf with the higher values on the bottom. Usually, though, higher on the top makes sense.

For numbers with three or more digits, you'll often decide to truncate (or round) the number to two places, using the first digit as the stem and the second as the leaf. So, if you had 432, 540, 571, and 638, you might display them as shown below with an indication that 6 | 3 means 630–639.

$$
\begin{array}{c|l}
6 & 3 \\
5 & 4\ 7 \\
4 & 3 \\
\end{array}
$$

When you make a stem-and-leaf by hand, make sure to give each digit the same width, in order to preserve the area principle. (That can lead to some fat 1's and thin 8's—but it makes the display honest.)

Do The Math

Stem-and-leaf displays are quick ways to make a picture for small sets of data, like these numbers posted by NFL quarterback Brett Favre during his record-breaking career with the Green Bay Packers, New York Jets, and Minnesota Vikings.

1. Here are the number of passes Favre completed each season:

    363, 343, 356, 343, 372, 346, 308, 341, 314,
    338, 341, 347, 304, 325, 359, 363, 318, 302

    Make a stem-and-leaf display using the first two digits as the stem and the last digit as the leaf. For example, 363 completions should look like this: 36 | 3.

2. Make a stem-and-leaf display showing the number of touchdown passes Favre threw each season:

    33, 22, 28, 18, 20, 30, 32, 27, 32, 20, 22, 31, 35, 39, 38, 33, 19, 18

3. You're probably not satisfied with that display of Brett's touchdown passes, because all of the data got crunched into just 3 lines—one for the teens, one for the 20s, and one for the 30s. Try again, this time splitting the stems. For example, that means you can make two bins for the 20s—one for 20–24 TDs and the other for 25–29.

(Check your answers on page 81.)

Dotplots

A **dotplot** is pretty simple. It just places a dot along an axis for each case in the data. It's like a stem-and-leaf display, but with dots instead of digits for all the leaves. Dotplots are a great way to display a small data set. On the next page is a dotplot of the time (in seconds) that the winning horse took to win the Kentucky Derby in each race between the first Derby in 1875 and the 2008 Derby.

Dotplots show basic facts about the distribution. We can find the slowest and quickest races by finding times for the topmost and bottommost dots. It's also clear that there are two clusters of points, one just below 160 seconds and the other at about 122 seconds. Something strange happened to the Derby times. Once we know to look for it, we can find out that in 1896 the distance of the Derby race was changed from 1.5 miles to the current 1.25 miles. That explains the two clusters of winning times.

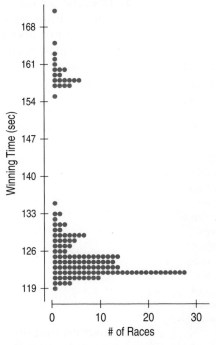

FIGURE 4.4 A dotplot of Kentucky Derby winning times plots each race as its own dot, showing the bimodal distribution.

Some dotplots stretch out horizontally, with the counts on the vertical axis, like a histogram. Others, such as the one shown here, run vertically, like a stem-and-leaf display. Newspapers sometimes offer dotplots using little pictures—pictographs—instead of dots.

Think Before You Draw, Again

Suddenly, we face a lot more options when it's time to invoke our first rule of data analysis and make a picture. You'll need to *Think* carefully to decide which type of graph to make. In the previous chapter you learned to check the Categorical Data Condition before making a pie chart or a bar chart. Now, before making a stem-and-leaf display, a histogram, or a dotplot, you need to check the

> **Quantitative Data Condition:** The data are values of a quantitative variable whose units are known.

Although a bar chart and a histogram may look somewhat similar, they're not the same. You can't display categorical data in a histogram or quantitative data in a bar chart. Always check the condition that confirms what type of data you have before proceeding with your display.

After you've made your histogram, stem-and-leaf, or dotplot, it's time to Tell what you see. When you describe a distribution, you should always Tell about three things: its **shape, center**, and **spread**.

The mode is sometimes defined as the single value that appears most often. That definition is fine for categorical variables because all we need to do is count the number of cases for each category. For quantitative variables, the mode is more ambiguous. What is the mode of the Kentucky Derby times? Well, seven races were timed at 122.2 seconds—more than any other race time. Should that be the mode? Probably not. For quantitative data, it makes more sense to use the term "mode" in the more general sense of the peak of the histogram rather than as a single summary value. In this sense, the important feature of the Kentucky Derby races is that there are two distinct modes, representing the two different versions of the race and warning us to consider those two versions separately.

The Shape of a Distribution

1. *Does the histogram have a single, central hump or several separate humps?* These humps are called **modes**.[5] The earthquake magnitudes have a single mode at just about 7. A histogram with one peak, such as the earthquake magnitudes, is dubbed **unimodal**; histograms with two peaks are **bimodal**, and those with three or more are called **multimodal**.[6] For example, here's a bimodal histogram.

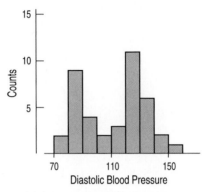

FIGURE 4.5 A bimodal histogram has two apparent peaks.

A histogram that doesn't appear to have any mode and in which all the bars are approximately the same height is called **uniform**.

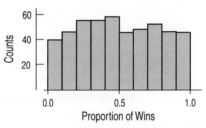

FIGURE 4.6 A uniform histogram. The bars are all about the same height. The histogram doesn't appear to have a mode.

2. *Is the histogram **symmetric**?* Can you fold it along a vertical line through the middle and have the edges match pretty closely, or are more of the values on one side?

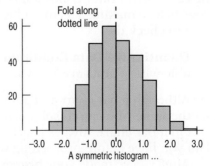

A symmetric histogram ...

FIGURE 4.7

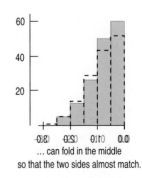

... can fold in the middle so that the two sides almost match.

You've heard of pie à la mode. Is there a connection between pie and the mode of a distribution? Actually, there is! The mode of a distribution is a *popular* value near which a lot of the data values gather. And "à la mode" means "in style"—*not* "with ice cream." That just happened to be a *popular* way to have pie in Paris around 1900.

---

[5]Well, technically, it's the value on the horizontal axis of the histogram that is the mode, but anyone asked to point to the mode would point to the hump.
[6]Apparently, statisticians don't like to count past two.

The (usually) thinner ends of a distribution are called the **tails**. If one tail stretches out farther than the other, the histogram is said to be **skewed** to the side of the longer tail. Here are two skewed histogram displaying one year's data about female heart attack patients in New York City. The distribution of the women's ages (blue) is *skewed to the left,* and the distribution of hospital charges (pink) is *skewed to the right*.

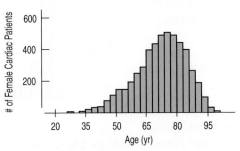

FIGURE 4.8 **Skewed distributions.**

3. *Do any unusual features stick out?* Often gaps and outliers can tell us something interesting or exciting about the data.

   Gaps help us see multiple modes and encourage us to notice when the data may come from different sources or contain more than one group. The Kentucky Derby data that we saw in the dotplot on page 51 has a large gap between two groups of times, one near 120 seconds and one near 160. That alerted us to the fact that the race distance has changed.

   You should also always mention any stragglers, or **outliers**, that stand off away from the body of the distribution. If you're collecting data on nose lengths and Pinocchio is in the group, you'd probably notice him, and you'd certainly want to mention it.

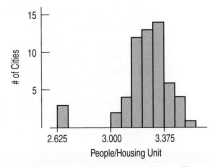

FIGURE 4.9 **A histogram with outliers.**
There are three cities in the leftmost bar. We wonder why these cities are different.

Outliers can affect almost every method we discuss in this course. So we'll always be on the lookout for them. An outlier can be the most informative part of your data. Or it might just be an error. But don't throw it away without comment. Treat it specially and discuss it when you tell about your data. Or find the error and fix it if you can. Be sure to look for outliers. Always.

For Example Describing Histograms

A credit card company wants to see how much customers in a particular segment of their market use their credit card. They have provided you with data[7] on the amount spent by 500 selected customers during a 3-month period and have asked you to summarize the expenditures. Of course, you begin by making a histogram.

Question: How would you describe the shape, center, and spread of this distribution?

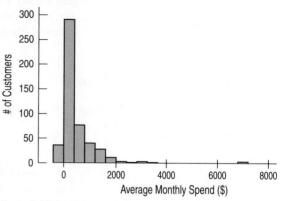

The distribution of credit card expenditures is unimodal and skewed to the high end. Customers typically spent around $500, but over half spend less than $400. There is an extraordinarily large value at about $7000, and some of the expenditures are negative.

Toto, I've a feeling we're not in math class anymore . . . When Dorothy and her dog Toto land in Oz, everything is more vivid and colorful, but also more dangerous and exciting. Dorothy has new choices to make. She can't always rely on the old definitions, and the yellow brick road has many branches. You may be coming to a similar realization about Statistics.

When we summarize data, our goal is usually more than just developing a detailed knowledge of the data we have at hand. We want to know what the data say about the world, so we'd like to know whether the patterns we see in histograms and summary statistics generalize to other individuals and situations. Scientists generally don't care about the particular guinea pigs they've treated, but rather about what their reactions say about how animals (and, perhaps, humans) would respond.

Because we want to see broader patterns rather than focus on the details of the data set we're looking at, many of the most important concepts in Statistics are not precisely defined. Whether a histogram is symmetric or skewed, whether it has one or more modes, whether a point is far enough from the rest of the data to be considered an outlier—these are all somewhat vague concepts. They all require judgment. You may be used to finding a single correct and precise answer, but in Statistics, there may be more than one interpretation. That may make you a little uncomfortable at first, but soon you'll see that this room for judgment brings you enormous power and responsibility. It means that your own knowledge about the world and your judgment matter. Supporting your findings with statistical evidence and justifications entitles you to your own opinions about what you see.

Just Checking

It's often a good idea to think about what the distribution of a data set might look like before we collect the data. What do you think the distribution of each of the following data sets will look like? Make a rough sketch of a possible histogram. Be sure to discuss its shape. Where do you think the center might be? How spread out do you think the values will be?

1. Number of miles run by Saturday morning joggers at a park.

2. Hours spent by U.S. adults watching football on Thanksgiving Day.

3. Amount of winnings of all people playing a particular state's lottery last week.

4. Ages of the faculty members at your school.

5. Last digit of phone numbers of your classmates.

(Check your answers on page 82.)

---

[7]These data are real, but cannot be further identified for obvious privacy reasons.

The Center of the Distribution: The Median

Let's think some more about the tsunami earthquakes. But this time, we'll look at just 25 years of data: 176 earthquakes that occurred from 1981 through 2005. These measurements should be more accurate because seismographs were in wide use. When we think of a typical value, we usually look for the **center** of the distribution. Where do you think the center of this distribution is? For a unimodal, symmetric distribution such as these earthquake data, it's easy. We'd all agree on the center of symmetry, where we would fold the histogram to match the two sides. But when the distribution is skewed or possibly multimodal, it's not immediately clear what we even mean by the center.

One reasonable choice of typical value is the value that is literally in the middle, with half the values below it and half above it.

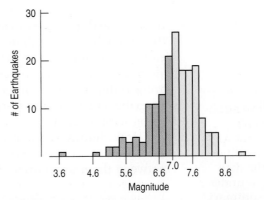

FIGURE   4.10   Tsunami-causing   earthquakes (1981–2005) The median splits the histogram into two halves of equal area.

Histograms follow the area principle, and the middle value that divides the histogram into two equal areas is called the **median.**

For the recent tsunamis, there are 176 earthquakes, so each half has 88 earthquakes. That makes the median the average of the two values on either side of the middle: the 88th and the 89th. The median earthquake magnitude is 7.0.

Be sure to include the units whenever you discuss the median. Knowing the median, we could say that a typical tsunami-causing earthquake, worldwide, was about 7.0 on the Richter scale. But there's more to say. After all, not every earthquake has a Richter scale value of 7.0. Whenever we find the center of data, the next step is always to look at how spread out the data are.

We always use n to indicate the number of values. Some people even say, "How big is the n?" when they mean the number of data values.

> **How do medians work?** Finding the median of a batch of n numbers is easy as long as you remember to put the numbers in order first. If n is odd, the median is the middle value.
>
> Suppose the data are these values: 14.1, 3.2, 25.3, 2.8, −17.5, 13.9, 45.8.
>
> First we order the values: −17.5, 2.8, 3.2, 13.9, 14.1, 25.3, 45.8.
>
> Since there are 7 values, the median is the $(7 + 1)/2 = $ 4th value, counting from the top or bottom: 13.9. Notice that 3 values are lower, 3 higher.
>
> When n is even, the median is the average of the two middle values.
>
> Suppose we had the same data plus another value at 35.7. Then the ordered values are −17.5, 2.8, 3.2, 13.9, 14.1, 25.3, 35.7, 45.8.
>
> Now the median is the average of the 4th and the 5th values. So the median is $(13.9 + 14.1)/2 = $ 14.0. Four data values are lower, and four higher.

Do The Math

4. Here again are the number of passes Brett Favre completed during each season. What's the median?

302, 304, 308, 314, 318, 325, 338, 341, 341,
343, 343, 346, 347, 359, 356, 363, 363, 372

5. And here are the number of touchdown passes Favre threw during his seasons with the Green Bay Packers. Find the median. (Don't forget to arrange the data in order first!)

28, 18, 20, 30, 32, 27, 32, 20, 22, 31, 35, 39, 38, 33, 19, 18

(Check your answers on page 81.)

Spread: Home on the Range

Statistics pays close attention to what we *don't* know as well as what we do know. Understanding how spread out the data are is a first step in understanding what a summary *cannot* tell us about the data. It's the beginning of telling us what we don't know.

If every earthquake that caused a tsunami registered 7.0 on the Richter scale, then knowing the median would tell us everything about the distribution of earthquake magnitudes. The more the data vary, however, the less the median alone can tell us. So we need to measure how much the data values vary around the center. In other words, how spread out are they? When we describe a distribution numerically, we always report a measure of its **spread** along with its center.

How should we measure the spread? We could simply ask: How far apart are the two extremes? The **range** of the data is defined as the *difference* between the maximum and minimum values:

$$Range = max - min.$$

Notice that the range is a *single number, not* an interval of values, as you might think from its use in common speech. The maximum magnitude of these earthquakes is 9.0 and the minimum is 3.7, so the *range* is $9.0 - 3.7 = 5.3$.

The range has the disadvantage that a single extreme value can make it very large, giving a value that doesn't really describe the data very well

Spread: The Interquartile Range

A better way to describe the spread of a variable ignores the extremes and concentrates on the middle half of the data. What do we mean by the middle half? Divide the data in half at the median. Now divide both halves in half again, cutting the data into four quarters. We call these new dividing points **quartiles**. One quarter of the data lies below the **lower quartile,** and one quarter of the data lies above the **upper quartile,** so half the data lies between them. The quartiles border the middle half of the data.

The difference between the quartiles tells us how much territory the middle half of the data covers and is called the **interquartile range.** It's commonly abbreviated IQR (and pronounced "eye-cue-are," not "ikker"):

$$IQR = upper\ quartile - lower\ quartile.$$

Even if a distribution itself is skewed or has some outliers, the IQR provides useful information about the spread.

For the earthquakes, there are 88 values below the median and 88 values above the median. The midpoint of the lower half is the average of the 44th and 45th values in the ordered data; that turns out to be 6.6. In the upper half we average the 132nd and 133rd values, finding a magnitude of 7.6 as the third quartile. The *difference* between the quartiles gives the IQR:

$$IQR = 7.6 - 6.6 = 1.0.$$

Now we know that the middle half of the earthquake magnitudes extends across a (interquartile) range of 1.0 Richter scale units. This seems like a reasonable summary of the spread of the distribution, as we can see from this histogram:

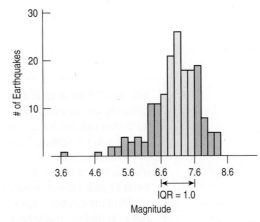

FIGURE 4.11 IQR. The quartiles bound the middle 50% of the values of the distribution. This gives a visual indication of the spread of the data. Here we see that the IQR is 1.0 Richter scale units.

The lower and upper quartiles are also known as the 25th and 75th **percentiles** of the data, respectively, since the lower quartile falls above 25% of the data and the upper quartile falls above 75% of the data. If we count this way, the median is the 50th percentile. We could, of course, define and calculate any percentile that we want. For example, the 10th percentile would be the number that falls above the lowest 10% of the data values.

How do quartiles work? A simple way to find the quartiles is to start by splitting the batch into two halves at the median. (When *n* is odd, some statisticians include the median in both halves; others omit it.) The lower quartile is the median of the lower half, and the upper quartile is the median of the upper half.

Here are our two examples again.

The ordered values of the first batch were −17.5, 2.8, 3.2, 13.9, 14.1, 25.3, and 45.8, with a median of 13.9. Excluding the median, the two halves of the list are −17.5, 2.8, 3.2 and 14.1, 25.3, 45.8.

Each half has 3 values, so the median of each is the middle one. The lower quartile is 2.8, and the upper quartile is 25.3.

The second batch of data had the ordered values −17.5, 2.8, 3.2, 13.9, 14.1, 25.3, 35.7, and 45.8.

Here *n* is even, so the two halves of 4 values are −17.5, 2.8, 3.2, 13.9 and 14.1, 25.3, 35.7, 45.8.

Now the lower quartile is $(2.8 + 3.2)/2 = 3.0$, and the upper quartile is $(25.3 + 35.7)/2 = 30.5$.

Do The Math

Back to Brett Favre once more . . . This time, find the quartiles and the interquartile range for each data set.

6. Pass completions:

302, 304, 308, 314, 318, 325, 338, 341, 341, 343, 343, 346, 347, 356, 359, 363, 363, 372

7. Touchdown passes for Green Bay:

28, 18, 20, 30, 32, 27, 32, 20, 22, 31, 35, 39, 38, 33, 19, 18

(Check your answers on page 81.)

So, what is a quartile anyway? Finding the quartiles sounds easy, but surprisingly, the quartiles are not well defined. It's not always clear how to find a value such that exactly one quarter of the data lies above or below that value. We offered a simple rule for Finding Quartiles in the box on page 57: Find the median of each half of the data split by the median. When n is odd, we (and your TI calculator) omit the median from each of the halves. Some other texts include the median in both halves before finding the quartiles. Both methods are commonly used.

All of the methods agree pretty closely. Actually we know of at least six different rules for finding quartiles! So don't worry too much about getting the "exact" value for a quartile. Remember, Statistics is about understanding the world, not about calculating the right number. The "answer" to a statistical question is a sentence about the issue raised in the question.

5-Number Summary

NOTATION ALERT

We always use Q1 to label the lower (25%) quartile and Q3 to label the upper (75%) quartile. We skip the number 2 because the median would, by this system, naturally be labeled Q2—but we don't usually call it that.

The **5-number summary** of a distribution reports its median, quartiles, and extremes (maximum and minimum). The 5-number summary for the recent tsunami earthquake *Magnitudes* looks like this:

Max	9.0
Q3	7.6
Median	7.0
Q1	6.6
Min	3.7

It's good idea to report the number of data values and the identity of the cases (the *Who*). Here there are 176 earthquakes.

The 5-number summary provides a good overview of the distribution of magnitudes of these tsunami-causing earthquakes. For a start, we can see that the median magnitude is 7.0. Because the IQR is only $7.6 - 6.6 = 1$, we see that many quakes are close to the median magnitude. Indeed, the quartiles show us that the middle half of these earthquakes had magnitudes between 6.6 and 7.6. One quarter of the earthquakes had magnitudes above 7.6, although one tsunami was caused by a quake measuring only 3.7 on the Richter scale.

Do The Math

Of course in real life, where data sets are often large and full of "messy" numbers, you'll use a calculator or computer to find the 5-number summary. But for now, let's get a feel for doing it without a calculator or computer by finding the 5-number summary for each of the following small data sets. We'll use the calculator after this, we promise!

8. The birth weights of the Bellflower, California octuplets born on January 26, 2009 are:

   2.7 lbs, 2.8 lbs, 3.3 lbs, 2.5 lbs, 1.5 lbs,
   2.8 lbs, 1.9 lbs, and 2.7 lbs

9. The weights of eleven cats (in pounds) are:

   5.1, 6.5, 7.1, 7.2, 7.6, 8.1, 8.1, 9.2, 10.5, 12.9, 14.2

10. The top ten times for U.S. women in the 400 meter dash (in seconds) are:

    47.60, 47.99, 48.25, 48.27, 48.59, 48.63, 48.70, 48.83, 48.89, 49.05

    (http://en.wikipedia.org/wiki/400_metres)

11. The number of calories in nine different candy bars are:

    280, 250, 290, 240, 210, 220, 190, 220, 230

    (http://webct.sic.edu/mthsci/Skaggs.pdf)

    (Check your answers on page 81.)

STEP-BY-STEP EXAMPLE

Shape, Center, and Spread: Flight Cancellations

The U.S. Bureau of Transportation Statistics (www.bts.gov) reports data on airline flights. Let's look at data giving the percentage of flights cancelled each month between 1995 and 2005.

Question: How often are flights cancelled?

WHO	Months
WHAT	Percentage of flights cancelled at U.S. airports
WHEN	1995–2005
WHERE	United States

THINK

Variable Identify the *variable,* and decide how you wish to display it.

To identify a variable, report the W's.

Select an appropriate display based on the nature of the data and what you want to know.

I want to learn about the monthly percentage of flight cancellations at U.S airports.

I have data from the U.S. Bureau of Transportation Statistics giving the percentage of flights cancelled at U.S. airports each month between 1995 and 2005.

✔ **Quantitative Data Condition:** Percentages are quantitative. It's okay to make a histogram and find numerical summaries.

(continued)

Mechanics We usually make histograms with a computer or graphing calculator.

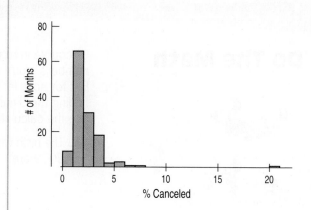

The histogram shows a distribution skewed to the high end and one extreme outlier, a month in which more than 20% of flights were cancelled.

In most months, fewer than 5% of flights are cancelled and usually only about 2% or 3%. That seems reasonable.

REALITY CHECK It's always a good idea to think about what you expect to see so that you can check whether the histogram looks like what you expected.

With 132 cases, we probably have more data than you'd choose to work with by hand. The results given here are from technology.

Count	132
Max	20.240
Q3	2.615
Median	1.755
Q1	1.445
Min	0.770
IQR	1.170

Interpretation Describe the shape, center, and spread of the distribution. Report on the symmetry, number of modes, and any gaps or outliers. You should also mention any concerns you may have about the data.

The distribution of flight cancellations is skewed to the right, and this makes sense: The values can't fall below 0%, but can increase almost arbitrarily due to bad weather or other events.

The median is 1.76% and the IQR is 1.17%. The low IQR indicates that in most months the cancellation rate is close to the median. In fact, it's between 1.4% and 2.6% in the middle 50% of all months, and in only 1/4 of the months were more than 2.6% of flights cancelled.

There is one extraordinary value: 20.2%. Looking it up, I find that the extraordinary month was September 2001. The attacks of September 11 shut down air travel for several days, accounting for this outlier.

Summarizing Symmetric Distributions: The Mean

In Algebra you used letters to represent values in a problem, but it didn't matter what letter you picked. You could call the width of a rectangle X or you could call it w (or *Fred*, for that matter). But in Statistics, the notation is part of the vocabulary.

We have already begun to point out examples of such special notation: n, Q1, and Q3. Think of them as part of the terminology you need to learn in this course.

Here's another one: Whenever we put a bar over a symbol, it means "find the mean."

In everyday language, sometimes "average" *does* mean what we want it to mean. We don't talk about your grade point mean or a baseball player's batting mean or the Dow Jones Industrial mean. So we'll continue to say "average" when that seems most natural. When we do, though, you may assume that what we mean is the mean.

Medians do a good job of summarizing the center of a distribution, even when the shape is skewed or when there is an outlier, as with the flight cancellations. But when we have symmetric data, there's another alternative. You already know how to average values, but this is a good place to introduce notation that we'll use throughout the book. We use the Greek capital letter sigma, Σ, to mean "sum" (sigma is "S" in Greek), and we'll write:

$$\bar{y} = \frac{Total}{n} = \frac{\sum y}{n}.$$

The formula says to add up all the values of the variable and divide that sum by the number of data values, n—just as you've always done.[8]

Once we've averaged the data, you'd expect the result to be called the *average*, but that would be too easy. A median is also a kind of average. To be clear, the value we calculated is called the **mean**, written $\bar{y}$ and pronounced "*y*-bar."

The earthquake magnitudes are pretty close to symmetric, so we can also summarize their center with a mean. The mean tsunami earthquake magnitude is 6.96—about what we might expect from the histogram. The **mean** feels like the center because it is the point where the histogram balances:

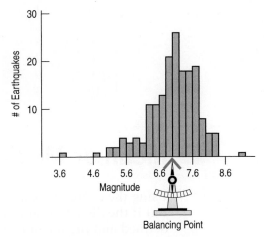

FIGURE 4.12 The mean. The mean is located at the *balancing point* of the histogram.

[8]You may also see the variable called x and the equation written $\bar{x} = \dfrac{Total}{n} = \dfrac{\sum x}{n}$.

Don't let that throw you. You are free to name the variable anything you want, but we'll generally use y for variables like this that we want to summarize, model, or predict. (Later we'll talk about variables that are used to explain, model, or predict y. We'll call them x.)

Do The Math

Ten neighborhood kids went out to score some candy on Halloween night. Here's a list of the number of treats they collected:

45, 34, 56, 32, 10, 32, 62, 11, 55, 34

12. Find the mean and median number of treats.
13. The kid who at first came home with 62 treats got even greedier and went back out. At the end of the night he ended up with 262 treats! Find the new mean and median for these ten children.
14. How do the new mean and median compare to the original values?
15. Which does a better job of describing the typical number of treats for the new data—the mean or median? Why?

The poor kid who got only 10 candy bars had 6 small bars weighing 2 ounces each, 3 large bars weighing 4 ounces each, and one jumbo bar weighing 8 ounces. To find the mean and the median weight of these candy bars, we can't simply look at the weights of 2, 4, and 8 ounces; we have to imagine the entire list of data values: 2, 2, 2, 2, 2, 2, 4, 4, 4, 8. Now we see that the median candy bar weight, the middle value, is 2 ounces. To calculate the mean weight, first we find total weight of all the bars: $6(2) + 3(4) + 1(8) = 32$ ounces. Now we'll divide that by the number of candy bars: $6 + 3 + 1 = 10$. That makes the mean weight of these candy bars $32/10 = 3.2$ ounces.

Bar weight	Number of bars
2	6
4	3
8	1

Find the mean and median values in each of these frequency distributions:

16.

Quiz score	Number of students
0	1
2	0
4	5
6	7
8	9
10	3

17.

Bathrooms in the house	Number of homes
1	11
2	21
3	5
4	2
5	1

(Check your answers on page 81.)

Mean or Median?

Using the center of balance makes sense when the data are symmetric. But if the distribution is skewed or has outliers, the center is not so well defined and the mean may not be what we want. For example, the mean of the flight cancellations doesn't give a very good idea of the typical percentage of cancellations.

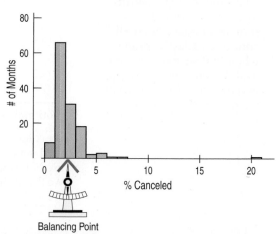

FIGURE 4.13 Mean vs. median. The median splits the area of the histogram in half at 1.755%. Because the distribution is skewed to the right, the mean (2.28%) is *higher* than the median. The points at the right have pulled the mean toward them away from the median.

The mean is 2.28%, but nearly 70% of months had cancellation rates below that, so the mean doesn't feel like a good overall summary. Why is the balancing point so high? The large outlying value pulls it to the right. For data like these, the median is a better summary of the center.

For the tsunami earthquake magnitudes, it doesn't seem to make much difference—the mean is 6.96; the median is 7.0. When the data are symmetric, the mean and median will be close, but when the data are skewed, the median is likely to be a better choice.

To choose between mean and median, we'll start (of course) by looking at the data. If the histogram is symmetric and there are no outliers, we'll prefer the mean. However, if the histogram is skewed or has outliers, we're usually better off with the median.

For Example Describing Center

Recap: You want to summarize the expenditures of 500 credit card company customers, and have looked at a histogram.

Question: Suppose you are told the mean expenditure is $478.19 and the median is $216.28. Which is the more appropriate measure of center, and why?

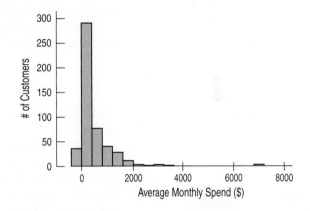

Because the distribution of expenditures is skewed, the median is the more appropriate measure of center. Unlike the mean, it's not affected by the large outlying value or by the skewness. Half of these credit card customers had average monthly expenditures less than $216.28 and half more.

> **When to expect skewness** Even without making a histogram, we can expect some variables to be skewed. When values of a quantitative variable are bounded on one side but not the other, the distribution may be skewed. For example, incomes and waiting times can't be less than zero, so they are often skewed to the right. Amounts of things (dollars, employees) are often skewed to the right for the same reason. If a test is too easy, the distribution will be skewed to the left because many scores will bump against 100%. and combinations of things are often skewed. In the case of the cancelled flights, flights are more likely to be cancelled in January (due to snowstorms) and in August (thunderstorms). Combining values across months leads to a skewed distribution.

What About Spread? The Standard Deviation

The IQR is always a reasonable summary of spread, but because it uses only the two quartiles of the data, it ignores much of the information about how individual values vary. A more powerful approach uses the **standard deviation**, which takes into account how far *each* value is from the mean. Like the mean, the standard deviation is appropriate only for symmetric data.

One way to think about spread is to examine how far each data value is from the mean. This difference is called a *deviation*. We could just average the deviations, but the positive and negative differences always cancel each other out. To keep them from canceling out, we *square* each deviation. Squaring eliminates negative values, so the sum won't be zero. That's great. Squaring also emphasizes larger differences—a feature that turns out to be both good and bad.

When we add up these squared deviations and find their average (almost), we call the result the **variance**:

$$s^2 = \frac{\sum (y - \bar{y})^2}{n - 1}.$$

Why almost? It *would* be a mean if we divided the sum by n. Instead, we divide by $n - 1$. Why? The simplest explanation is "to drive you crazy." But there are good reasons, some of which we'll see later.

There is one problem with the variance, though: Whatever the units of the original data are, the variance is in *squared* units. We want measures of spread to have the same units as the data. And we probably don't want to talk about square dollars or mpg^2. So, to get back to the original units, we take the square root of the variance. The result, s, is the **standard deviation.**

Putting it all together, the standard deviation of the data is found by the following formula:

$$s = \sqrt{\frac{\sum (y - \bar{y})^2}{n - 1}}.$$

Luckily you'll almost always rely on a calculator or computer to do the calculating.

Understanding what the standard deviation really means will take some time, and we'll revisit the concept in later chapters. For now, have a look at this histogram of resting pulse rates. The distribution is roughly symmetric, so it's okay to choose the mean and standard deviation as our summaries of center and spread. The mean pulse rate is 72.7 beats per minute, and we can see that's a typical heart rate. We also see that some heart rates are higher and some lower—but how much? Well, the standard deviation of 6.5 beats per minute indicates that, on average, we might expect people's heart rates to differ from the mean rate by about 6.5 beats per minute. Looking at the histogram, we can see that 6.5 beats above or below the mean appears to be a typical deviation.

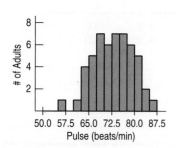

WHO	52 adults
WHAT	Resting heart rates
UNITS	Beats per minute

How does standard deviation work? To find the standard deviation, start with the mean, $\bar{y}$. Then find the *deviations* by taking $\bar{y}$ from each value: $(y - \bar{y})$

Square each deviation: $(y - \bar{y})^2$.

Now you're nearly home. Just add these up and divide by $n - 1$. That gives you the variance, s^2. To find the standard deviation, s, take the square root. Here we go:

Suppose the batch of values is 14, 13, 20, 22, 18, 19, and 13.

The mean is $\bar{y} = 17$. So the deviations are found by subtracting 17 from each value:

Original Values	Deviations	Squared Deviations
14	$14 - 17 = -3$	$(-3)^2 = 9$
13	$13 - 17 = -4$	$(-4)^2 = 16$
20	$20 - 17 = 3$	9
22	$22 - 17 = 5$	25
18	$18 - 17 = 1$	1
19	$19 - 17 = 2$	4
13	$13 - 17 = -4$	16

Add up the squared deviations: $9 + 16 + 9 + 25 + 1 + 4 + 16 = 80$.

Now divide by $n - 1$: $80/6 = 13.33$.

Finally, take the square root: $s = \sqrt{13.33} = 3.65$

Do The Math

We promised you'd get to use a calculator to find standard deviations . . . soon. But first, trying a few by hand will help you understand how the standard deviation works to measure the spread in a set of data. Here's what you should do for the following data sets:

Step 1: Make a 3-column table like we did on page 64, listing the data values (the y's) in the first column.

Step 2: Find the mean: $\bar{y}$.

Step 3: Find out how far each data value is from the mean: $(y - \bar{y})$. Write these deviations in the second column.

Step 4: Square the deviations—that's the third column: $(y - \bar{y})^2$.

Step 5: Add up the squared deviations: $\sum (y - \bar{y})^2$.

Step 6: Divide that sum by $n - 1$, one less than the number of data values, to find the variance: $s^2 = \dfrac{\sum (y - \bar{y})^2}{n - 1}$.

Step 7: And finally, take the square root: $s = \sqrt{\dfrac{\sum (y - \bar{y})^2}{n - 1}}$. Voila—that's the standard deviation!

18. Data: 100, 105, 90, 85, 70
19. Data: 26, 16, 18, 19, 23, 22, 18, 18
20. Look at both of your tables. What do you notice about the sum of the deviations (2nd column) in each? Why does that happen?

(Check your answers on pages 81–82.)

Thinking About Variation

Why do banks favor a single line that feeds several teller windows rather than separate lines for each teller? The average waiting time is the same. But the time you can expect to wait is less variable when there is a single line, and people prefer consistency.

Statistics is about variation, so spread is an important fundamental concept in Statistics. Measures of spread help us to be precise about what we *don't* know. If many data values are scattered far from the center, the IQR and the standard deviation will be large. If the data values are close to the center, then these measures of spread will be small. If all our data values were exactly the same, we'd have no question about summarizing the center, and all measures of spread would be zero—and that would make for a boring world. In the real world data do vary, and that's why we need Statistics!

Measures of spread tell how well other summaries describe the data. That's why we always (always!) report a spread along with any summary of the center.

Just Checking

6. The U.S. Census Bureau reports the median family income in its summary of census data. Why do you suppose they use the median instead of the mean? What might be the disadvantages of reporting the mean?

7. You've just bought a new car that claims to get a highway fuel efficiency of 31 miles per gallon. Of course, your mileage will "vary." If you had to guess, would you expect the IQR of gas mileage attained by all cars like yours to be 30 mpg, 3 mpg, or 0.3 mpg? Why?

8. A company selling a new MP3 player advertises that the player has a mean lifetime of 5 years. If you were in charge of quality control at the factory, would you prefer that the standard deviation of lifespans of the players you produce be 2 years or 2 months? Why?

(Check your answers on page 82.)

What to *Tell* About a Quantitative Variable

What should you *Tell* about a quantitative variable?

- Start by making a histogram or stem-and-leaf display, and describe the shape of the distribution.
- Next, discuss the center *and* spread.
  - **(1)** Always report *both* center and spread.
  - **(2)** Always pair the median with the IQR or the mean with the standard deviation.
  - **(3)** If the shape is skewed, report the median and IQR. You may want to include the mean and standard deviation as well, but you should point out why the mean and median differ.
  - **(4)** If the shape is symmetric, report the mean and standard deviation (and possibly the median and IQR as well). For unimodal symmetric data, the IQR is usually a bit larger than the standard deviation. If that's not true of your data set, look again to make sure that the distribution isn't skewed and there are no outliers.
- Also, discuss any unusual features.
- If there are multiple modes, try to understand why. If you can identify a reason for separate modes (for example, women and men typically have heart attacks at different ages), it may be a good idea to split the data into separate groups.
- If there are any clear outliers, point them out. If you are reporting the mean and standard deviation, you may want to report them with the outliers present and with the outliers removed. The differences may be revealing. (Of course, the median and IQR won't be affected very much by the outliers.)

HOW "ACCURATE" SHOULD WE BE?

Don't think you should report means and standard deviations to a zillion decimal places; such implied accuracy is really meaningless. Although there is no ironclad rule, statisticians commonly report summary statistics to one or two decimal places more than the original data have.

STEP-BY-STEP EXAMPLE **Summarizing a Distribution**

One of the authors owned a 1989 Nissan Maxima for 8 years. Being a statistician, he recorded the car's fuel efficiency (in mpg) each time he filled the tank. He wanted to know what fuel efficiency to expect as "ordinary" for his car. (Hey, he's a statistician. What would you expect?[9]) Knowing this, he was able to predict when he'd need to fill the tank again and to notice if the fuel efficiency suddenly got worse, which could be a sign of trouble.

Question: How would you describe the distribution of *Fuel efficiency* for this car?

Plan State what you want to find out.

I want to summarize the distribution of Nissan Maxima fuel efficiency.

Variable Identify the variable and report the W's.

The data are the fuel efficiency values in miles per gallon for the first 100 fill-ups of a 1989 Nissan Maxima between 1989 and 1992.

---

[9]He also recorded the time of day, temperature, price of gas, and phase of the moon. (OK, maybe not phase of the moon.) His data are on the DVD.

Be sure to check the appropriate condition.

✔ **Quantitative Data Condition:** The fuel efficiencies are quantitative with units of miles per gallon. Histograms and boxplots are appropriate displays for displaying the distribution. Numerical summaries are appropriate as well.

Mechanics Make a histogram and boxplot. Based on the shape, choose appropriate numerical summaries.

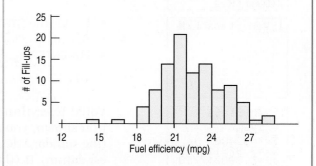

A value of 22 mpg seems reasonable for such a car. The spread is reasonable, although the range looks a bit large.

A histogram of the data shows a fairly symmetric distribution with a low outlier.

Count	100
Mean	22.4 mpg
StdDev	2.45
Q1	20.8
Median	22.0
Q3	24.0
IQR	3.2

The mean and median are close, so the outlier doesn't seem to be a problem. I can use the mean and standard deviation.

Conclusion Summarize and interpret your findings in context. Be sure to discuss the distribution's shape, center, spread, and unusual features (if any).

The distribution of mileage is unimodal and roughly symmetric with a mean of 22.4 mpg. There are one or two low outliers that should be investigated, but it does not influence the mean very much. The standard deviation suggests that from tankful to tankful, I can expect the car's fuel economy to differ from the mean by an average of about 2.45 mpg.

TI Tips   Calculating the statistics

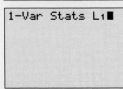

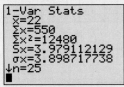

Your calculator can easily find all the numerical summaries of data. To try it out, you simply need a set of values in one of your datalists. We'll illustrate using the boys' agility test results from this chapter's earlier TI Tips (still in L1), but you can use any data currently stored in your calculator.

- Under the **STAT CALC** menu, select 1-Var Stats and hit **ENTER**.
- Specify the location of your data, creating a command like 1-Var Stats L1.
- Hit **ENTER** again.

Voilà! Everything you wanted to know, and more. Among all of the information shown, you are primarily interested in these statistics: x̄ (the mean), Sx (the standard deviation), n (the count), and—scrolling down—minX (the smallest datum), Q_1 (the first quartile), Med (the median), Q_3 (the third quartile), and maxX (the largest datum).

Sorry, but the TI doesn't explicitly tell you the range or the IQR. Just subtract: IQR = $Q_3 - Q_1$ = 25 − 19.5 = 5.5. What's the range?

By the way, if the data come as a frequency table with the values stored in, say, L4 and the corresponding frequencies in L5, all you have to do is ask for 1-Var Stats L4, L5.

🚫 WHAT CAN GO WRONG?

A data display should tell a story about the data. To do that, it must speak in a clear language, making plain what variable is displayed, what any axis shows, and what the values of the data are. And it must be consistent in those decisions.

A display of quantitative data can go wrong in many ways. The most common failures arise from only a few basic errors:

- **Don't make a histogram of a categorical variable.** Just because the variable contains numbers doesn't mean that it's quantitative. Here's a histogram of the insurance policy numbers of some workers. It's not very informative because these numbers are just labels indicating the type of policy. A histogram or stem-and-leaf display of a categorical variable makes no sense. A bar chart or pie chart would be more appropriate.

- **Don't look for shape, center, and spread of a bar chart.** A bar chart showing the sizes of the piles displays the distribution of a categorical variable, but the bars could be arranged in any order left to right. Concepts like symmetry, center, and spread make sense only for quantitative variables.

- **Don't use bars in every display—save them for histograms and bar charts.** In both bar charts and histograms, the bars represent counts of categories or bins. Some people create other

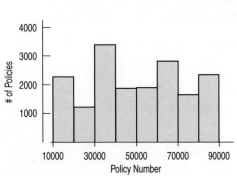

FIGURE 4.14 It's not appropriate to display these data with a histogram.

displays that use bars to represent individual data values. Beware: Such graphs are neither bar charts nor histograms. For example, a student was asked to make a histogram from data showing the number of juvenile bald eagles seen during each of the 13 weeks in the winter of 2003–2004 at a site in Rock Island, IL. Instead, he made this plot:

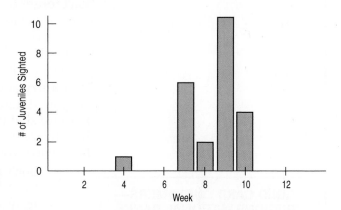

FIGURE 4.15 Bad graph. This isn't a histogram or a bar chart. It's an ill-conceived graph that uses bars to represent individual data values (number of eagles sighted) week by week.

Look carefully. That's not a histogram. This student used bars to show counts of birds for each week. We need counts of weeks. A correct histogram should have a tall bar at "0" to show there were many weeks when no eagles were seen, like this:

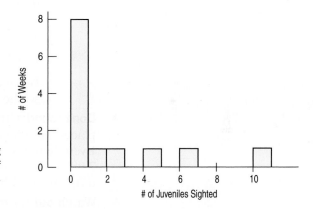

FIGURE 4.16 Good graph. A histogram of the eagle-sighting data shows the number of weeks in which different counts of eagles occurred. This display shows the distribution of juvenile-eagle sightings.

■ **Choose a bin width appropriate to the data.** Here are the tsunami earthquakes with two rather ineffective choices for the bin size:

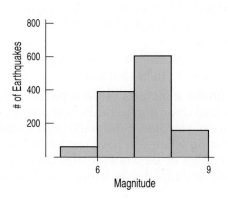

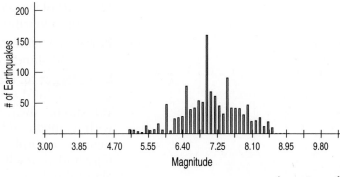

(continued)

- **Don't forget to do a reality check.** Don't let the computer or calculator do your thinking for you. Make sure the calculated summaries make sense. For example, does the mean look like it is in the center of the histogram? Think about the spread: An IQR of 50 mpg would clearly be wrong for gas mileage. And no measure of spread can be negative.

- **Don't forget to sort the values before finding the median or percentiles.** It seems obvious, but when you work by hand, it's easy to forget to sort the data first before counting in to find medians, quartiles, or other percentiles. Don't report that the median of the five values 194, 5, 1, 17, and 893 is 1 just because 1 is the middle number.

- **Don't worry about small differences when using different methods.** If you compare different computer outputs or calculators, you may find that they give slightly different answers for the same data. These differences, though, are unlikely to be important in interpreting the data, so don't let them worry you.

- **Don't compute numerical summaries of a categorical variable.** Neither the mean zip code nor the standard deviation of social security numbers is meaningful. If the variable is categorical, you should instead report summaries such as percentages of individuals in each category.

- **Don't report too many decimal places.** Statistical programs and calculators often report a ridiculous number of digits. A general rule for numerical summaries is to report one or two more digits than the number of digits in the data. For example, earlier we saw a dotplot of the Kentucky Derby race times. The mean and standard deviation of those times could be reported as:

$$\bar{y} = 130.63401639344262\,\text{sec} \qquad s = 13.66448201942662\,\text{sec}$$

But we knew the race times only to the nearest quarter second, so the extra digits are meaningless.

- **Don't round in the middle of a calculation.** Don't *report* too many decimal places, but it's best not to do any rounding until the end of your calculations. Even though you might report the mean of the earthquakes as 7.08, it's really 7.08339. Use that more precise number in any calculations to avoid differences in the final result.

- **Watch out for multiple modes.** The summaries of the Kentucky Derby times are meaningless for another reason. As we saw in the dotplot, the Derby was initially a longer race. It would make much more sense to report that the old 1.5 mile Derby had a mean time of 159.6 seconds, while the current Derby has a mean time of 124.6 seconds. If the distribution has multiple modes, consider separating the data into different groups and summarizing each group separately.

- **Beware of outliers.** The median and IQR are resistant to outliers, but the mean and standard deviation are not. To help spot outliers . . .

- **Don't forget to: Make a picture (make a picture, make a picture).** The sensitivity of the mean and standard deviation to outliers is one reason you should always make a picture of the data. Summarizing a variable with its mean and standard deviation when you have not looked at a histogram or dotplot to check for outliers or skewness invites disaster. You may find yourself drawing absurd or dangerously wrong conclusions about the data.

GOLD CARD CUSTOMERS—REGIONS NATIONAL BANKS

Month	April 2007	May 2007
Average Zip Code	45,034.34	38,743.34

QUANTITATIVE DATA **IN YOUR WORLD**

Bolt Strikes Twice, with a Second World Record

BEIJING (AP)—Arms churning high, face twisted in pain as he sprinted toward the finish line, Usain Bolt kept glancing at the clock. The win in the Olympic 200 meters was a given, his second gold medal of the Beijing Games assured. This was now about a world record. About racing against history.

Showing just what he can do when he goes all out start to finish, Bolt forged the greatest race ever run Wednesday night under the hazy lights at the Bird's Nest, heaving his chest toward the finish line—not simply to beat someone for the gold, but to become a part of track's glorious, and sometimes troubled, lore.

He finished in 19.30 seconds to break Michael Johnson's 12-year-old world record, one of the most venerable in the books.

Officially, he won by an astounding 0.66 second over American Shawn Crawford, the defending

Olympic champion . . . [by] about four body lengths, the biggest margin in an Olympic 200.

After the unrelenting effort with a slight headwind in his face, Bolt sprawled out on the ground, arms and legs outstretched, basking in the roar of the Bird's Nest crowd and the glow of becoming, quite possibly, the greatest sprinter ever.

Bolt's Feat
On Wednesday, Usain Bolt won the 200 meters in 19.30 seconds, breaking Michael Johnson's 1996 record by two-hundredths of a second. Both times are far better than the 250 next fastest times.

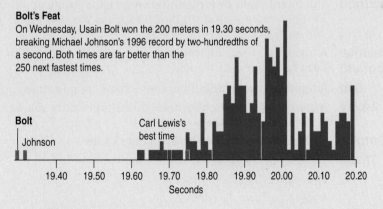

http://www.nbcolympics.com/trackandfield/news/newsid=239399.html

WHAT HAVE WE LEARNED?

We've learned how to make a picture of quantitative data to help us see the story the data have to *Tell*.

▶ We can display the distribution of quantitative data with a *histogram*, a *stem-and-leaf* display, or a *dotplot*.

▶ We *Tell* what we see about the distribution by talking about *shape, center, spread*, and any *unusual features*.

We've learned how to summarize distributions of quantitative variables numerically.

▶ Measures of center for a distribution include the median and the mean.
We write the formula for the mean as $\bar{y} = \dfrac{\sum y}{n}$.

▶ Measures of spread include the range, IQR, and standard deviation.
The standard deviation is computed as $s = \sqrt{\dfrac{\sum (y - \bar{y})^2}{n - 1}}$.

▶ We'll report the median and IQR when the distribution is skewed. If it's symmetric, we'll summarize the distribution with the mean and standard deviation (and possibly the median and IQR as well). Always pair the median with the IQR and the mean with the standard deviation.

We've learned to *Think* about the type of variable we're summarizing.

▶ All the methods of this chapter assume that the data are quantitative.

▶ The **Quantitative Data Condition** serves as a check that the data are, in fact, quantitative. One good way to be sure is to know the measurement units. You'll want those as part of the *Think* step of your answers.

Terms

Distribution
The distribution of a quantitative variable slices up all the possible values of the variable into equal-width bins and gives the number of values (or counts) falling into each bin.

Histogram (relative frequency histogram)
A histogram uses adjacent bars to show the distribution of a quantitative variable. Each bar represents the frequency (or relative frequency) of values falling in each bin.

Gap
A region of the distribution where there are no values.

Stem-and-leaf display
A stem-and-leaf display shows quantitative data values in a way that sketches the distribution of the data.

Dotplot
A dotplot graphs a dot for each data value.

Shape
To describe the shape of a distribution, look for

▶ single vs. multiple modes.

▶ symmetry vs. skewness.

▶ outliers and gaps.

Center
The place in the distribution of a variable that you'd point to if you wanted to indicate a typical value. Measures of center include the mean and median.

Spread
A numerical summary of how tightly the values cluster around the center. Measures of spread include the IQR and standard deviation.

Mode
A hump or local high point in the shape of the distribution of a variable.

Unimodal (Bimodal)
Having one mode. This is a useful term for describing the shape of a histogram when it's generally mound-shaped. Distributions with two modes are called **bimodal.** Those with more than two are **multimodal.**

Uniform
A distribution that stays roughly the same height is said to be uniform.

Symmetric
A distribution is symmetric if the two halves on either side of the center look approximately like mirror images of each other.

Tails	The tails of a distribution are the parts that typically trail off on either side. Distributions can be characterized as having long tails (if they straggle off for some distance) or short tails (if they don't).
Skewed	A distribution is skewed if it's not symmetric and one tail stretches out farther than the other. Distributions are said to be **skewed left** when the longer tail stretches to the left, and **skewed right** when it goes to the right.
Outliers	Outliers are extreme values that don't appear to belong with the rest of the data. They may be unusual values that deserve further investigation, or they may just be mistakes; there's no obvious way to tell. Don't delete outliers automatically—you have to think about them. Outliers can affect many statistical analyses, so you should always be alert for them.
Median	The median is the middle value, with half of the data above and half below it. If n is even, it is the average of the two middle values. It is usually paired with the IQR.
Range	The difference between the lowest and highest values in a data set. *Range = max − min.*
Quartile	The lower quartile (Q1) is the value with a quarter of the data below it. The upper quartile (Q3) has three quarters of the data below it. The median and quartiles divide data into four parts with equal numbers of data values.
Interquartile range (IQR)	The IQR is the difference between the first and third quartiles. *IQR = Q3 − Q1.* It is usually reported along with the median.
Percentile	The ith percentile is the number that falls above $i\%$ of the data.
5-Number summary	The 5-number summary of a distribution reports the minimum value, Q1, the median, Q3, and the maximum value.
Mean	The mean is found by summing all the data values and dividing by the count:

$$\bar{y} = \frac{Total}{n} = \frac{\sum y}{n}.$$

Report it with the standard deviation.

Variance	The variance is the sum of squared deviations from the mean, divided by the count minus 1:

$$s^2 = \frac{\sum (y - \bar{y})^2}{n - 1}.$$

Standard deviation	The standard deviation is the square root of the variance:

$$s = \sqrt{\frac{\sum (y - \bar{y})^2}{n - 1}}$$

Report it along with the mean.

Skills

- ▶ Be able to identify an appropriate display for any quantitative variable.
- ▶ Be able to guess the shape of the distribution of a variable by knowing something about the data.
- ▶ Be able to select a suitable measure of center and a suitable measure of spread for a variable based on information about its distribution.
- ▶ Know the basic properties of the median: The median divides the data into the half of the data values that are below the median and the half that are above.
- ▶ Know the basic properties of the mean: The mean is the point at which the histogram balances.
- ▶ Know that the standard deviation summarizes how spread out all the data are around the mean.
- ▶ Understand that the median and IQR resist the effects of outliers, while the mean and standard deviation do not.
- ▶ Understand that in a skewed distribution, the mean is pulled in the direction of the skewness (toward the longer tail) relative to the median.

(continued)

▶ Know how to display the distribution of a quantitative variable with a stem-and-leaf display (drawn by hand for smaller data sets), a dotplot, or a histogram (made by computer for larger data sets).

▶ Know how to compute the median and IQR of a set of data.

▶ Know how to compute the mean and standard deviation of a set of data.

▶ Be able to describe the distribution of a quantitative variable in terms of its shape, center, and spread.

▶ Be able to describe any unusual features revealed by the display of a variable.

▶ Know how to describe summary statistics in a sentence using appropriate units.

DISPLAYING AND SUMMARIZING QUANTITATIVE VARIABLES ON THE COMPUTER

Displays and summaries of quantitative variables are among the simplest things you can do in most statistics packages. Always start by making a picture.

The vertical scale may be counts or proportions. Sometimes it isn't clear which. But the shape of the histogram is the same either way.

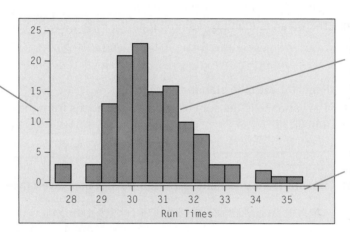

Most packages choose the number of bars for you automatically. Often you can adjust that choice.

The axis should be clearly labeled so you can tell what "pile" each bar represents.

Many statistics packages offer a prepackaged collection of summary measures. The result might look like this:

```
Variable: Weight
N = 234
Mean = 143.3          Median = 139
St. Dev =11.1         IQR = 14
```

Alternatively, a package might make a table for several variables and summary measures:

Variable	N	mean	median	stdev	IQR
Weight	234	143.3	139	11.1	14
Height	234	68.3	68.1	4.3	5
Score	234	86	88	9	5

Packages often offer many more summary statistics than you need. Of course, some of these may not be appropriate when the data are skewed or have outliers. It is your responsibility to check a histogram or stem-and-leaf display and decide which summary statistics to use.

It is common for packages to report summary statistics to many decimal places of "accuracy." Of course, that doesn't mean that those digits have any meaning. Generally it's a good idea to round the values off, allowing perhaps one more digit of precision than was given in the original data.

EXERCISES

1. **Thinking about shape.** Would you expect distributions of these variables to be uniform, unimodal, or bimodal? Symmetric or skewed? Explain why.
   a) The number of speeding tickets each student in the senior class of a college has ever had.
   b) Players' scores (number of strokes) at the U.S. Open golf tournament in a given year.

2. **More shapes.** Would you expect distributions of these variables to be uniform, unimodal, or bimodal? Symmetric or skewed? Explain why.
   a) Ages of people at a Little League game.
   b) Number of siblings of people in your class.

3. **About shapes III.** Would you expect distributions of these variables to be uniform, unimodal, or bimodal? Symmetric or skewed? Explain why.
   a) Weights of female babies born in a particular hospital over the course of a year.
   b) The length of the average hair on the heads of students in a large class.

4. **Shapes again.** Would you expect distributions of these variables to be uniform, unimodal, or bimodal? Symmetric or skewed? Explain why.
   a) Pulse rates of college-age males.
   b) Number of times each face of a die shows in 100 tosses.

5. **Sugar in cereals.** The histogram displays the sugar content (as a percent of weight) of 49 brands of breakfast cereals.

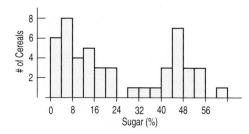

   a) Describe this distribution.
   b) What do you think might account for this shape?

6. **Singers.** The display shows the heights of some of the singers in a chorus, collected so that the singers could be positioned on stage with shorter ones in front and taller ones in back.

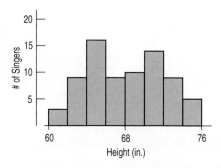

a) Describe the distribution.
b) Can you account for the features you see here?

7. **Vineyards.** The histogram shows the sizes (in acres) of 36 vineyards in the Finger Lakes region of New York. Write a brief description of this distribution (shape, center, spread, unusual features).

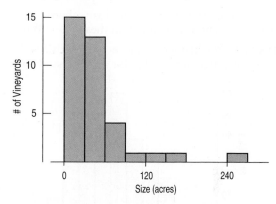

8. **Run times.** One of the authors collected the times (in minutes) it took him to run 4 miles on various courses during a 10-year period. Here is a histogram of the times.

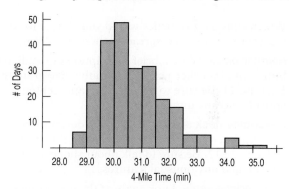

Describe the distribution and summarize the important features. What is it about running that might account for the shape you see?

9. **Vineyards revisited.** Look again at the histogram in Exercise 7. Suppose you wanted to describe the acreages of these vineyards by finding summary statistics.
   a) Would you describe the center using the median or the mean. Why?
   b) Would you describe the spread using the IQR or the standard deviation? Why?

10. **Run times revisited.** Look again at the histogram in Exercise 8. Suppose you wanted to describe this runner's times by finding summary statistics.
   a) Would you describe the center using the median or the mean. Why?
   b) Would you describe the spread using the IQR or the standard deviation? Why?

11. **Pizza prices.** The histogram shows the distribution of the prices of plain pizza slices (in $) for 156 weeks in Dallas, TX.

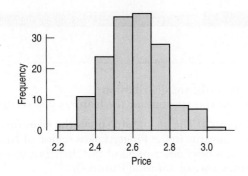

Which summary statistics would you choose to summarize the center and spread in these data? Why?

T 12. Neck size. The histogram shows the neck sizes (in inches) of 250 men recruited for a health study in Utah.

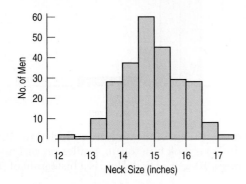

Which summary statistics would you choose to summarize the center and spread in these data? Why?

13. Another pizza. Write a brief description of the distribution of pizza slice prices seen in the histogram in Exercise 11. Be sure to talk about shape, center, spread, and any unusual features.

14. Describing neck sizes. Write a brief description of the distribution of men's neck sizes seen in the histogram in Exercise 12. Be sure to talk about shape, center, spread, and any unusual features.

T 15. Pizza prices again. Look again at the histogram of the pizza prices in Exercise 11.
a) Is the mean closer to $2.40, $2.60, or $2.80? Why?
b) Is the standard deviation closer to $0.15, $0.50, or $1.00? Explain.

T 16. Neck sizes again. Look again at the histogram of men's neck sizes in Exercise 12.
a) Is the mean closer to 14, 15, or 16 inches? Why?
b) Is the standard deviation closer to 1 inch, 3 inches, or 5 inches? Explain.

17. Super Bowl points. How many points do football teams score in the Super Bowl? Here are the total numbers of points scored by both teams in each of the first 43 Super Bowl games. Display these data in a stem-and-leaf plot.

45, 47, 23, 30, 29, 27, 21, 31, 22, 38, 46, 37, 66, 50, 37, 47, 44, 47, 54, 56, 59, 52, 36, 65, 39, 61, 69, 43, 75, 44, 56, 55, 53, 39, 41, 37, 69, 61, 45, 31, 46, 31, 50

18. Super Bowl wins. In the Super Bowl, by how many points does the winning team outscore the losers?

Here are the winning margins for the first 43 Super Bowl games. Display these data in a stem-and-leaf plot.

25, 19, 9, 16, 3, 21, 7, 17, 10, 4, 18, 17, 4, 12, 17, 5, 10, 29, 22, 36, 19, 32, 4, 45, 1, 13, 35, 17, 23, 10, 14, 7, 15, 7, 27, 3, 27, 3, 3, 11, 12, 3, 4

19. Points again. Look again at the Super Bowl total points data in Exercise 17 and at the stem-and-leaf plot you made.
a) Find the median.
b) Find the quartiles.

20. Wins again. Look again at the Super Bowl winning points margin data in Exercise 18 and at the stem-and-leaf plot you made.
a) Find the median.
b) Find the quartiles.

21. Describing Super Bowl points. Using your stem-and-leaf plot from Exercise 17 and your summary statistics from Exercise 19, write a brief description of the total number of points scored by both teams in Super Bowls. Be sure to talk about shape, center, spread, and any unusual features.

22. Describing Super Bowl wins. Using your stem-and-leaf plot from Exercise 18 and your summary statistics from Exercise 20, write a brief description of how many points Super Bowl winners win by. Be sure to talk about shape, center, spread, and any unusual features.

23. Final grades. A professor (of something other than Statistics!) distributed the following histogram to show the distribution of grades on his 200-point final exam. Comment on the display.

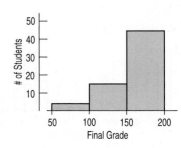

24. Final grades revisited. After receiving many complaints about his final-grade histogram from students currently taking a Statistics course, the professor from Exercise 43 distributed the following revised histogram:

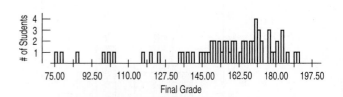

a) Comment on this display.
b) Describe the distribution of grades.

B

T **25. Movie lengths.** The histogram shows the running times in minutes of 122 feature films released in 2005.

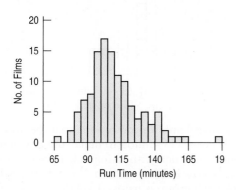

a) You plan to see a movie this weekend. Based on these movies, how long do you expect a typical movie to run?

b) Would you be surprised to find that your movie ran for $2\frac{1}{2}$ hours (150 minutes)?

c) Which would you expect to be higher: the mean or the median run time for all movies? Why?

T **26. Golf drives.** The display shows the average drive distance (in yards) for 202 professional golfers on the men's PGA tour.

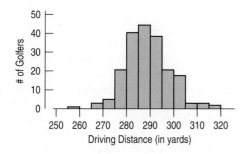

a) Describe this distribution.

b) Approximately what proportion of professional male golfers drive, on average, less than 280 yards?

c) Estimate the mean by examining the histogram.

d) Do you expect the mean to be smaller than, approximately equal to, or larger than the median? Why?

27. Movie lengths II. Exercise 25 looked at the running times of movies released in 2005. The standard deviation of these running times is 19.6 minutes, and the quartiles are $Q_1 = 97$ minutes and $Q_3 = 119$ minutes.

a) Write a sentence or two describing the spread in running times based on
  i) the quartiles.
  ii) the standard deviation.

b) Do you have any concerns about using either of these descriptions of spread? Explain.

28. Golf drives II. Exercise 26 looked at distances PGA golfers can hit the ball. The standard deviation of these average drive distances is 9.3 yards, and the quartiles are $Q_1 = 282$ yards and $Q_3 = 294$ yards.

a) Write a sentence or two describing the spread in distances based on
  i) the quartiles.
  ii) the standard deviation.

b) Do you have any concerns about using either of these descriptions of spread? Explain.

29. Heart attack stays. The histogram shows the lengths of hospital stays (in days) for all the female patients admitted to hospitals in New York during one year with a primary diagnosis of acute myocardial infarction (heart attack).

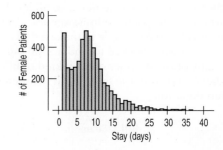

a) From the histogram, would you expect the mean or median to be larger? Explain.

b) Write a few sentences describing this distribution (shape, center, spread, unusual features).

c) Which summary statistics would you choose to summarize the center and spread in these data? Why?

T **30. E-mails.** A university teacher saved every e-mail received from students in a large Introductory Statistics class during an entire term. He then counted, for each student who had sent him at least one e-mail, how many e-mails each student had sent.

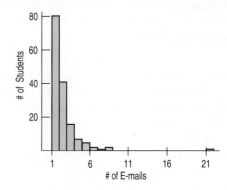

a) From the histogram, would you expect the mean or the median to be larger? Explain.

b) Write a few sentences describing this distribution (shape, center, spread, unusual features).

c) Which summary statistics would you choose to summarize the center and spread in these data? Why?

31. Stars. During 2009 the film critic for the *Ithaca Journal* reviewed a new movie every week, rating them from 0 to 4 stars. She awarded 4 stars to only 8 movies that year; the rest of her ratings are

summarized in the frequency distribution shown.
Find her mean and median ratings for the year.

Stars	Number of films
0	3
1	9
2	13
3	19
4	8

32. Donations. The History Center, a small local museum, doesn't charge a specific admission price; rather, a box at the entrance asks people to make a donation. One day the staff was shocked to find a $100 bill there. Their tally of the number of each kind of bill donated that day is shown in the table. Find the mean and median donation

Bill	Count
1	38
5	27
10	11
20	3
50	0
100	1

33. Standard deviation I. Consider these six data values:

2, 6, 6, 9, 11, 14.

a) Find the mean and standard deviation *by hand*.
b) Now look at these data values: 82, 86, 86, 89, 91, 94. Without doing any calculations, explain why the standard deviation will be the same even though the mean is obviously different.

34. Standard deviation II. Consider these six data values:

10, 16, 18, 20, 22, 28.

a) Find the mean and standard deviation *by hand*.
b) Now look at these data values: 110, 116, 118, 120, 122, 128. Without doing any calculations, explain why the standard deviation will be the same even though the mean is obviously different.

35. Standard deviation III. Consider these five data values:

10, 14, 15, 16, 20.

a) Find the mean and standard deviation *by hand*.
b) Now look at these data values: 10, 11, 15, 19, 20. Without doing any calculations explain why the standard deviation will be larger even though the mean is still the same.

36. Standard deviation IV. Consider these five data values:

4, 6, 7, 8, 10.

a) Find the mean and standard deviation *by hand*.
b) Now look at these data values: 4, 7, 7, 7, 10. Without doing any calculations explain why the standard deviation will be smaller even though the mean is still the same.

37. Movie budgets. The histogram shows the budgets (in millions of dollars) of major release movies in 2005.

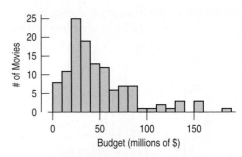

An industry publication reports that the average movie costs $35 million to make, but a watchdog group concerned with rising ticket prices says that the average cost is $46.8 million. What statistic do you think each group is using? Explain.

38. Sick days. During contract negotiations, a company seeks to change the number of sick days employees may take, saying that the annual "average" is 7 days of absence per employee. The union negotiators counter that the "average" employee misses only 3 days of work each year. Explain how both sides might be correct, identifying the measure of center you think each side is using and why the difference might exist.

39. Gasoline. In March 2006, 16 gas stations in Grand Junction, CO, posted these prices for a gallon of regular gasoline:

2.22	2.21	2.45	2.24
2.27	2.28	2.27	2.23
2.26	2.46	2.29	2.32
2.36	2.38	2.33	2.27

a) Make a stem-and-leaf display of these gas prices. Use split stems; for example, use two 2.2 stems—one for prices between $2.20 and $2.24 and the other for prices from $2.25 to $2.29.
b) Describe the shape, center, and spread of this distribution.
c) What unusual feature do you see?
d) How do the two high prices affect the mean and standard deviation? Calculate these statistics for all the data, and again with the high prices removed. (Use your calculator!) Compare the results.

40. The Great One. During his 20 seasons in the NHL, Wayne Gretzky scored 50% more points than anyone who ever played professional hockey. He accomplished this amazing feat while playing in 280 fewer games than Gordie Howe, the previous record holder. Here are the number of games Gretzky played during each season:

79, 80, 80, 80, 74, 80, 80, 79, 64, 78,
73, 78, 74, 45, 81, 48, 80, 82, 82, 70

a) Create a stem-and-leaf display for these data, using split stems.
b) Describe the shape of the distribution.
c) Describe the center and spread of this distribution.
d) What unusual feature do you see? What might explain this?

e) A standard deviation could be used to describe Gretzky's season-to-season consistency. How do the two low seasons affect that statistic? Using your calculator, calculate the standard deviation both with and without those data. Compare the results.

C

41. Mistake. A clerk entering salary data into a company spreadsheet accidentally put an extra "0" in the boss's salary, listing it as $2,000,000 instead of $200,000. Explain how this error will affect these summary statistics for the company payroll:
a) measures of center: median and mean.
b) measures of spread: range, IQR, and standard deviation.

42. Cold weather. A meteorologist preparing a talk about global warming compiled a list of weekly low temperatures (in degrees Fahrenheit) he observed at his southern Florida home last year. The coldest temperature for any week was 36°F, but he inadvertently recorded the Celsius value of 2°. Assuming that he correctly listed all the other temperatures, explain how this error will affect these summary statistics:
a) measures of center: mean and median.
b) measures of spread: range, IQR, and standard deviation.

43. Payroll. A small warehouse employs a supervisor at $1200 a week, an inventory manager at $700 a week, six stock boys at $400 a week, and four drivers at $500 a week.
a) Find the mean and median wage.
b) How many employees earn more than the mean wage?
c) Which measure of center best describes a typical wage at this company: the mean or the median?
d) Which measure of spread would best describe the payroll: the range, the IQR, or the standard deviation? Why?

44. Singers. The frequency table shows the heights (in inches) of 130 members of a choir.

Height	Count	Height	Count
60	2	69	5
61	6	70	11
62	9	71	8
63	7	72	9
64	5	73	4
65	20	74	2
66	18	75	4
67	7	76	1
68	12		

a) Find the median and IQR.
b) Find the mean.
c) What percent of the singers are taller than the mean height?

45. States. The stem-and-leaf display shows populations of the 50 states and Washington, DC, in millions of people, according to the 2000 census.

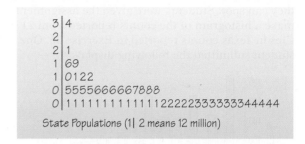

State Populations (1| 2 means 12 million)

a) What measures of center and spread are most appropriate?
b) Without doing any calculations, which must be larger: the median or the mean? Explain how you know.
c) From the stem-and-leaf display, find the 5-number summary.
d) Write a few sentences describing this distribution.

46. LeBron. In the 2007–08 NBA season, basketball superstar LeBron James led his team into the NBA finals. This stem-and-leaf plot displays the number of points he scored in his 75 regular season games.

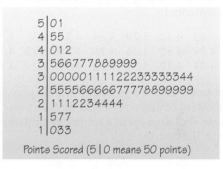

Points Scored (5 | 0 means 50 points)

a) Without doing any calculations, how might the mean compare to the median?
b) Using the display, find the 5-number summary.
c) Write a few sentences describing this distribution.

T 47. Home runs again. Students were asked to make a histogram of the number of home runs hit by Mark McGwire from 1986 to 2001. One student submitted the following display:

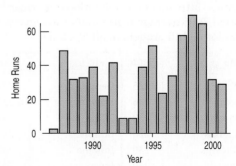

a) Comment on this graph.
b) Sketch your own histogram of the data.

48. Bird species. The Cornell Lab of Ornithology holds an annual Christmas Bird Count (www.birdsource.org), in which bird watchers at various locations around the country see how many different species of birds they can spot. Students were given the assignment to make a histogram of the counts reported from 21 sites in Texas counts reported in Exercise 36. One student submitted the following display:

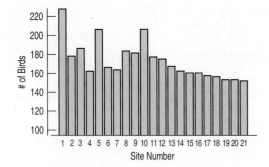

a) Comment on this graph.
b) Sketch your own histogram of the data.

49. Acid rain. Two researchers measured the pH (a scale on which a value of 7 is neutral and values below 7 are acidic) of water collected from rain and snow over a 6-month period in Allegheny County, PA. Describe their data with a graph, appropriate statistics, and a few sentences:

4.57	5.62	4.12	5.29	4.64	4.31	4.30	4.39	4.45
5.67	4.39	4.52	4.26	4.26	4.40	5.78	4.73	4.56
5.08	4.41	4.12	5.51	4.82	4.63	4.29	4.60	

T 50. Horsepower. Create a stem-and-leaf display for these horsepowers of autos reviewed by *Consumer Reports* one year, find appropriate summary statistics, and describe the distribution:

155	103	130	80	65
142	125	129	71	69
125	115	138	68	78
150	133	135	90	97
68	105	88	115	110
95	85	109	115	71
97	110	65	90	
75	120	80	70	

T 51. Math scores 2005. The National Center for Education Statistics (http://nces.ed.gov/nationsreportcard/) reported 2005 average mathematics achievement scores for eighth graders in all 50 states:

State	Score	State	Score
Alabama	225	Georgia	234
Alaska	236	Hawaii	230
Arizona	230	Idaho	242
Arkansas	236	Illinois	233
California	230	Indiana	240
Colorado	239	Iowa	240
Connecticut	242	Kansas	246
Delaware	240	Kentucky	231
Florida	239	Louisiana	230
Maine	241	Ohio	242
Maryland	238	Oklahoma	234
Massachusetts	247	Oregon	238
Michigan	238	Pennsylvania	241
Minnesota	246	Rhode Island	233
Mississippi	227	South Carolina	238
Missouri	235	South Dakota	242
Montana	241	Tennessee	232
Nebraska	238	Texas	242
Nevada	230	Utah	239
New Hampshire	246	Vermont	244
New Jersey	244	Virginia	240
New Mexico	224	Washington	242
New York	238	West Virginia	231
North Carolina	241	Wisconsin	241
North Dakota	243	Wyoming	243

Using appropriate graphical displays and summary statistics, write a report about the performance of eighth graders nationwide.

T 52. Gasoline usage 2004. The California Energy Commission (www.energy.ca.gov/gasoline/) collects data on the amount of gasoline sold in each state. The following data show the per capita (gallons used per person) consumption in the year 2004. Using appropriate graphical displays and summary statistics, write a report on the gasoline use by state in the year 2004.

State	Gallons per Capita	State	Gallons per capita
Alabama	529.4	Montana	544.4
Alaska	461.7	Nebraska	470.1
Arizona	381.9	Nevada	367.9
Arkansas	512.0	New Hampshire	544.4
California	414.4	New Jersey	488.2
Colorado	435.7	New Mexico	508.8
Connecticut	435.7	New York	293.4
Delaware	541.6	North Carolina	505.0
Florida	496.0	North Dakota	553.7
Georgia	537.1	Ohio	451.1
Hawaii	358.7	Oklahoma	614.2
Idaho	454.8	Oregon	418.4
Illinois	408.3	Pennsylvania	386.8
Indiana	491.7	Rhode Island	454.6
Iowa	555.1	South Carolina	578.6
Kansas	511.8	South Dakota	564.4
Kentucky	526.6	Tennessee	552.5
Louisiana	507.8	Texas	532.7
Maine	576.3	Utah	460.6
Maryland	447.5	Vermont	545.5
Massachusetts	458.5	Virginia	526.9
Michigan	482.0	Washington	423.6
Minnesota	527.7	West Virginia	426.7
Mississippi	558.5	Wisconsin	449.8
Missouri	550.5	Wyoming	615.0

53. Histogram. Find a histogram that shows the distribution of a variable in a newspaper, a magazine, or the Internet.
a) Does the article identify the W's?
b) Discuss whether the display is appropriate.
c) Discuss what the display reveals about the variable and its distribution.
d) Does the article accurately describe and interpret the data? Explain.

54. Not a histogram. Find a graph other than a histogram that shows the distribution of a quantitative variable in a newspaper, a magazine, or the Internet.
a) Does the article identify the W's?
b) Discuss whether the display is appropriate for the data.
c) Discuss what the display reveals about the variable and its distribution.
d) Does the article accurately describe and interpret the data? Explain.

55. In the news. Find an article in a newspaper, a magazine, or the Internet that discusses an "average."
a) Does the article discuss the W's for the data?
b) What are the units of the variable?
c) Is the average used the median or the mean? How can you tell?
d) Is the choice of median or mean appropriate for the situation? Explain.

56. In the news II. Find an article in a newspaper, a magazine, or the Internet that discusses a measure of spread.
a) Does the article discuss the W's for the data?
b) What are the units of the variable?
c) Does the article use the range, IQR, or standard deviation?
d) Is the choice of measure of spread appropriate for the situation? Explain.

Answers

Do the Math

1.

```
37 | 2
36 | 33
35 | 69
34 | 113367
33 | 8
32 | 5
31 | 48
30 | 248
```

Pass Completions (37 | 2 means 372)

2.

```
3 | 012233589
2 | 002278
1 | 889
```

TD Passes (3 | 0 means 30)

3.

```
3 | 589
3 | 012233
2 | 78
2 | 0022
1 | 889
```

TD Passes (3 | 0 means 30)

4. 342 completions

5. 29 TD passes

6. Q1 = 318, Q3 = 356, IQR = 38

7. Q1 = 20, Q3 = 32.5, IQR = 12.5

8. Min = 1.5, Q1 = 2.2, Median = 2.7, Q3 = 2.8, Max = 3.3 pounds

9. Min = 5.1, Q1 = 7.1, Median = 8.1, Q3 = 10.5, Max = 14.2 pounds

10. Min = 47.60, Q1 = 48.25, Median = 48.61, Q3 = 48.83, Max = 49.05 seconds

11. Min = 190, Q1 = 215, Median = 230, Q3 = 265, Max = 290 calories

12. Median = 34, mean = 37.1 treats

13. Median = 34, mean = 57.1 treats

14. The median stayed the same; the mean increased a lot.

15. The median is better, because it's still a typical value. All but one of the kids got fewer treats than the mean, so it's hardly typical.

16. Mean = 6.56, median = 6

17. Mean = 2.025, median = 2

18. Mean = 90, $n = 5$

Original values	Deviations	Squared Deviations
100	10	100
105	15	225
90	0	0
85	−5	25
70	−20	400
Sum of squared deviations: 750		
Divided by 4 (variance)		187.5
Square root (st. dev.):		13.69

19. Mean = 20, n = 8

Original values	Deviations	Squared Deviations
26	6	36
16	−4	16
18	−2	4
19	−1	1
23	3	9
22	2	4
18	−2	4
18	−2	4
Sum of squared deviations:		78
Divided by 7 (variance)		11.143
Square root (st. dev.):		3.34

20. For each set of data, the sum of the non-squared deviations from the mean is 0. This is because the mean is the balance point of all the data values.

Just Checking

(Thoughts will vary.)

1. Roughly symmetric, slightly skewed to the right. Center around 3 miles? Few over 10 miles.

2. Bimodal. Center between 1 and 2 hours? Many people watch no football; others watch most of one or more games. Probably only a few values over 5 hours.

3. Strongly skewed to the right, with almost everyone at $0; a few small prizes, with the winner an outlier.

4. Fairly symmetric, somewhat uniform, perhaps slightly skewed to the right. Center in the 40s? Few ages below 25 or above 70.

5. Uniform, symmetric. Center near 5. Roughly equal counts for each digit 0–9.

6. Incomes are probably skewed to the right and not symmetric, making the median the more appropriate measure of center. The mean will be influenced by the high end of family incomes and not reflect the "typical" family income as well as the median would. It will give the impression that the typical income is higher than it is.

7. An IQR of 30 mpg would mean that only 50% of the cars get gas mileages in an interval 30 mpg wide. Fuel economy doesn't vary that much. 3 mpg is reasonable. It seems plausible that 50% of the cars will be within about 3 mpg of each other. An IQR of 0.3 mpg would mean that the gas mileage of half the cars varies little from the estimate. It's unlikely that cars, drivers, and driving conditions are that consistent.

8. We'd prefer a standard deviation of 2 months. Making a consistent product is important for quality. Customers want to be able to count on the MP3 player lasting somewhere close to 5 years, and a standard deviation of 2 years would mean that lifespans were highly variable.

Stories Quantitative Data Tell

The Hopkins Memorial Forest is a 2500-acre reserve in Massachusetts, New York, and Vermont managed by the Williams College Center for Environmental Studies (CES). CES monitors forest conditions, posting daily measurements at their website.[1] You can go there, download, and analyze data for any range of days. We'll focus for now on 1989. As we'll see, some interesting things happened that year.

One of the variables CES measures in the forest is wind speed. Wind speeds can vary greatly during a day and from day to day, but if we step back a bit farther, we can see patterns. We'll discover new insights

as we change from viewing the whole year's data at one glance, to comparing seasons, to looking for patterns across months, and, finally, to looking at *Average Wind Speed* day by day.

---

[1] www.williams.edu/CES/hopkins.htm

The Big Picture

WHO	Days during 1989
WHAT	Average daily wind speed (mph), Average barometric pressure (mb), Average daily temperature (deg Celsius)
WHEN	1989
WHERE	Hopkins Forest, in Western Massachusetts
WHY	Long-term observations to study ecology and climate

Let's start with the "big picture." Here's a histogram and 5-number summary of the *Average Wind Speed* for every day in 1989. Because of the skewness, we'll report the median and IQR. We can see that the distribution of *Average Wind Speed* is unimodal and skewed to the right. Median daily wind speed is about 1.90 mph, and on half of the days, the average wind speed is between 1.15 and 2.93 mph. We also see a rather windy 8.67-mph day. Was that unusually windy or just the windiest day that year? To answer that, we'll need to work with the summaries a bit more.

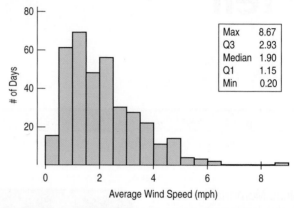

Max	8.67
Q3	2.93
Median	1.90
Q1	1.15
Min	0.20

FIGURE 5.1 A histogram of daily *Average Wind Speed* for 1989. The distribution is unimodal and skewed to the right, with a possible high outlier.

Boxplots and 5-Number Summaries

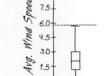

Once we have a 5-number summary of a (quantitative) variable, we can display that information in a **boxplot**. To make a boxplot of the average wind speeds, follow these steps:

1. Draw a single vertical axis spanning the extent of the data.[2] Draw short horizontal lines at the lower and upper quartiles and at the median. Then connect them with vertical lines to form a box. The box can have any width that looks OK.[3]

2. To help us construct the boxplot, we erect "fences" around the main part of the data. We place the fences 1.5 IQRs from the quartiles. For the wind speed data, we compute 1.5 IQR = 1.5 × 1.78 = 2.67 mph, so

$$Upper\ fence = Q3 + 1.5\ IQR = 2.93 + 2.67 = 5.60 \text{ mph}$$

and

$$Lower\ fence = Q1 - 1.5\ IQR = 1.15 - 2.67 = -1.52 \text{ mph}$$

(Note that the fences are not part of the display. We show them here with dotted lines for illustration. You should not include them in your boxplot.)

3. We use the fences to grow "whiskers." Draw lines from the ends of the box up and down to *the most extreme data values found within the fences.*

---

[2]The axis could also run horizontally.

[3]Some computer programs draw wider boxes for larger data sets. That can be useful when comparing groups.

4.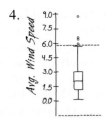

If a data value falls outside one of the fences, we do *not* connect it with a whisker.

4. Finally, we add the **outliers** by displaying any data values beyond the fences with special symbols.

Do The Math

Eventually, of course, we'll rely on a calculator or computer to make boxplots for us, but first let's be sure these plots make sense by creating a couple by hand. Here are some data about Kellogg's breakfast cereals from www.statcrunch.com.

Cereal Name	Calories	Fiber (g)
All-Bran	70	9
All-Bran with Extra Fiber	50	14
Apple Jacks	110	1
Corn Flakes	100	1
Corn Pops	110	1
Cracklin' Oat Bran	110	4
Crispix	110	1
Froot Loops	110	1
Frosted Flakes	110	1
Frosted Mini-Wheats	100	3
Fruitful Bran	120	5
Just Right Crunchy Nuggets	110	1
Just Right Fruit & Nut	140	2
Mueslix Crispy Blend	160	3
Nut & Honey Crunch	120	0
Nutri-Grain Almond-Raisin	140	3
Nutri-grain Wheat	90	3
Product 19	100	1
Raisin Bran	120	5
Raisin Squares	90	2
Rice Krispies	110	0
Smacks	110	1
Special K	110	1

1. Let's start with *Calories*.
 a) Find the median and the quartiles. (Remember to arrange the data in order first!)
 b) Find the IQR.
 c) Look for outliers. Remember, the fences are 1.5 IQRs outside the quartiles.
 d) Make and label a boxplot for *Calories*.
2. Make and label a boxplot for *Fiber*.

(Check your answers on page 113.)

The prominent statistician John W. Tukey, the originator of the boxplot, was asked by one of the authors why the outlier nomination rule cut at 1.5 IQRs beyond each quartile. He answered that the reason was that 1 IQR would be too small and 2 IQRs would be too large. That works for us.

What does a boxplot show? The center of a boxplot is (remarkably enough) a box that shows the middle half of the data, between the quartiles. The height of the box is equal to the IQR. If the median is roughly centered between the quartiles, then the middle half of the data is roughly symmetric. If the median is not centered, the distribution is skewed. If the whiskers are not roughly the same length they can show skewness as well. Displaying outliers individually keeps them out of the way when you judge skewness and encourages you to give them special attention. They may be mistakes, or they may be the most interesting cases in your data.

For the Hopkins Forest data, the central box contains each day whose *Average Wind Speed* is between 1.15 and 2.93 miles per hour (see Figure 5.2). The off-center median and the longer upper whisker indicate that the distribution of wind speeds is skewed high. We also see a few very windy days. Boxplots are particularly good at pointing out outliers. These extraordinarily windy days deserve more attention. We'll get to that shortly.

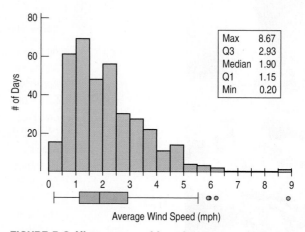

Max	8.67
Q3	2.93
Median	1.90
Q1	1.15
Min	0.20

FIGURE 5.2 Histogram and boxplot. By turning the boxplot and putting it on the same scale as the histogram, we can compare both displays of the daily wind speeds and see how each represents the distribution.

Just Checking

1. Here are boxplots showing the distributions of 5 different variables. Match each boxplot to one of the following histograms and justify your choice.

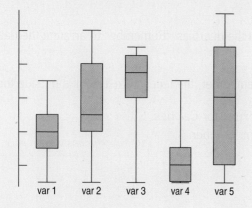

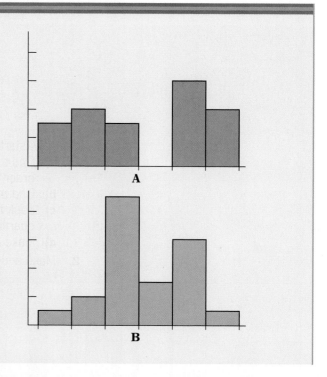

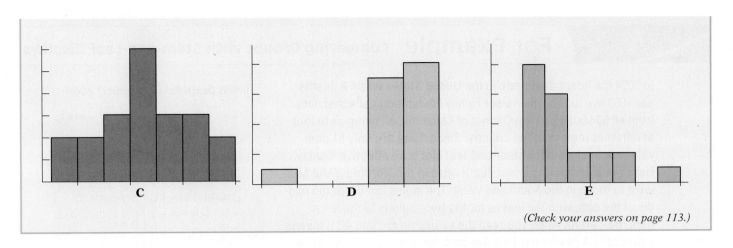

(*Check your answers on page 113.*)

Comparing Groups with Histograms

Is it windier in the winter or the summer? Are any months particularly windy? Questions like these can be more interesting, and we answer them by comparing groups of data.

First, let's split the year into two groups: April through September (Spring/Summer) and October through March (Fall/Winter). To compare the groups, we create two histograms using the same scale. Here are displays of the average daily wind speed for Spring/Summer (on the left) and Fall/Winter (on the right):

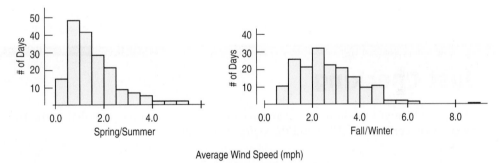

FIGURE 5.3 Histograms of *Average Wind Speed.* These tell very different stories for days in Spring/Summer (left) and Fall/Winter (right)

The shapes, centers, and spreads of these two distributions are strikingly different. During spring and summer (histogram on the left), the distribution is skewed to the right. A typical day during these warmer months has an average wind speed of only 1 to 2 mph, and few have average speeds above 3 mph. In the colder months (histogram on the right), however, the shape is less strongly skewed. It's also more spread out, telling us that wind speeds are more variable in the Fall and Winter. The typical wind speed is also higher then, and days with average wind speeds above 3 mph are not unusual; several are noticeably higher than any Spring/Summer wind speeds.

Summaries for *Average Wind Speed* by Season				
Group	Mean	StdDev	Median	IQR
Fall/Winter	2.71	1.36	2.47	1.87
Spring/Summer	1.56	1.01	1.34	1.32

For Example   Comparing Groups with Stem-and-Leaf Displays

In 2004 the infant death rate in the United States was 6.8 deaths per 1000 live births. The Kaiser Family Foundation collected data from all 50 states and the District of Columbia, allowing us to look at different regions of the country. Since there are only 51 data values, a back-to-back stem-and-leaf plot is an effective display. Here's one comparing infant death rates in the Northeast and Midwest to those in the South and West. The stems run down the middle of the plot, with the leaves for the two regions to the left or right. Be careful when you read the values on the left: 4|11| means a rate of 11.4 deaths per 1000 live birth for one of the southern or western states.

Infant Death Rates (by state) 2004

South and West		North and Midwest
4	11	
3 0	10	
0 0	9	
0 4 1 6 9 5 8	8	1 0
0 5 0 3	7	5 8 0 7 4 1
4 1 0 4 9 1 1 6 4	6	3 1 5 4 4
6 3 6 2	5	8 4 0 6
	4	8 8 9 7
	3	

(4 |11| means 11.4 deaths per 1000 live births)

Question: How do infant death rates compare for these regions?

In general, infant death rates were generally higher for states in the South and West than in the Northeast and Midwest. The distribution for the northeastern and midwestern states is roughly uniform, varying from a low of 4.8 to a high of 8.1 deaths per 1000 live births. Ten southern and western states had higher infant death rates than any in the Northeast or Midwest, with one state over 11 deaths per 1000. Rates varied more widely in the South and West, where the distribution is skewed to the right and possibly bimodal. We should investigate further to see which states represent the cluster of high death rates.

Just Checking

Here are histograms showing the average annual temperatures recorded in the Hopkins Memorial Forest for the years between 1890 to 1959 and the years 1960 to 2005.

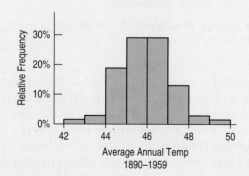

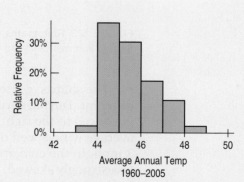

2. Describe fully what the tallest bar in the 1960–2005 histogram represents.

3. Write a few sentences comparing the distributions of annual average temperatures for the years before and years after 1960. Remember to discuss the similarities and differences in shape, center, and spread (in the context, of course).

4. Do these data offer evidence of global warming? Explain.

(Check your answers on page 113.)

Comparing Groups with Boxplots

Are some months windier than others? Even local residents may not have a good idea of which parts of the year are the most windy. (Do you know for your hometown?) Remember: we're not interested just in the centers, but also in the spreads. Are wind speeds equally consistent from month to month, or do some months show more variation?

Earlier, we compared histograms of the wind speeds for two halves of the year. Now we'll group the daily observations by month.

Histograms or stem-and-leaf displays are a fine way to look at one distribution or two. But it would be hard to see patterns by comparing 12 histograms. When we want to compare several groups we often plot boxplots side by side. That way we can easily see which groups have higher medians, which have the greater IQRs, where the central 50% of the data is located in each group, and which have the greater overall range. And, when the boxes are in an order, we can get a general idea of patterns in both the centers and the spreads. Equally important, we can see past any outliers in making these comparisons because they've been displayed separately.

Here are boxplots of the *Average Daily Wind Speed* by month:

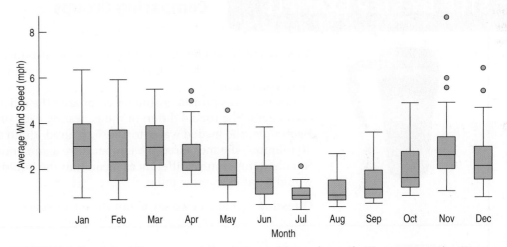

FIGURE 5.4 **Boxplots of the average daily wind speed for each month.** We see seasonal patterns in both the centers and spreads.

Now we see that wind speeds tend to decrease in the summer. Winds are both strongest and most variable in November through March. And there was one remarkably windy day in November.

When we looked at a boxplot of wind speeds for the entire year, there were only 5 outliers. Now, when we group the days by *Month,* the boxplots display more days as outliers. That's because some days that seemed ordinary among the entire year's data are outliers for their month. That windy day in July certainly wouldn't stand out in November or December, but for July, it was remarkable. And that really windy day in November? Stay tuned to see what happened then.

For Example   Comparing Distributions

Roller coasters[4] are a thrill ride in many amusement parks worldwide. And thrill seekers want a coaster that goes fast. There are two main types of roller coasters: those with wooden tracks and those with steel tracks. Do they typically run at different speeds? Here are boxplots:

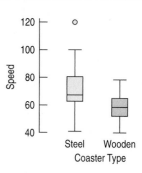

Question: How do the speeds of wood and steel roller coasters compare?

Overall, wooden-track roller coasters are slower than steel-track coasters. In fact, the fastest half of the steel coasters are faster than three-quarters of the wooden coasters. Although the IQRs of the two groups are similar, the range of speeds among steel coasters is larger than the range for wooden coasters. The distribution of speeds of wooden coasters appears to be roughly symmetric, but the speeds of the steel coasters are skewed to the right, and there is a high outlier at 120 mph. We should look into why that steel coaster is so fast.

STEP-BY-STEP EXAMPLE   Comparing Groups

Most scientific studies compare two or more groups. It is almost always a good idea to start an analysis of data from such studies by comparing boxplots. Here's an example:

For her class project, a student compared the efficiency of various coffee containers. She used 4 different mugs, testing each of them 8 different times. Each time, she heated water to 180°F, poured it into a mug, and sealed it. After 30 minutes, she measured the temperature again and recorded the difference. Smaller temperature differences mean that the liquid stayed hot—just what we would want in a coffee mug.

Question: What can we say about the effectiveness of these four mugs?

Plan State what you want to find out.

Variables Identify the *variables* and report the W's.

Be sure to check the appropriate condition.

I want to compare the effectiveness of the different mugs in maintaining temperature. I have 8 measurements of Temperature Change for each of the mugs.

✔ **Quantitative Data Condition:** The *Temperature Changes* are quantitative, with units of °F. Boxplots are appropriate displays for comparing the groups. Numerical summaries of each group are appropriate as well.

Mechanics Report the 5-number summaries of the four groups. Including the IQR is a good idea as well.

	Min	Q1	Median	Q3	Max	IQR
CUPPS	6°F	6	8.25	14.25	18.50	8.25
Nissan	0	1	2	4.50	7	3.50
SIGG	9	11.50	14.25	21.75	24.50	10.25
Starbucks	6	6.50	8.50	14.25	17.50	7.75

---

[4]See the Roller Coaster Data Base at www.rcdb.com.

Make a picture. Because we want to compare the distributions for four groups, boxplots are an appropriate choice.

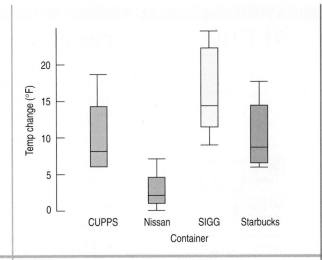

Conclusion Interpret what the boxplots and summaries say about the ability of these mugs to retain heat. Compare the shapes, centers, and spreads, and note any outliers.

The individual distributions of temperature changes are all slightly skewed to the high end. The Nissan cup does the best job of keeping liquids hot, with a median loss of only 2°F, and the SIGG cup does the worst, typically losing 14°F. The difference is large enough to be important: A coffee drinker would be likely to notice a 14° drop in temperature. And the mugs are clearly different: 75% of the Nissan tests showed less heat loss than any of the other mugs in the study. The IQR of results for the Nissan cup is also the smallest of these test cups, indicating that it is a consistent performer.

Just Checking

The Bureau of Transportation Statistics of the U.S. Department of Transportation collects and publishes statistics on airline travel (www.transtats.bts.gov). Here are three displays of the % of flights arriving late each month from 1995 through 2005:

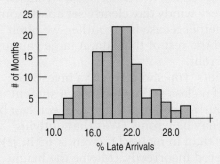

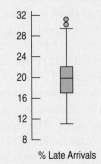

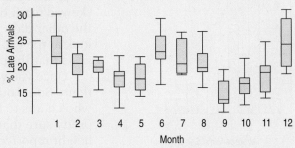

5. Describe what the histogram says about late arrivals.

6. What does the boxplot of late arrivals suggest that you can't see in the histogram?

7. Describe the patterns shown in the boxplots by month. At what time of year are flights least likely to be late? Can you suggest reasons for this pattern?

(Check your answers on page 113.)

TI Tips

Comparing groups with boxplots

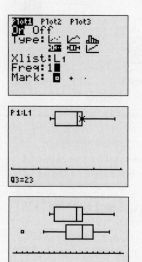

In the last chapter we looked at the performances of fourth-grade students on an agility test. Now let's make comparative boxplots for the boys' scores and the girls' scores:

> *Boys:* 22, 17, 18, 29, 22, 22, 23, 24, 23, 17, 21
> *Girls:* 25, 20, 12, 19, 28, 24, 22, 21, 25, 26, 25, 16, 27, 22

Enter these data in L1 (*Boys*) and L2 (*Girls*).

Set up STATPLOT's Plot1 to make a boxplot of the boys' data:

- Turn the plot On;
- Choose the first boxplot icon (you want your plot to indicate outliers);
- Specify Xlist:L1 and Freq:1, and select the Mark you want the calculator to use for displaying any outliers.

Use ZoomStat to display the boxplot for *Boys*. You can now TRACE to see the statistics in the 5-number summary. Try it!

As you did for the boys, set up Plot2 to display the girls' data. This time when you use ZoomStat with both plots turned on, the display shows the parallel boxplots. See the outlier?

This is a great opportunity to practice your Tell skills. How do these fourth graders compare in terms of agility?

Outliers

When we looked at boxplots for the *Average Wind Speed* by *Month*, we noticed that several days stood out as possible outliers and that one very windy day in November seemed truly remarkable. What should we do with such outliers?

First, try to understand outliers in the context of the data. A good place to start is with a histogram to get an idea of how the outlier fits (or doesn't fit) in with the rest of the data.

A histogram of the *Average Wind Speed* in November shows a slightly skewed main body of data and that very windy day clearly set apart from the other days. When considering whether a case is an outlier, we often look at the gap between that case and the rest of the data. A large gap suggests that the case really is quite different.

Once you've identified likely outliers, you should always investigate them. Some outliers are just errors. If a class survey includes a student who claims to be 170 inches tall (about 14 feet, or 4.3 meters), you can be pretty sure that's an error. The units may be wrong. (Was that outlying height reported in centimeters rather than in inches [170 cm = 65 in.]?) Or a number may just have been entered incorrectly. (Maybe the height was 70 inches and someone hit the 1 by mistake.) If you can identify the correct value, then you should certainly fix it. One important reason to look into outliers is to correct errors in your data.

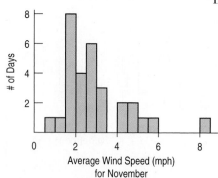

FIGURE 5.5 The *Average Wind Speed* in November is slightly skewed with a high outlier.

In fact, though, many outliers are not wrong; they're just different. Such cases often repay the effort to understand them. Sometimes you can learn more from the extraordinary cases than from summaries of the overall data set.

There are two things we should *never* do with outliers. The first is to drop an outlier from the analysis without comment just because it's unusual. If you want to exclude an outlier, you must discuss your decision and, to the extent you can, justify your decision. The other is to silently leave an outlier in place and proceed as if nothing were unusual. Analyses of data with outliers are very likely to be influenced by those outliers—sometimes to a large and misleading degree. Our best path is to report summaries and analyses with *and* without the outlier. In this way a reader can judge for him- or herself what influence the outlier has and decide what to think about the data.

What about that windy November day in the Hopkins Forest? Was it really that windy, or could there have been a problem with the anemometers? A quick Internet search for weather on November 21, 1989, finds that there was a severe storm:

WIND, SNOW, COLD GIVE N.E. A TASTE OF WINTER

Published on November 22, 1989
Author: Andrew Dabilis, Globe Staff

An intense storm roared like the Montreal Express through New England yesterday, bringing frigid winds of up to 55 m.p.h., 2 feet of snow in some parts of Vermont and a preview of winter after weeks of mild weather. Residents throughout the region awoke yesterday to an icy vortex that lifted an airplane off the runway in Newark and made driving dangerous in New England because of rapidly shifting winds that seemed to come from all directions.

For Example   Checking Out the Outliers

Recap: We've looked at the speeds of roller coasters and found a difference between steel- and wooden-track coasters. We also noticed an extraordinary value.

Question: The fastest coaster in this collection turns out to be the "Top Thrill Dragster" at Cedar Point amusement park. What might make this roller coaster unusual? You'll have to do some research, but that's often what happens with outliers.

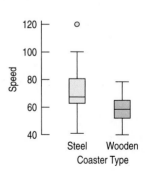

The Top Thrill Dragster is easy to find in an Internet search. We learn that it is a "hydraulic launch" coaster. That is, it doesn't get its remarkable speed just from gravity, but rather from a kick-start by a hydraulic piston. That could make it different from the other roller coasters.

(You might also discover that it is no longer the fastest roller coaster in the world.)

Timeplots: Order, Please!

The Hopkins Forest wind speeds are reported as daily averages. Previously, we grouped the days into months or seasons, but we could look at the wind speed values day by day. Whenever we have data measured over time, it is a good idea to look for patterns by plotting the data in time order. Here are the daily average wind speeds plotted over time:

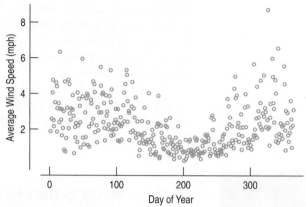

FIGURE 5.6 A timeplot of *Average Wind Speed* shows the overall pattern and changes in variation.

A display of values against time is called a **timeplot**. This timeplot reflects the pattern that we saw when we plotted the wind speeds by month. But without the arbitrary divisions between months, we can see a calm period during the summer, starting around day 200 (the middle of July), when the wind is relatively mild and doesn't vary greatly from day to day. We can also see that the wind becomes both more variable and stronger during the early and late parts of the year.

Looking into the Future

It is always tempting to try to extend what we see in a timeplot into the future. Sometimes that makes sense. Most likely, the Hopkins Forest climate follows regular seasonal patterns. It's probably safe to predict a less windy June next year and a windier November. But we certainly wouldn't predict another storm on November 21.

Other patterns are riskier to extend into the future. If a stock has been rising, will it continue to go up? No stock has ever increased in value indefinitely, and no stock analyst has consistently been able to forecast when a stock's value will turn around. Stock prices, unemployment rates, and other economic, social, or psychological concepts are much harder to predict. The path a ball will follow when thrown from a certain height at a given speed is well understood. The path the stock market will take is much less clear.

If you had tracked stock prices by plotting the Dow Jones average monthly as it rose from 2003 through 2007, you might have decided it was a good time to invest some money. Unless we have strong (nonstatistical) reasons for doing otherwise, we should resist the temptation to think that any trend we see will continue, even into the near future. The trend in stock prices may have looked promising:

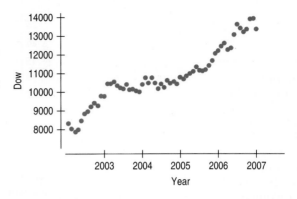

Bad call, though. By March 2009 the Dow had fallen by over 50%.

Shifting Data

All quantitative data have units—feet, quarts, pounds, horsepower, and so on. Sometimes it's helpful to change the units. For example, if we're given temperatures in degrees Celsius, it might help us to think about them in degrees Fahrenheit instead. A change like that is called *rescaling* the data. Or maybe rather than looking at the actual temperatures we just want to see how many degrees above or below freezing they are. That kind of change is called *shifting* the data. Shifting and rescaling can be very useful, so we need to understand how they work. In particular, how do shifting or rescaling data affect the shape, center, and spread of the distribution? Let's tackle shifting first.

Since the 1960s, the Centers for Disease Control's National Center for Health Statistics has been collecting health and nutritional information on people of all ages and backgrounds. A recent survey, the National Health and Nutrition Examination Survey (NHANES) 2001–2002,[5] recorded body measurements for more than 11,000 people. Among them were 80 men between 19 and 24 years old and of average height (between 5′8″ and 5′10″ tall). Here are a histogram and boxplot of their weights:

WHO	80 male participants of the NHANES survey between the ages of 19 and 24 who measured between 68 and 70 inches tall
WHAT	Their weights
UNIT	Kilograms
WHEN	2001–2002
WHERE	United States
WHY	To study nutrition, and health issues and trends
HOW	National survey

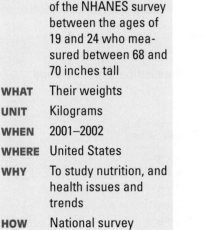

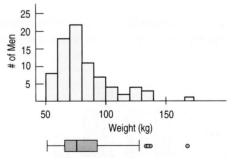

FIGURE 5.7 **Histogram and boxplot for the men's weights.** The shape is skewed to the right with several high outliers.

Their mean weight is 82.36 kg. For men this age and height, the National Institutes of Health recommends a maximum healthy weight of 74 kg. We can see that some of these men are heavier than that. To compare their weights to the recommended maximum, we'll subtract 74 kg from each of their weights.

---

[5]www.cdc.gov/nchs/nhanes.htm

Doctors' height and weight charts sometimes give ideal weights for various heights that include 2-inch heels. If the mean height of adult women is 66 inches including 2-inch heels, what is the mean height of women without shoes? Each woman is shorter by 2 inches when barefoot, so the mean is decreased by 2 inches, to 64 inches.

What does that do to the center, shape, and spread of the histogram? Here's the picture:

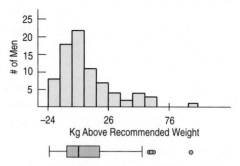

FIGURE 5.8 **After the shift.** Subtracting 74 kilograms shifts the entire histogram down but leaves the spread and the shape exactly the same.

On average, these men weigh 82.36 kg, so on average they're 82.36 − 74 = 8.36 kg overweight. That's the mean of the new distribution. When we **shift** the data by adding (or subtracting) a constant to each value, all measures of position (center, percentiles, min, max) will increase (or decrease) by the same amount.

What about the shape and spread? Look at the two histograms again. Adding or subtracting a constant changes each data value equally, so the entire distribution just shifts. Its shape doesn't change and neither does the spread. None of the measures of spread we've discussed—not the range, not the IQR, not the standard deviation—changes.

> *Adding (or subtracting) a constant to every data value adds (or subtracts) the same constant to measures of position, but leaves measures of spread unchanged.*

For Example   Shifting Temperatures

Remember that windy November in the Hopkins Forest? Was it cold, too? Here's the distribution of *Average Daily Temperature* for the month, along with the summary statistics.

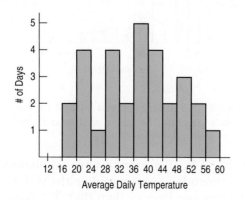

Summary	Avg Daily Temp
Min	19.8°F
Q1	29.1°F
Median	37.1°F
Q3	46.8°F
Max	59.7°F
Range	39.9°F
IQR	17.7°F
Mean	37.0°F
StdDev	11.3°F

Question: Let's investigate how far above or below freezing these November days were by subtracting 32 degrees from each of the temperatures. How will that shift affect the shape, center, and spread of the *Average Daily Temperature*?

Subtracting the same amount from each temperature will not change the shape of the distribution. Each new temperature will simply be 32 degrees less than it was, shifting the entire distribution downward, as we see in these "before" and "after" boxplots.

The maximum temperature was 59.7°F, so the maximum distance above freezing is now $59.7 - 32 = 27.7°F$. The minimum is also 32°F lower than it was, as are both measures of center (mean and median), and the quartiles.

However, because the entire distribution shifted the same amount, the spread did not change, as we can see clearly in the boxplots. All the measures of spread—range, IQR, and the standard deviation—remain exactly the same.

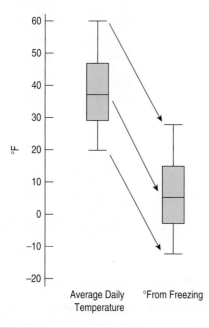

Summary	°*FromFreezing*
Min	−12.2°F
Q1	−2.9°F
Median	5.1°F
Q3	14.8°F
Max	27.7°F
Range	39.9°F
IQR	17.7°F
Mean	5.0°F
StdDev	11.3°F

Rescaling Data

Not everyone thinks naturally in metric units. Suppose we want to look at the men's weights in pounds instead. We have to **rescale** the data. Because there are about 2.2 pounds in every kilogram, we'll convert the weights by multiplying each value by 2.2. Multiplying or dividing each value by a constant changes the measurement units. Here are histograms of the two weight distributions, plotted on the same scale, so you can see the effect of multiplying:

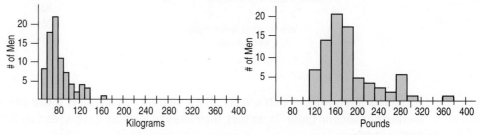

FIGURE 5.9 Men's weights in both kilograms and pounds. How do the distributions and numerical summaries change?

What happened to the shape of the distribution? Although the histograms don't look exactly alike, we see that the shape really hasn't changed: Both are unimodal and skewed to the right.

What happened to the mean? The men weigh 82.36 kg on average, which is 181.19 pounds. Not too surprisingly, the mean multiplied by 2.2 as well. As the boxplots and 5-number summaries show, all measures of position act the same way. They all get multiplied by this same constant.

What happened to the spread? Take a look at the boxplots. The spread in pounds (on the right) is larger. How much larger? If you guessed 2.2 times, you've figured out how measures of spread get rescaled.

FIGURE 5.10 The boxplots (drawn on the same scale) show the weights measured in kilograms (on the left) and pounds (on the right). Because 1 kg is 2.2 lb, all the points in the right box are 2.2 times larger than the corresponding points in the left box. So each measure of position and spread is 2.2 times as large when measured in pounds rather than kilograms.

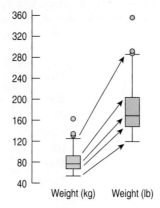

Summary	Weight (kg)	Weight (lb)
Min	54.3	119.46
Q1	67.3	148.06
Median	76.85	169.07
Q3	92.3	203.06
Max	161.5	355.30
IQR	25	55
SD	22.27	48.99

When we multiply (or divide) all the data values by any constant, all measures of position (such as the mean, median, and percentiles) and measures of spread (such as the range, the IQR, and the standard deviation) are multiplied (or divided) by that same constant.

For Example   Rescaling Wind Speeds

Here again is the histogram of average daily Hopkins Forest wind speeds in miles per hour during November 1989, along with the summary statistics.

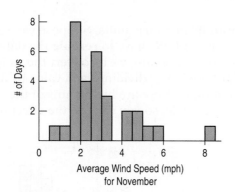

Summary	AverageWindSpeed
Min	1.01 mph
Q1	2.01 mph
Median	2.63 mph
Q3	3.42 mph
Max	8.67 mph
Range	7.66 mph
IQR	1.41 mph
Mean	3.12 mph
StdDev	1.63 mph

Scientists often work in metric units. Suppose they convert all of these speeds from miles per hour (mph) to meters per second (m/s). To do that, they need to multiply each speed by 0.45. (For example, 4 mph = 4 × 0.45 = 1.80 m/s.)

Question: How will this rescaling affect the shape, center, and spread of *Average Wind Speed*?

Although multiplying all the wind speeds by 0.45 makes them all smaller, it does not change the shape of the distribution. The rescaled speeds are still skewed high with three outliers, as we see in these "before" and "after" boxplots.

The median wind speed was 2.63 mph, so the new median is 0.45 × 2.63 = 1.18 m/s. The mean wind speed is also multiplied by the same amount, as are the min, the max, and the quartiles.

Because multiplying each speed by 0.45 made the distribution shrink, the spread also changed. The new IQR is 0.45 × 1.41 = 0.63 m/s. The range and standard deviation were also multiplied by 0.45.

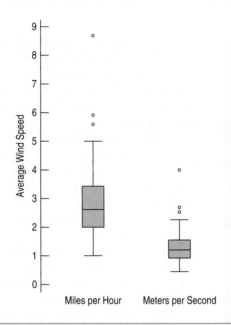

Summary	AvgWindSpeed
Min	0.45 m/s
Q1	0.90 m/s
Median	1.18 m/s
Q3	1.54 m/s
Max	3.90 m/s
Range	3.45 m/s
IQR	0.63 m/s
Mean	1.40 m/s
StdDev	0.73 m/s

Just Checking

8. In 1995 the Educational Testing Service (ETS) adjusted the scores of SAT tests. Before ETS recentered the SAT Verbal test, the mean of all test scores was 450.

a) How would adding 50 points to each score affect the mean?

b) The standard deviation was 100 points. What would the standard deviation be after adding 50 points?

c) Suppose we drew boxplots of test takers' scores a year before and a year after the recentering. How would the boxplots of the two years differ?

d) Which of these statistics would not change: Q1, IQR, or max?

9. A company manufactures wheels for in-line skates. The diameters of the wheels have a mean of 3 inches and a standard deviation of 0.1 inches. Because so many of their customers use the metric

system, the company decided to report their production statistics in millimeters (1 inch = 25.4 mm).

a) What's the mean wheel diameter in millimeters?

b) What's the standard deviation of the wheel diameters in millimeters?

c) Which of these statistics would not change: Q1, IQR, or max?

(Check your answers on page 113.)

🚫 WHAT CAN GO WRONG?

- **Avoid inconsistent scales.** No fair changing scales in the middle or plotting two variables on different scales but on the same display. When comparing two groups, be sure to compare them on the same scale.

- **Label clearly.** Variables should be identified clearly and axes labeled so a reader knows what the plot displays.

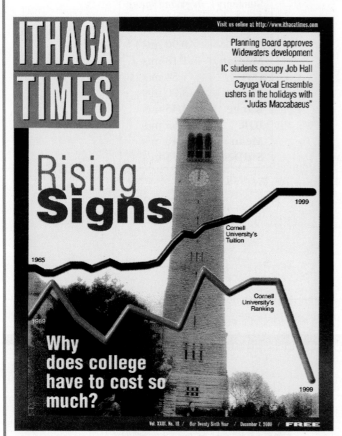

Here's a remarkable example of a plot gone wrong. It illustrated a news story about rising college costs. It uses timeplots, but it gives a misleading impression. First think about the story you're being told by this picture. Then try to figure out what has gone wrong.

What's wrong? Just about everything.

- The horizontal scales are inconsistent. Both lines show trends over time, but exactly for what years? The tuition sequence starts in 1965, but rankings are graphed from 1989. Plotting them on the same (invisible) scale makes it seem that they're for the same years.

- The vertical axis isn't labeled. That hides the fact that it's inconsistent. Does it graph dollars (of tuition) or ranking (of Cornell University)?

This display violates three of the rules. And it's even worse than that: It violates another rule that we didn't even bother to mention.

- The two inconsistent scales for the vertical axis don't point in the same direction! The line for Cornell's rank shows that it has "plummeted" from 15th place to 6th place in academic rank. Most of us think that's an *improvement*, but that's not the message of this graph.

- **Beware of outliers.** If the data have outliers and you can correct them, you should do so. If they are clearly wrong or impossible, you should remove them and report on them. Otherwise, consider summarizing the data both with and without the outliers.

COMPARING DISTRIBUTIONS **IN YOUR WORLD**

Go To College, Earn More $$$

Getting a college degree takes years of study. Will that effort pay off in the job market? It probably will, according to the U.S. Bureau of Labor Statistics (BLS).

More people are going to college now than ever before, in part because of the advantages that a college degree confers. College-educated workers' higher earnings and lower unemployment are good reasons to go to college, and these benefits are also evidence of the demand for college graduates. Higher earnings show that employers are willing to pay more to have college graduates work for them. And lower unemployment means college graduates are more likely to find a job when they want one.

As a whole, college-educated workers earn more money than workers who have less education. In 2005, workers who had a bachelor's degree had median weekly earnings of $937 [$48,724 per year], compared with $583 [$30,316 a year] a week for high school graduates— that's a difference of $354 per week [$18,408 a year], or a 61-percent jump in median earnings. For workers who had a master's, doctoral, or professional degree, median earnings were even higher.

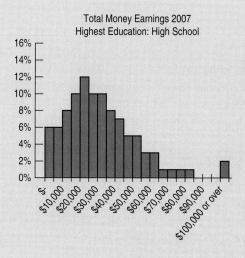

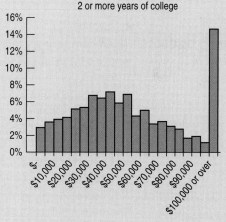

In addition to earning more money, workers who have more education are also less likely to be unemployed. . . . [T]he 2005 unemployment rate for workers who have a bachelor's or higher degree [was] half the rate for high school graduates . . . and less than a third of the rate for dropouts . . . .

Higher earnings and less unemployment combine to give graduates substantially higher incomes over a lifetime compared with their less-educated counterparts.

(The 2004-14 Job Outlook, http://goliath.ecnext.com/coms2/gi_0199-6046145/The-2004-14-job-outlook.html)

WHAT HAVE WE LEARNED?

▶ We've learned the value of comparing groups and looking for patterns among groups and over time.

▶ We've seen that boxplots are very effective for comparing groups graphically. When we compare groups, we discuss their shape, center, and spreads, and any unusual features.

▶ We've seen why we need to identify and investigate outliers.

▶ We've graphed data that have been measured over time against a time axis and looked for long-term trends.

We've learned that the story data can tell may be easier to understand after shifting or rescaling the data.

▶ Shifting data by adding or subtracting the same amount from each value affects measures of center and position but not measures of spread.

▶ Rescaling data by multiplying or dividing every value by a constant changes all the summary statistics—center, position, and spread.

Terms

Boxplot A boxplot displays the 5-number summary as a central box with whiskers that extend to the non-outlying data values. Boxplots are particularly effective for comparing groups and for displaying outliers.

Outlier Any point more than 1.5 IQR from either end of the box is nominated as an outlier.

Comparing distributions When comparing the distributions of several groups using histograms or stem-and-leaf displays, consider their:

▶ Shape

▶ Center

▶ Spread

Comparing boxplots When comparing groups with boxplots:

▶ Compare the shapes. Do the boxes look symmetric or skewed? Are there differences between groups?

▶ Compare the medians. Which group has the higher center? Is there any pattern to the medians?

▶ Compare the IQRs. Which group is more spread out? Is there any pattern to how the IQRs change?

▶ Check for possible outliers. Identify them if you can and discuss why they might be unusual. Of course, correct them if you find that they are errors.

Timeplot A timeplot displays data that change over time. Often, successive values are connected with lines to show trends more clearly. Sometimes a smooth curve is added to the plot to help show long-term patterns and trends.

Shifting Adding a constant to each data value adds the same constant to the mean, the median, and the quartiles, but does not change the standard deviation or IQR.

Rescaling Multiplying each data value by a constant multiplies both the measures of position (mean, median, and quartiles) and the measures of spread (standard deviation and IQR) by that constant.

Skills

- ► Be able to select a suitable display for comparing groups. Understand that histograms show distributions well, but are difficult to use when comparing more than two or three groups. Boxplots are more effective for comparing several groups.
- ► Understand that how you group data can affect what kinds of patterns and relationships you are likely to see. Know how to select groupings to show the information that is important for your analysis.
- ► Be aware of the effects of skewness and outliers on measures of center and spread. Know how to select appropriate measures for comparing groups based on their displayed distributions.
- ► Understand that outliers, whatever their source, deserve special attention.
- ► Recognize when it is appropriate to make a timeplot.
- ► Understand how adding (subtracting) a constant or multiplying (dividing) by a constant changes the center and/or spread of a variable.

- ► Know how to make side-by-side histograms on comparable scales to compare the distributions of two groups.
- ► Know how to make side-by-side boxplots to compare the distributions of two or more groups.
- ► Know how to describe differences among groups in terms of patterns and changes in their center, spread, shape, and unusual values.
- ► Know how to make a timeplot of data that have been measured over time.

- ► Know how to compare the distributions of two or more groups by comparing their shapes, centers, and spreads. Be prepared to explain your choice of measures of center and spread for comparing the groups.
- ► Be able to describe trends and patterns in the centers and spreads of groups—especially if there is a natural order to the groups, such as a time order.
- ► Be prepared to discuss patterns in a timeplot in terms of both the general trend of the data and the changes in how spread out the pattern is.
- ► Be cautious about assuming that trends over time will continue into the future.
- ► Be able to describe the distribution of a quantitative variable in terms of its shape, center, and spread.
- ► Be able to describe any anomalies or extraordinary features revealed by the display of a variable.
- ► Know how to compare the distributions of two or more groups by comparing their shapes, centers, and spreads.
- ► Know how to describe patterns over time shown in a timeplot.
- ► Be able to discuss any outliers in the data, noting how they deviate from the overall pattern of the data.

COMPARING DISTRIBUTIONS ON THE COMPUTER

Most programs for displaying and analyzing data can display plots to compare the distributions of different groups. Typically these are boxplots displayed side-by-side.

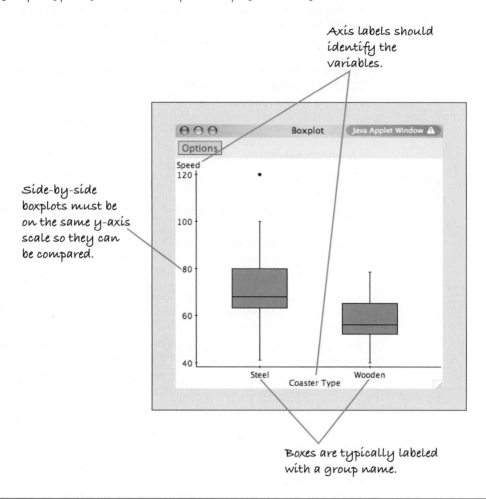

Axis labels should identify the variables.

Side-by-side boxplots must be on the same y-axis scale so they can be compared.

Boxes are typically labeled with a group name.

EXERCISES

A

1. **Test scores.** Three Statistics classes all took the same test. Histograms and boxplots of the scores for each class are shown below. Match each class with the corresponding boxplot.

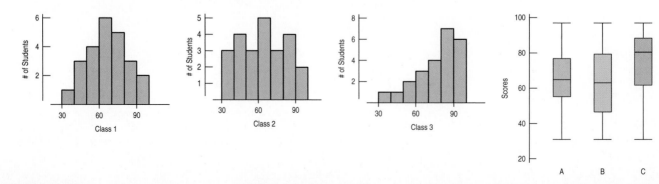

2. Matching. Match each boxplot to the corresponding histogram and justify your choices.

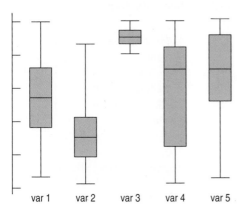

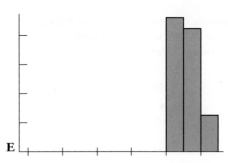

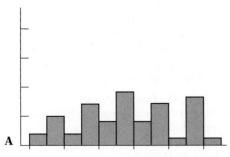

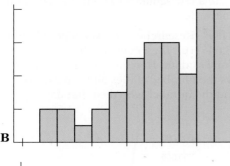

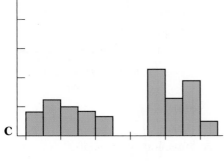

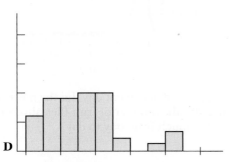

T **3. Still rockin'.** Crowd Management Strategies monitors accidents at rock concerts. In their database, they list the names and other variables of victims whose deaths were attributed to "crowd crush" at rock concerts. Here are the histogram and boxplot of the victims' ages for data from 1999 to 2000:

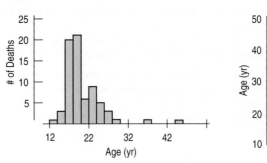

a) What features of the distribution can you see in both the histogram and the boxplot?
b) What features of the distribution can you see in the histogram that you could not see in the boxplot?
c) What summary statistic would you choose to summarize the center of this distribution? Why?
d) What summary statistic would you choose to summarize the spread of this distribution? Why?

T **4. Slalom times.** The Men's Combined skiing event consists of a downhill and a slalom. Here are two displays of the slalom times in the Men's Combined at the 2006 Winter Olympics:

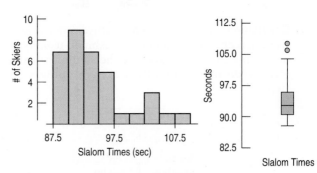

a) What features of the distribution can you see in both the histogram and the boxplot?
b) What features of the distribution can you see in the histogram that you could not see in the boxplot?
c) What summary statistic would you choose to summarize the center of this distribution? Why?
d) What summary statistic would you choose to summarize the spread of this distribution? Why?

T **5. Women's basketball.** Here are boxplots of the points scored during the first 10 games of the season for both Scyrine and Alexandra:

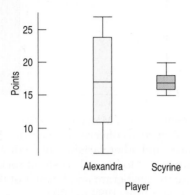

a) Summarize the similarities and differences in their performance so far.
b) The coach can take only one player to the state championship. Which one should she take? Why?

T **6. Marriage age.** In 1975, did men and women marry at the same age? Here are boxplots of the age at first marriage for a sample of U.S. citizens then. Write a brief report discussing what these data show.

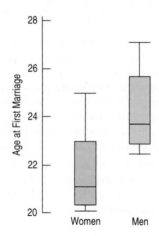

7. Graduation? A survey of major universities asked what percentage of incoming freshmen usually graduate "on time" in 4 years. Use the summary statistics given to decide whether there are any outliers. Explain.

	% on Time
Count	48
Mean	68.35
Median	69.90
StdDev	10.20
Min	43.20
Max	87.40
Range	44.20
25th %tile	59.15
75th %tile	74.75

T **8. Vineyards.** Here are summary statistics for the sizes (in acres) of Finger Lakes vineyards. Are any of the vineyards outliers? Explain.

Count	36
Mean	46.50 acres
StdDev	47.76
Median	33.50
IQR	36.50
Min	6
Q1	18.50
Q3	55
Max	250

9. Graduation revisited. Look again at the on-time graduation summaries in Exercise 7.
a) Create a boxplot of these data.
b) Would you describe this distribution as symmetric or skewed? Explain.
c) Write a few sentences about the graduation rates.

10. Vineyards again. Look again at the sizes of Finger Lakes vineyards summarized in Exercise 8.
a) Create a boxplot of these data.
b) Would you describe this distribution as symmetric or skewed? Explain.
c) Write a few sentences about the sizes of the vineyards.

11. Fruit flies. Researchers tracked a population of 1,203,646 fruit flies, counting how many died each day for 171 days. Here are two timeplots. The first shows the number of flies alive on each day and the second the number who died that day.

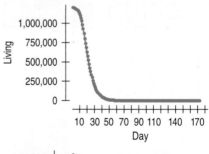

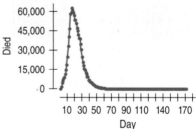

a) On approximately what day did the most flies die?
b) When did the number of fruit flies alive stop changing very much from day to day?

12. Fruit flies II. Researchers tracked a population of 1,203,646 fruit flies for 171 days. The timeplot below shows the daily mortality rate—the fraction of the number alive who died each day.

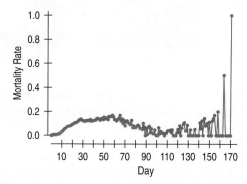

a) During the first 100 days, when was the death rate highest?
b) Over what time interval did the death rate remain fairly stable?
c) What happened on the last day?

13. Cell phones. A cell phone company charges all users a basic monthly fee and then adds on extra charges based on usage and other services. They compile data recording each user's bill. Suppose they raise the basic fee by $3 a month. Which of the statistics in each group will stay the same? Which will change—and how?
a) Min, max, range
b) Median, quartiles, IQR
c) Mean, standard deviation

14. Basketball players. Basketball teams list players' heights in the programs. Suppose one team measured their players while they were wearing their sneakers, adding one inch to everyone's height. You've found summaries based on the listed heights. If you want to describe the true heights, which of the statistics in each group should you change—and how? Which will stay the same?
a) Min, max, range
b) Median, quartiles, IQR
c) Mean, standard deviation

15. Cell phones II. Suppose the cell phone company in Exercise 13 had decided to raise their rates by 10% for everything—basic fee, usage charges, and other services. How will that increase affect each of these statistics summarizing customers' monthly bills?
a) Min, max, range
b) Median, quartiles, IQR
c) Mean, standard deviation

16. Basketball players II. The basketball team in Exercise 14 will be playing in an international tournament, so they need to change the program to list all the heights in centimeters instead of inches (1 inch = 2.54 cm). How will that change affect each of these statistics summarizing players' heights?
a) Min, max, range
b) Median, quartiles, IQR
c) Mean, standard deviation

17. Shipments. A company selling clothing on the Internet reports that the packages it ships have a median weight of 68 ounces and an IQR of 40 ounces.

a) The company plans to include a sales flyer weighing 4 ounces in each package. What will the new median and IQR be?
b) If the company recorded the shipping weights of these new packages in pounds instead of ounces, what would the median and IQR be? (1 lb. = 16 oz.)

18. Hotline. A company's customer service hotline handles many calls relating to orders, refunds, and other issues. The company's records indicate that the median length of calls to the hotline is 4.4 minutes with an IQR of 2.3 minutes.
a) If the company were to describe the duration of these calls in seconds instead of minutes, what would the median and IQR be?
b) In an effort to speed up the customer service process, the company decides to streamline the series of pushbutton menus customers must navigate, cutting the time by 24 seconds. What will the median and IQR of the length of hotline calls become?

B

19. Pizza prices. A company that sells frozen pizza to stores in four markets in the United States (Denver, Baltimore, Dallas, and Chicago) wants to examine the prices that the stores charge for pizza slices. Here are boxplots comparing data from a sample of stores in each market:

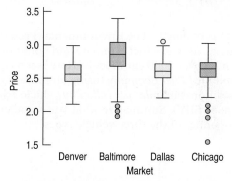

a) Do prices appear to be the same in the four markets? Explain.
b) Does the presence of any outliers affect your overall conclusions about prices in the four markets?

20. Costs. To help travelers know what to expect, researchers collected the prices of commodities in 16 cities throughout the world. Here are boxplots comparing the prices of a ride on public transportation, a newspaper, and a cup of coffee in the 16 cities (prices are all in $US).

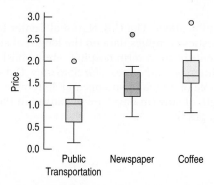

a) On average, which commodity is the most expensive?
b) Is a newspaper always more expensive than a ride on public transportation? Explain.
c) Does the presence of outliers affect your conclusions in a) or b)?

T **21. Fuel economy.** Describe what these boxplots tell you about the relationship between the number of cylinders a car's engine has and the car's fuel economy (mpg):

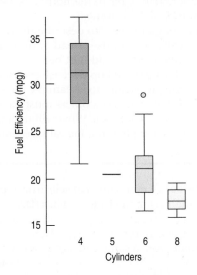

22. Fuel economy II. The Environmental Protection Agency provides fuel economy and pollution information on over 2000 car models. Here is a boxplot of *Combined Fuel Economy* (using an average of driving conditions) in *miles per gallon* by vehicle *Type* (car, van, or SUV). Summarize what you see about the fuel economies of the three vehicle types.

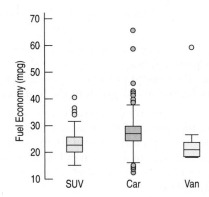

23. Hospital stays. The U.S. National Center for Health Statistics compiles data on the length of stay by patients in short-term hospitals and publishes its findings in *Vital and Health Statistics*. Data from a sample of 39 male patients and 35 female patients on length of stay (in days) are displayed in the following histograms.

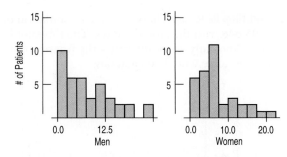

a) What would you suggest be changed about these histograms to make them easier to compare?
b) Describe these distributions by writing a few sentences comparing the duration of hospitalization for men and women.
c) Can you suggest a reason for the peak in women's length of stay?

24. Deaths 2003. A National Vital Statistics Report (www.cdc.gov/nchs/) indicated that nearly 300,000 black Americans died in 2003, compared with just over 2 million white Americans. Here are histograms displaying the distributions of their ages at death:

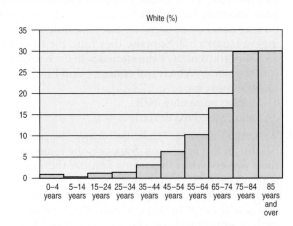

a) Describe the overall shapes of these distributions.
b) How do the distributions differ?
c) Look carefully at the bar definitions. Where do these plots violate the rules for statistical graphs?

25. Cereals. Sugar is a major ingredient in many breakfast cereals. The histogram displays the sugar content as a percentage of weight for 49 brands of cereal. The boxplot compares sugar content for adult and children's cereals.

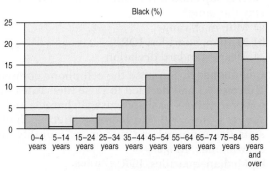

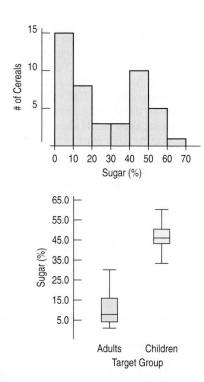

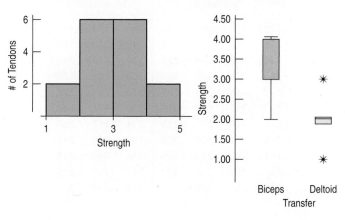

a) What is the range of the sugar contents of these cereals.
b) Describe the shape of the distribution.
c) What aspect of breakfast cereals might account for this shape?
d) Are all children's cereals higher in sugar than adult cereals?
e) Which group of cereals varies more in sugar content? Explain.

26. Tendon transfers. People with spinal cord injuries may lose function in some, but not all, of their muscles. The ability to push oneself up is particularly important for shifting position when seated and for transferring into and out of wheelchairs. Surgeons compared two operations to restore the ability to push up in children. The histogram shows scores rating pushing strength for all patients two years after surgery. The boxplots compare results for the two surgical methods. (Mulcahey, Lutz, Kozen, Betz, "Prospective Evaluation of Biceps to Triceps and Deltoid to Triceps for Elbow Extension in Tetraplegia," *Journal of Hand Surgery*, 28, 6, 2003)

a) Describe the shape of this distribution.
b) What is the range of the strength scores?
c) What fact about results of the two procedures is hidden in the histogram?
d) Which method had the higher (better) median score?
e) Was that method always best?
f) Which method produced the most consistent results? Explain.

27. Caffeine. A student study of the effects of caffeine asked volunteers to take a memory test 2 hours after drinking soda. Some drank caffeine-free cola, some drank regular cola (with caffeine), and others drank a mixture of the two (getting a half-dose of caffeine). Here are the 5-number summaries for each group's scores (number of items recalled correctly) on the memory test:

	n	Min	Q1	Median	Q3	Max
No caffeine	15	16	20	21	24	26
Low caffeine	15	16	18	21	24	27
High caffeine	15	12	17	19	22	24

a) Create parallel boxplots to display these results as best you can with this information.
b) Write a few sentences comparing the performances of the three groups.

28. SAT scores. Here are the summary statistics for Verbal SAT scores for a high school graduating class:

	n	Mean	Median	SD	Min	Max	Q1	Q3
Male	80	590	600	97.2	310	800	515	650
Female	82	602	625	102.0	360	770	530	680

a) Create parallel boxplots comparing the scores of boys and girls as best you can from the information given.
b) Write a brief report on these results. Be sure to discuss the shape, center, and spread of the scores.

29. Bread. Clarksburg Bakery is trying to predict how many loaves to bake each day. In the last 100 days, they have sold between 95 and 140 loaves per day. Here is a histogram of the number of loaves they sold for the last 100 days.

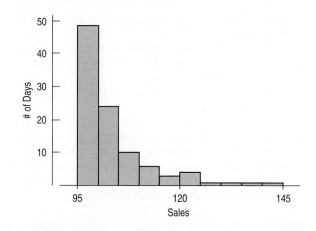

a) Describe the distribution.
b) Which should be larger, the mean number of sales or the median? Explain.
c) Here are the summary statistics for Clarksburg Bakery's bread sales. Use these statistics and the histogram above to create a boxplot. You may approximate the values of any outliers.

Summary of Sales	
Median	100
Min	95
Max	140
25th %tile	97
75th %tile	105.5

30. Camp sites. Shown below are the histogram and summary statistics for the number of camp sites at public parks in Vermont.

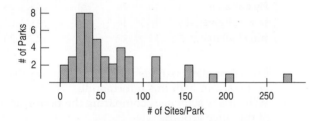

Count	46
Mean	62.8 sites
Median	43.5
StdDev	56.2
Min	0
Max	275
Q1	28
Q3	78

a) Which statistics would you use to identify the center and spread of this distribution? Why?
b) How many parks would you classify as outliers? Explain.
c) Create a boxplot for these data.
d) Write a few sentences describing the distribution.

31. Payroll. Here are the summary statistics for the weekly payroll of a small company: lowest salary = $300, mean salary = $700, median = $500, range = $1200, IQR = $600, first quartile = $350, standard deviation = $400.
a) Do you think the distribution of salaries is symmetric, skewed to the left, or skewed to the right? Explain why.
b) Between what two values are the middle 50% of the salaries found?
c) Suppose business has been good and the company gives every employee a $50 raise. Tell the new value of each of the summary statistics given above.
d) Instead, suppose the company gives each employee a 10% raise. Tell the new value of each of the summary statistics given above.

32. Hams. A specialty foods company sells "gourmet hams" by mail order. The hams vary in size from 4.15 to 7.45 pounds, with a mean weight of 6 pounds and standard deviation of 0.65 pounds. The quartiles and median weights are 5.6, 6.2, and 6.55 pounds.
a) Find the range and the IQR of the weights.
b) Do you think the distribution of the weights is symmetric or skewed? If skewed, which way? Why?
c) If these weights were expressed in ounces (1 pound = 16 ounces) what would the mean, standard deviation, quartiles, median, IQR, and range be?
d) When the company ships these hams, the box and packing materials add 30 ounces. What are the mean, standard deviation, quartiles, median, IQR, and range of weights of boxes shipped (in ounces)?
e) One customer made a special order of a 10-pound ham. Which of the summary statistics of part d might *not* change if that data value were added to the distribution?

C

33. Rainmakers? In an experiment to determine whether seeding clouds with silver iodide increases rainfall, 52 clouds were randomly assigned to be seeded or not. The amount of rain they generated was then measured (in acre-feet). Here are the summary statistics:

	n	Mean	Median	SD	IQR	Q1	Q3
Unseeded	26	164.59	44.20	278.43	138.60	24.40	163
Seeded	26	441.98	221.60	650.79	337.60	92.40	430

a) Which of the summary statistics are most appropriate for describing these distributions. Why?
b) Do you see any evidence that seeding clouds may be effective? Explain.

34. Cholesterol. A study examining the health risks of smoking measured the cholesterol levels of people who had smoked for at least 25 years and people of similar ages who had smoked for no more than 5 years and then stopped. Create appropriate graphical displays for both groups, and write a brief report comparing their cholesterol levels. Here are the data:

Smokers				Ex-Smokers		
225	211	209	284	250	134	300
258	216	196	288	249	213	310
250	200	209	280	175	174	328
225	256	243	200	160	188	321
213	246	225	237	213	257	292
232	267	232	216	200	271	227
216	243	200	155	238	163	263
216	271	230	309	192	242	249
183	280	217	305	242	267	243
287	217	246	351	217	267	218
200	280	209		217	183	228

T **35. Population growth.** Here is a "back-to-back" stem-and-leaf display that shows two data sets at once—one going to the left, one to the right. The display compares the percent change in population for two regions of the United States (based on census figures for 1990 and 2000). The fastest-growing states were Nevada at 66% and Arizona at 40%. To show the distributions better, this display breaks each stem into two lines, putting leaves 0–4 on one stem and leaves 5–9 on the other.

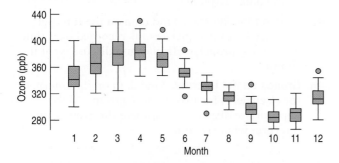

```
   NE/MW States  │  S/W States
                6│6
                6│
                5│
                5│
                4│
                4│0
                3│
                3│001
                2│6
                2│001134
                1│578
          2100  1│001134444
99998876655   0│6999
       4431   0│1
```
Population Growth rate
(|6|6 means 66%)

a) Use the data displayed in the stem-and-leaf display to construct comparative boxplots.
b) Write a few sentences describing the difference in growth rates for the two regions of the United States.

T **36. Ozone.** Ozone levels (in parts per billion, ppb) were recorded at sites in New Jersey monthly between 1926 and 1971. Here are boxplots of the data for each month (over the 46 years), lined up in order (January = 1):

a) In what month was the highest ozone level ever recorded?
b) Which month has the largest IQR?
c) Which month has the smallest range?
d) Write a brief comparison of the ozone levels in January and June.
e) Write a report on the annual patterns you see in the ozone levels.

T **37. Bike safety 2003.** The Bicycle Helmet Safety Institute website includes a report on the number of bicycle fatalities per year in the United States. The table below shows the counts for the years 1994–2003.

Year	Bicycle fatalities
1994	796
1995	828
1996	761
1997	811
1998	757
1999	750
2000	689
2001	729
2002	663
2003	619

a) What are the W's for these data?
b) Display the data in a stem-and-leaf display.
c) Display the data in a timeplot.
d) What is apparent in the stem-and-leaf display that is hard to see in the timeplot?
e) What is apparent in the timeplot that is hard to see in the stem-and-leaf display?
f) Write a few sentences about bicycle fatalities in the United States.

T **38. Drunk driving 2005.** Accidents involving drunk drivers account for about 40% of all deaths on the nation's highways. The table tracks the number of alcohol-related fatalities for 24 years. (www.madd.org)

Year	Deaths (thousands)	Year	Deaths (thousands)
1982	26.2	1994	17.3
1983	24.6	1995	17.7
1984	24.8	1996	17.7
1985	23.2	1997	16.7
1986	25.0	1998	16.7
1987	24.1	1999	16.6
1988	23.8	2000	17.4
1989	22.4	2001	17.4
1990	22.6	2002	17.5
1991	20.2	2003	17.1
1992	18.3	2004	16.9
1993	17.9	2005	16.9

a) Create a stem-and-leaf display or a histogram of these data.
b) Create a timeplot.
c) Using features apparent in the stem-and-leaf display (or histogram) and the timeplot, write a few sentences about deaths caused by drunk driving.

39. SAT or ACT? Each year thousands of high school students take either the SAT or the ACT, standardized tests used in the college admissions process. Combined SAT Math and Verbal scores go as high as 1600, while the maximum ACT composite score is 36. Since the two exams use very different scales, comparisons of performance are difficult. A convenient rule of thumb is $SAT = 40 \times ACT + 150$; that is, multiply an ACT score by 40 and add 150 points to estimate the equivalent SAT score. An admissions officer reported the following statistics about the ACT scores of 2355 students who applied to her college one year. Find the summaries of equivalent SAT scores.

Lowest score = 19   Mean = 27
Standard deviation = 3

Q3 = 30   Median = 28   IQR = 6

40. Cold U? A high school senior uses the Internet to get information on February temperatures in the town where he'll be going to college. He finds a website with some statistics, but they are given in degrees Celsius. The conversion formula is $°F = 9/5 \, °C + 32$. Determine the Fahrenheit equivalents for the following summary information.

Maximum temperature = 11°C   Range = 33°
Mean = 1°   Standard deviation = 7°
Median = 2°   IQR = 16°

T 41. Derby speeds 2007. How fast do horses run? Kentucky Derby winners top 30 miles per hour, as shown in this graph. The graph shows the percentage of Derby winners that have run *slower* than each given speed. Note that few have won running less than 33 miles per hour, but about 86% of the winning horses have run less than 37 miles per hour. (A cumulative frequency graph like this is called an "ogive.")

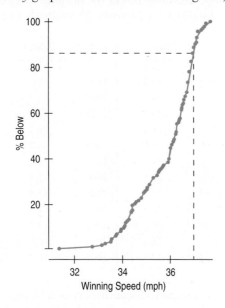

a) Estimate the median winning speed.
b) Estimate the quartiles.
c) Estimate the range and the IQR.
d) Create a boxplot of these speeds.
e) Write a few sentences about the speeds of the Kentucky Derby winners.

T 42. Cholesterol. The Framingham Heart Study recorded the cholesterol levels of more than 1400 men. Here is an ogive of the distribution of these cholesterol measures. (An ogive shows the percentage of cases at or below a certain value.) Construct a boxplot for these data, and write a few sentences describing the distribution.

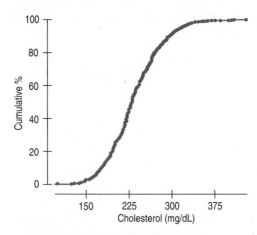

43. In the news. Find an article in a newspaper, magazine, or the Internet that compares two or more groups of data.
a) Does the article discuss the W's?
b) Is the chosen display appropriate? Explain.
c) Discuss what the display reveals about the groups.
d) Does the article accurately describe and interpret the data? Explain.

44. In the news. Find an article in a newspaper, magazine, or the Internet that shows a timeplot.
a) Does the article discuss the W's?
b) Is the timeplot appropriate for the data? Explain.
c) Discuss what the timeplot reveals about the variable.
d) Does the article accurately describe and interpret the data? Explain.

45. Time on the Internet. Find data on the Internet (or elsewhere) that give results recorded over time. Make an appropriate display and discuss what it shows.

46. Groups on the Internet. Find data on the Internet (or elsewhere) for two or more groups. Make appropriate displays to compare the groups, and interpret what you find.

Answers

Do The Math Answers

1. a) Median = 110, Q1 = 100, Q3 = 120
   b) IQR = 20
   c) Outliers: 50, 160
   d)

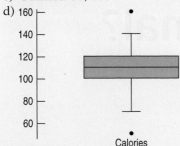

Calories

2. Note that the median = Q1.

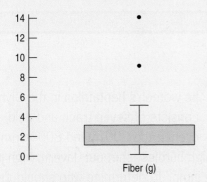

Fiber (g)

Just Checking

1. • The boxplot of var1 matches histogram C. The graphs are symmetrical and most tightly clustered around the middle value.
   • The boxplot of var2 matches histogram B. The lower half of these data is less spread out than the upper half of the data.
   • The boxplot of var3 matches histogram D. These data are skewed to the left.
   • The boxplot of var4 matches histogram E. This distribution is skewed to the right.
   • The boxplot of var5 matches histogram A. This distribution is fairly uniform.

2. The tallest bar in the histogram for 1960–2005 indicates that the average annual temperature in the Hopkins Memorial Forest was between 44 and 45 degrees for over 35% of those years.

3. The typical annual average temperature for both time periods was about 45 to 46 degrees (center), with most of the averages between 44 and 48 degrees. The distribution of average temperatures before 1960 is symmetric, and there were about the same number of years where the average temperature was higher or lower than typical. In the period after 1960, the distribution is skewed to the right, with fewer years of higher than average temperatures.

4. These data do not show evidence that average annual temperatures in the Hopkins Memorial Forest have become higher after 1960.

5. The % late arrivals have a unimodal, symmetric distribution centered at about 20%. In most months between 16% and 23% of the flights arrived late.

6. The boxplot of % late arrivals makes it easier to see that the median is just below 20%, with quartiles at about 17% and 22%. It nominates two months as high outliers.

7. The boxplots by month show a strong seasonal pattern. Flights are more likely to be late in the winter and summer and less likely to be late in the spring and fall. One likely reason for the pattern is snowstorms in the winter and thunderstorms in the summer.

8. a) The mean would increase to 500.
   b) The standard deviation is still 100 points.
   c) The two boxplots would look nearly identical (the shape of the distribution would remain the same), but the later one would be shifted 50 points higher.
   d) IQR

9. a) 76.2 mm
   b) 2.54 mm
   c) None (Each is multiplied by 25.4.)

What's Normal?

The women's heptathlon in the Olympics consists of seven track and field events: the 200-m and 800-m runs, 100-m high hurdles, shot put, javelin, high jump, and long jump. To determine who should get the gold medal, somehow the performances in all seven events have to be combined into one score. How can performances in such different events be compared? They don't even have the same units; the races are recorded in minutes and seconds and the throwing and jumping events in meters. In the 2004 Olympics, Austra Skujyté of Lithuania put the shot 16.4 meters, about 3 meters farther than the average of all contestants. Carolina Klüft won the long jump with a 6.78-m jump, about a meter better than the average. Which performance deserves more points? Even though both events are measured in meters, it's not clear how to compare them. The solution to this problem turns out to be a powerful method for comparing all sorts of values whether they have the same units or not. We *can* compare apples and oranges!

The Standard Deviation as a Ruler

GRADING ON A CURVE

If you score 79% on an exam, what grade should you get? One teaching philosophy looks only at the raw percentage, 79, and bases the grade on that alone. Another looks at your *relative* performance and bases the grade on how you did compared with the rest of the class. Teachers and students often debate which method is better.

The trick in comparing very different-looking values is to use standard deviations. The standard deviation tells us how much the whole collection of values varies, so it's a natural ruler for comparing an individual value to the group. Over and over during this course, we will ask questions such as "How far is this value from the mean?" or "How different are these two statistics?" The answer in every case will be to measure using the standard deviation as a ruler.

In order to compare the two heptathlon events, let's start with a picture. We'll use stem-and-leaf displays so we can see the individual distances.

Long Jump			Shot Put	
Stem	Leaf		Stem	Leaf
67	8		16	4
66			15	
65	1		15	
64	2		14	56778
63	0566		14	24
62	11235		13	5789
61	0569		13	012234
60	2223		12	55
59	0278		12	0144
58	4		11	59
57	0		11	23

FIGURE 6.1 Stem-and-leaf displays for both the long jump and the shot put in the 2004 Olympic Heptathlon. Carolina Klüft (green scores) won the long jump, and Austra Skujyté (red scores) won the shot put. Which heptathlete did better for both events *combined*?

The two winning performances at the top of each stem-and-leaf display appear to be about the same distance from the center of the pack. But look again carefully. What do we mean by the *same distance*? The two displays have different scales. For the shot put each line represents half a meter, but for the long jump each line is only a tenth of a meter. Our eyes naturally adjust to the scales and we suspect each is about the same distance from the center of the data. How can we make this hunch more precise? Let's see how many standard deviations each performance is from the mean.

	Event	
	Long Jump	**Shot Put**
Mean (all contestants)	6.16 m	13.29 m
SD	0.23 m	1.24 m
n	26	28
Klüft	6.78 m	14.77 m
Skujyté	6.30 m	16.40 m

The mean long jump is 6.16 m. Klüft's 6.78-m jump is 6.78 − 6.16 = 0.62 meters longer. How many *standard deviations* better than the mean is that? The standard deviation for this event was 0.23 m, so her jump was 0.62/0.23 = 2.70 *standard deviations better* than the mean.

Skujyté's winning shot put was 16.40 − 13.29 = 3.11 meters longer than the mean shot put distance, and that's 3.11/1.24 = 2.51 standard deviations better than the mean. That's a great performance but not quite as impressive as Klüft's long jump, which was farther above the mean when measured in *standard deviations*.

Standardizing with z-Scores

To compare these athletes' performances, we figured out how many standard deviations each was from the event's mean.

Expressing the distance in standard deviations **standardizes** the performances. To standardize a value, we simply subtract the mean performance in that event and then divide this difference by the standard deviation. We can write the calculation as

$$z = \frac{y - \bar{y}}{s}.$$

These values are called **standardized values,** and are commonly denoted with the letter z. Usually, we just call them **z-scores.**

Standardized values have *no units*. z-scores measure the distance of each data value from the mean in standard deviations. A z-score of 2 tells us that a data value is 2 standard deviations above the mean. It doesn't matter whether the original variable was measured in inches, dollars, or seconds. Data values below the mean have negative z-scores, so a z-score of -1.6 means that the data value was 1.6 standard deviations below the mean. The farther a data value is from the mean, the more unusual it is. Looking at the z-scores, we can see that even though both were winning performances, Klüft's long jump with a z-score of 2.70 is slightly more impressive than Skujyté's shot put with a z-score of 2.51.

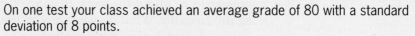

Do The Math

On one test your class achieved an average grade of 80 with a standard deviation of 8 points.

1. If you got a 96, what's your z-score?
2. Your best friend got a 76. What's her z-score?
3. What test grade has a z-score of $+1.5$?
4. The teacher calls home whenever a student's z-score is worse than -2.0. What grade earns you that phone call?

In Chapter 5 you made boxplots for the number of calories and the fiber content in 23 kinds of Kellogg's cereals. Now let's think about the sugar content. Those cereals average 7.6 grams of sugar per serving with a standard deviation of 4.5 grams.

5. Find the z-scores for the following cereals and describe what the z-score tells you about that cereal.
   a) Frosted Flakes: 11 grams of sugar per serving
   b) Apple Jacks: 14 grams of sugar per serving
   c) Crispix: 3 grams of sugar per serving
6. The z-score for Honey Smacks' sugar content is a very high 3.87! How many grams of sugar are in one serving?
7. Product 19 is very low in sugar, with a z-score of -0.8. How many grams are in a serving of this cereal?

(Check your answers on pages 135–136.)

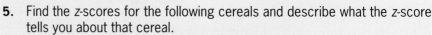

For Example   Standardizing Skiing Times

The men's combined skiing event in the winter Olympics consists of two races: a downhill and a slalom. Times for the two events are added together, and the skier with the lowest total time wins. In the 2006 Winter Olympics, the mean slalom time was 94.2714 seconds with a standard deviation of 5.2844 seconds. The mean downhill time was 101.807 seconds with a standard deviation of 1.8356 seconds. Ted Ligety of the United States, who won the gold medal with a combined time of 189.35 seconds, skied the slalom in 87.93 seconds and the downhill in 101.42 seconds.

Question: On which race did he do better compared with the competition?

For the slalom, Ligety's z-score is found by subtracting the mean time from his time and then dividing by the standard deviation:

$$z_{Slalom} = \frac{87.93 - 94.2714}{5.2844} = -1.2$$

Similarly, his z-score for the downhill is:

$$z_{Downhill} = \frac{101.42 - 101.807}{1.8356} = -0.21$$

The z-scores show that Ligety's time in the slalom is farther below the mean than his time in the downhill. His performance in the slalom was more remarkable.

Just Checking

1. Your Statistics teacher has announced that the lower of your two tests will be dropped. You got a 90 on test 1 and an 80 on test 2. You're all set to drop the 80 until she announces that she grades "on a curve." She standardized the scores in order to decide which is the lower one. If the mean on the first test was 88 with a standard deviation of 4 and the mean on the second was 75 with a standard deviation of 5,

   a) Which one will be dropped?
   b) Does this seem "fair"?

2. The calorie content for 23 varieties of Kellogg's cereals averages 109 calories per serving with a standard deviation of 22.2 calories. The mean fiber content for these cereals is 2.7 grams per serving with a standard deviation of 3.2 grams. A serving of Kellogg's All-Bran with Extra Fiber has a very low 50 calories and a very high 14 grams of fiber. Which is more remarkable—the calorie content or the fiber content? Explain.

(Check your answers on page 136.)

More about z-Scores

When we standardize data to get a z-score, we do two things. First, we shift the data by subtracting the mean. Then, we rescale the values by dividing by their standard deviation. Let's think about how standardizing affects the distribution.

When we subtract the mean of the data from every data value, we shift the mean to zero. Remember, though: such a shift doesn't change the standard deviation.

When we *divide* each of these shifted values by s, however, the standard deviation should be divided by s as well. Since the standard deviation was s to start with, the new standard deviation becomes 1.

How, then, does standardizing affect the distribution of a variable? Let's consider the three aspects of a distribution: the shape, center, and spread.

z-scores have mean 0 and standard deviation 1.

- *Standardizing into z-scores does not change the **shape** of the distribution of a variable.*
- *Standardizing into z-scores changes the **center** by making the mean 0.*
- *Standardizing into z-scores changes the **spread** by making the standard deviation 1.*

When Is a z-Score BIG?

A *z*-score gives us an indication of how unusual a value is because it tells us how far it is from the mean. If the data value sits right at the mean, it's not unusual at all and its *z*-score is 0. A *z*-score of 1 tells us that the data value is 1 standard deviation above the mean, while a *z*-score of −1 tells us that the value is 1 standard deviation below the mean. How far from 0 does a *z*-score have to be in order to be interesting or unusual? There is no universal rule, but the larger the score is (negative or positive), the more unusual it is. For symmetric data, it's not uncommon for at least half of the data to have *z*-scores between −1 and 1. No matter what the shape of the distribution, a *z*-score of 3 (plus or minus) or more is rare, and a *z*-score of 6 or 7 shouts out for attention.

To say more about how big we expect a *z*-score to be, we need to *model* the data's distribution. Like all models of the real world, the model will be wrong—wrong in the sense that it can't match reality exactly. But it can still be useful. Just as a model of an airplane in a wind tunnel can give insights even though it doesn't show every rivet,[1] models of data give us insights about the real world. It's important to remember, though, that they're only *models* of reality and not reality itself.

"All models are wrong—but some are useful."

—George Box,
famous statistician

There is no universal standard for *z*-scores, but there is a model that shows up over and over in Statistics. You may have heard of "bell-shaped curves." Statisticians call them Normal models. **Normal models** are appropriate for distributions whose shapes are unimodal and roughly symmetric.

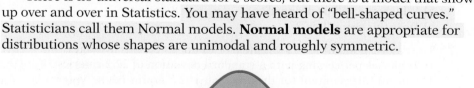

---

[1] In fact, the model is useful *because* it doesn't have every rivet. It is because models offer a simpler view of reality that they are so useful as we try to understand reality.

$N(\mu, \sigma)$ always denotes a Normal model. The μ, pronounced "mew," is the Greek letter for "m" and always represents the mean in a model. The σ, sigma, is the lowercase Greek letter for "s" and always represents the standard deviation in a model. Latin letters $\bar{y}$ and s represent the mean and standard deviation of actual data.

IS THE STANDARD NORMAL A STANDARD?

Yes. We call it the "Standard Normal" because it models standardized values. It is also a "standard" because this is the particular Normal model that we almost always use.

IS NORMAL NORMAL?

Don't be misled. The name "Normal" doesn't mean that these are the *usual* shapes for histograms. The name follows a tradition of positive thinking in Mathematics and Statistics in which functions, equations, and relationships that are easy to work with or have other nice properties are called "normal," "common," "regular," "natural," or similar terms. It's as if by calling them ordinary, we could make them actually occur more often and simplify our lives.

We write $N(\mu, \sigma)$ to represent a Normal model with a mean of μ and a standard deviation of σ. Why the Greek? Well, *this* mean and standard deviation are not numerical summaries of data. They are part of the model. They don't come from the data. Rather, they are numbers that we choose to help specify the model. Such numbers are called **parameters** of the model.

We don't want to confuse the parameters with summaries of the data such as $\bar{y}$ and s, so we use special symbols. In Statistics, we almost always use Greek letters for parameters. By contrast, summaries of data are called **statistics** and are usually written with Latin letters.

If we model data with a Normal model and standardize them using the corresponding μ and σ, we still call the standardized value a z-**score,** and we write

$$z = \frac{y - \mu}{\sigma}.$$

Usually it's easier to standardize data first (using its mean and standard deviation). Then we need only the model $N(0,1)$. The Normal model with mean 0 and standard deviation 1 is called the **standard Normal model** (or the **standard Normal distribution**).

But be careful. You shouldn't use a Normal model for just any data set. Remember that standardizing won't change the shape of the distribution. If the distribution is not unimodal and symmetric to begin with, standardizing won't make it Normal.

When we use the Normal model, we assume that the distribution of the data is, well, Normal. There's no way to check whether this **Normality Assumption** is true. In fact, it almost certainly isn't. Real data don't behave like perfect mathematical models. Models are idealized; real data are real. The good news, however, is that to use a Normal model, it's sufficient to check the following condition:

Nearly Normal Condition. The shape of the data's distribution is unimodal and symmetric. Check this by making a histogram.

Don't model data with a Normal model without checking whether this condition is satisfied.

The First Three Rules for Working with Normal Models

1. Make a picture.
2. Make a picture.
3. Make a picture.

Although we're thinking about models, not histograms of data, the three rules don't change. To help you think clearly, a simple hand-drawn sketch is all you need. Even experienced statisticians sketch pictures to help them think about Normal models. You should too.

Of course, when we have data, we'll also need to make a histogram to check the **Nearly Normal Condition** to be sure we can use the Normal model to model the data's distribution. Other times, we may be told that a Normal model is appropriate based on prior knowledge of the situation or on theoretical considerations.

How to Sketch a Normal Curve That Looks Normal To sketch a good Normal curve, you need to remember only three things:

- The Normal curve is bell-shaped and symmetric around its mean. Start at the middle, and sketch to the right and left from there.
- Even though the Normal model extends forever on either side, you need to draw it only for 3 standard deviations. After that, there's so little left that it isn't worth sketching.
- The place where the bell shape changes from curving downward to curving back up—the *inflection point*—is exactly one standard deviation away from the mean.

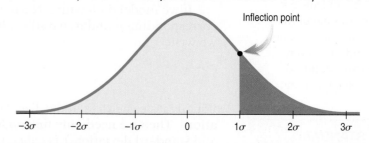

The 68–95–99.7 Rule

Normal models give us an idea of how extreme a value is by telling us how likely it is to find one that far from the mean. We'll soon show how to find these numbers precisely—but one simple rule is usually all we need.

It turns out that in a Normal model, about 68% of the values fall within 1 standard deviation of the mean, about 95% of the values fall within 2 standard deviations of the mean, and about 99.7%—almost all—of the values fall within 3 standard deviations of the mean. These facts are summarized in a rule that we call (let's see . . .) the **68–95–99.7 Rule.**

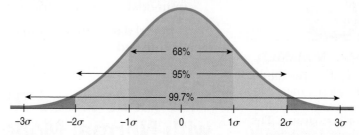

FIGURE 6.2 Normal model percentages. Reaching out one, two, and three standard deviations on a Normal model gives the 68–95–99.7 Rule, seen as proportions of the area under the curve.

TI-*nspire*

Normal models. Watch the Normal model react as you change the mean and standard deviation.

STEP-BY-STEP EXAMPLE   Working with the 68–95–99.7 Rule

The SAT Reasoning Test has three parts: Writing, Math, and Critical Reading (Verbal). Each part has a distribution that is roughly unimodal and symmetric and is designed to have an overall mean of about 500 and a standard deviation of 100 for all test takers. In any one year, the mean and standard deviation may differ from these target values by a small amount, but they are a good overall approximation.

Question: Suppose you earned a 600 on one part of your SAT. Where do you stand among all students who took that test?

You could calculate your *z*-score and find out that it's $z = (600 - 500)/100 = 1.0$, but what does that tell you about your percentile? You'll need the Normal model and the 68–95–99.7 Rule to answer that question.

THINK	**Plan** State what you want to know.	I want to see how my SAT score compares with the scores of all other students. To do that, I'll need to model the distribution.
	Variables Identify the variable.	Let y = my SAT score. Scores are quantitative but have no meaningful units other than points.
	Be sure to check the appropriate conditions.	✔ **Nearly Normal Condition:** If I had data, I would check the histogram. I have no data, but I am told that the SAT scores are roughly unimodal and symmetric.
	Specify the parameters of your model.	I will model SAT score with a N(500, 100) model.
SHOW	**Mechanics** Make a picture of this Normal model. (A simple sketch is all you need.)	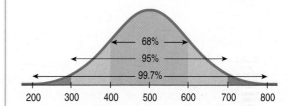
	Locate your score.	My score of 600 is 1 standard deviation above the mean. That corresponds to one of the points of the 68–95–99.7 Rule.
TELL	**Conclusion** Interpret your result in context.	About 68% of those who took the test had scores that fell no more than 1 standard deviation from the mean, so 100% − 68% = 32% of all students had scores more than 1 standard deviation away. Only half of those were on the high side, so about 16% (half of 32%) of the test scores were better than mine. My score of 600 is higher than about 84% of all scores on this test.

The bounds of SAT scoring at 200 and 800 can also be explained by the 68–95–99.7 Rule. Since 200 and 800 are three standard deviations from 500, it hardly pays to report scores any farther on either side. We'd get more information only on 100 − 99.7 = 0.3% of students.

Just Checking

3. As a group, the Dutch are among the tallest people in the world. The average Dutch man is 184 cm tall—just over 6 feet! The standard deviation of men's heights is about 8 cm. Assuming the distribution is approximately Normal, use the 68–95–99.7 Rule to sketch a model for the heights of Dutch men. Label the axis clearly and indicate appropriate percentages.

(Check your answers on page 136.)

For Example   Using the 68–95–99.7 Rule

Question: In the 2006 Winter Olympics men's combined event, Jean-Baptiste Grange of France skied the slalom in 88.46 seconds—about 1 standard deviation faster than the mean. If a Normal model is useful in describing slalom times, about how many of the 35 skiers finishing the event would you expect skied the slalom *faster* than Jean-Baptiste?

From the 68–95–99.7 Rule, we expect 68% of the skiers to be within one standard deviation of the mean. Of the remaining 32%, we expect half on the high end and half on the low end. 16% of 35 is 5.6, so, conservatively, we'd expect about 5 skiers to do better than Jean-Baptiste.

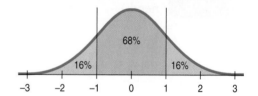

Just Checking

4. Recall that the average Dutch man is 184 cm tall, and a Normal model is appropriate. Based on the 68–95–99.7 Rule, the standard deviation is about 8 cm. What percentage of all Dutch men should be over 2 meters (6'6") tall?

5. Let's say it takes you 20 minutes, on average, to drive to school, with a standard deviation of 2 minutes. Suppose a Normal model is appropriate for the distributions of driving times. Based on the 68–95–99.7 Rule:

   a) About how often will it take you between 18 and 22 minutes to get to school?
   b) How often will you arrive at school in less than 22 minutes?
   c) How often will it take you more than 24 minutes?

(Check your answers on page 136.)

Finding Normal Percentiles

With a mean of 500 and standard deviation of 100, an SAT score of 600 is easy to assess, because we can think of it as one standard deviation above the mean. If your score was 680, though, where do you stand among the rest of the people tested? Your z-score is 1.80, so you're somewhere between 1 and 2 standard deviations above the mean. We figured out that no more than 16% of people score better than 600. By the same logic, no more than 2.5% of people score better than 700. Can we be more specific than "between 16% and 2.5%"?

When the value doesn't fall exactly 1, 2, or 3 standard deviations from the mean, we can look it up in a table of **Normal percentiles** or use technology.[2] Either way, we first convert our data to z-scores before using the table. Your SAT score of 680 has a z-score of $(680 - 500)/100 = 1.80$.

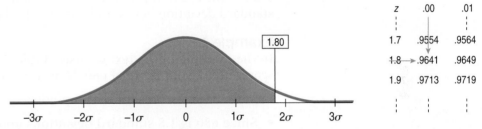

FIGURE 6.3 Cutpoints and percentiles. A table of Normal percentiles (Table Z in Appendix G) lets us find the percentage of individuals in a Standard Normal distribution falling below any specified z-score value.

In the piece of the table shown, we find your z-score by looking down the left column for the first two digits, 1.8, and across the top row for the third digit, 0. The table gives the percentile as 0.9641. That means that 96.4% of the z-scores are less than 1.80. Only 3.6% of people, then, scored better than 680 on the SAT.

Most of the time, though, you'll do this with your calculator.

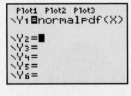

Normal percentiles. Explore the relationship between z-scores and areas in a Normal model.

TI Tips | Finding Normal percentages

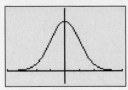

Your calculator knows the Normal model. Have a look under 2nd DISTR. There you will see three "norm" functions, normalpdf (, normalcdf (, and invNorm (. Let's play with the first two.

- normalpdf(calculates y-values for graphing a Normal curve. You probably won't use this very often, if at all. If you want to try it, graph Y1 = normalpdf(X) in a graphing WINDOW with Xmin = −4, Xmax = 4, Ymin = −0.1, and Ymax = 0.5.
- normalcdf(finds the proportion of area under the curve between two z-score cut points, by specifying normalcdf (zLeft,zRight). Make friends with this function; you will use it often!

(continued)

---

[2]See Table Z in Appendix G, if you're curious. But your calculator (and any statistics computer package) does this, too—and more easily!

Example 1

The Normal model shown shades the region between $z = -0.5$ and $z = 1.0$.

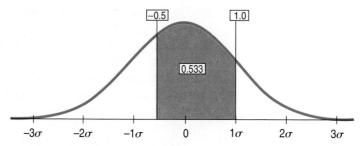

To find the shaded area:

> Under 2nd DISTR select normalcdf(and then hit ENTER.
> Specify the cut points: normalcdf(−.5, 1.0) and hit ENTER again.

There's the area. Approximately 53% of a Normal model lies between half a standard deviation below and one standard deviation above the mean.

Example 2

In the example in the text we used Table Z to determine the fraction of SAT scores above your score of 680. Now let's do it again, this time using your TI.

First we need z-scores for the cutpoints:

- Since 680 is 1.8 standard deviations above the mean, your z-score is 1.8; that's the left cut point.
- Theoretically the standard Normal model extends rightward forever, but you can't tell the calculator to use infinity as the right cut point. Recall that for a Normal model almost all the area lies within ± 3 standard deviations of the mean, so any upper cut point beyond, say, $z = 5$ does not cut off anything very important. We suggest you always use 99 (or −99) when you really want infinity as your cut point—it's easy to remember and way beyond any meaningful area.

Now you're ready. Use the command normalcdf(1.8,99).

There you are! The Normal model estimates that approximately 3.6% of SAT scores are higher than 680.

Do The Math 8. Find the shaded area of each Normal Model.

a)

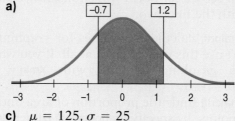

b)

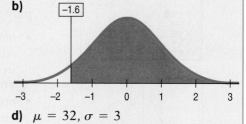

c) $\mu = 125, \sigma = 25$

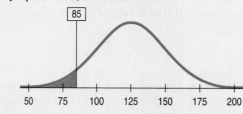

d) $\mu = 32, \sigma = 3$

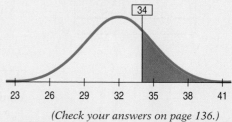

(Check your answers on page 136.)

STEP-BY-STEP EXAMPLE   Working with Normal Models Part I

The Normal model is our first model for data. It's the first in a series of modeling situations where we step away from the data at hand to make more general statements about the world. We'll become more practiced in thinking about and learning the details of models as we progress through the book. To give you some practice in thinking about the Normal model, let's find a percentile in detail.

Question: What proportion of SAT scores fall between 450 and 600?

THINK

Plan State the problem.

Variables Name the variable.

Check the appropriate conditions and specify which Normal model to use.

I want to know the proportion of SAT scores between 450 and 600.

Let y = SAT score.

✔ **Nearly Normal Condition:** We are told that SAT scores are nearly Normal.

I'll model SAT scores with a N(500, 100) model, using the mean and standard deviation specified for them.

SHOW

Mechanics Make a picture of this Normal model. Locate the desired values and shade the region of interest.

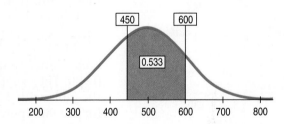

Find z-scores for the cut points 450 and 600. Use technology to find the desired proportions, represented by the area under the curve. (This was Example 1 in the TI Tips—take another look.)

Standardizing the two scores, I find that

$$z = \frac{(y - \mu)}{\sigma} = \frac{(600 - 500)}{100} = 1.00$$

and

$$z = \frac{(450 - 500)}{100} = -0.50$$

So,

Area$(450 < y < 600)$ = Area$(-0.5 < z < 1.0)$
$$= 0.5328$$

(If you use a table, then you need to subtract the two areas to find the area *between* the cut points.)

(**OR:** From Table Z, the area $(z < 1.0) = 0.8413$ and area $(z < -0.5) = 0.3085$, so the proportion of z-scores between them is $0.8413 - 0.3085 = 0.5328$, or 53.28%.)

TELL

Conclusion Interpret your result in context.

The Normal model estimates that about 53.3% of SAT scores fall between 450 and 600.

Just Checking

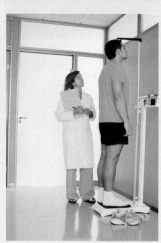

6. Remember those tall Dutch men? Their mean height was 184 cm and the standard deviation was 8 cm. Answer each of these questions by sketching a Normal model, shading the appropriate area, finding the z-scores, and using your calculator (or the table) to determine the percentage.

a) What fraction of Dutch men should be less than 190 cm tall?

b) What fraction of Dutch men should be between 170 and 180 cm tall?

c) What fraction of Dutch men should be over 6'6" (198 cm) tall?

(Check your answers on page 136.)

From Percentiles to Scores: z in Reverse

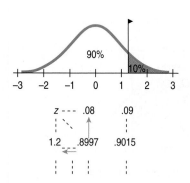

Finding areas from z-scores is the simplest way to work with the Normal model. But sometimes we start with areas and are asked to work backward to find the corresponding z-score or even the original data value. For instance, what z-score cuts off the top 10% in a Normal model?

Make a picture like the one shown, shading the rightmost 10% of the area. Notice that this is the 90th percentile. Look in Table Z for an area of 0.900. The exact area is not there, but 0.8997 is pretty close. That shows up in the table with 1.2 in the left margin and 0.08 in the top margin. The z-score for the 90th percentile, then, is approximately z = 1.28.

Computers and calculators will determine the cut point more precisely (and more easily).

TI Tips | Finding Normal Cutpoints

To find the z-score at the 25th percentile, go to **2nd DISTR** again. This time we'll use the third of the "norm" functions, **invNorm(**.

Just specify the desired percentile with the command **invNorm(.25)** and hit **ENTER**. The calculator says that the cut point for the leftmost 25% of a Normal model is approximately z = −0.674.

One more example: What z-score cuts off the highest 10% of a Normal model? That's easily done—just remember to specify the *percentile*. Since we want the cut point for the *highest* 10%, we know that the other 90% must be *below* that z-score. The cut point, then, must stand at the 90th percentile, so specify **invNorm(.90)**.

Only 10% of the area in a Normal model is more than about 1.28 standard deviations above the mean.

Do The Math

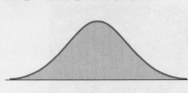

9. For each, sketch the standard normal distribution, shade the area described, and find the *z*-score cutpoints.
  a) the lowest 40% of the distribution
  b) the highest 30% of the distribution
  c) the highest 2% of the distribution
  d) the middle 30% of the distribution

(Check your answers on page 136.)

STEP-BY-STEP EXAMPLE   Working with Normal Models Part II

Question: Suppose a college says it admits only people with SAT Verbal test scores among the top 10%. How high a score does it take to be eligible?

Plan State the problem.

Variable Define the variable.

Check to see if a Normal model is appropriate, and specify which Normal model to use.

How high an SAT Verbal score do I need to be in the top 10% of all test takers?

Let *y* = my SAT score.

✔ **Nearly Normal Condition:** I am told that SAT scores are nearly Normal. I'll model them with $N(500, 100)$.

Mechanics Make a picture of this Normal model. Locate the desired percentile approximately by shading the rightmost 10% of the area.

The college takes the top 10%, so its cutoff score is the 90th percentile. Find the corresponding *z*-score using your calculator as shown in the TI Tips. (**OR:** Use Table Z as shown in Appendix G.)

Convert the *z*-score back to the original units.

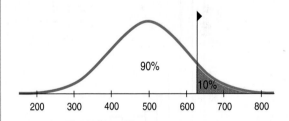

The cut point is $z = 1.28$.

A *z*-score of 1.28 is 1.28 standard deviations above the mean. Since the SD is 100, that's $1.28 \times 100 = 128$ SAT points. The cutoff is 128 points above the mean of 500, or $500 + 128 = 628$.

Conclusion Interpret your results in the proper context.

Because the school wants SAT Verbal scores in the top 10%, the cutoff is 628. (Actually, since SAT scores are reported only in multiples of 10, I'd have to score at least a 630.)

Just Checking

7. Let's think about the Normal model for the heights of Dutch men one more time. Remember that their mean height was 184 cm and the standard deviation was 8 cm. Answer each of these questions by sketching a Normal model, shading the appropriate area, finding the cutpoint z-score, and then determining the cutpoint height.

a) How tall are the tallest 10% of all Dutch men?
b) How tall are the shortest 20% of Dutch men?
c) How tall are the middle 50% of Dutch men?

(Check your answers on page 136.)

🚫 WHAT CAN GO WRONG?

- **Don't use a Normal model when the distribution is not unimodal and symmetric.** Normal models are so easy and useful that it is tempting to use them even when they don't describe the data very well. That can lead to wrong conclusions. Don't use a Normal model without first checking the **Nearly Normal Condition.** Look at a picture of the data to check that it is unimodal and symmetric. A histogram, can help you tell whether a Normal model is appropriate.

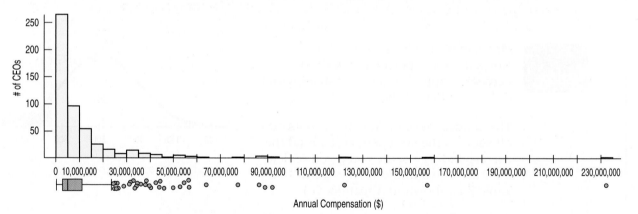

In 2005 CEOs of Fortune 500 companies had a mean total compensation of $10,307,311.87 with a standard deviation of $17,964,615.16. Using the Normal model rule, we should expect about 68% of the CEOs to have compensations between −$7,657,303.29 and $28,271,927.03. In fact, more than 90% of the CEOs have annual compensations in this range. What went wrong? The distribution is skewed, not symmetric. Using the 68–95–99.7 Rule for data like these will lead to silly results.

- **Don't use the mean and standard deviation when outliers are present.** Both means and standard deviations can be distorted by outliers, and no model based on distorted values will do a good job. A z-score calculated from a distribution with outliers may be misleading. It's always a good idea to check for outliers. How? Make a picture.

- **Don't round your results in the middle of a calculation.** We *reported* the mean of the heptathletes' long jump as 6.16 meters. More precisely, it was 6.16153846153846 meters.

  You should use all the precision available in the data for all the intermediate steps of a calculation. Using the more precise value for the mean (and also carrying 15 digits for the SD), the *z*-score calculation for Klüft's long jump comes out to

$$z = \frac{6.78 - 6.16153846153846}{0.2297597407326585} = 2.691775053755667700$$

  We'd report that as 2.692, as opposed to the rounded-off value of 2.70 we got earlier from the table.

- **Don't worry about minor differences in results.** Because various calculators and programs may carry different precision in calculations, your answers may differ slightly from those we show in the text and in the Step-By-Steps, or even from the values given in the answers in the back of the book. Those differences aren't anything to worry about. They're not the main story Statistics tries to tell.

STANDARDIZED SCORES IN YOUR WORLD

Choosing a college to attend is a huge decision that impacts the rest of a student's life. To help with this decision, many students and parents turn to ratings and rankings compiled by news organizations such as *US News and World Reports*. But how do they go about ranking colleges?

On their website, *US News* describes their process. They gather data from each college on indicators of academic excellence such as student retention, faculty resources, student selectivity, and graduation rates. Weighing those factors according to how much *US News* believes each matters, colleges are ranked against their peers.

Then *US News* identifies **Best Value Schools**. They measure three additional variables: (1) the ratio of each school's quality scores to their tuition, (2) the percentage of students receiving financial aid, and (3) the average percentage of the school's total costs covered by that financial aid. And then? US News says:

Best Values in Colleges

The schools' overall Best Value ranks were determined by first standardizing the scores achieved by every school in each of the three above variables and weighting those scores. The ratio of quality to price accounted for 60 percent of the overall score; the percentage of all undergraduates receiving need-based grants accounted for 25 percent; and the average discount accounted for 15 percent. The school with the highest total weighted points became No. 1 in its category. The other schools were then ranked in descending order.

http://colleges.usnews.rankingsandreviews.com/college

WHAT HAVE WE LEARNED?

We've learned the power of standardizing data.

▶ Standardizing uses the standard deviation as a ruler to measure distance from the mean, creating *z*-scores.

▶ Using these *z*-scores, we can compare apples and oranges—values from different distributions or values based on different units.

▶ And a *z*-score can identify unusual or surprising values among data.

We've learned that the 68–95–99.7 Rule can be a useful rule of thumb for understanding distributions.

▶ For data that are unimodal and symmetric, about 68% fall within 1 SD of the mean, 95% fall within 2 SDs of the mean, and 99.7% fall within 3 SDs of the mean (see p. 120).

Again we've seen the importance of *Thinking* about whether a method will work.

▶ **Normality Assumption:** We sometimes work with Normal tables (Table Z). Those tables are based on the Normal model.

▶ Data can't be exactly Normal, so we check the **Nearly Normal Condition** by making a histogram (is it unimodal, symmetric, and free of outliers?).

Terms

Standardizing We standardize to eliminate units. Standardized values can be compared and combined even if the original variables had different units and magnitudes.

Standardized value A value found by subtracting the mean and then dividing by the standard deviation.

Normal model A useful family of models for unimodal, symmetric distributions.

Parameter A numerically valued attribute of a model. For example, the values of μ and σ in a $N(\mu, \sigma)$ model are parameters.

Statistic A value calculated from data to summarize aspects of the data. For example, the mean, $\bar{y}$, and standard deviation, s, are statistics.

z-score A *z*-score tells how many standard deviations a value is from the mean; *z*-scores have a mean of 0 and a standard deviation of 1. When working with data, use the statistics $\bar{y}$, and s:

$$z = \frac{y - \bar{y}}{s}.$$

When working with models, use the parameters μ and σ:

$$z = \frac{y - \mu}{\sigma}.$$

Standard Normal model A Normal model, $N(\mu, \sigma)$ with mean $\mu = 0$ and standard deviation $\sigma = 1$. Also called the **standard Normal distribution.**

Nearly Normal Condition A distribution is nearly Normal if it is unimodal and symmetric. We can check by looking at a histogram.

68–95–99.7 Rule In a Normal model, about 68% of values fall within 1 standard deviation of the mean, about 95% fall within 2 standard deviations of the mean, and about 99.7% fall within 3 standard deviations of the mean.

Normal percentile The Normal percentile corresponding to a *z*-score gives the percentage of values in a standard Normal distribution found at that *z*-score or below.

Skills

▶ Recognize when standardization can be used to compare values.

▶ Understand that standardizing uses the standard deviation as a ruler.

▶ Recognize when a Normal model is appropriate.

▶ Know how to calculate the z-score of an observation.

▶ Know how to compare values of two different variables using their z-scores.

▶ Be able to use Normal models and the 68–95–99.7 Rule to estimate the percentage of observations falling within 1, 2, or 3 standard deviations of the mean.

▶ Know how to find the percentage of observations falling below any value in a Normal model using a Normal table or appropriate technology.

▶ Know how to check whether a variable satisfies the **Nearly Normal Condition** by making a histogram.

▶ Know what z-scores mean.

▶ Be able to explain how extraordinary a standardized value may be by using a Normal model.

NORMAL MODELS ON THE COMPUTER

Many software packages allow you to work with Normal percentiles. You can specify the model's mean and standard deviation, and then find percentiles or cutoffs.

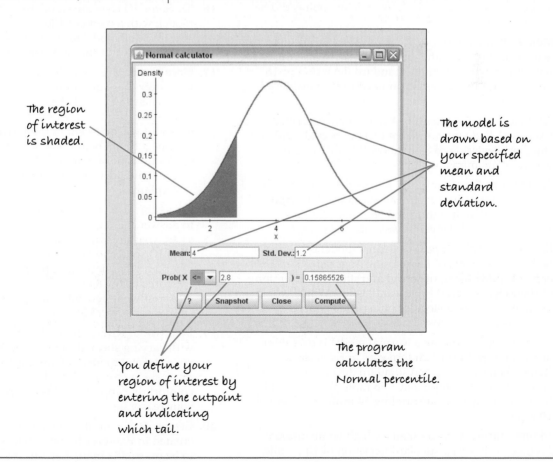

The region of interest is shaded.

The model is drawn based on your specified mean and standard deviation.

You define your region of interest by entering the cutpoint and indicating which tail.

The program calculates the Normal percentile.

EXERCISES

A

1. **Cattle.** The Virginia Cooperative Extension reports that the mean weight of yearling Angus steers is 1152 pounds. Suppose that the standard deviation is 84 pounds. How many standard deviations from the mean would a steer weighing 1000 pounds be?

 2. **Car speeds.** John Beale of Stanford, CA, recorded the speeds of cars driving past his house, where the speed limit read 20 mph. The mean of 100 readings was 23.84 mph, with a standard deviation of 3.56 mph. How many standard deviations from the mean would a car going under the speed limit be?

3. **Stats test.** Suppose your Statistics professor reports test grades as z-scores, and you got a score of 2.20 on an exam. Write a sentence explaining what that means.

4. **Checkup.** One of the authors has an adopted grandson whose birth family members are very short. After examining him at his 2-year checkup, the boy's pediatrician said that the z-score for his height relative to American 2-year-olds was −1.88. Write a sentence explaining what that means.

5. **Stats test, part II.** The mean score on the Stats exam was 75 points with a standard deviation of 5 points, and Gregor's z-score was −2. How many points did he score?

6. **Mensa.** People with z-scores above 2.5 on an IQ test are sometimes classified as geniuses. If IQ scores have a mean of 100 and a standard deviation of 16 points, what IQ score do you need to be considered a genius?

7. **Another steer.** The yearling Angus steers described in Exercise 1 had a mean weight of 1152 pounds with a standard deviation of 84 pounds. What's the actual weight of an animal whose z-score is −1.5?

8. **Another car.** The Stanford, CA, cars described in Exercise 2 were traveling at a mean speed of 23.84 mph, with a standard deviation of 3.56 mph. One car's z-score was +3.2. How fast was that one going?

9. **Unusual steers?** Suppose a Normal model is useful for describing weights of the yearling Angus steers in Exercise 1. Remember, the mean and standard deviation of the weights were 1152 lbs and 84 lbs. Which would be more unusual, a steer weighing 1000 pounds or one weighing 1250 pounds?

10. **Unusual cars?** Suppose a Normal model is useful for describing speeds of Stanford cars in Exercise 2. Remember, the mean and standard deviation of the speeds were 23.84 mph and 3.56 mph. Which would be more unusual, a car traveling 34 mph or one going 10 mph?

11. **Temperatures.** A town's January high temperatures average 36°F with a standard deviation of 10°, while in July the mean high temperature is 74° and the standard deviation is 8°. In which month is it more unusual to have a day with a high temperature of 55°? Explain.

12. **Placement exams.** An incoming freshman took her college's placement exams in French and mathematics. In French, she scored 82 and in math 86. The overall results on the French exam had a mean of 72 and a standard deviation of 8, while the mean math score was 68, with a standard deviation of 12. On which exam did she do better compared with the other freshmen?

13. **Small steer.** In Exercise 9 we suggested the model $N(1152, 84)$ for weights in pounds of yearling Angus steers. What weight would you consider to be unusually low for such an animal? Explain.

14. **Slow cars.** Based on the Normal model $N(23.84, 3.56)$ for the speeds of cars, what speeds would you consider to be unusually slow in that neighborhood? Explain your reasoning.

15. **Rivets.** A company that manufactures rivets believes the shear strength (in pounds) is modeled by $N(800, 50)$. Draw the Normal model for these rivets. Clearly label it to show what the 68–95–99.7 Rule predicts.

16. **IQ.** Some IQ tests are standardized to a Normal model, with a mean of 100 and a standard deviation of 16. Draw the model for these IQ scores. Clearly label it, showing what the 68–95–99.7 Rule predicts.

17. **More rivets.** Use your Normal model from Exercise 15 to answer these questions.
    a) In what interval of shear strengths would you expect to find 99.7% of these rivets?
    b) About what percent of these rivets would you expect to fail below 900 pounds?
    c) Would it be safe to use these rivets in a situation requiring a shear strength of 750 pounds? Explain.

18. **IQs again.** Use your Normal model from Exercise 16 to answer these questions.
    a) In what interval would you expect the central 95% of IQ scores to be found?
    b) About what percent of people should have IQ scores below 84?
    c) About what percent of people should have IQ scores above 132?

19. **Safe rivets.** Rivets are used in a variety of applications with varying shear strength requirements. Based on your Normal model in Exercise 15, what is the maximum shear strength for which you would feel comfortable approving this company's rivets? Explain your reasoning.

20. **Genius.** Based on the Normal model for IQs that you created in Exercise 16, what IQs would you consider to be unusually high? Explain your reasoning.

B

21. Combining test scores. The first Stats exam had a mean of 65 and a standard deviation of 10 points; the second had a mean of 80 and a standard deviation of 5 points. Derrick scored an 80 on both tests. Julie scored a 70 on the first test and a 90 on the second. They both totaled 160 points on the two exams, but Julie claims that her total is better. Explain.

22. Combining scores again. The first Stat exam had a mean of 80 and a standard deviation of 4 points; the second had a mean of 70 and a standard deviation of 15 points. Reginald scored an 80 on the first test and an 85 on the second. Sara scored an 88 on the first but only a 65 on the second. Although Reginald's total score is higher, Sara feels she should get the higher grade. Explain her point of view.

23. Professors. A friend tells you about a recent study dealing with the number of years of teaching experience among current college professors. He remembers the mean but can't recall whether the standard deviation was 6 months, 6 years, or 16 years. Tell him which one it must have been, and why.

24. Rock concerts. A popular band on tour played a series of concerts in large venues. They always drew a large crowd, averaging 21,359 fans. While the band did not announce (and probably never calculated) the standard deviation, which of these values do you think is most likely to be correct: 20, 200, 2000, or 20,000 fans? Explain your choice.

25. Trees. A forester measured 27 of the trees in a large woods that is up for sale. He found a mean diameter of 10.4 inches and a standard deviation of 4.7 inches. Suppose that these trees provide an accurate description of the whole forest and that a Normal model applies.
a) Draw the Normal model for tree diameters.
b) What size would you expect the central 95% of all trees to be?
c) About what percent of the trees should be less than an inch in diameter?
d) About what percent of the trees should be between 5.7 and 10.4 inches in diameter?
e) About what percent of the trees should be over 15 inches in diameter?

26. Guzzlers? Environmental Protection Agency (EPA) fuel economy estimates for automobile models tested recently predicted a mean of 24.8 mpg and a standard deviation of 6.2 mpg for highway driving. Assume that a Normal model can be applied.
a) Draw the model for auto fuel economy. Clearly label it, showing what the 68–95–99.7 Rule predicts.
b) In what interval would you expect the central 68% of autos to be found?
c) About what percent of autos should get more than 31 mpg?
d) About what percent of cars should get between 31 and 37.2 mpg?
e) Describe the gas mileage of the worst 2.5% of all cars.

27. Trees, part II. Later on, the forester in Exercise 25 shows you a histogram of the tree diameters he used in analyzing the woods that was for sale. Do you think he was justified in using a Normal model? Explain, citing some specific concerns.

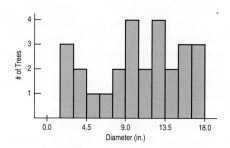

T 28. Car speeds, the picture. For the car speed data of Exercise 2, here is the histogram and boxplot of the 100 readings. Do you think it is appropriate to apply a Normal model here? Explain.

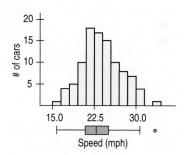

29. Normal cattle. Using $N(1152, 84)$, the Normal model for weights of Angus steers in Exercise 9, what percent of steers weigh
a) over 1250 pounds?
b) under 1200 pounds?
c) between 1000 and 1100 pounds?

30. IQs revisited. Based on the Normal model $N(100, 16)$ describing IQ scores, what percent of people's IQs would you expect to be
a) over 80?
b) under 90?
c) between 112 and 132?

31. More cattle. Based on the model $N(1152, 84)$ describing Angus steer weights, what are the cutpoint values for
a) the highest 10% of the weights?
b) the lowest 20% of the weights?
c) the middle 40% of the weights?

32. More IQs. In the Normal model $N(100, 16)$, what cutpoint value bounds
a) the highest 5% of all IQs?
b) the lowest 30% of the IQs?
c) the middle 80% of the IQs?

33. Cattle, percentiles. Consider the Angus weights model $N(1152, 84)$ again.
a) What weight represents the 40th percentile?
b) What weight represents the 99th percentile?

34. IQ percentiles. Consider the IQ model $N(100, 16)$ again.
a) What IQ represents the 15th percentile?
b) What IQ represents the 98th percentile?

35. Kindergarten. Companies that design furniture for elementary school classrooms produce a variety of sizes for kids of different ages. Suppose the heights of kindergarten children can be described by a Normal model with a mean of 38.2 inches and standard deviation of 1.8 inches.
a) What fraction of kindergarten kids should the company expect to be less than 3 feet tall?
b) In what height interval should the company expect to find the middle 80% of kindergartners?
c) At least how tall are the biggest 10% of kindergartners?

36. Body temperatures. Most people think that the "normal" adult body temperature is 98.6°F. That figure, based on a 19th-century study, has recently been challenged. In a 1992 article in the *Journal of the American Medical Association*, researchers reported that a more accurate figure may be 98.2°F. Furthermore, the standard deviation appeared to be around 0.7°F. Assume that a Normal model is appropriate.
a) In what interval would you expect most people's body temperatures to be? Explain.
b) What fraction of people would be expected to have body temperatures above 98.6°F?
c) Below what body temperature are the coolest 20% of all people?

T 37. Receivers. NFL data from the 2006 football season reported the number of yards gained by each of the league's 167 wide receivers:

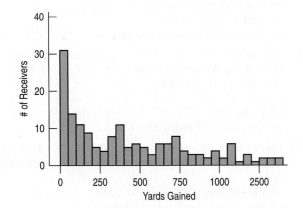

The mean is 435 yards, with a standard deviation of 384 yards.
a) According to the Normal model, what percent of receivers would you expect to gain fewer yards than 2 standard deviations below the mean number of yards?
b) For these data, what does that mean?
c) Explain the problem in using a Normal model here.

38. Customer database. A large philanthropic organization keeps records on the people who have contributed to their cause. In addition to keeping records of past giving, the organization buys demographic data on

neighborhoods from the U.S. Census Bureau. Eighteen of these variables concern the ethnicity of the neighborhood of the donor. Here are a histogram and summary statistics for the percentage of whites in the neighborhoods of 500 donors:

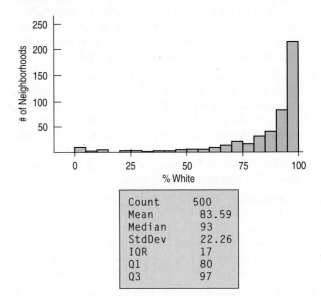

Count	500
Mean	83.59
Median	93
StdDev	22.26
IQR	17
Q1	80
Q3	97

a) Which is a better summary of the percentage of white residents in the neighborhoods, the mean or the median? Explain.
b) Which is a better summary of the spread, the IQR or the standard deviation? Explain.
c) From a Normal model, about what percentage of neighborhoods should have a percent white within one standard deviation of the mean?
d) What percentage of neighborhoods actually have a percent white within one standard deviation of the mean?
e) Explain the discrepancy between parts c and d.

C

39. Final exams. Anna, a language major, took final exams in both French and Spanish and scored 83 on each. Her roommate Megan, also taking both courses, scored 77 on the French exam and 95 on the Spanish exam. Overall, student scores on the French exam had a mean of 81 and a standard deviation of 5, and the Spanish scores had a mean of 74 and a standard deviation of 15.
a) To qualify for language honors, a major must maintain at least an 85 average for all language courses taken. So far, which student qualifies?
b) Which student's overall performance was better?

40. MP3s. Two companies market new batteries targeted at owners of personal music players. DuraTunes claims a mean battery life of 11 hours, while RockReady advertises 12 hours.
a) Explain why you would also like to know the standard deviations of the battery lifespans before deciding which brand to buy.
b) Suppose those standard deviations are 2 hours for DuraTunes and 1.5 hours for RockReady. You are headed for 8 hours at the beach. Which battery is most likely to last all day? Explain.

c) If your beach trip is all weekend, and you probably will have the music on for 16 hours, which battery is most likely to last? Explain.

41. Winter Olympics 2006 downhill. Fifty-three men qualified for the men's alpine downhill race in Torino. The gold medal winner finished in 1 minute, 48.8 seconds. All competitors' times (in seconds) are found in the following list:

108.80	109.52	109.82	109.88	109.93	110.00
110.04	110.12	110.29	110.33	110.35	110.44
110.45	110.64	110.68	110.70	110.72	110.84
110.88	110.88	110.90	110.91	110.98	111.37
111.48	111.51	111.55	111.70	111.72	111.93
112.17	112.55	112.87	112.90	113.34	114.07
114.65	114.70	115.01	115.03	115.73	116.10
116.58	116.81	117.45	117.54	117.56	117.69
118.77	119.24	119.41	119.79	120.93	

a) The mean time was 113.02 seconds, with a standard deviation of 3.24 seconds. If the Normal model is appropriate, what percent of times will be less than 109.78 seconds?
b) What is the actual percent of times less than 109.78 seconds?
c) Why do you think the two percentages don't agree?
d) Create a histogram of these times. What do you see?

42. Winter Olympics 2006 speed skating. The top 25 women's 500-m speed skating times are listed in the table below:

Skater	Country	Time
Svetlana Zhurova	Russia	76.57
Wang Manli	China	76.78
Hui Ren	China	76.87
Tomomi Okazaki	Japan	76.92
Lee Sang-Hwa	South Korea	77.04
Jenny Wolf	Germany	77.25
Wang Beixing	China	77.27
Sayuri Osuga	Japan	77.39
Sayuri Yoshii	Japan	77.43
Chiara Simionato	Italy	77.68
Jennifer Rodriguez	United States	77.70
Annette Gerritsen	Netherlands	78.09
Xing Aihua	China	78.35
Sanne van der Star	Netherlands	78.59
Yukari Watanabe	Japan	78.65
Shannon Rempel	Canada	78.85
Amy Sannes	United States	78.89
Choi Seung-Yong	South Korea	79.02
Judith Hesse	Germany	79.03
Kim You-Lim	South Korea	79.25
Kerry Simpson	Canada	79.34
Krisy Myers	Canada	79.43
Elli Ochowicz	United States	79.48
Pamela Zoellner	Germany	79.56
Lee Bo-Ra	South Korea	79.73

a) The mean finishing time was 78.21 seconds, with a standard deviation of 1.03 second. If the Normal model is appropriate, what percent of the times should be within 0.5 second of 78.21?
b) What percent of the times actually fall within this interval?
c) Explain the discrepancy between a and b.

43. Cattle, finis. Consider the Angus weights model $N(1152, 84)$ one last time. What does the model predict to be the IQR of the weights of these steers?

44. IQ, finis. Consider the IQ scores model $N(100, 16)$ one last time. What does the model predict to be the IQR of IQs?

45. Cholesterol. Assume the cholesterol levels of adult American women can be described by a Normal model with a mean of 188 mg/dL and a standard deviation of 24.
a) Draw and label the Normal model.
b) What percent of adult women do you expect to have cholesterol levels over 200 mg/dL?
c) What percent of adult women do you expect to have cholesterol levels between 150 and 170 mg/dL?
d) Estimate the IQR of the cholesterol levels.
e) Above what value are the highest 15% of women's cholesterol levels?

46. Tires. A tire manufacturer believes that the treadlife of its snow tires can be described by a Normal model with a mean of 32,000 miles and standard deviation of 2500 miles.
a) If you buy a set of these tires, would it be reasonable for you to hope they'll last 40,000 miles? Explain.
b) Approximately what fraction of these tires can be expected to last less than 30,000 miles?
c) Approximately what fraction of these tires can be expected to last between 30,000 and 35,000 miles?
d) Estimate the IQR of the treadlives.
e) In planning a marketing strategy, a local tire dealer wants to offer a refund to any customer whose tires fail to last a certain number of miles. However, the dealer does not want to take too big a risk. If the dealer is willing to give refunds to no more than 1 of every 25 customers, for what mileage can he guarantee these tires to last?

Answers
Do the Math

1. You're 16 points above the mean; that's $z = 2$ standard deviations.

2. 76 is 4 points below the mean; that's $z = -0.5$ standard deviations.

3. $z = 1.5$ means one and a half standard deviations above the mean. That's $1.5 \times 8 = 12$ points, so the test score is $80 + 12 = 92$.

4. $z = -2.0$ means 2 standard deviations below the mean; $80 - 2(8) = 64$ points.

5. a) Frosted Flakes: $z = \dfrac{y - \bar{y}}{s} = \dfrac{11 - 7.6}{4.5} = 0.76$; the sugar content of Frosted Flakes is 0.76 standard deviations above the mean for Kellogg's cereals.

b) Apple Jacks: $z = \dfrac{y - \bar{y}}{s} = \dfrac{14 - 7.6}{4.5} = 1.42$; Apple Jacks' sugar content is 1.42 standard deviations above the mean for Kellogg's cereals.

c) Crispix: $z = \dfrac{y - \bar{y}}{s} = \dfrac{3 - 7.6}{4.5} = -1.02$; the sugar content of Crispix is just over one standard deviation lower than the mean sugar content of Kellogg's cereals.

6. 3.87 standard deviations above the mean would be $7.6 + 3.87(4.5) = 25$ grams of sugar per serving.

7. 0.8 standard deviations below the mean would be $7.6 - 0.8(4.5) = 4$ grams of sugar per serving.

8. a) 0.643   b) 0.945
c) Area$(z < -1.6) = 0.055$
d) Area$(z > 0.67) = 0.25$

9. a) The 40th percentile cutpoint is $z = -0.25$.
b) The 70th percentile cutpoint is $z = 0.52$
c) The 98th percentile cutpoint is $z = 2.05$
d) The middle 30% leaves 70% in the two tails—that's 35% in each tail. The 35th percentile cutpoint is $z = -0.39$, so the middle 30% of the distribution lies in the interval $-0.39 < z < 0.39$.

Just Checking

1. a) On the first test, the mean is 88 and the SD is 4, so $z = (90 - 88)/4 = 0.5$. On the second test, the mean is 75 and the SD is 5, so $z = (80 - 75)/5 = 1.0$. The first test has the lower z-score, so it is the one that will be dropped.

b) No. The second test is 1 standard deviation above the mean, farther away than the first test, so it's the better score relative to the class.

2. For calories, $z = \dfrac{50 - 109}{22.2} = -2.66$; for fiber,

$z = \dfrac{14 - 2.7}{3.2} = 3.53$.

The fiber content is farther from the mean, so it's more remarkable.

3.

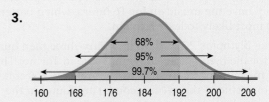

4. The mean is 184 centimeters, with a standard deviation of 8 centimeters. 2 meters is 200 centimeters, which is 2 standard deviations above the mean. We expect 5% of the men to be more than 2 standard deviations below or above the mean, so half of those, 2.5%, are likely to be above 2 meters.

5. a) The interval 18–22 minutes represents 1 standard deviation above and below the mean. We expect that to happen on about 68% of the days.

b) We know that 68% of the time we'll be within 1 standard deviation (2 min) of 20. So 32% of the time we'll arrive in less than 18 or more than 22 minutes. Half of those times (16%) will be greater than 22 minutes, so about 84% should be less than 22 minutes.

c) 24 minutes is 2 standard deviations above the mean. Because of the 95% rule, we know about 2.5% of the times will be more than 24 minutes.

6. a) Area $(z < 0.75) = 77\%$
b) Area $(-1.75 < z < -0.5) = 26.8\%$
c) Area $(z > 1.75) = 4\%$

7. a) The 90th percentile cutpoint is $z = 1.28$ standard deviations above the mean. $184 + 1.28(8) = 194.24$, so we expect the tallest 10% of Dutch men to be over 194.24 cm tall.

b) The 20th percentile cutpoint is $z = -0.84$; $184 - 0.84(8) = 177.28$. We expect the shortest 20% of Dutch men to be less than 177.28 cm tall.

c) The middle 50% lies between the quartiles, and the 25th and 75th percentile cutpoints are $z = \pm 0.67$. We expect the middle 50% of Dutch men to be between $184 - 0.67(8) = 178.64$ and $184 + 0.67(8) = 189.36$ cm tall.

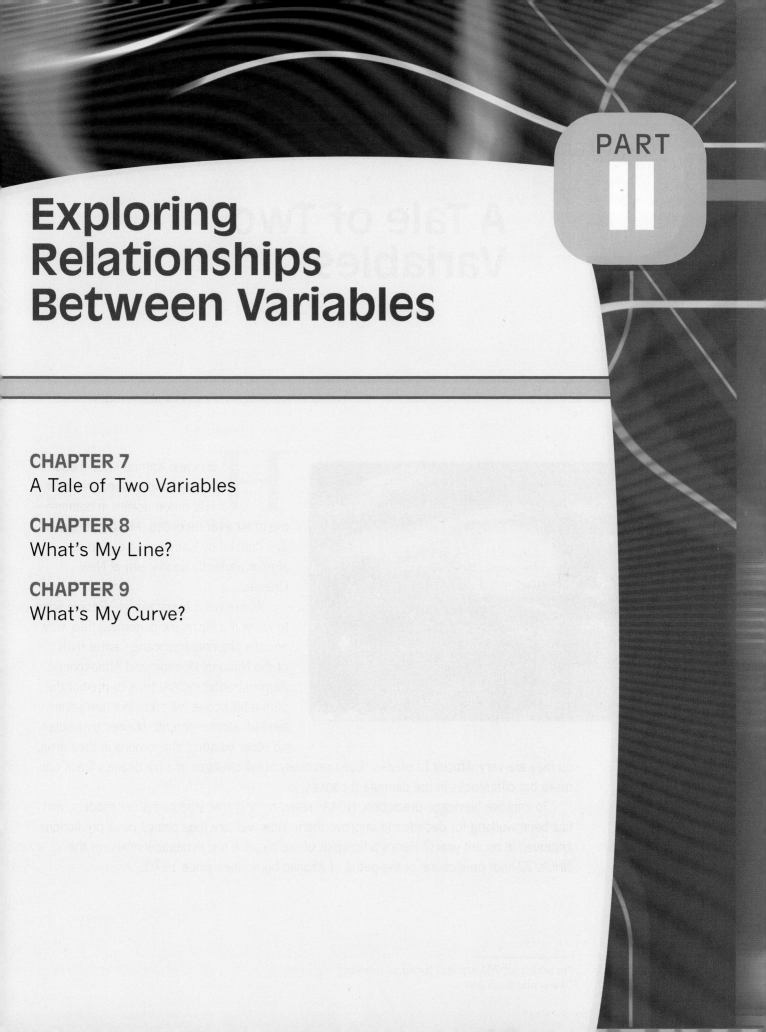

Exploring Relationships Between Variables

A Tale of Two Variables

Hurricane Katrina killed 1836 people[1] and caused well over 100 billion dollars in damage—the most ever recorded. Much of the damage caused by Katrina was due to its almost perfectly deadly aim at New Orleans.

Where will a hurricane go? People want to know if a hurricane is coming their way, and the National Hurricane Center (NHC) of the National Oceanic and Atmospheric Administration (NOAA) tries to predict the path a hurricane will take. But hurricanes tend to wander around, pushed by fronts and other weather phenomena in their area, so they are very difficult to predict. Even relatively small changes in a hurricane's track can make big differences in the damage it causes.

To improve hurricane prediction, NOAA[2] relies on sophisticated computer models, and has been working for decades to improve them. How well are they doing? Have predictions improved in recent years? Here's a timeplot of the mean error, in nautical miles, of the NHC's 72-hour predictions of the paths of Atlantic hurricanes since 1970:

---

[1]In addition, 705 are still listed as missing.
[2]www.nhc.noaa.gov

WHO	Years 1970–2005
WHAT	Mean error in the position of Atlantic hurricanes as predicted 72 hours ahead by the NHC
UNITS	nautical miles
WHEN	1970–2005
WHERE	Atlantic and Gulf of Mexico
WHY	NHC wants to improve prediction models.

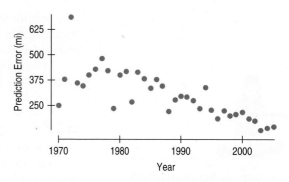

FIGURE 7.1 Prediction errors. A scatterplot of the average error in nautical miles of the predicted position of Atlantic hurricanes for predictions made by the National Hurricane Center of NOAA, plotted against the *Year* in which the predictions were made.

LOOK, MA, NO ORIGIN!

Scatterplots usually don't—and shouldn't—show the origin, because often neither variable has values near 0. The display should focus on the part of the coordinate plane that actually contains the data. In our example about hurricanes, none of the prediction errors or years were anywhere near 0, so the computer drew the scatterplot with axes that don't quite meet.

Clearly, predictions have improved. The plot shows a fairly steady decline in the average error, from almost 500 nautical miles in the late 1970s to about 150 nautical miles in 2005. We can also see that in some years predictions were unusually good and that 1972 was a really bad year for predicting hurricane tracks.

This timeplot is an example of a more general kind of display called a **scatterplot.** Scatterplots may be the most common displays for data. By just looking at them, you can see patterns, trends, relationships, and even the occasional extraordinary value sitting apart from the others. As the great philosopher Yogi Berra[3] once said, "You can observe a lot by watching."[4] Scatterplots are the best way to start observing the relationship between two *quantitative* variables.

Relationships between variables are often at the heart of what we'd like to learn from data:

- Are grades higher now than they used to be?
- Do people tend to reach puberty at a younger age than in previous generations?
- Does applying magnets to parts of the body relieve pain? If so, are stronger magnets more effective?
- Do students learn better with the use of computers?

Questions such as these relate two quantitative variables and ask whether there is an **association** between them. Scatterplots are the ideal way to *picture* such associations.

Looking at Scatterplots

Look for **Direction**: What's my sign—positive, negative, or neither?

Look for **Form**: straight, curved, something weird, or no pattern?

How would you describe the association of hurricane *Prediction Error* and *Year*? What should we look for in a scatterplot? What do *you* see? Try to describe the scatterplot of *Prediction Error* against *Year*.

The **direction** of the association is important. Over time, the NHC's prediction errors have decreased. A pattern like this that runs from the upper left to the lower right is said to be **negative.** A pattern running the other way is called **positive.**

The second thing to look for in a scatterplot is its **form.** Is there a **linear** relationship, a cloud or swarm of points stretched out in a generally consistent, straight form? For example, the scatterplot of *Prediction Error* vs. *Year* has such an underlying linear form, although some points stray away from it.

---

[3]Hall of Fame catcher and manager of the New York Mets and Yankees.
[4]But then he also said, "I really didn't say everything I said." So we can't really be sure.

Scatterplots can reveal many kinds of patterns. Often they will not be straight, but straight line patterns are both the most common and the most useful for statistics.

What about relationships that aren't straight? In Chapter 9 we'll see what to do if it curves gently, while still increasing or decreasing steadily,

Look for Strength: how much scatter?

. But if it curves sharply—up and then down, for example there is much less we can say about it with the methods of this book.

The third feature to look for in a scatterplot is the **strength** of the relationship.

At one extreme, do the points appear tightly clustered in a single stream

(whether straight, curved, or bending all over the place)? Or, at the other extreme, does the swarm of points seem to form a vague cloud through which

Look for Unusual Features: are there outliers or subgroups?

we can barely discern any trend or pattern? The *Prediction error* vs. *Year* plot shows moderate scatter around a generally straight form. This indicates that the linear trend of improving prediction is pretty consistent and moderately strong.

Finally, always look for the unexpected. Often the most interesting thing to see in a scatterplot is something you never thought to look for. One example of such a surprise is an **outlier** standing away from the overall pattern of the scatterplot. Such a point is almost always interesting and always deserves special attention. In the scatterplot of prediction errors, the year 1972 stands out as a year with very high prediction errors. An Internet search shows that it was a relatively quiet hurricane season. However, it included the very unusual—and deadly—Hurricane Agnes, which combined with another low-pressure center to ravage the northeastern United States, killing 122 and causing 1.3 billion 1972 dollars in damage. Possibly, Agnes was also unusually difficult to predict.

You should also look for clusters or subgroups that stand away from the rest of the plot or that show a trend in a different direction. You should question why they are different. This may be a clue that you should split the data into subgroups instead of looking at them all together.

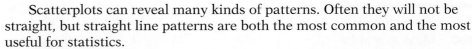

For Example   Describing the Scatterplot of Hurricane Winds and Pressure

Hurricanes develop low pressure at their centers. This pulls in moist air, pumps up their rotation, and generates high winds. Standard sea-level pressure is around 1013 millibars (mb), or 29.9 inches of mercury. Hurricane Katrina had a central pressure of 920 mb and sustained winds of 110 knots.

Here's a scatterplot of *Maximum Wind Speed* (kts) vs. *Central Pressure* (mb) for 163 hurricanes that have hit the United States since 1851.

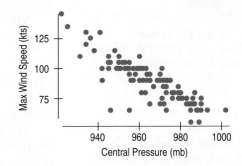

Question: Describe what this plot shows.

The scatterplot shows a negative direction; in general, lower central pressure is found in hurricanes that have higher maximum wind speeds. This association is linear and moderately strong.

Just Checking

1. Here are some scatterplots of associations between variables. Describe the **direction** ("positive," "negative"), **form** ("linear," "curved"), and **strength** ("weak," "moderate," "strong") of each relationship—in context, of course! For example:

Answer: There is a ***strong***, ***positive***, ***linear*** relationship between the price of a pizza and the number of toppings on it.

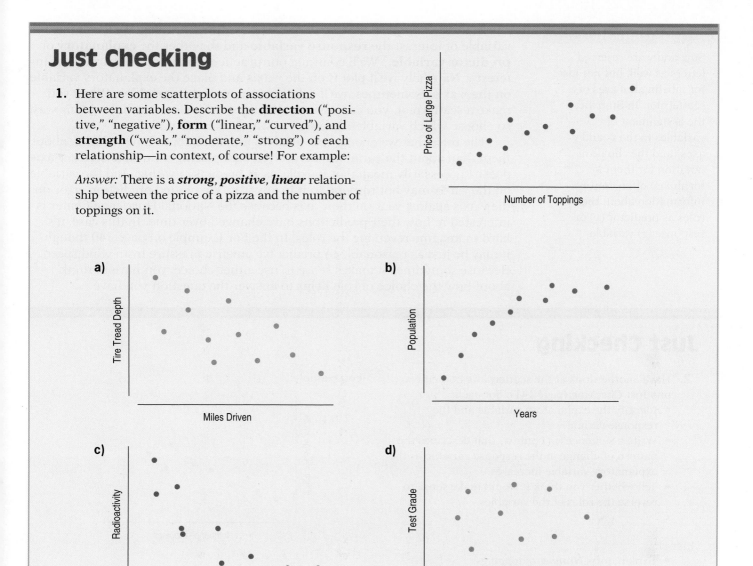

a)

Tire Tread Depth

Miles Driven

b)

Population

Years

c)

Radioactivity

Age of Isotope

d)

Test Grade

Hours Studied

Price of Large Pizza

Number of Toppings

(Check your answers on page 162.)

Roles for Variables

Which variable should go on the *x*-axis and which on the *y*-axis? What we want to know about the relationship can tell us how to make the plot. We often have questions such as:

- Do baseball teams that score more runs sell more tickets to their games?
- Do people with more years of education earn higher salaries?
- Do students who score higher on their SAT or ACT tests have higher grade point averages in college?
- Can we estimate a person's percent body fat more simply by just measuring waist or wrist size?

So x and y are reserved letters as well, but not just for labeling the axes of a scatterplot. In Statistics, the assignment of variables to the x- and y-axes (and the choice of notation for them in formulas) often conveys information about their roles as predictor (x) or response (y) variable.

In these examples, the two variables play different roles. We'll call the variable of interest the **response variable** and the other the **explanatory** or **predictor variable.**[5] We'll continue our practice of naming the variable of interest y. Naturally we'll plot it on the y-axis and place the explanatory variable on the x-axis. Sometimes, we'll call them the **x-** and **y-variables.** When you make a scatterplot, you can assume that those who view it will think this way, so choose which variables to assign to which axes carefully.

The roles that we choose for variables are more about how we *think* about them than about the variables themselves. Just placing a variable on the x-axis doesn't necessarily mean that it explains or predicts *anything.* And the variable on the y-axis may not respond to it in any way. We plotted prediction error on the y-axis against year on the x-axis because the National Hurricane Center is interested in how their predictions have changed over time. In this case, it's hard to imagine reversing the roles. In the For Example on page 140 though, it may be just as reasonable to predict barometric pressure from wind speed. Because sometimes it makes sense to use either choice, you have to think about how the choice of role helps to answer the question you have.

Just Checking

2. Have another look at the scatterplots in the previous Just Checking (page 141). For each:
   - Identify the explanatory variable and the response variable.
   - Write a sentence (in context) that describes the pattern of change in the response variable as the explanatory variable increases.
   - Tell whether you think it would make sense to reverse the roles of the variables.

For example:

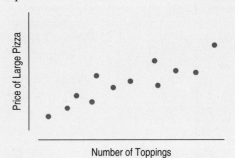

Answer:
   - Explanatory: *Number of toppings*
   - Response: *Price of large pizza*
   - As the number of toppings on a large pizza increases, the price generally increases.
   - Yes; we might want to try to predict the number of toppings knowing the price.

(Check your answers on page 162.)

TI Tips   Creating a scatterplot

Let's use your calculator to make a scatterplot. First you need some data. It's okay to just enter the data in any two lists, but let's get fancy. When you are handling lots of data and several variables (as you will be soon), remembering what you stored in L1, L2, and so on can become confusing. You can—and should—give your variables meaningful names. To see how, let's store some data that you will use several times in this chapter and the next. They show the change in tuition costs at Arizona State University during the 1990s.

---

[5]In algebra the x- and y-variables are sometimes referred to as the *independent* and *dependent* variables, respectively. These names, however, can be misleading, so we recommend you don't use them.

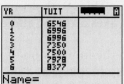

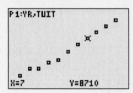

Naming the Lists

- Go into **STAT Edit**, place the cursor on one of the list names (**LI**, say), and use the arrow key to move to the right across all the lists until you encounter a blank column.
- Type **YR** to name this first variable, then hit **ENTER**.
- Often when we work with years it makes sense to use values like "90" (or even "0") rather than big numbers like "1990." For these data enter the years 1990 through 2000 as 0, 1, 2, . . . , 10.
- Now go to the next blank column, name this variable **TUIT**, and enter these values: 6546, 6996, 6996, 7350, 7500, 7978, 8377, 8710, 9110, 9411, 9800.

Making the Scatterplot

- Set up the **STATPLOT** by choosing the scatterplot icon (the first option).
- Identify which lists you want as **Xlist** and **Ylist**. If the data are in L1 and L2, that's easy to do—but your data are stored in lists with special names. To specify your **Xlist**, go to **2nd LIST NAMES**, scroll down the list of variables until you find **YR**, then hit **ENTER**.
- Use **LIST NAMES** again to specify **Ylist:TUIT**.
- Pick a symbol for displaying the points.
- Now **ZoomStat** to see your scatterplot. (Didn't work? **ERR:DIM MISMATCH** means you don't have the same number of *x*'s and *y*'s. Go to **STAT Edit** and look carefully at your two datalists. You can easily fix the problem once you find it.)
- Notice that if you **TRACE** the scatterplot the calculator will tell you the *x*- and *y*-value at each point.

What can you Tell about the trend in tuition costs at ASU? (Remember: direction, form, and strength!)

Correlation

WHO	Students
WHAT	Height (inches), weight (pounds)
WHERE	Ithaca, NY
WHY	Data for class
HOW	Survey

Data collected from students in Statistics classes included their *Height* (in inches) and *Weight* (in pounds). It's no great surprise to discover that there is a positive association between the two. As you might suspect, taller students tend to weigh more. (If we had reversed the roles and chosen height as the explanatory variable, we might say that heavier students tend to be taller.)[6] And the form of the scatterplot is fairly straight as well, although there seems to be a high outlier. The second plot shows the relationship is the same when we measure heights in centimeters and weights in kilograms.

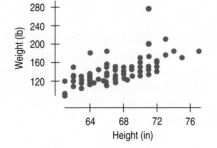
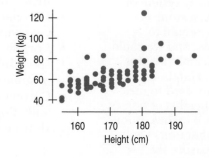

FIGURE 7.2 Weight vs. Height of Statistics students. Plotting *Weight* vs. *Height* in different units doesn't change the shape of the pattern.

[6]The son of one of the authors, when told (as he often was) that he was tall for his age, used to point out that, actually, he was young for his height.

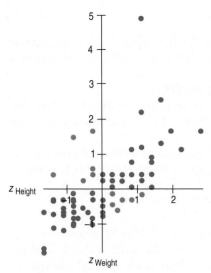

FIGURE 7.3 In this scatterplot of z-scores, points are colored according to how they affect the association: green for positive, red for negative, and blue for neutral.

HEIGHT AND WEIGHT, AGAIN

We could have measured the students' weights in stones. In the now outdated UK system of measures, a stone is a measure equal to 14 pounds. And we could have measured heights in hands. Hands are still commonly used to measure the heights of horses. A hand is 4 inches. But no matter what *units* we use to measure the two variables, the *correlation* stays the same.

The pattern in the scatterplots looks straight and is clearly a positive association, but how strong is it? If you had to put a number (say, between 0 and 1) on the strength, what would it be?

Since the units shouldn't matter to our measure of strength, we can remove them by standardizing each variable. Now, for each point, instead of the values (x, y) we'll have the standardized coordinates (z_x, z_y).

Because standardizing makes the means of both variables 0, the center of the new scatterplot is at the origin. The scales on both axes are now standard deviation units.

Which points in the scatterplot of the z-scores give the impression of a positive association? In a positive association, y tends to increase as x increases, so the points in the upper right and lower left (colored green) vote for a positive association. The red points in the upper left and lower right quadrants vote against that positive association. Points on either axis (colored blue) abstain from voting either way. We could just count points to see who wins, but we do something fancier—and more precise.

For each point, we multiply the standardized coordinates $z_x z_y$, then add up these products for all the points in the scatterplot. After dividing by $n - 1$[7] we have the famous **correlation coefficient:**

$$r = \frac{\sum z_x z_y}{n - 1}.$$

Why does this work? For the green points, z_x and z_y have the same sign so the product $z_x z_y$ is positive. For the red points, z_x and z_y have opposite signs so the product $z_x z_y$ is negative. For the blue points, either z_x or z_y is 0, so they don't contribute to the correlation. If most of the points are in the green quadrants, the sum will tend to be positive. If most are in the red quadrants, it will tend to be negative. Points far from the origin, evidence of a stronger association, have bigger products. Dividing by $n - 1$ keeps the sum from getting larger just because there are more points. These properties allow the correlation coefficient to summarize the direction *and* strength of the association found in all the points—just what we wanted!

Correlation Properties

For the students' heights and weights, $r = 0.644$. What does that mean?

Here's a useful list of facts about the correlation coefficient:

- The sign of a correlation coefficient gives the direction of the association.
- Correlation is always between -1 and $+1$. Correlation *can* be exactly equal to -1.0 or $+1.0$, but these values are unusual in real data because they mean that all the data points fall *exactly* on a single straight line.
- Correlation is commutative. The correlation of x with y is the same as the correlation of y with x.
- Correlation has no units, because it's based on z-scores. (That means you shouldn't give correlation as a percentage, because it suggests a percentage of *something*—and correlation, lacking units, has no "something.")
- Correlation is not affected by changes in the center or units of either variable. Correlation depends only on the z-scores, and they are unaffected by shifting or rescaling.

---

[7] Yes, the same $n - 1$ as in the standard deviation calculation.

- Correlation measures the strength of the *linear* association between the two variables. Variables can be strongly associated but still have a small correlation if the association isn't linear. As always, *make a picture*.
- Correlation is sensitive to outliers. A single outlying value can make a small correlation large or make a large one small. As always, beware of outliers.

TI-*nspire*

Correlation and Scatterplots.
See how the correlation changes as you drag data points around in a scatterplot.

> **How strong is strong?** You'll often see correlations characterized as "weak," "moderate," or "strong," but be careful. There's no agreement on what those terms mean. The same numerical correlation might be strong in one context and weak in another. You might be thrilled to discover a correlation of 0.7 between the new summary of the economy you've come up with and stock market prices, but you'd consider it a design failure if you found a correlation of "only" 0.7 between two tests intended to measure the same thing. Deliberately vague terms like "weak," "moderate," or "strong" that describe a linear association can be useful additions to the numerical summary that correlation provides. But be sure to include the correlation and show a scatterplot, so others can judge for themselves.

For Example   Changing Scales

Recap: We've seen the scatterplot displaying hurricane wind speeds in knots and their central pressures in millibars.

Question: Suppose we wanted to consider the wind speeds in miles per hour (1 mile per hour = 0.869 knots) and central pressures in inches of mercury (1 inch of mercury = 33.86 millibars). How would that conversion affect the value of *r*?

Not at all! Correlation is based on standardized values (z-scores), so the value of r will be unaffected by changes in units.

Just Checking

Your Statistics teacher tells you that the correlation between the scores (points out of 50) on Exam 1 and Exam 2 was 0.75.

3. Before answering any questions about the correlation, what would you like to see? Why?

4. If she adds 10 points to each Exam 1 score, how will this change the correlation?

5. If she standardizes scores on each exam, how will this affect the correlation?

6. In general, if someone did poorly on Exam 1, are they likely to have done poorly or well on Exam 2? Explain.

7. If someone did poorly on Exam 1, can you be sure that they did poorly on Exam 2 as well? Explain.

8. Here and on the next page are five scatterplots; the five correlations are listed below. Match each value of *r* with the appropriate graph.

  I. *r* = 0.87
  II. *r* = −0.50
  III. *r* = −0.68
  IV. *r* = 0.52
  V. *r* = 0.07

a)
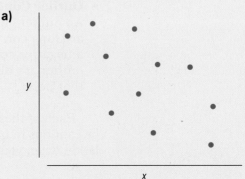

(continued)

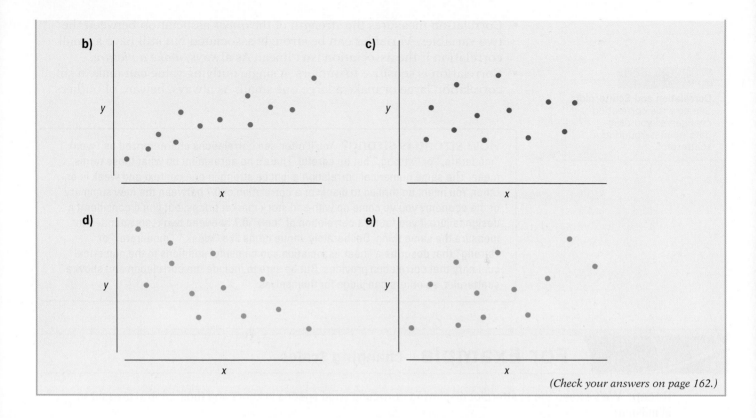

(Check your answers on page 162.)

Correlation Conditions

Correlation measures the strength of the *linear* association between two *quantitative* variables. Before you use correlation, you must check several *conditions:*

- **Quantitative Variables Condition:** Are both variables quantitative? Correlation applies only to quantitative variables. Don't apply correlation to categorical data masquerading as quantitative. Check that you know the variables' units and what they measure.
- **Straight Enough Condition:** Is the form of the scatterplot straight enough that a linear relationship makes sense? Sure, you can *calculate* a correlation coefficient for any pair of variables. But correlation measures the strength only of the *linear* association, and will be misleading if the relationship is not linear. What is "straight enough"? This is a judgment call. Do you think that the underlying relationship is curved? If so, then summarizing its strength with a correlation would be misleading.
- **Outlier Condition:** Outliers can distort the correlation dramatically. An outlier can make an otherwise weak correlation look big or hide a strong correlation. It can even give an otherwise positive association a negative correlation coefficient (and vice versa). When you see an outlier, it's often a good idea to report the correlation with and without that point.

Each of these conditions is easy to check with a scatterplot. You should be cautious in interpreting (or accepting others' interpretations of) the correlation when you can't check the conditions for yourself.

For Example   Correlating Wind Speed and Pressure

Recap: Here again is the scatterplot displaying hurricane wind speeds and central pressures. The correlation coefficient for these wind speeds and pressures is $r = -0.879$.

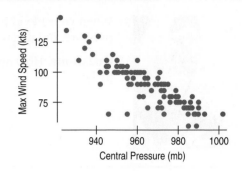

Question: Check the conditions for using correlation. If you feel they are satisfied, interpret this correlation.

✔ Quantitative Variables Condition: Both wind speed and central pressure are quantitative variables, measured (respectively) in knots and millibars.

✔ Straight Enough Condition: The pattern in the scatterplot is quite straight.

✔ Outlier Condition: A few hurricanes seem to straggle away from the main pattern, but they don't appear to be extreme enough to be called outliers. It may be worthwhile to check on them, however.

The conditions for using correlation are satisfied. The correlation coefficient of $r = -0.879$ indicates quite a strong negative linear association between the wind speeds of hurricanes and their central pressures.

Just Checking

9. Take a look at these four scatterplots. Do you think it's appropriate to find the correlation? If so, see if you can guess the value of r. If not, explain why not.

a)

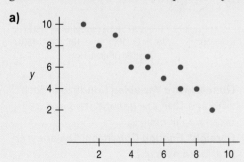

b)

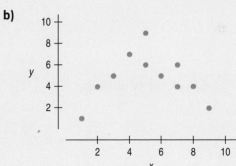

c)

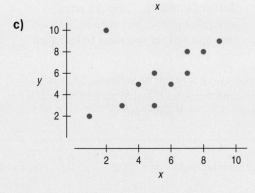

d)

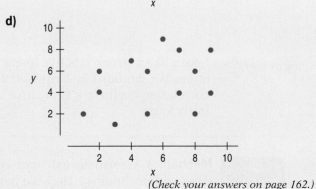

(Check your answers on page 162.)

STEP-BY-STEP EXAMPLE   Looking at Association

When your blood pressure is measured, it is reported as two values: systolic blood pressure and diastolic blood pressure.

Questions: How are these two measurements related to each other? Do they tend to be both high or both low? How strongly associated are they?

 THINK

Plan State what you are trying to investigate.

Variables Identify the two quantitative variables whose relationship we wish to examine. Report the W's, and be sure both variables are recorded for the same individuals.

Plot Make the scatterplot. Use a computer program or graphing calculator if you can.

Check the conditions.

REALITY CHECK Looks like a strong positive linear association. We shouldn't be surprised if the correlation coefficient is positive and fairly large.

I'll examine the relationship between two measures of blood pressure.

The variables are systolic and diastolic blood pressure (SBP and DBP), recorded in millimeters of mercury (mm Hg) for each of 1406 participants in the Framingham Heart Study, a famous health study in Framingham, MA.[8]

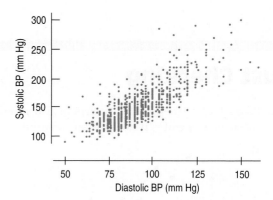

✔ **Quantitative Variables Condition:** Both SBP and DBP are quantitative and measured in mm Hg.
✔ **Straight Enough Condition:** The scatterplot looks straight.
✔ **Outlier Condition:** There are a few straggling points, but none far enough from the body of the data to be called outliers.

I have two quantitative variables that satisfy the conditions, so it's okay to use correlation as a measure of association.

 SHOW

Mechanics We usually calculate correlations with technology. Here we have 1406 cases, so we'd never try it by hand.

The correlation is $r = 0.792$.

---

[8]www.nhlbi.nih.gov/about/framingham

Conclusion Describe the direction, form, and strength you see in the plot, along with any unusual points or features. Be sure to state your interpretations in the proper context.

The scatterplot shows a positive direction, with higher SBP associated with higher DBP. The plot is generally straight, with a moderate amount of scatter. The correlation of 0.792 is consistent with what I saw in the scatterplot. A few cases stand out with unusually high SBP compared with their DBP. It seems far less common for the DBP to be high by itself.

TI Tips

Finding the correlation

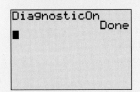

Now let's use the calculator to find a correlation. Unfortunately, the statistics package on your TI calculator does not automatically do that. Correlations are one of the most important things we might want to do, so here's how to fix that, once and for all.

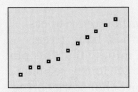

- Hit **2nd CATALOG** (on the zero key). You now see a list of everything the calculator knows how to do. Impressive, huh?
- Scroll down until you find **DiagnosticOn**. Hit **ENTER**. Again. It should say **Done**.

Now and forevermore (or perhaps until you change batteries) your calculator will find correlations.

Finding the Correlation

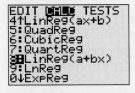

- *Always* check the conditions first. Look at the scatterplot for the Arizona State tuition data again. Does this association look linear? Are there outliers? This plot looks fine, but remember that correlation can be used to describe the strength of *linear* associations only, and outliers can distort the results. Eyeballing the scatterplot is an essential first step. (You should be getting used to checking on assumptions and conditions before jumping into a statistical procedure—it's always important.)
- Under the **STAT CALC** menu, select **8:LinReg (a+bx)** and hit **ENTER**.

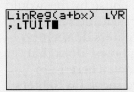

- Now specify *x* and *y* by importing the names of your variables from the **LIST NAMES** menu. First name your *x*-variable followed by a comma, then your *y*-variable, creating the command

$$\text{LinReg(a+bx)LYR,LTUIT}$$

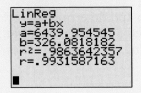

Wow! A lot of stuff happened. If you suspect all those other numbers are important, too, you'll really enjoy the next chapter. But for now, it's the value of *r* you care about. What does this correlation, *r* = 0.993, say about the trend in tuition costs?

Warning: Correlation ≠ Causation

Whenever we have a strong correlation, it's tempting to try to explain it by imagining that the predictor variable has *caused* the response to change. Humans are like that; we tend to see causes and effects in everything.

Sometimes this tendency can be amusing. A scatterplot of the human population (*y*) of Oldenburg, Germany, in the beginning of the 1930s plotted against the number of storks nesting in the town (*x*) shows a tempting pattern.

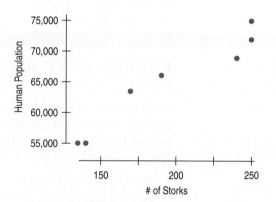

FIGURE 7.4 Storks and population. The number of storks in Oldenburg, Germany, plotted against the population of the town for 7 years in the 1930s. The association is clear. How about the causation? (*Ornithologishe Monatsberichte*, 44, no. 2)

Anyone who has seen the beginning of the movie *Dumbo* remembers Mrs. Jumbo anxiously waiting for the stork to bring her new baby. Even though you know it's silly, you can't help but think for a minute that this plot shows that storks are the culprits. The two variables are obviously related to each other (the correlation is 0.97!), but that doesn't prove that storks bring babies.

It turns out that storks nest on house chimneys. More people means more houses, more nesting sites, and so more storks. The causation is actually in the *opposite* direction, but you can't tell from the scatterplot or correlation. You need additional information—not just the data—to determine the real mechanism.

A scatterplot of the damage (in dollars) caused to a house by fire would show a strong correlation with the number of firefighters at the scene. Surely the damage doesn't cause firefighters. And firefighters do seem to cause damage, spraying water all around and chopping holes. Does that mean we shouldn't call the fire department? Of course not. There is an underlying variable that leads to both more damage and more firefighters: the size of the blaze.

A hidden variable that stands behind a relationship and determines it by simultaneously affecting the other two variables is called a **lurking variable.** You can often debunk claims made about data by finding a lurking variable behind the scenes.

Scatterplots and correlation coefficients *never* prove causation. That's one reason it took so long for the U.S. Surgeon General to get warning labels on cigarettes. Although there was plenty of evidence that increased smoking was *associated* with increased levels of lung cancer, it took years to provide evidence that smoking actually *causes* lung cancer.

Does cancer cause smoking? Even if the correlation of two variables is due to a causal relationship, the correlation itself cannot tell us what causes what.

Sir Ronald Aylmer Fisher (1890–1962) was one of the greatest statisticians of the 20th century. Fisher testified in court (in testimony paid for by the tobacco companies) that a causal relationship might underlie the correlation of smoking and cancer:

> "Is it possible, then, that lung cancer . . . is one of the causes of smoking cigarettes? I don't think it can be excluded . . . the pre-cancerous condition is one involving a certain amount of slight chronic inflammation . . . .
>
> A slight cause of irritation . . . is commonly accompanied by pulling out a cigarette, and getting a little compensation for life's minor ills in that way. And . . . is not unlikely to be associated with smoking more frequently."

Ironically, the proof that smoking indeed is the cause of many cancers came from experiments conducted following the principles of experiment design and analysis that Fisher himself developed—and that we'll see in Chapter 11.

WHAT CAN GO WRONG?

Did you know that there's a strong correlation between playing an instrument and drinking coffee? No? One reason might be that the statement doesn't make sense. Correlation is a statistic that's valid only for *quantitative* variables.

- **Don't say "correlation" when you mean "association."** How often have you heard the word "correlation"? Chances are pretty good that when you've heard the term, it's been misused. When people want to sound scientific, they often say "correlation" when talking about the relationship between two variables. It's one of the most widely misused Statistics terms, and given how often statistics are misused, that's saying a lot. One of the problems is that many people use the specific term *correlation* when they really mean the more general term *association*. "Association" is a deliberately vague term describing the relationship between two variables.

  "Correlation" is a precise term that measures the strength and direction of the linear relationship between *quantitative* variables.

- **Don't correlate categorical variables.** People who misuse the term "correlation" to mean "association" often fail to notice whether the variables they discuss are quantitative. There may be an association between people's religions and what area of the country they live in, but both variables

(continued)

are categorical, so there can't be any correlation. Be sure to check the **Quantitative Variables Condition.**

■ **Don't confuse correlation with causation.** One of the most common mistakes people make in interpreting statistics occurs when they observe a high correlation between two variables and jump to the perhaps tempting conclusion that one thing must be causing the other. Scatterplots and correlations *never* demonstrate causation. At best, these statistical tools can only reveal an association between variables, and that's a far cry from establishing cause and effect. While it's true that some associations may be causal, the nature and direction of the causation can be very hard to establish, and there's always the risk of overlooking lurking variables.

■ **Make sure the association is linear.** Not all associations between quantitative variables are linear. Correlation can miss even a strong nonlinear association. A student project evaluating the quality of brownies baked at different temperatures reports a correlation of −0.05 between judges' scores and baking temperature. That seems to say there is no relationship—until we look at the scatterplot:

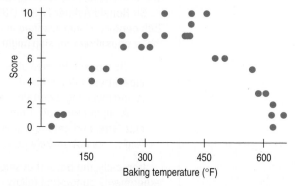

FIGURE 7.5 Low correlation. The relationship between brownie taste *Score* and *Baking Temperature* is strong, but not at all linear.

There is a strong association, but the relationship is not linear. Don't forget to check the Straight Enough Condition.

■ **Don't assume the relationship is linear just because the correlation coefficient is high.** The relationship between camera f/stops and shutter speeds plotted below is clearly not straight and yet the correlation is 0.979. Although the relationship must be straight for the correlation to be an appropriate measure, a high correlation is no guarantee of straightness. It's always important to look at the scatterplot.

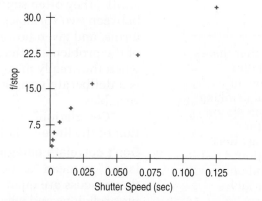

FIGURE 7.6 A scatterplot of *f/stop* vs. *Shutter Speed.* We see a bent relationship even though the correlation is $r = 0.979$.

■ **Beware of outliers.** You can't interpret a correlation coefficient safely without a background check for outliers. Here's a silly example:

The relationship between IQ and shoe size among comedians shows a surprisingly strong positive correlation of 0.50. To check assumptions, we look at the scatterplot:

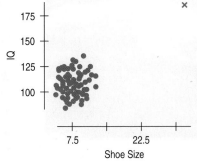

FIGURE 7.7 A scatterplot of *IQ* vs. *Shoe Size.* From this "study," what is the relationship between the two? The correlation is 0.50. Who does that point (the green x) in the upper right-hand corner belong to?

The outlier is Bozo the Clown, known for his large shoes, and widely acknowledged to be a comic "genius." Without Bozo, the correlation is near zero.

Even a single outlier can dominate the correlation value. That's why you need to check the Outlier Condition.

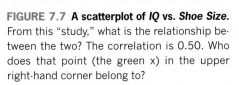

ASSOCIATION AND CAUSATION **IN THE NEWS**

Kids who smoke watch more R-rated films, study shows

by Carrie Stetler
The Star-Ledger Wednesday March 11, 2009,
2:04 PM

. . . [R]esearchers have found evidence that sixth-graders who watch R-rated movies are more likely to smoke.

A study at the University of Massachusetts Medical School found that permission to watch R-rated movies was right up there with peer pressure when it came to leading kids to tobacco.

The findings were based on a four-year study on nicotine dependency. Children were asked questions about cigarette use among family members and friends, in addition to seemingly unrelated questions like, "How often do your parents let you watch movies and videos that are rated R?"

The kids who watched R-rated movies were almost twice as likely to perceive cigarettes as being "readily accessible," according to the study, which was published last week in the American Journal of Preventive Medicine.

Researchers couldn't explain the connection. Was it lax parenting or were the children emulating on-screen characters who smoke?

"We don't know why this is so. It may have to do with a parenting style that is permissive and activities that are not age appropriate," said lead author of the study, Chyke Doubeni, of the university's medical school. "Or it may be an outcome of all the smoking scenes."

18. **More correlation errors.** Students in the Economics class discussed in Exercise 17 also wrote these conclusions. Explain the mistakes they made.
    a) "There was a very strong correlation of 1.22 between *Life Expectancy* and *GDP*."
    b) "The correlation between *Literacy Rate* and *GDP* was 0.83. This shows that countries wanting to increase their standard of living should invest heavily in education."

B

19. **Hard water.** In a study of streams in the Adirondack Mountains, the following relationship was found between the water's pH and its hardness (measured in grains):

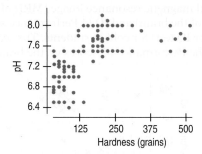

    a) Write a few sentences describing the association between *Hardness* and *pH*.
    b) Is it appropriate to summarize the strength of association with a correlation? Explain.

20. **Kentucky Derby 2006.** The fastest horse in Kentucky Derby history was Secretariat in 1973. The scatterplot shows speed (in miles per hour) of the winning horses each year.

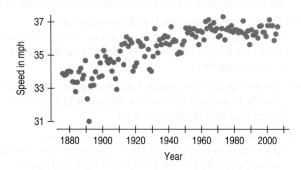

    a) Write a few sentences describing the association between *Year* and *Speed*.
    b) Explain why we shouldn't find the correlation between *Year* and *Speed*.

21. **Prediction units.** The errors in predicting hurricane tracks (examined in this chapter) were given in nautical miles. An ordinary mile is 0.86898 nautical miles. Most people living on the Gulf Coast of the United States would prefer to know the prediction errors in miles rather than nautical miles. Explain why converting the errors to miles would not change the correlation between *Prediction Error* and *Year*.

22. **More predictions.** Hurricane Katrina's hurricane force winds extended 120 miles from its center. Katrina was a big storm, and that affects how we think about the prediction errors. Suppose we add 120 miles to each error to get an idea of how far from the predicted track we might still find damaging winds. Explain what would happen to the correlation between *Prediction Error* and *Year,* and why.

23. **Height and reading.** A researcher studies children in elementary school and finds a strong positive linear association between height and reading scores.
    a) Does this mean that taller children are generally better readers?
    b) What might explain the strong correlation?

24. **Cellular telephones and life expectancy.** A survey of the world's nations in 2004 shows a strong positive correlation between percentage of the country using cell phones and life expectancy in years at birth.
    a) Does this mean that cell phones are good for your health?
    b) What might explain the strong correlation?

25. **Correlation conclusions I.** The correlation between *Age* and *Income* as measured on 100 people is $r = 0.75$. Explain whether or not each of these possible conclusions is justified:
    a) When *Age* increases, *Income* increases as well.
    b) The form of the relationship between *Age* and *Income* is straight.
    c) There are no outliers in the scatterplot of *Income* vs. *Age*.
    d) Whether we measure *Age* in years or months, the correlation will still be 0.75.

26. **Correlation conclusions II.** The correlation between *Fuel Efficiency* (as measured by miles per gallon) and *Price* of 150 cars at a large dealership is $r = -0.34$. Explain whether or not each of these possible conclusions is justified:
    a) The more you pay, the lower the fuel efficiency of your car will be.
    b) The form of the relationship between *Fuel Efficiency* and *Price* is moderately straight.
    c) There are several outliers that explain the low correlation.
    d) If we measure *Fuel Efficiency* in kilometers per liter instead of miles per gallon, the correlation will increase.

27. **Income and housing.** The Office of Federal Housing Enterprise Oversight (www.ofheo.gov) collects data on various aspects of housing costs around the United States. Here is a scatterplot of the *Housing Cost Index* versus the *Median Family Income* for each of the 50 states. The correlation is 0.65.

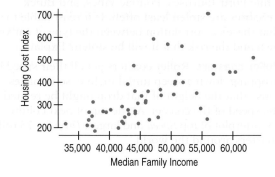

a) Describe the relationship between the *Housing Cost Index* and the *Median Family Income* by state.
b) Do these data provide proof that by raising the median income in a state, the Housing Cost Index will rise as a result? Explain.

28. Interest rates and mortgages. Since 1980, average mortgage interest rates have fluctuated from a low of under 6% to a high of over 14%. Is there a relationship between the amount of money people borrow and the interest rate that's offered? Here is a scatterplot of *Total Mortgages* in the United States (in millions of 2005 dollars) versus *Interest Rate* at various times over the past 26 years. The correlation is −0.84.

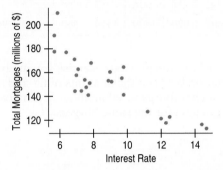

a) Describe the relationship between *Total Mortgages* and *Interest Rate*.
b) Do these data provide proof that if mortgage rates are lowered, people will take out more mortgages? Explain.

29. Income and housing revisited. In Exercise 27 we saw that there was a correlation of 0.65 between *Median Family Income* and *Housing Cost Index* for the 50 states.
a) If we standardized both variables, what would the correlation coefficient between the standardized variables be?
b) If we had measured *Median Family Income* in thousands of dollars instead of dollars, how would the correlation change?
c) Washington, DC, has a Housing Cost Index of 548 and a median income of about $45,000. If we were to include DC in the data set, how would that affect the correlation coefficient?

30. Interest rates and mortgages revisited. In Exercise 28 we saw that there was a correlation of 0.65 between *Interest Rate* and *Total Mortgages* since 1980.
a) If we standardized both variables, what would the correlation coefficient between the standardized variables be?
b) If we were to measure *Total Mortgages* in thousands of dollars instead of millions of dollars, how would the correlation coefficient change?
c) Suppose in another year, interest rates were 11% and mortgages totaled $250 million. How would including that year with these data affect the correlation coefficient?

31. Fuel economy 2007. Here are advertised horsepower ratings and expected gas mileage for several 2007 vehicles. (http://www.kbb.com/KBB/ReviewsAndRatings)

Vehicle	Horsepower	Highway Gas Mileage (mpg)
Audi A4	200	32
BMW 328	230	30
Buick LaCrosse	200	30
Chevy Cobalt	148	32
Chevy TrailBlazer	291	22
Ford Expedition	300	20
GMC Yukon	295	21
Honda Civic	140	40
Honda Accord	166	34
Hyundai Elantra	138	36
Lexus IS 350	306	28
Lincoln Navigator	300	18
Mazda Tribute	212	25
Toyota Camry	158	34
Volkswagen Beetle	150	30

a) Make a scatterplot for these data.
b) Describe the direction, form, and strength of the plot.
c) Find the correlation between horsepower and miles per gallon.
d) Write a few sentences telling what the plot says about fuel economy.

32. Manatees 2005. Marine biologists warn that the growing number of powerboats registered in Florida threatens the existence of manatees. The following data come from the Florida Fish and Wildlife Conservation Commission (www.floridamarine.org) and the National Marine Manufacturers Association (www.nmma.org/facts).

Year	Manatees Killed	Powerboat Registrations (in 1000s)
1982	13	447
1983	21	460
1984	24	481
1985	16	498
1986	24	513
1987	20	512
1988	15	527
1989	34	559
1990	33	585
1992	33	614
1993	39	646
1994	43	675
1995	50	711
1996	47	719
1997	53	716
1998	38	716
1999	35	716
2000	49	735
2001	81	860
2002	95	923
2003	73	940
2004	69	946
2005	79	974

a) In this context, which is the explanatory variable?
b) Make a scatterplot of these data.
c) Describe the association you see.
d) Find the correlation between *Boat Registrations* and *Manatee Deaths*.
e) Does your analysis prove that powerboats are killing manatees? Explain.

33. Firing pottery. A ceramics factory can fire eight large batches of pottery a day. Sometimes a few of the pieces break in the process. In order to understand the problem better, the factory records the number of broken pieces in each batch for 3 days and then creates the scatterplot shown.

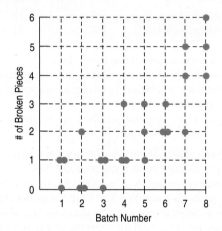

a) Make a histogram showing the distribution of the number of broken pieces in the 24 batches of pottery examined.
b) Describe the distribution as shown in the histogram. What feature of the problem is more apparent in the histogram than in the scatterplot?
c) What aspect of the company's problem is more apparent in the scatterplot?

34. Coffee sales. Owners of a new coffee shop tracked sales for the first 20 days and displayed the data in a scatterplot (by day).

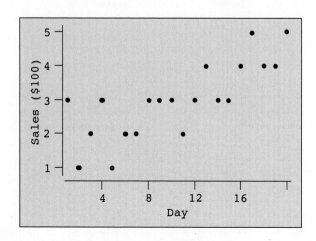

a) Make a histogram of the daily sales since the shop has been in business.
b) State one fact that is obvious from the scatterplot, but not from the histogram.
c) State one fact that is obvious from the histogram, but not from the scatterplot.

35. Burgers. Fast food is often considered unhealthy because much of it is high in both fat and sodium. But are the two related? Here are the fat and sodium contents of several brands of burgers. Analyze the association between fat content and sodium.

Fat (g)	19	31	34	35	39	39	43
Sodium (mg)	920	1500	1310	860	1180	940	1260

36. Burgers II. In the previous exercise you analyzed the association between the amounts of fat and sodium in fast food hamburgers. What about fat and calories? Here are data for the same burgers:

Fat (g)	19	31	34	35	39	39	43
Calories	410	580	590	570	640	680	660

37. Attendance 2006. American League baseball games are played under the designated hitter rule, meaning that pitchers, often weak hitters, do not come to bat. Baseball owners believe that the designated hitter rule means more runs scored, which in turn means higher attendance. Is there evidence that more fans attend games if the teams score more runs? Data collected from American League games during the 2006 season indicate a correlation of 0.667 between runs scored and the number of people at the game. (http://mlb.mlb.com)

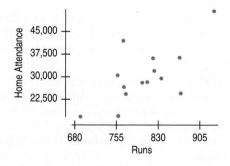

a) Does the scatterplot indicate that it's appropriate to calculate a correlation? Explain.
b) Describe the association between attendance and runs scored.
c) Does this association prove that the owners are right that more fans will come to games if the teams score more runs?

38. Second inning 2006. Perhaps fans are just more interested in teams that win. The displays below are based on American League teams for the 2006 season. (http://espn.go.com) Are the teams that win necessarily those which score the most runs?

Correlations			
	Wins	**Runs**	**Attend**
Wins	1.000		
Runs	0.605	1.000	
Attend	0.697	0.667	1.000

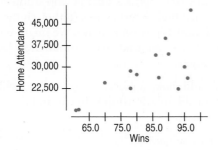

a) Do winning teams generally enjoy greater attendance at their home games? Describe the association.
b) Is attendance more strongly associated with winning or scoring runs? Explain.
c) How strongly is scoring more runs associated with winning more games?

39. Thrills. People who responded to a July 2004 Discovery Channel poll named the 10 best roller coasters in the United States. The table below shows the length of the initial drop (in feet) and the duration of the ride (in seconds). What do these data indicate about the height of a roller coaster and the length of the ride you can expect?

Roller Coaster	State	Drop (ft)	Duration (sec)
Incredible Hulk	FL	105	135
Millennium Force	OH	300	105
Goliath	CA	255	180
Nitro	NJ	215	240
Magnum XL-2000	OH	195	120
The Beast	OH	141	65
Son of Beast	OH	214	140
Thunderbolt	PA	95	90
Ghost Rider	CA	108	160
Raven	IN	86	90

40. Cramming. One Thursday, researchers gave students enrolled in a section of basic Spanish a set of 50 new vocabulary words to memorize. On Friday the students took a vocabulary test. When they returned to class the following Monday, they were retested—without advance warning. Here are the test scores for the 25 students. What do these data indicate about learning by studying for a test at the last minute?

Fri.	Mon.	Fri.	Mon.	Fri.	Mon.
42	36	48	37	39	41
44	44	43	41	46	32
45	46	45	32	37	36
48	38	47	44	40	31
44	40	50	47	41	32
43	38	34	34	48	39
41	37	38	31	37	31
35	31	43	40	36	41
43	32				

41. Cold nights. Is there an association between time of year and the nighttime temperature in North Dakota? A researcher assigned the numbers 1–365 to the days January 1–December 31 and recorded the temperature at 2 a.m. for each. What might you expect the correlation between *DayNumber* and *Temperature* to be? Explain.

42. Association. A researcher investigating the association between two variables collected some data and was surprised when he calculated the correlation. He had expected to find a fairly strong association, yet the correlation was near 0. Discouraged, he didn't bother making a scatterplot. Explain to him how the scatterplot could still reveal the strong association he anticipated.

43. Planets (more or less). On August 24, 2006, the International Astronomical Union voted that Pluto is not a planet. Some members of the public have been reluctant to accept that decision. Let's look at some of the data. Is there any pattern to the locations of the planets? The table shows the average distance of each of the traditional nine planets from the sun.

Planet	Position Number	Distance from Sun (million miles)
Mercury	1	36
Venus	2	67
Earth	3	93
Mars	4	142
Jupiter	5	484
Saturn	6	887
Uranus	7	1784
Neptune	8	2796
Pluto	9	3666

a) Make a scatterplot and describe the association. (Remember: direction, form, and strength!)
b) Why would you not want to talk about the correlation between a planet's *Position* and *Distance* from the sun?

44. Pendulum. A student experimenting with a pendulum counted the number of full swings the pendulum made in 20 seconds for various lengths of string. Here are her data:

Length (in.)	6.5	9	11.5	14.5	18	21	24	27	30	37.5
Number of Swings	22	20	17	16	14	13	13	12	11	10

a) Make a scatterplot and describe the association. (Remember: form, direction, and strength!)
b) Why would you not want to talk about the correlation between the *Length* of a pendulum and the *Number of Swings* it makes in 20 seconds?

Answers
Just Checking

1. a) There is a moderate, negative, linear relationship between the number of miles these tires are driven and the depth of their tread.
   b) There is a strong positive, curved, relationship between years and population.
   c) There is a moderately strong, curved, negative relationship between the age of these isotopes and their radioactivity.
   d) There is a very weak, linear, positive relationship between the numbers of hours studied and the grade on this test.

2. a) Explanatory: *Miles driven*; response: *Tire tread depth*. In general, the more miles tires are driven the lower the tread depth. We could also use tread depth to estimate miles driven.
   b) Explanatory: *Years*; response: *Population*. As time goes on, the population grows rapidly at first, and then the rate of growth slows. We would not use the population to figure out what year it is.
   c) Explanatory: *Age of isotope*; response: *Radioactivity*. The level of radioactivity of this isotope decreases sharply at first, and then more slowly as the isotope ages. We could use the level of radioactivity to estimate the age of the isotope; indeed, this is how carbon dating works.
   d) Explanatory: *Hours studied*; response: *Test grade*. In general, students who studied longer earned higher test scores (although the relationship is very weak). We might use students' test grades to estimate how long they studied.

3. We should make a picture. A scatterplot will show whether the relationship is linear and alert us to any outliers.

4. It won't change.

5. It won't change.

6. They are likely to have done poorly. The positive correlation means that low scores on Exam 1 are associated with low scores on Exam 2 (and similarly for high scores).

7. No. The general association is positive, but individual performances may vary.

8. a) II b) I c) V d) III e) IV

9. a) Yes; $r = -0.93$ (Any guess stronger than -0.8 is pretty good.)
   b) No; there's a curve. This violates the Straight Enough Condition.
   c) No; there's an outlier. This violates the Outlier Condition.
   d) Yes; $r = 0.31$ (Any guess between 0.2 and 0.4 is pretty good.)

What's My Line?

The Whopper™ has been Burger King's signature sandwich since 1957. One Double Whopper with cheese provides 53 grams of protein—all the protein you need in a day. It also supplies 1020 calories and 65 grams of fat. The Daily Value (based on a 2000-calorie diet) for fat is 65 grams. So after a Double Whopper you'll want the rest of your calories that day to be fat-free.[1]

Of course, the Whopper isn't the only item Burger King sells. How are fat and protein related for the entire BK menu? The scatterplot of the *Fat* (in grams) versus the *Protein* (in grams) for foods sold at Burger King shows a positive, moderately strong, linear relationship.

If you want to have 25 grams of protein in your Burger King lunch, how much fat should you expect to consume? The correlation between *Fat* and *Protein* is 0.83, a sign that the linear association seen in the scatterplot is fairly strong. But *strength* of the relationship is only part of the picture.

[1]Sorry about the fries.

WHO	Items on the Burger King menu
WHAT	Protein content and total fat content
UNITS	Grams of protein Grams of fat
HOW	Supplied by BK on request or at their Web site

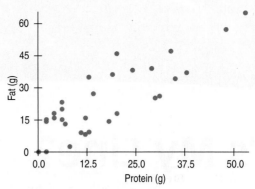

FIGURE 8.1 Total *Fat* versus *Protein* for 30 items on the BK menu. The Double Whopper is in the upper-right corner. It's extreme, but is it out of line?

"Statisticians, like artists, have the bad habit of falling in love with their models."

—George Box, famous statistician

Now we can say more. We can **model** the relationship with a line and give its equation. The equation will let us predict the fat content for any Burger King food, given its amount of protein. For the Burger King foods, we'd choose a linear model to describe the relationship between *Protein* and *Fat*. The **linear model** is just an equation of a straight line through the data. Of course, no line can go through all the points, but a linear model can summarize the general pattern with only a couple of parameters: the slope and *y*-intercept. (Sound familiar?) Like all models of the real world, the line will be wrong—wrong in the sense that it can't match reality *exactly*. But it can help us understand how the variables are associated.

Do The Math

As you can probably tell, we're going to be writing equations of lines that model what we see in a scatterplot, and then using those equations to help us understand the relationship between the two variables.

So let's review what you learned about lines in your Algebra class. We'll start off with a few equations in the familiar $y = mx + b$ form, and then look at others written as $y = a + bx$, the way you'll be seeing them in Statistics. (Those may look a bit different, but they're really the same.) For each line,

a) identify the slope,
b) identify the intercept, and
c) find the value of y for $x = 10$.

1. $y = 3x - 5$
2. $y = \frac{1}{2}x + 3$
3. $y = -2x + 33$
4. $y = 7 + 4x$

5. $y = 22 - 0.5x$
6. $y = -17 + 2x$
7. $y = 13 - 0.25x$
8. $y = -0.72 + 0.88x$

(Check your answers on page 190.)

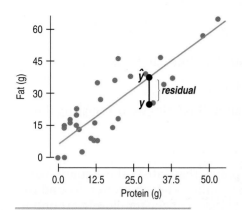

Residuals

Not only can't we draw a line through all the points in the scatterplot, the best line might not even hit *any* of the points. Then what makes it the "best" line? We want to find the line that somehow comes *closer* to all the points than any other line, so the key is finding out how far from the line each point would be.

Some of the points will be above the line and some below. For example, the line might estimate that a BK Broiler chicken sandwich with 30 grams of protein should have 36 grams of fat when, in fact, it actually has only 25 grams of fat. We call the estimate made from a model the **predicted value,** and write it as $\hat{y}$ (called *y-hat*) to distinguish it from the true value y (called, uh, y). The difference between the true value and the line's predicted value is called the **residual.** The residual tells us how far off the model's prediction is at that point. The BK Broiler chicken's residual would be $y - \hat{y} = 25 - 36 = -11$ g of fat.

> *residual = observed value −*
>   *predicted value*
> A *negative* residual means the observed value is smaller than predicted. And a *positive* residual shows that the model underestimates the true value.

To find residuals, we always subtract the predicted value from the observed one. The negative residual tells us that the actual fat content of the BK Broiler chicken is about 11 grams *less* than the model predicts for a typical Burger King menu item with 30 grams of protein.

Our challenge now is how to find the best line.

Just Checking

Suppose we've come up with the line that allows us to use the number of grams of *Protein* in a Burger King sandwich to estimate the *Fat* content.

1. For a certain sandwich, the model predicts 15 grams of fat but it actually has 17 grams. What's the residual?

2. For another sandwich the model predicts 40 grams of fat and we see that the residual is −8 grams. What's the actual fat content of this sandwich?

3. There's one Burger King sandwich that has a residual of 0. Explain what that means.

(Check your answers on page 190.)

"Best Fit" Means Least Squares

When we draw a line through a scatterplot, some residuals are positive and some negative. We can't see how well the line fits by just adding them up— the positive and negative residuals would just cancel each other out. We faced the same issue when we calculated a standard deviation to measure spread. And we deal with it the same way here: by squaring the residuals. Squaring makes them all positive so we can add them up. Squaring also emphasizes the large residuals. After all, we're more worried about points far from the line. When we add all the squared residuals together, that sum tells us how well this line fits the data—the smaller the sum, the better the fit. Different lines produce different sums, maybe bigger, maybe smaller. The **line of best fit** is the line for which the sum of the squared residuals is smallest of all; we call it the **least squares** line.

> **TI-*nspire***
> **Least Squares.** Try to minimize the sum of areas of residual squares as you drag a line across a scatterplot.

You might think that finding this line would be pretty hard. Surprisingly, it's not.

The Linear Model

"Putting a hat on it" is standard Statistics notation to indicate that something has been predicted by a model. Whenever you see a hat over a variable name or symbol, you can assume it is the predicted version of that variable or symbol (and look around for the model).

In a linear model, we use b for the slope and a for the y-intercept.

You may remember from Algebra that a straight line can be written as

$$y = mx + b.$$

We'll use this form for our linear model, but in Statistics we use slightly different notation:

$$\hat{y} = a + bx.$$

We write $\hat{y}$ (y-hat) to emphasize that the points that satisfy this equation are just our *predicted* values, not the actual data values (which scatter around the line). If the model is a good one, the data values will cluster closely around it.

Statisticians write a for the intercept and b for the slope of the line. These are called the **coefficients** of the linear model. The coefficient b is the **slope,** which tells how rapidly y changes with respect to x. The coefficient a is the **intercept,** which tells where the line hits (intercepts) the y-axis.

For the Burger King menu items, the best fit line is

$$\widehat{Fat} = 6.8 + 0.97\, Protein.$$

What does this mean? The slope, 0.97, says that an additional gram of protein is associated with an additional 0.97 grams of fat, on average. Less formally, we might say that Burger King foods pack about 0.97 grams of fat per gram of protein. Slopes are always expressed in y-units per x-unit. They tell how the y-variable changes (in its units) for a one-unit change in the x-variable. When you see a phrase like "students per teacher" or "kilobytes per second," think slope.

How about the intercept, 6.8? Algebraically, that's the value the line takes when x is zero. Here, our model predicts that even a BK item with no protein would have, on average, about 6.8 grams of fat. Is that reasonable? Well, the apple pie, with 2 grams of protein, has 14 grams of fat, so it's not impossible. But often 0 is not a plausible value for x (the year 0, a baby born weighing 0 grams, . . .). Then the intercept serves only as a starting value for our predictions, and you shouldn't try to interpret it as a meaningful value.

UNITS OF Y PER UNIT OF X

Get into the habit of identifying the units by writing down "y-units per x-unit," with the unit names put in place. You'll find it'll really helps you to Tell about the line in context.

For Example   A Linear Model for Hurricanes

In Chapter 7 we looked at the relationship between the central pressure and maximum wind speed of Atlantic hurricanes. We saw that the scatterplot was straight enough, and then found a correlation of −0.879, but we had no model to describe how these two important variables are related or to allow us to predict wind speed from pressure. Now we use technology to find the linear model. It looks like this:

$$\widehat{MaxWindSpeed} = 955.27 - 0.897 CentralPressure$$

Questions: Interpret this model. What does the slope mean in this context? Does the intercept have a meaningful interpretation?

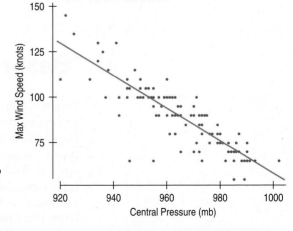

The negative slope says that as *CentralPressure* falls, *MaxWindSpeed* increases. That makes sense from our general understanding of how hurricanes work: Low central pressure pulls in moist air, driving the rotation and the resulting destructive winds. The slope's value says that, on average, the maximum wind speed increases by about 0.897 knots for every 1-millibar drop in central pressure.

It's not meaningful, however, to interpret the intercept as the wind speed predicted for a central pressure of 0—that would be a vacuum. Instead, it is merely a starting value for the model.

Do The Math

When we use a linear model to describe the relationships between two variables, the model's equation often reveals important details about that relationship. Take for example the cost of a pizza in dollars, c, modeled by the equation, $\hat{c} = 8 + 1.5t$, where t is the number of toppings. This intercept suggests that the cost of a pizza with no toppings is $8 (the intercept of the graph). And the slope reveals that the cost increases by $1.50 for each additional topping.

9. For each model described below, explain what the slope and intercept mean (in context, as always!).
   a) $\hat{c} = 25 + 2w$, where c is the cost of shipping a package (in dollars) and w is the weight of the package (in pounds).
   b) $\hat{F} = 40 + \frac{1}{4}c$, where F is temperature in degrees Fahrenheit and c is the number of chirps a cricket makes in 1 minute. (Believe it or not, this generally is true for crickets in temperatures between 55° and 100°F.)
   c) $\hat{P} = 15 + 0.1m$, where P is your cell phone plan's monthly charge (in dollars) and m is the number of minutes you used.
   d) $\hat{C} = 11 - 0.5h$, where C is how tall a candle is (in inches) after it has been burning for h hours.

10. For each of the following situations, create a linear equation that models the relationship between the variables. (Be sure to define your explanatory and response variables.)
   a) Renting a car costs $19.99 plus 25 cents for each mile you drive the car. Write an equation that models your rental cost.
   b) A football player breaks free and has an open field to the goal line 80 yards away. He can run 9 yards per second. Write an equation that models how far he has left to run for the touchdown.
   c) On Tuesdays a family restaurant hosts "Kid's Night" where an adult who purchases a full-price meal for $12 can buy children's dinners for an additional $2.50 per kid. Write an equation that models the total cost for an adult who brings several children to Kid's Night.
   d) You invest $500 to purchase supplies to bake pies. You sell each pie you bake for $8. Write an equation to model your profit.

(Check your answers on page 190.)

Just Checking

A scatterplot of house *Price* (in thousands of dollars) vs. house *Size* (in thousands of square feet) for houses sold recently in Saratoga, NY shows a relationship that is straight, with only moderate scatter and no outliers. The correlation between house *Price* and house *Size* is 0.77, and the equation of the linear model is

$$\widehat{Price} = -3.117 + 94.45\,Size.$$

4. What does the slope of 94.45 mean?

5. What are the units of the slope?

6. Your house is 2000 sq ft bigger than your neighbor's house. How much more do you expect it to be worth?

7. Is the intercept of −3.117 meaningful? Explain.

(Check your answers on page 190.)

The Least Squares Line

How do we find the actual values of slope and intercept for the least squares line? The formulas are simple. All we need are the correlation (to tell us the strength of the linear association), the standard deviations (to give us the units), and the means (to tell us where to put the line).

The slope of the line is

$$b = \frac{rs_y}{s_x}$$

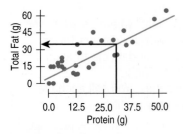

SLOPE
$$b = \frac{rs_y}{s_x}$$

Protein	Fat
$\bar{x} = 17.2$ g	$\bar{y} = 23.5$ g
$s_x = 14.0$ g	$s_y = 16.4$ g
$r = 0.83$	

INTERCEPT
$$a = \bar{y} - b\bar{x}.$$

For the Burger King foods, the slope is $b = \frac{rs_y}{s_x} = \frac{0.83 \times 16.4\,\text{g fat}}{14\,\text{g protein}} = 0.97$ grams of fat per gram of protein.

What about the intercept? If you had to predict the y-value for a data point whose x-value was average, what would you say? The best fit line always predicts $\bar{y}$ for points whose x-value is $\bar{x}$. Putting that into our equation, we find

$$\bar{y} = a + b\bar{x},$$

or, rearranging the terms,

$$a = \bar{y} - b\bar{x}.$$

So knowing the slope and the fact that the line goes through the point $(\bar{x}, \bar{y})$ tells us how to find the intercept. For the Burger King foods, that comes out to

$$a = 23.5\,\text{g fat} - 0.97\frac{\text{g fat}}{\text{g protein}} \times 17.2\,\text{g protein} = 6.8\,\text{g fat}.$$

And now we know the equation of the least squares line:

$$\widehat{fat} = 6.8 + 0.97\,protein.$$

With this linear model, it's easy to predict fat content for any menu item we want. For example, for the BK Broiler chicken sandwich with 30 grams of *protein*, we can plug in 30 grams for the amount of *protein* and see that the *predicted fat* content is

$$\widehat{fat} = 6.8 + 0.97(30) = 35.9\,\text{grams of fat}.$$

Because the BK Broiler chicken sandwich actually has only 25 grams of fat, its residual is

$$fat - \widehat{fat} = 25 - 35.9 = -10.9\,\text{g}.$$

To use a linear model, we should check the same conditions as we did for correlation: the **Quantitative Variables Condition,** the **Straight Enough Condition,** and the **Outlier Condition.**

FIGURE 8.2 Burger King menu items in their natural units with the regression line.

STEP-BY-STEP EXAMPLE   Calculating a Linear Model

Wildfires are an ongoing source of concern shared by several government agencies. In 2004, the Bureau of Land Management, Bureau of Indian Affairs, Fish and Wildlife Service, National Park Service, and USDA Forest Service spent a combined total of $890,233,000 on fire suppression, down from nearly twice that much in 2002. These government agencies join together in the National Interagency Fire Center, whose Web site (www.nifc.gov) reports statistics about wildfires.

Question: Has the annual number of wildfires been changing, on average? If so, how fast and in what way?

Plan State the problem.

I want to know how the number of wildfires in the continental United States has changed in the past two decades.

Variables Identify the variables and report the W's.

I have data giving the number of wildfires for each year (in thousands of fires) from 1982 to 2005.

✔ **Quantitative Variables Condition:** Both the number of fires and the year are quantitative.

Just as we did for correlation, check the conditions by making a picture. Never fit a linear model without looking at the scatterplot first.

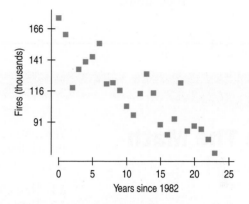

Note: It's common (and usually simpler) not to use four-digit numbers to identify years. Here we have chosen to number the years beginning in 1982, so 1982 is represented as year 0 and 2005 as year 23.

✔ **Straight Enough Condition:** The scatterplot shows a strong linear relationship with a negative association.

✔ **Outlier Condition:** No outliers are evident in the scatterplot.

Because these conditions are satisfied, it is OK to model the relationship with a line.

Mechanics Find the equation of the best fit line. Summary statistics give the building blocks of the calculation.

(We generally report summary statistics to one more digit of accuracy than the data. We do the same for intercept and predicted values, but for slopes we usually report an additional digit. Remember, though, not to round off until you finish computing an answer.)[2]

Find the slope, b.

Find the intercept, a.

Year: $\bar{x} = 11.5$ (representing 1993.5)

$s_x = 7.07$ years

Fires: $\bar{y} = 114.098$ fires

$s_y = 28.342$ fires

Correlation: $r = -0.862$

$$b = \frac{rs_y}{s_x} = \frac{-0.862(28.342)}{7.07}$$

$$= -3.4556 \text{ fires per year}$$

$$a = \bar{y} - b\bar{x} = 114.098 - (-3.4556)11.5$$

$$= 153.837$$

(continued)

---

[2] We rounded the numbers off a bit to show the steps more clearly. If you repeat these calculations yourself using a calculator or statistics program, you may get somewhat different results. When calculated with more precision, the intercept is 153.809 and the slope is −3.453.

	So the least squares line is $$\hat{y} = 153.837 - 3.4556x, \text{ or}$$ $$\widehat{Fires} = 153.837 - 3.4556\,year$$
Write the equation of the model, using meaningful variable names.	

 TELL	**Conclusion** Interpret what you have found in the context of the question. Discuss in terms of the variables and their units.	During the period from 1982 to 2005, the annual number of fires declined at an average rate of about 3,456 (3.456 thousand) fires per year. For prediction, the model uses a base estimate of 153,837 fires in 1982.

Do The Math

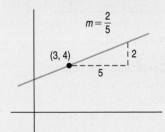

When we create linear models to fit our scatterplots, we'll always know two things: the slope of the line, and a point it goes through (the mean-mean point). Try writing the equations of these lines:

11. Through the point (3,20) with a slope of 4
12. Through the point (4,7) with a slope of −2.5
13. Through the point (80,12) with a slope of 0.2
14. Through the point (10,3.3) with a slope of −0.66

(Check your answers on page 190.)

Just Checking

8. A student collected data about the heights and weights of ten classmates, hoping to develop a model that would use a teenager's height to estimate his weight. Here are the scatterplot and the summary statistics.

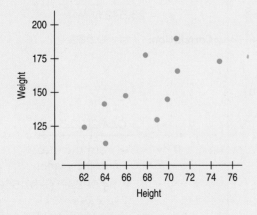

	Height (x)	Weight (y)
Mean	$\bar{x} = 68$	$\bar{y} = 150$
St. Dev.	$s_x = 4$	$s_y = 25$
Correlation	$r = 0.7$	

a) Does the scatterplot indicate it's okay to create a linear model? Explain.
b) Find the slope of the best fit line.
c) What point must lie on the line?
d) Find the intercept.

e) Write the equation of the linear model. (Use meaningful variable names.)
f) One student who was absent the day the data were collected is 60" tall. Use the model to estimate her weight.

(Check your answers on page 190.)

Correlation and the Line

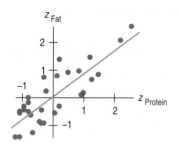

FIGURE 8.3 **The Burger King scatterplot in z-scores.**

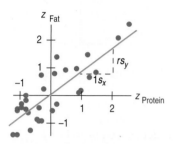

FIGURE 8.4 **Standardized Fat vs. standardized** *Protein* **with the linear model.** Each one-standard-deviation change in *Protein* results in a predicted change of *r* standard deviations in *Fat*.

Sir Francis Galton was the first to speak of "regression," although others had fit lines to data by the same method.

In Chapter 7, we learned a lot about how correlation worked by looking at a scatterplot of the standardized variables. Let's see what else standardizing can tell us. Here's a scatterplot of z_y (standardized *Fat*) vs. z_x (standardized *Protein*) for the BK items along with their least squares line. What's the slope of this line?

We know that the slope $b = \dfrac{rs_y}{s_x}$ but here we are working with standardized variables whose standard deviations are both 1. So the slope of the linear model for z_y and z_x, is $b = r$.

What about the intercept? Look at the plot. We know that $a = \bar{y} - b\bar{x}$. But standardized variables have zero means, so we just get $a = 0$; the line goes through the origin.

Wow! This line has an equation that's about as simple as we could possibly hope for:

$$\hat{z}_y = rz_x.$$

Great. It's simple, but what does it tell us? It says that if we move one standard deviation from the mean in x, we can expect to move about r standard deviations away from the mean in y. Now that we're thinking about linear models, the correlation is more than just a vague measure of strength of association: It's a great way to think about what the line tells us.

Let's be more specific. For the BK menu items, the correlation is 0.83. If we standardize both protein and fat, we can write

$$\hat{z}_{Fat} = 0.83z_{Protein}.$$

This equation tells us that for every standard deviation above (or below) the mean a menu item is in protein, we'd predict that its fat content is 0.83 standard deviations above (or below) the mean fat content.

In general, moving any number of standard deviations in x moves r times that number of standard deviations in y. Let's think some more about what that means.

Suppose you were told that a new male student was about to join the class, and you were asked to guess his height in inches. What would you guess? A safe guess would be the mean height of male students. Now suppose you are also told that this student has a grade point average (*GPA*) of 3.9—about 2 SDs above the mean *GPA*. Would that change your guess? Probably not. The correlation between *GPA* and *height* is near 0, so knowing the *GPA* value doesn't tell you anything and doesn't move your guess. (And the equation tells us that as well, since it says that we should move 0×2 SDs from the mean.)

On the other hand, suppose you were told that, measured in centimeters, the student's height was 2 SDs above the mean. There's a perfect correlation between *height in inches* and *height in centimeters*, so you'd know he's 2 SDs above mean height in inches as well. (The equation would tell us to move 1.0×2 SDs from the mean.)

What if you're told that the student is 2 SDs above the mean in *shoe size*? Now you might guess that he's taller than average, since there's a positive correlation between *height* and *shoe size*. But would you guess that he's 2 SDs above the mean? When there was no correlation, we didn't move away from the mean at all. With a perfect correlation, we moved our guess the full 2 SDs. Any correlation between these extremes should lead us to move somewhere between 0 and 2 SDs above the mean. (To be exact, the equation tells us to move $r \times 2$ standard deviations away from the mean.)

THE FIRST REGRESSION

Sir Francis Galton related the heights of sons to the heights of their fathers with a regression line. The slope of his line was less than 1. That is, sons of tall fathers were tall, but not as much above the average height as their fathers had been above their mean. Sons of short fathers were short, but generally not as far from their mean as their fathers. Galton interpreted the slope correctly as indicating a "regression" toward the mean height. He labeled the regression factor r, and called it the "index of correlation."

Notice that if x is 2 SDs above its mean, we won't ever guess more than 2 SDs away for y, since r can't be bigger than 1.0. So, each predicted y tends to be closer to its mean (in standard deviations) than its corresponding x was. This property of the linear model is called **regression to the mean,** and the line is called the **regression line.**

Just Checking

Remember the houses that sold recently in Saratoga, NY? The correlation between house *Price* and house *Size* is 0.77.

9. You go to an open house and find that the home is 1 standard deviation above the mean in size. What would you guess about its price?

10. You read an ad for a house priced 2 standard deviations below the mean. What would you guess about its size?

11. A friend tells you about a house whose size in square meters (he's European) is 1.5 standard deviations above the mean. What would you guess about its size in square feet?

12. Remember the student heights and weights? Here are the summary statistics again. In the last *Just*

Checking you used the linear model to estimate that a student 60" tall should weigh about 115 pounds. Let's try that estimate again, this time using the regression concept.

	Height (x)	Weight (y)
Mean	$\bar{x} = 68$	$\bar{y} = 150$
St. Dev.	$s_x = 4$	$s_y = 25$
Correlation	$r = 0.7$	

a) How many standard deviations below the mean is this student's height of 60"?
b) Using the regression factor r, how many standard deviations below the mean weight should we expect this student's weight to be?
c) How many pounds is that?
d) What's the predicted weight?

(Check your answers on page 190.)

Residuals Revisited

A linear model assumes that the relationship between the two variables is a perfect straight line. The residuals are the part of the data that the model misses. We can write

$$Data = Model + Residual$$

or, equivalently,

$$Residual = Data - Model.$$

Or, in symbols,

$$e = y - \hat{y}.$$

For Example Katrina's Residual

Recap: The linear model relating hurricanes' wind speeds to their central pressures was

$$\widehat{MaxWindSpeed} = 955.27 - 0.897 CentralPressure$$

Let's use this model to make predictions and see how those predictions do.

Question: Hurricane Katrina had a central pressure measured at 920 millibars. What does our regression model predict for her maximum wind speed? How good is that prediction, given that Katrina's actual wind speed was measured at 110 knots?

Substituting 920 for the central pressure in the linear model equation gives

$$\widehat{MaxWindSpeed} = 955.27 - 0.897(920) = 130.03$$

The model predicts a maximum wind speed of 130 knots for Hurricane Katrina.

The residual for this prediction is the observed value minus the predicted value:

$$110 - 130 = -20 \text{ kts.}$$

In the case of Hurricane Katrina, the model predicts a wind speed 20 knots higher than was actually observed.

Just Checking

Our linear model for Saratoga homes uses the *Size* (in thousands of square feet) to estimate the *Price* (in thousands of dollars): $\widehat{Price} = -3.117 + 94.45Size$. Suppose you're thinking of buying a home there.

13. Would you prefer to find a home with a negative or a positive residual? Explain.

14. You plan to look for a home of about 3000 square feet. How much should you expect to have to pay?

15. You find a nice home that size selling for $300,000. What's the residual?

Your linear model for the class heights and weights was $\widehat{Weight} = -149 + 4.4Height$.

16. A student is 70" tall.
   a) Estimate her weight.
   b) She actually weighs 144 pounds. What's the residual?

17. Another student is 65" tall and has a residual of 18 pounds. How much does he weigh?

(Check your answers on page 190.)

When we want to know how well the model fits, we can ask instead what the model missed. To see that, we look at the residuals. Residuals help us to see whether the model makes sense. When a regression model is appropriate, it should capture the underlying relationship. Nothing interesting should be left behind. So after we fit a regression model, we usually plot the residuals in the hope of finding . . . nothing.

Residuals plots then are a great way to check the **Straight Enough Condition**. A scatterplot of the residuals versus the *x*-values should be the most boring scatterplot you've ever seen. It shouldn't have any interesting features, like a direction or shape. It should stretch horizontally, with about the same amount of scatter throughout. It should show no bends, and it should have no outliers.

MAKE A PICTURE

To use regression, first check that
• the scatterplot is straight enough.

After you've fit the regression, make a residual plot and check that there are no obvious patterns. In particular, check that
• there are no obvious bends,
• the spread of the residuals is about the same throughout, and
• there are no obvious outliers.

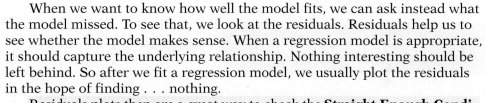

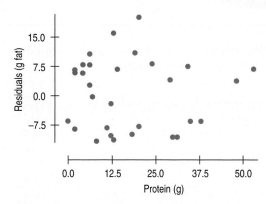

FIGURE 8.5 No problem. The residuals for the BK menu regression look appropriately boring.

What if there *is* a bend? Stick around for Chapter 9!

A Tale of Two Regressions

Our regression model for the Burger King sandwiches was $\widehat{fat} = 6.8 + 0.97\ protein$. That equation allowed us to estimate that a sandwich with 30 grams of protein would have 35.9 grams of fat. Suppose, though, that we knew the fat content and wanted to predict the amount of protein. It might seem natural to think that by solving our equation for *protein* we'd get a model for predicting *protein* from *fat*. But that doesn't work.

Our original model is $\hat{y} = a + bx$, but the new one needs to evaluate an $\hat{x}$ based on a value of y. There's no y in our original model, only $\hat{y}$, and that makes all the difference. Our model doesn't fit the BK data values perfectly, and we based it on the vertical errors the model makes in estimating y—not on horizontal errors related to x.

If we want to predict *protein* from *fat*, we need to create a new model.

Now the slope is $b = \dfrac{rs_x}{s_y} = \dfrac{(0.83)(14.0)}{16.4} = 0.709$ grams of protein per gram of fat. The equation turns out to be $\widehat{protein} = 0.54 + 0.709\ fat$, so we'd predict that a sandwich with 35.9 grams of fat should have 26.0 grams of protein—not the 30 grams that we used in the first equation.

Moral of the story: *Think*. (Where have you heard *that* before?) Decide which variable you want to use (x) to predict values for the other (y). Then find the model that does that. If, later, you want to make predictions in the other direction, you'll need to start over and create the other model from scratch. That's easy on a calculator or computer—just reverse the roles of the variables.

Protein	Fat
$\bar{x} = 17.2\,g$	$\bar{y} = 23.5\,g$
$s_x = 14.0\,g$	$s_y = 16.4\,g$
	$r = 0.83$

STEP-BY-STEP EXAMPLE   Regression

Even if you hit the fast-food joints for lunch, you should first have a good breakfast. Nutritionists, concerned about "empty calories" in breakfast cereals, recorded facts about 77 cereals, including their *Calories* per serving and *Sugar* content (in grams).

Question: How are calories and sugar content related in breakfast cereals?

THINK

Plan State the problem and determine the role of the variables.

Variables Name the variables and report the W's.

I am interested in the relationship between sugar content and calories in cereals. I'll use *Sugar* to estimate *Calories*.

✔ **Quantitative Variables Condition:** I have two quantitative variables, *Calories* and *Sugar* content per serving, measured on 77 breakfast cereals. The units of measurement are calories and grams of sugar, respectively.

Check the conditions for regression by making a picture. Never fit a regression line without looking at the scatterplot first.

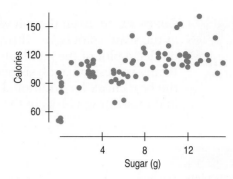

✔ **Outlier Condition:** There are no obvious outliers or groups.
✔ The **Straight Enough Condition** is satisfied; I will fit a regression model to these data.

 Mechanics If there are no clear violations of the conditions, fit a straight line model of the form $\hat{y} = a + bx$ to the data.

To use *Sugar* to estimate *Calories*, let $x = $ *Sugar* and $y = $ *Calories*

Calories
$\bar{y} = 107$ calories
$s_y = 19.5$ calories

Sugar
$\bar{x} = 7$ grams
$s_x = 4.4$ grams

Correlation
$r = 0.564$

Find the slope.

$$b = \frac{rs_y}{s_x} = \frac{0.564(19.5)}{4.4}$$
$$= 2.50 \text{ calories per gram of sugar.}$$

Find the intercept.

$a = \bar{y} - b\bar{x} = 107 - 2.50(7) = 89.5$ calories.

So the least squares line is

Write the equation, using meaningful variable names.

$$\hat{y} = 89.5 + 2.50\, x \text{ or}$$
$$\widehat{Calories} = 89.5 + 2.50\, Sugar.$$

 Conclusion Describe what the model says in words and numbers. Be sure to use the names of the variables and their units.

The key to interpreting a regression model is to start with the phrase "*b* *y*-units per *x*-unit," substituting the estimated value of the slope for *b* and the names of the respective units. The intercept is then a starting or base value.

The scatterplot shows a positive, linear relationship and no outliers. The slope of the least squares regression line suggests that cereals have about 2.50 more *Calories* per additional gram of *Sugar*.

The intercept predicts that sugarless cereals would average about 89.5 calories.

(continued)

Check Again Even though we looked at the scatterplot *before* fitting a linear model, a plot of the residuals is essential to any regression analysis because it is the best check for additional patterns and interesting quirks in the data.

THINK

AGAIN

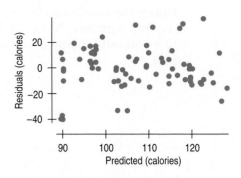

TI-*nspire*

Residuals plots. See how the residuals plot changes as you drag points around in a scatterplot.

The residuals show random scatter and a shapeless form so our linear model appears to be appropriate.

TI Tips — Regression lines and residuals plots

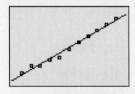

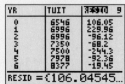

By now you will not be surprised to learn that your calculator can do it all: scatterplot, regression line, and residuals plot. Let's try it using the Arizona State tuition data from the last chapter. (TI Tips, page 142) You should still have those saved in lists named **YR** and **TUIT**. First, recreate the scatterplot.

1. Find the equation of the regression line.

Actually, you already found the line when you used the calculator to get the correlation. But this time we'll go a step further so that we can display the line on our scatterplot. We want to tell the calculator to do the regression and save the equation of the model as a graphing variable.

- Under **STAT CALC** choose **LinReg(a+bx)**.
- Specify that *x* and *y* are **YR** and **TUIT**, as before, but . . .
- Now add a comma and one more specification. Press **VARS**, go to the **Y-VARS** menu, choose **1:Function**, and finally(!) choose **Y1**.
- Hit **ENTER**.

There's the equation. The calculator tells you that the regression line is $\widehat{tuit} = 6440 + 326year$. Can you explain what the slope and intercept mean?

2. Add the line to the plot.

When you entered this command, the calculator automatically saved the equation as **Y1**. Just hit **GRAPH** to see the line drawn across your scatterplot.

3. Check the residuals.

Remember, you are not finished until you check to see if a linear model is appropriate. That means you need to see if the residuals appear to be randomly distributed. To do that, you need to look at the residuals plot.

This is made easy by the fact that the calculator has already placed the residuals in a list named **RESID**. Want to see them? Go to **STAT EDIT** and look through the lists. (If **RESID** is not already there, go to the first blank list and import the name **RESID** from your **LIST NAMES** menu. The residuals should appear.) Every time you have the calculator compute a regression analysis, it will automatically save this list of residuals for you.

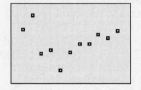

4. Now create the residuals plot.

- Set up STAT PLOT Plot2 as a scatterplot with Xlist:YR and Ylist:RESID.
- Before you try to see the plot, go to the Y= screen. By moving the cursor around and hitting ENTER in the appropriate places you can turn off the regression line and Plot1, and turn on Plot2.
- ZoomStat will now graph the residuals plot.

Uh-oh! See the curve? The residuals are high at both ends, low in the middle. Looks like a linear model may not be appropriate after all. Notice that the residuals plot makes the curvature much clearer than the original scatterplot did.

Moral: Always check the residuals plot!

What now? The next chapter provides techniques for dealing with data like these.

Just Checking

Let's look at the student height/weight relationship one last time. Earlier you created the linear model $\widehat{Weight} = -149 + 4.4Height$ to use students' heights to estimate their weights.

    Now suppose we know a student weighs 200 pounds and want to estimate his height. That earlier model won't work, because it uses *actual heights* to find *estimated weights*. Now we know the *actual weight* and want to *estimate height*. We need a new model.

18. In this new model, which is the explanatory variable and which is the response variable?

19. Create the appropriate model. You may use the summary statistics, or enter the actual data into your calculator to find the equation of the regression line. Either way, be sure to think carefully about which variable is *x* and which is *y*.

20. How tall do you estimate the 200-pound student is?

Height (inches)		Weight (pounds)
66		147
62		124
71		189
64		141
75		172
70		144
64		112
71		165
69		129
68		177
68	**mean**	150
4	**SD**	25
	$r = 0.7$	

(Check your answers on page 190.)

Beware of Extrapolation

Linear models can be very useful, but don't try to push them too far. Predictions extending far beyond the data we used to build the model can be untrustworthy. The farther a new *x*-value is from $\bar{x}$, the less trust we should place in the predicted value. Once we venture into new *x* territory, such a prediction is called an **extrapolation**. Extrapolations are dubious because they require the very questionable assumption that nothing about the relationship between *x* and *y* changes even at extreme values of *x* and beyond.

    Extrapolations can get us into deep trouble. When the *x*-variable is *Time*, extrapolation becomes an attempt to peer into the future. People have always

"Prediction is difficult, especially about the future."

—Niels Bohr, Danish physicist

wanted to see into the future, and it doesn't take a crystal ball to foresee that they always will. In the past, seers, oracles, and wizards were called on to predict the future. Today mediums, fortune-tellers, and Tarot card readers still find many customers.

Those with a more scientific outlook may use a linear model as their digital crystal ball. Linear models are based on the *x*-values of the data at hand and cannot be trusted beyond that span. Some physical phenomena do exhibit a kind of "inertia" that allows us to guess their behavior will continue, but we can't count on that in phenomena such as stock prices, sales figures, hurricane tracks, or public opinion.

Extrapolating from current trends is so tempting that even professional forecasters make this mistake, and sometimes the errors are striking.

In the period from 1982 to 1998 oil prices went down so much that by 1998, prices (adjusted for inflation) were the lowest they'd been since before World War II. Of course, these decreases clearly couldn't continue, or oil would be free by now. The Energy Information Administration offered two *different* 20-year forecasts for oil prices after 1998, and both called for relatively modest increases in oil prices. So, how accurate have these forecasts been? Here's a timeplot of the EIA's predictions and the actual prices (in 2005 dollars).

Oops! They seemed to have missed the sharp run-up in oil prices in the past few years.

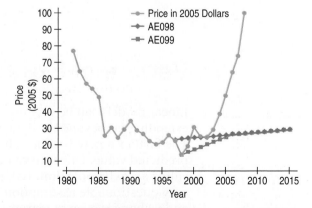

FIGURE 8.6 Predicting the future? Here are the EIA forecasts with the actual prices from 1981 to 2008. Neither forecast predicted the sharp run-up in the past few years.

Where do you think oil prices will go in the next decade? *Your* guess may be as good as anyone's!

Of course, knowing that extrapolation is dangerous doesn't stop people. The temptation to see into the future is hard to resist. So our more realistic advice is this:

> *If you must extrapolate into the future, at least don't believe that the prediction will come true.*

Lurking Variables and Causation

One common way to interpret a regression slope is to say that "a change of 1 unit in *x* results in a change of *b* units in *y*." This way of saying things encourages causal thinking. Beware.

In Chapter 7, we tried to make it clear that no matter how strong the correlation is between two variables, there's no simple way to show that one variable causes the other. Putting a regression line through a cloud of points just increases the temptation to think and to say that the *x*-variable *causes* the *y*-variable. So we'll say it again: No matter how strong the association, no matter how straight the line, there is no way to conclude from a regression alone that one variable *causes* the other. Unless you design an experiment, there is no way to be sure that a **lurking variable** is not the source of any apparent association.

Here's an example: The scatterplot shows the *Life Expectancy* (average of men and women, in years) for each of 41 countries of the world, plotted against the square root[3] of the number of *Doctors* per person in the country.

The strong positive association ($r = 0.79$) seems to confirm our hunch that more *Doctors* per person improves healthcare, leading to longer lifetimes and a greater *Life Expectancy*. The strength of the association would *seem* to argue that we should send more doctors to developing countries to increase life expectancy.

But do doctors *cause* greater life expectancy? Perhaps, but there may be another explanation for the association.

The next similar-looking scatterplot's *x*-variable is the square root of the number of *Televisions* per person in each country. The positive association in this scatterplot is even *stronger* than the association in the previous plot ($r = 0.87$). Should we conclude that increasing the number of TVs actually extends lifetimes? If so, we should send TVs instead of doctors to developing countries. Not only is the correlation with life expectancy higher, but TVs are much cheaper than doctors.

What's wrong with this reasoning? Maybe there's a lurking variable here. Countries with higher standards of living have both longer life expectancies *and* more doctors (and more TVs). Could higher living standards cause changes in the other variables? If so, then improving living standards might be expected to prolong lives, increase the number of doctors, and increase the number of TVs.

From this example, you can see how easy it is to mistakenly fall into the causation trap. For all we know, doctors (or TVs!) *do* increase life expectancy. But we can't tell that from data like these, no matter how much we'd like to. Resist the temptation to conclude that *x* causes *y* from a regression, no matter how obvious that conclusion seems to you.

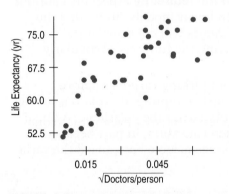

FIGURE 8.7 The relationship between *Life Expectancy* (years) and availability of *Doctors* (measured as √doctors per person) for countries of the world is strong, positive, and linear.

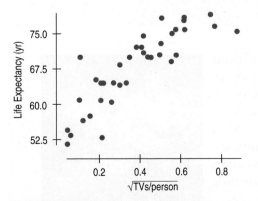

FIGURE 8.8 To increase life expectancy, don't send doctors, send TVs; they're cheaper and more fun. Or maybe that's not the right interpretation of this scatterplot of *life expectancy* against availability of TVs (as √TVs per person).

---

[3]Don't worry about the square root thing. It just makes the relationship satisfy the Straight Enough Condition.

WHAT CAN GO WRONG?

There are many ways in which data that appear at first to be good candidates for regression analysis may be unsuitable. And there are ways that people use regression that can lead them astray.

- **Make sure the relationship is straight.** Check the Straight Enough Condition. Always examine the residuals for evidence that the Linearity Assumption has failed. It's often easier to see deviations from a straight line in the residuals plot than in the scatterplot of the original data.

- **Don't fit a straight line to a nonlinear relationship.** Linear regression is suited only to relationships that are, well, *linear*. We'll come back to this topic in Chapter 9.

- **Don't extrapolate beyond the data.** A linear model will often do a reasonable job of summarizing a relationship for the observed *x*-values. But beware of predicting *y*-values for *x*-values that lie outside the range of the original data. The model may no longer hold there, so such *extrapolations* too far from the data are dangerous.

- **Beware especially of extrapolating into the future!** Be especially cautious about extrapolating into the future with linear models. To predict the future, you must assume that future changes will continue at the same rate you've observed in the past. Predicting the future is particularly tempting and particularly dangerous.

- **Beware of lurking variables.** Think about lurking variables before interpreting a linear model. It's particularly tempting to explain a strong regression by thinking that the *x*-variable *causes* the *y*-variable. A linear model alone can never demonstrate such causation, in part because it cannot eliminate the chance that a lurking variable has caused the variation in both *x* and *y*.

REGRESSION IN YOUR WORLD

Under the rules of the Federation of International Football Associations (FIFA), international soccer matches may not be played at altitudes above 2500 meters (about 8200 feet), in the fear that the thinner air gives an advantage to teams used to playing at high altitudes. But is that true? University at Oxford researcher Patrick McSherry investigated. He looked at the scores of 1460 soccer games played in ten South American countries. By taking into account each team's overall winning percentage and goals both scored and allowed, he was able to calculate the effect of altitude differences on home and away outcomes. What did he find?

High Altitude Soccer Teams Have Significant Advantage

ScienceDaily (Jan. 4, 2008) — Soccer teams from high altitude countries have a significant advantage when playing at both low and high altitudes, finds a study in the . . . British Medical Journal. Altitude difference had a significant negative impact on performance. High altitude teams scored more and conceded fewer goals as altitude difference increased. Each additional 1,000m of altitude difference increased the goal difference by about half of a goal.

http://www.sciencedaily.com/releases/2007/12/071221094837.htm

WHAT HAVE WE LEARNED?

We've learned that when the relationship between quantitative variables is fairly straight, a linear model can help summarize that relationship and give us insights about it:

▶ The regression (best fit) line doesn't pass through all the points, but it is the best compromise in the sense that the sum of squares of the residuals is the smallest possible.

We've learned several things the correlation, *r*, tells us about the regression:

▶ The slope of the line is based on the correlation, adjusted for the units of *x* and *y*:

$$b = \frac{rs_y}{s_x}$$

We've learned to interpret that slope in context:

▶ The slope describes predicted changes in *y*-units per one unit change in *x*.

▶ For each SD of *x* that we are away from the *x* mean, we expect to be *r* SDs of *y* away from the *y* mean.

▶ Because *r* is always between −1 and +1, each predicted *y* is fewer SDs away from its mean than the corresponding *x* was, a phenomenon called regression to the mean.

The residuals also reveal how well the model works:

▶ If a plot of residuals against predicted values shows a pattern, we should re-examine the data to see why.

Of course, the linear model makes no sense unless the **Linearity Assumption** is satisfied. We check the **Straight Enough Condition** and **Outlier Condition** with a scatterplot, as we did for correlation, and also with a plot of residuals.

And we've learned that even a good regression doesn't mean we should believe that the model says more than it really does.

▶ Extrapolation far from $\bar{x}$ can lead to silly and useless predictions.

▶ Even an *r* near ±1.0 doesn't indicate that *x* causes *y* (or the other way around). Watch out for lurking variables that may affect both *x* and *y*.

Terms

Model	An equation or formula that simplifies and represents reality.
Linear model	A linear model is an equation of a line. To interpret a linear model, we need to know the variables (along with their W's) and their units.
Predicted value	The value of $\hat{y}$ found for a given *x*-value in the data. A predicted value is found by substituting the *x*-value in the regression equation. The predicted values are the values on the fitted line; the points $(x, \hat{y})$ all lie exactly on the fitted line.

Residuals	Residuals are the differences between data values and the corresponding values predicted by the regression model—or, more generally, values predicted by any model.

$$\text{Residual} = \text{observed value} - \text{predicted value} = e = y - \hat{y}$$

Least squares	The least squares criterion specifies the unique line that minimizes the variance of the residuals or, equivalently, the sum of the squared residuals.
Regression line (Line of best fit)	The particular linear equation

$$\hat{y} = a + bx$$

that satisfies the least squares criterion is called the least squares regression line. Casually, we often just call it the regression line, or the line of best fit.

Slope	The slope, b, gives a value in "y-units per x-unit." Changes of one unit in x are associated with changes of b units in predicted values of y. The slope can be found by

$$b = \frac{rs_y}{s_x}.$$

Intercept	The intercept, a, gives a starting value in y-units. It's the $\hat{y}$-value when x is 0. You can find it from $a = \bar{y} - b\bar{x}$.
Regression to the mean	Because the correlation is always less than 1.0 in magnitude, each predicted $\hat{y}$ tends to be fewer standard deviations from its mean than its corresponding x was from its mean.
Extrapolation	Although linear models provide an easy way to predict values of y for a given value of x, it is unsafe to predict for values of x far from the ones used to find the linear model equation. Such extrapolation may pretend to see into the future, but the predictions should not be trusted.
Lurking variable	A variable that is not explicitly part of a model but affects the way the variables in the model appear to be related. Because we can never be certain that observational data are not hiding a lurking variable that influences both x and y, it is never safe to conclude that a linear model demonstrates a causal relationship, no matter how strong the linear association.

Skills

▶ Be able to identify response (y) and explanatory (x) variables in context.

▶ Understand how a linear equation summarizes the relationship between two variables.

▶ Recognize when a regression model should be used to summarize a linear relationship between two quantitative variables.

▶ Know how to examine your data for violations of the **Straight Enough Condition** that would make it inappropriate to use a linear model.

▶ Know that residuals are the differences between the data values and the corresponding values predicted by the line and that the *least squares criterion* finds the line that minimizes the sum of the squared residuals.

▶ Know how to use a plot of residuals to check the **Straight Enough Condition** and the **Outlier Condition.**

▶ Know the danger of extrapolating beyond the range of the x-values used to find the linear model, especially when the extrapolation tries to predict into the future.

▶ Look for lurking variables whenever you consider the association between two variables. Understand that a strong association does not mean that the variables are causally related.

▶ Know how to find a regression equation from the summary statistics for each variable and the correlation between the variables.

▶ Know how to find a regression equation using your calculator or statistics software and how to find the slope and intercept values in the regression output.

(continued)

▶ Know how to use a linear model to predict a value of *y* for a given *x*.

▶ Know how to compute the residual for each data value and how to plot the residuals.

▶ Be able to write a sentence explaining what a linear equation says about the relationship between *y* and *x*, basing it on the fact that the slope is given in *y*-units per *x*-unit.

▶ Be able to describe a prediction made from a regression model, relating the predicted value to the specified *x*-value.

▶ Include appropriate cautions about extrapolation when reporting predictions from a linear model.

▶ Discuss possible lurking variables.

REGRESSION ON THE COMPUTER

All statistics packages make a table of results for regression. These tables may differ slightly from one package to another, but all are essentially the same—and all include much more than we need to know for now.

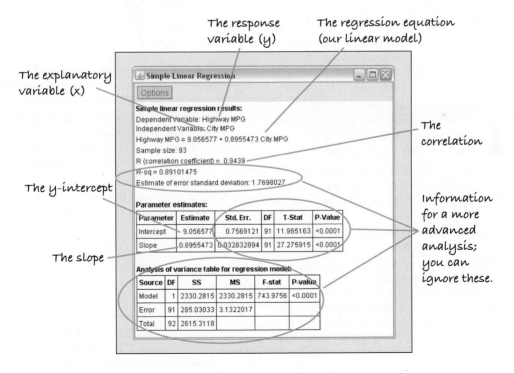

The slope and intercept coefficient are given in a table such as this one. Usually the slope is labeled with the name of the *x*-variable, and the intercept is labeled "Intercept" or "Constant." So the regression equation shown here is

$$\widehat{Highway} = 9.057 + 0.896City$$

It is not unusual for statistics packages to give many more digits of the estimated slope and intercept than could possibly be estimated from the data. Ordinarily, you should round most of the reported numbers to one digit more than the precision of the data, and the slope to two more.

EXERCISES

A

1. **Cereals.** For many people, breakfast cereal is an important source of fiber in their diets. Cereals also contain potassium, a mineral shown to be associated with maintaining a healthy blood pressure. An analysis of the amount of fiber (in grams) and the potassium content (in milligrams) in servings of 77 breakfast cereals produced the linear model $\widehat{Potassium} = 38 + 27Fiber$. If your cereal provides 9 grams of fiber per serving, how much potassium does the model estimate you will get?

2. **Horsepower.** In Chapter 7's Exercise 31 we examined the relationship between the fuel economy (mpg) and horsepower for 15 models of cars. Further analysis produces the linear model $\widehat{mpg} = 46.87 - 0.084HP$. If the car you are thinking of buying has a 200-horsepower engine, what does this model suggest your gas mileage would be?

3. **More cereal.** Exercise 1 describes a regression model that estimates a cereal's potassium content from the amount of fiber it contains. In this context, what does it mean to say that a cereal has a negative residual?

4. **Horsepower again.** Exercise 2 describes a regression model that uses a car's horsepower to estimate its fuel economy. In this context, what does it mean to say that a certain car has a positive residual?

5. **Another bowl.** In Exercise 1, the regression model $\widehat{Potassium} = 38 + 27Fiber$ relates fiber (in grams) and potassium content (in milligrams) in servings of breakfast cereals. Explain what the slope means.

6. **More horsepower.** In Exercise 2, the regression model $\widehat{mpg} = 46.87 - 0.084HP$ relates cars' horsepower to their fuel economy (in mpg). Explain what the slope means.

7. **Human Development Index.** The United Nations Development Programme (UNDP) uses the Human Development Index (HDI) in an attempt to summarize in one number the progress in health, education, and economics of a country. The gross domestic product per capita (GDPPC), by contrast, is often used to summarize the *overall* economic strength of a country. Is the HDI related to the GDPPC? Here is a scatterplot of *HDI* against *GDPPC*.

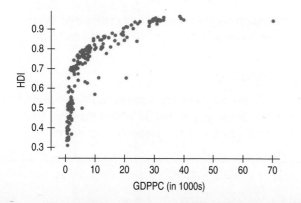

a) Explain why fitting a linear model to these data might be misleading.
b) If you fit a linear model to the data, what do you think a scatterplot of residuals will look like?

8. **HDI revisited.** The United Nations Development Programme (UNDP) uses the Human Development Index (HDI) in an attempt to summarize in one number the progress in health, education, and economics of a country. The number of cell phone subscribers per 1000 people is positively associated with economic progress in a country. Can the number of cell phone subscribers be used to predict the HDI? Here is a scatterplot of HDI against cell phone subscribers:

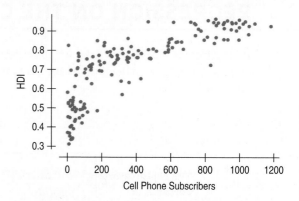

a) Explain why fitting a linear model to these data might be misleading.
b) If you fit a linear model to the data, what do you think a scatterplot of residuals will look like?

9. **Residuals.** Tell what each of the residual plots below indicates about the appropriateness of the linear model that was fit to the data.

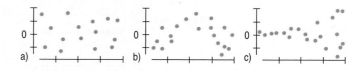

10. **Residuals.** Tell what each of the residual plots below indicates about the appropriateness of the linear model that was fit to the data.

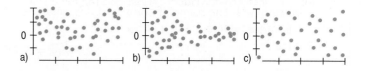

11. **Real estate.** A random sample of records of sales of homes from Feb. 15 to Apr. 30, 1993, from the files maintained by the Albuquerque Board of Realtors

gives the *Price* and *Size* (in square feet) of 117 homes. They hope to predict *Price* (in thousands of dollars) from *Size*. The residuals plot indicated that a linear model is appropriate.
a) What are the variables and units in this regression?
b) What units does the slope have?
c) Do you think the slope is positive or negative? Explain.

12. Roller coaster. People who responded to a July 2004 Discovery Channel poll named the 10 best roller coasters in the United States. A table in the last chapter's exercises shows the length of the initial drop (in feet) and the duration of the ride (in seconds). We want to predict *Duration* from *Drop*.
a) What are the variables and units in this regression?
b) What units does the slope have?
c) Do you think the slope is positive or negative? Explain.

13. Real estate redux. For *Price* and *Size* of homes in Albuquerque as described in Exercise 11, the correlation is $r = 0.845$.
a) What would you predict about the *Price* of a home 1 standard deviation above average in *Size*?
b) What would you predict about the *Price* of a home 2 standard deviations below average in *Size*?

14. Another ride. The correlation between the *Duration* of a roller coaster ride and the height of its initial *Drop*, described in Exercise 12, is $r = 0.35$.
a) What would you predict about the *Duration* of the ride on a coaster whose initial *Drop* was 1 standard deviation below the mean *Drop*?
b) What would you predict about the *Duration* of the ride on a coaster whose initial *Drop* was 3 standard deviations above the mean *Drop*?

15. More real estate. Consider the Albuquerque home sales from Exercise 11 again. The regression analysis gives the model $\widehat{Price} = 47.82 + 0.061\,Size$.
a) Explain what the slope of the line says about housing prices and house size.
b) What price would you predict for a 3000-square-foot house in this market?
c) A real estate agent shows a potential buyer a 1200-square-foot home, saying that the asking price is $6000 less than what one would expect to pay for a house of this size. What is the asking price, and what is the $6000 called?

16. Last ride. Consider the roller coasters described in Exercise 12 again. The regression analysis gives the model $\widehat{Duration} = 91.033 + 0.242\,Drop$.
a) Explain what the slope of the line says about how long a roller coaster ride may last and the height of the coaster.
b) A new roller coaster advertises an initial drop of 200 feet. How long would you predict the rides last?
c) Another coaster with a 150-foot initial drop advertises a 2-minute ride. Is this longer or shorter than you'd expect? By how much? What's that called?

17. Regression equations. Fill in the missing information in the following table.

	$\bar{x}$	s_x	$\bar{y}$	s_y	r	$\hat{y} = a + bx$
a)	10	2	20	3	0.5	
b)	2	0.06	7.2	1.2	−0.4	
c)	12	6			−0.8	$\hat{y} = 200 − 4x$
d)	2.5	1.2		100		$\hat{y} = −100 + 50x$

18. More regression equations. Fill in the missing information in the following table.

	$\bar{x}$	s_x	$\bar{y}$	s_y	r	$\hat{y} = a + bx$
a)	30	4	18	6	−0.2	
b)	100	18	60	10	0.9	
c)		0.8	50	15		$\hat{y} = −10 + 15x$
d)			18	4	−0.6	$\hat{y} = 30 − 2x$

B

19. What's the cause? Suppose a researcher studying health issues measures blood pressure and the percentage of body fat for several adult males and finds a strong positive association. Describe three different possible cause-and-effect relationships that might be present.

20. What's the effect? A researcher studying violent behavior in elementary school children asks the children's parents how much time each child spends playing computer games and has their teachers rate each child on the level of aggressiveness they display while playing with other children. Suppose that the researcher finds a moderately strong positive correlation. Describe three different possible cause-and-effect explanations for this relationship.

21. Cigarettes. Is the nicotine content of a cigarette related to the "tars"? A collection of data (in milligrams) on 29 cigarettes produced the scatterplot and model shown:

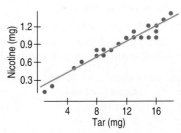

a) Do you think a linear model is appropriate here? Explain.
b) The correlation between *Tar* and *Nicotine* is $r = 0.96$. What would you predict about the average *Nicotine* content of cigarettes that are 2 standard deviations below average in *Tar* content?
c) If a cigarette is 1 standard deviation above average in *Nicotine* content, what do you suspect is true about its *Tar* content?

22. Attendance 2006. In the previous chapter you looked at the relationship between the number of wins by American League baseball teams and the average attendance at their home games for the 2006 season. Here are the scatterplot and the regression line:

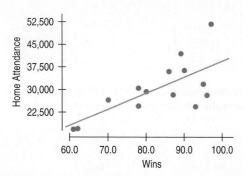

a) Do you think a linear model is appropriate here? Explain.
b) The point in the upper right of the plots is the New York Yankees. What can you say about the residual for the Yankees?
c) The correlation between *Wins* and *Average Attendance* is $r = 0.70$. What would you predict about the *Average Attendance* for a team that is 2 standard deviations above average in *Wins*?
d) If a team is 1 standard deviation below average in attendance, what would you predict about the number of games the team has won?

23. Last cigarette. Here is the regression analysis of tar and nicotine content of the cigarettes in Exercise 21.

```
Dependent variable is: nicotine
Variable      Coefficient
Constant      0.154030
Tar           0.065052
```

a) Write the equation of the regression line.
b) Estimate the *Nicotine* content of cigarettes with 4 milligrams of *Tar*.
c) Interpret the meaning of the slope of the regression line in this context.
d) What does the *y*-intercept mean?
e) If a new brand of cigarette contains 7 milligrams of tar and a nicotine level whose residual is –0.5 mg, what is the nicotine content?

24. Last inning 2006. Here is the regression analysis for average attendance and games won by American League baseball teams, seen in Exercise 22.

```
Dependent variable is: Home Attendance
Variable      Coefficient
Constant      −14364.5
Wins          538.915
```

a) Write the equation of the regression line.
b) Estimate the *Average Attendance* for a team with 50 *Wins*.
c) Interpret the meaning of the slope of the regression line in this context.
d) In general, what would a negative residual mean in this context?
e) The St. Louis Cardinals, the 2006 World Champions, are not included in these data because they are a

National League team. During the 2006 regular season, the Cardinals won 83 games and averaged 42,588 fans at their home games. Calculate the residual for this team, and explain what it means.

25. Jumps 2004. How are Olympic performances in various events related? The plot shows winning long-jump and high-jump distances, in inches, for the Summer Olympics from 1912 through 2004.

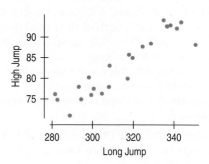

Here are the summary statistics:

Event	Mean	StdDev
Long Jump	316.04	20.85
High Jump	83.85	7.46
Correlation = 0.925		

a) Do the assumptions and conditions for linear regression appear to be met? Explain.
b) Write the equation of the line of regression for estimating *High Jump* from *Long Jump*.
c) Interpret the slope of the line.
d) In a year when the long jump is 350 inches, what high jump would you predict?

26. Online clothes. An online clothing retailer keeps track of its customers' purchases. For those customers who signed up for the company's credit card, the company also has information on the customer's *Income*. A random sample of 500 of these customers shows the following scatterplot of *Total Yearly Purchases* by *Income*:

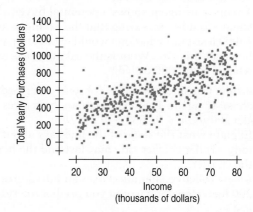

The correlation between *Total Yearly Purchases* and *Income* is 0.722. Summary statistics for the two variables are:

	Mean	SD
Income	$50,343.40	$16,952.50
Total Yearly Purchase	$572.52	$253.62

a) Do the assumptions and conditions for linear regression appear to be met? Explain.
b) What is the linear regression equation for predicting *Total Yearly Purchase* from *Income*?
c) Interpret the slope of the line.
d) What is the predicted average *Total Yearly Purchase* for someone with a yearly *Income* of $20,000? For someone with an annual *Income* of $80,000?

27. Used cars 2007. Classified ads in the *Ithaca Journal* offered several used Toyota Corollas for sale. Listed below are the ages of the cars and the advertised prices.

Age (yr)	Price Advertised ($)
1	13,990
1	13,495
3	12,999
4	9500
4	10,495
5	8995
5	9495
6	6999
7	6950
7	7850
8	6999
8	5995
10	4950
10	4495
13	2850

a) Make a scatterplot for these data.
b) Find the equation of the regression line.
c) Do you think a linear model is appropriate?
d) Explain the meaning of the slope of the line.
e) Explain the meaning of the *y*-intercept of the line.
f) You have a chance to buy one of two cars. They are about the same age and appear to be in equally good condition. Would you rather buy the one with a positive residual or the one with a negative residual? Explain.
g) You see a "For Sale" sign on a 10-year-old Corolla stating the asking price as $3500. What is the residual?
h) Would this regression model be useful in establishing a fair price for a 20-year-old car? Explain.

28. Birthrates 2005. The table shows the number of live births per 1000 women aged 15–44 years in the United States, starting in 1965. (National Center for Health Statistics, www.cdc.gov/nchs/)

Year	1965	1970	1975	1980	1985	1990	1995	2000	2005
Rate	19.4	18.4	14.8	15.9	15.6	16.4	14.8	14.4	14.0

a) Make a scatterplot and describe the general trend in *Birthrates*. (Enter *Year* as years since 1900: 65, 70, 75, etc.)
b) Find the equation of the regression line.
c) Check to see if the line is an appropriate model. Explain.
d) Interpret the slope of the line.
e) Is the intercept meaningful? Explain.
f) In 1978 the birthrate was actually 15.0. What's the residual?
g) Since 2005 residuals have been positive. What does that mean?
h) Predict the *Birthrate* for 2025. Comment on your faith in this prediction.

29. Body fat. It is difficult to determine a person's body fat percentage accurately without immersing him or her in water. Researchers hoping to find ways to make a good estimate immersed 20 male subjects, then measured their waists and recorded their weights.

Waist (in.)	Weight (lb)	Body Fat (%)	Waist (in.)	Weight (lb)	Body Fat (%)
32	175	6	33	188	10
36	181	21	40	240	20
38	200	15	36	175	22
33	159	6	32	168	9
39	196	22	44	246	38
40	192	31	33	160	10
41	205	32	41	215	27
35	173	21	34	159	12
38	187	25	34	146	10
38	188	30	44	219	28

a) Create a model to predict *%Body Fat* from *Weight*.
b) Do you think a linear model is appropriate? Explain.
c) Interpret the slope of your model.

30. Body fat again. Would a model that uses the person's *Waist* size be able to predict the *%Body Fat* accurately? Using the data in Exercise 29, create and analyze that model.

31. Burgers. In the last chapter, you examined the association between the amounts of *Fat* and *Calories* in fast-food hamburgers. Here are the data:

Fat (g)	19	31	34	35	39	39	43
Calories	410	580	590	570	640	680	660

a) Create a scatterplot of *Calories* vs. *Fat*.
b) Write the equation of the line of regression.
c) Use the residuals plot to explain whether your linear model is appropriate.
d) Explain the meaning of the *y*-intercept of the line.
e) Explain the meaning of the slope of the line.
f) A new burger containing 28 grams of fat is introduced. According to this model, its residual for calories is +33. How many calories does the burger have?

T **32. Modeling manatees 2005.** Marine biologists warn that the growing number of powerboats registered in Florida threatens the existence of manatees. Here again are the data we saw in Chapter 7, Exercise 32.

Year	Manatees Killed	Powerboat Registrations (in 1000s)
1982	13	447
1983	21	460
1984	24	481
1985	16	498
1986	24	513
1987	20	512
1988	15	527
1989	34	559
1990	33	585
1992	33	614
1993	39	646
1994	43	675
1995	50	711
1996	47	719
1997	53	716
1998	38	716
1999	35	716
2000	49	735
2001	81	860
2002	95	923
2003	73	940
2004	69	946
2005	79	974

a) Create a linear model of the association between *Manatee Deaths* and *Powerboat Registrations*.
b) Use the residuals plot to see if your model is appropriate.
c) Interpret the slope of your model.
d) Interpret the *y*-intercept of your model.
e) Find your model the residual for the high number of manatee deaths in 2005.
f) Which is better for the manatees, positive residuals or negative residuals? Explain.
g) A boating organization estimates that next year there will be 1,200,000 powerboats registered. Use your model to predict the number of manatees that will be killed. Why don't you trust this prediction?

33. A second helping of burgers. In Exercise 31 you created a model that can estimate the number of *Calories* in a burger when the *Fat* content is known.
a) Explain why you cannot use that model to estimate the fat content of a burger with 600 calories.
b) Using an appropriate model, estimate the fat content of a burger with 600 calories.

34. Manatees again. While listening to the news you hear a report that this year 88 manatees were killed.
a) Explain why you can't use the model you created in Exercise 32 to estimate the number of powerboats registered this year.
b) Create the proper model and make that estimate.

C

35. Misinterpretations. A Biology student used data from many species to create a regression model that uses a bird's *Height* when perched for predicting its *Wingspan*. Assuming the calculations were done correctly, explain what is wrong with his interpretations:
a) If a bird grows one inch taller, its wings will get 1.8 inches longer.
b) A bird 10 inches tall will have a wingspan of 17 inches.

36. More misinterpretations. A Sociology student investigated the association between a country's *Literacy Rate* and *Life Expectancy*, then drew the conclusions listed below. Explain why each statement is incorrect. (Assume that all the calculations were done properly.)
a) People living in countries where the literacy rate is 88% have a life expectancy of 68.3 years.
b) The slope of the line shows that an increase of 5% in *Literacy Rate* will produce a 2-year improvement in *Life Expectancy*.

37. ESP. People who claim to "have ESP" participate in a screening test in which they have to guess which of several images someone is thinking of. You and a friend both took the test. You scored 2 standard deviations above the mean, and your friend scored 1 standard deviation below the mean. The researchers offer everyone the opportunity to take a retest.
a) Should you choose to take this retest? Explain.
b) Now explain to your friend what his decision should be and why.

38. SI jinx. Players in any sport who are having great seasons, turning in performances that are much better than anyone might have anticipated, often are pictured on the cover of *Sports Illustrated*. Frequently, their performances then falter somewhat, leading some athletes to believe in a "*Sports Illustrated* jinx." Similarly, it is common for phenomenal rookies to have less stellar second seasons—the so-called "sophomore slump." While fans, athletes, and analysts have proposed many theories about what leads to such declines, a statistician might offer a simpler (statistical) explanation. Explain.

39. Heating. After keeping track of his heating expenses for several winters, a homeowner believes he can estimate the monthly cost from the average daily Fahrenheit temperature by using the model $\widehat{Cost} = 133 - 2.13$ *Temp*. Here is the residuals plot for his data:

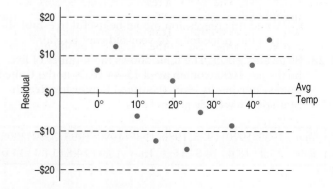

a) Interpret the slope of the line in this context.

b) Interpret the *y*-intercept of the line in this context.

c) During months when the temperature stays around freezing, would you expect cost predictions based on this model to be accurate, too low, or too high? Explain.

d) What heating cost does the model predict for a month that averages 10°?

e) During one of the months on which the model was based, the temperature did average 10°. What were the actual heating costs for that month?

f) Should the homeowner use this model? Explain.

g) Would this model be more successful if the temperature were expressed in degrees Celsius? Explain.

40. Speed. How does the speed at which you drive affect your fuel economy? To find out, researchers drove a compact car for 200 miles at speeds ranging from 35 to 75 miles per hour. From their data, they created the model $\overline{Fuel\ Efficiency} = 32 - 0.1\ Speed$ and created this residual plot:

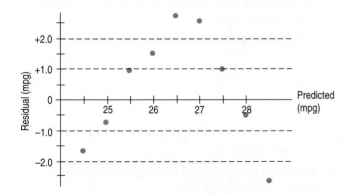

a) Interpret the slope of this line in context.

b) Explain why it's silly to attach any meaning to the *y*-intercept.

c) When this model predicts high *Fuel Efficiency*, what can you say about those predictions?

d) What *Fuel Efficiency* does the model predict when the car is driven at 50 mph?

e) What was the actual *Fuel Efficiency* when the car was driven at 45 mph?

f) Do you think there appears to be a strong association between *Speed* and *Fuel Efficiency*? Explain.

g) Do you think this is the appropriate model for that association? Explain.

41. SAT scores. The SAT is a test often used as part of an application to college. SAT scores are between 200 and 800, but have no units. Tests are given in both Math and Verbal areas. Doing the SAT-Math problems also involves the ability to read and understand the questions, but can a person's verbal score be used to predict the math score? Verbal and math SAT scores of a high school graduating class are displayed in the scatterplot, with the regression line added.

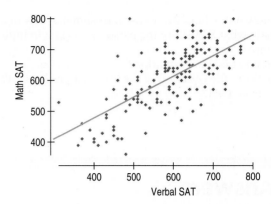

For these data, $r = 0.685$. The verbal scores averaged 596.3, with a standard deviation of 99.5, and the math scores averaged 612.2, with a standard deviation of 96.1.

a) Is a linear model appropriate?

b) Write the equation of the regression line.

c) Interpret the slope of this line.

d) Predict the math score of a student with a verbal score of 500.

e) Every year some student scores a perfect 1600. Based on this model, what would be that student's Math score residual?

42. Success in college. Colleges use SAT scores in the admissions process because they believe these scores provide some insight into how a high school student will perform at the college level. Suppose the entering freshmen at a certain college have mean combined *SAT Scores* of 1833, with a standard deviation of 123. In the first semester these students attained a mean *GPA* of 2.66, with a standard deviation of 0.56. A scatterplot showed the association to be reasonably linear, and the correlation between *SAT* score and *GPA* was 0.47.

a) Write the equation of the regression line.

b) Explain what the *y*-intercept of the regression line indicates.

c) Interpret the slope of the regression line.

d) Predict the GPA of a freshman who scored a combined 2100.

e) As a student, would you rather have a positive or a negative residual in this context? Explain.

43. SAT, take 2. Suppose we wanted to use SAT math scores to estimate verbal scores based on the information in Exercise 41.

a) What is the correlation?

b) Write the equation of the line of regression predicting verbal scores from math scores.

c) In general, what would a positive residual mean in this context?

d) A person tells you her math score was 500. Predict her verbal score.

44. Success, part 2. You'd like to use the statistics for college students given in Exercise 42 to create a model that would allow you to use a freshman's GPA to esti-mate what SAT score the person got when in high school.

a) What is the correlation?

b) Write the equation of the line that predicts *SAT* from *GPA*.

c) Your friend's freshman GPA was a solid 3.50. What would you guess her SAT score was?

d) When you tell her what you guessed, she says her residual is negative. What does that mean in this context?

Answers

Do The Math

1. slope = 3, intercept = −5, $y = 25$

2. slope = 1/2, intercept = 3, $y = 8$

3. slope = −2, intercept = 33, $y = 13$

4. slope = 4, intercept = 7, $y = 47$

5. slope = −0.5, intercept = 22, $y = 17$

6. slope = 2, intercept = −17, $y = 3$

7. slope = −0.25, intercept = 13, $y = 10.5$

8. slope = 0.88, intercept = −0.72, $y = 8.08$

9. a) Shipping a package costs $25 plus $2 per pound.

b) Intercept is 40°, the temperature when there's no chirping; temperature increases 1/4°F for each additional chirp per minute.

c) The monthly plan has a base charge of $15, plus an additional 10 cents for each minute of call time.

d) The intercept is 11 inches, the original height of the candle; as it burns it shortens 1/2 inch per hour.

10. a) $C = 19.99 + 0.25m$ (C = cost in dollars; m = number of miles driven)

b) $D = 80 - 9s$ (D = yards to the goal line; s = number of seconds running)

c) $C = 12 + 2.50k$ (C = cost in dollars; k = number of kids who came)

d) $P = 8s - 500$ (P = profit in dollars; s = number of pies sold)

11. $y = 8 + 4x$

12. $y = 17 - 2.5x$

13. $y = -4 + 0.2x$

14. $y = 9.9 - 0.66x$

Just Checking

1. 2 grams

2. 32 grams

3. The fat content is exactly what the model predicts.

4. An increase in home size of 1000 square feet is associated with an increase in price of $94,450, on average.

5. Units are thousands of dollars per thousand square feet.

6. About $188,900, on average

7. No. Even if it were positive, no one wants a house with 0 square feet!

8. a) Yes; the pattern looks linear and there are no outliers.

b) $b = \dfrac{rs_y}{s_x} = \dfrac{(0.7)(25)}{4} = 4.4$

c) (68, 150)

d) $150 = a + 4.4(68) \rightarrow a = -149$

e) $\widehat{Weight} = -149 + 4.4Height$

f) $-149 + 4.4(60) = 115$ pounds

9. You should expect the price to be 0.77 standard deviations above the mean.

10. You should expect the size to be 2(0.77) = 1.54 standard deviations below the mean.

11. The home is 1.5 standard deviations above the mean in size no matter how size is measured.

12. a) 2   b) 0.7(2) = 1.4
c) 1.4(25) = 35   d) 150 − 35 = 115 pounds

13. Negative; that indicates it's priced lower than a typical home of its size.

14. $280,233

15. $19,767 (positive!)

16. a) −149 + 4.4(70) = 159 pounds

b) 144 − 159 = −15 pounds

17. Estimate −149 + 4.4(65) = 137 pounds; 18 pounds heavier would be 155 pounds.

18. *Weight* is explanatory; *Height* is response.

19. $\widehat{Height} = 51.2 + 0.112Weight$

20. 51.2 + 0.112(200) = 73.6 inches tall

What's My Curve?

Jessica Meir and Paul Ponganis study emperor penguins at the Scripps Institution of Oceanography's Center for Marine Biotechnology and Biomedicine at the University of California at San Diego. Says Jessica:

Emperor penguins are the most accomplished divers among birds, making routine dives of 5–12 minutes, with the longest recorded dive over 27 minutes. These birds can also dive to depths of over 500 meters! Since air-breathing animals like penguins must hold their breath while submerged, the duration of any given dive depends on how much oxygen is in the bird's body at the beginning of the dive, how quickly that oxygen gets used, and the lowest level of oxygen the bird can tolerate. The rate of oxygen depletion is primarily determined by the penguin's heart rate. Consequently, studies of heart rates during dives can help us understand how these animals regulate their oxygen consumption in order to make such impressive dives.

The researchers equip emperor penguins with devices that record their heart rates during dives. Here's a scatterplot of the *Dive Heart Rate* (beats per minute) and the *Duration* (minutes) of dives by these high-tech penguins.

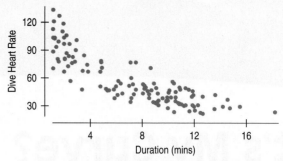

FIGURE 9.1 **Penguin dives.** The scatterplot of *Dive Heart Rate* in beats per minute (bpm) vs. *Duration* (minutes) shows a strong, roughly linear, negative association.

The scatterplot looks fairly linear with a moderately strong negative association ($r = -0.85$). The linear regression equation

$$\widehat{DiveHeartRate} = 96.9 - 5.47\,Duration$$

says that for longer dives, the average *Dive Heart Rate* is lower by about 5.47 beats per dive minute, starting from a value of 96.9 beats per minute.

The scatterplot of the residuals holds a surprise though. If a linear model is appropriate we should not see a pattern, but instead there's a bend, starting high on the left, dropping down in the middle of the plot, and rising again at the right.

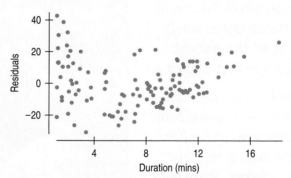

FIGURE 9.2 **Plotting the residuals reveals a bend.** It was also in the original scatterplot, but here it's easier to see.

Graphs of residuals often reveal patterns such as this that were easy to miss in the original scatterplot. Now looking back at that scatterplot, you may see that the pattern of points isn't really straight. There's a slight bend, so we shouldn't model the penguin data with a line. Now what?

Getting the "Bends": When the Residuals Aren't Straight

No attempt to model a relationship is complete until you have plotted the residuals. Because the residuals are what's "left over," they tell you what the model missed. Ideally, the model won't have missed anything meaningful, and that will show up as randomly scattered residuals. Sometimes, though, the residuals will reveal a pattern that was harder to see in the original scatterplot, as you just saw for the penguin dives. If there's a clear bend in the residuals, your model has missed something important about the relationship, and it's time to look for a better model.

Modeling bent relationships calls for a curve. There are a lot of possible curves, though, and finding a useful one can be a challenge. In this chapter, we'll consider two kinds of curved models: exponential curves and power curves. And, of course, each time we try fitting one of these curves, we'll check the residuals hoping to see . . . nothing.

Keep in mind, though, that the real world is a messy place, and our models won't be perfect. While we really hope to see nothing but randomly scattered residuals, even a little bit of curvature may represent a big improvement over a linear model if those residuals are small. And sometimes we'll strike out entirely, unable to find a model we think is useful. Let the games begin!

We can't *know* whether the **Linearity Assumption** is true, but we can see if it's *plausible* by checking the **Straight Enough Condition** by looking at the residuals.

Exponential Models

Linear models work well for things that grow, well, linearly. That means as the explanatory variable changes, the response variable grows (or shrinks) by adding (or subtracting) the same amount. If a crop of oats yields 80 bushels per acre, then planting one more acre should allow us to harvest 80 more bushels. If we plant 100 acres we expect 8000 bushels, 101 acres 8080 bushels, 102 acres 8160 bushels, and so on. Every additional acre gives us another 80 bushels of oats no matter how many acres we already have.

But population growth is different. Populations tend to increase by about the same *percentage* each year. If the annual growth rate is 2%, then in one year a city of 100,000 people will grow by 2% of 100,000—that's 2000 people—to a population of 102,000. The following year the city will add 2% of 102,000, now an additional 2040 people. The increase gets bigger every year, because the same 2% growth rate is based on an ever larger population. Linear growth stays constant, but exponential growth accelerates. If a larger farm increases from 500 to 501 acres of oats, that farmer still expects to harvest the same additional 80 bushels. But in a much larger city of 500,000 residents, a 2% growth rate would add 10,000 more people.

Equations of **exponential models** involve (surprise!) exponents, and look like this:

$$\hat{y} = a(b^x)$$

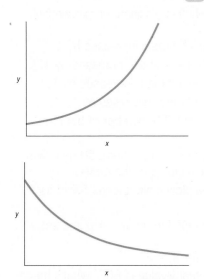

FIGURE 9.3 Exponential curves. In an exponential model, y grows (or shrinks) slowly when y is small, and more quickly when y is large.

Like a line, these curves have an intercept, and it tells us essentially the same thing. Remember that raising a number to the 0 power always results in 1, so $x = 0$ makes $\hat{y} = a(b^0) = a$. We see that once again a represents the model's starting value. And b, akin to slope for a line, describes the growth rate—but now that rate is given as a percentage. If $b = 1.02$, that's 102%. This represents 2% growth, perhaps annually as in the city we imagined before. If $\hat{y}$ decreases by 15% for each 1 unit of change in x, then the model's value of b would be $100\% - 15\% = 0.85$.

Do The Math

Let's explore some of the properties of exponential functions.

1. Here are two exponential functions. For each function, complete a table showing y values for values of x from 0 to 5, then plot the points on a grid and sketch the curve.
    a) $y = 4(3)^x$ b) $\hat{y} = 256\left(\frac{3}{4}\right)^x$

2. Look at each table of values. Is this function linear or exponential? Explain how you know.

a)

x	y
0	5
1	10
2	15
3	20
4	25
5	30

b)

x	y
0	5
1	10
2	20
3	40
4	80
5	160

c)

x	y
0	486
1	162
2	54
3	18
4	6
5	2

d)

x	y
0	20
1	17
2	14
3	11
4	8
5	5

3. Read each description of a function and tell whether it's linear or exponential. Write the equation.
    a) Initial value is 30 and y increases by 5 each time x increases by 1.
    b) Initial value is 30 and y is multiplied by 5 each time x increases by 1.
    c) Initial value is 64 and y decreases by 7 each time x increases by 1.
    d) Initial value is 64 and y is cut in half each time x increases by 1.
    e) There are initially 5000 bacteria on a wound. The number of bacteria increases by 15% each minute.
    f) At the start of a game there were 5000 fans in a stadium. 50 new fans entered each minute for the first several minutes of the game.
    g) A balloon holding 800 ml of helium has a slow leak; it loses 5% of its helium each minute.

4. The exponential function $\hat{y} = 300(1.5)^x$ models the relationship in a set of data.
    a) What value of y does this model predict for x = 2?
    b) In the observed data when x = 2 the actual y-value is 600. What's the residual?
    c) At x = 4 there's an actual data value with a residual of 12. What is y?

(Check your answers on page 214.)

Exponential growth shows up in many situations. Populations, be they people or bacteria, often grow exponentially. Savings accounts earning compound interest grow exponentially. Other things *shrink* exponentially. That's how your body assimilates medications. You metabolize a certain percentage of the drug remaining in your system each hour, so the amount in your bloodstream decreases rapidly at first and then more slowly as time passes. The same principle is at work in radioactive decay and carbon dating of ancient artifacts. Any time you suspect that the size of a change may depend on the current value of the variable, an exponential model is worth a try.

Obviously, working with curves is messier than working with a straight line, but here's the good news: your calculator or computer can create exponential models, and you won't even know it's harder!

A Model for Penguin Dives

The researchers studying emperor penguins set out to understand how the duration of a dive is related to the penguin's heart rate. We saw that fitting a linear model to their data left us with clear curvature in the residuals, a telltale sign that we haven't found the right model. Now we'll let our computer try an exponential model instead. Here are the scatterplot showing the curve and the resulting residuals plot:

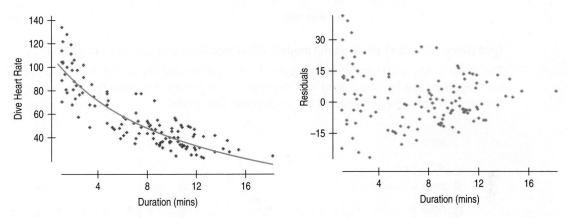

FIGURE 9.4 The curved model. An exponential curve nicely fits the bend in the relationship, and now there's no apparent bend in the residuals.

That's clearly an improvement. The residuals look much more random, suggesting that this model might be useful. The equation for our exponential model is:

$$\widehat{DiveHeartRate} = 102.32(0.91)^{Duration}$$

The value of a tells us our model estimates that penguins' heart rates should average around 102.32 beats per minute when they're not diving. The value of b indicates that heart rates tend to decrease exponentially about $1.00 - 0.91 = 9\%$ for each minute a dive lasts.

We can use this model to make predictions, just as we did with linear models. For example, let's estimate a typical heart rate for penguins during dives that last for 10 minutes:

$$\widehat{DiveHeartRate} = 102.32(0.91)^{10} = 39.85 \text{ beats per minute.}$$

For Example Spotting Exponential Growth

In 1960 you could have purchased a brand-new Cadillac for about $5000. Today that might buy you a golf cart!

Just how fast have car prices risen? Here's a scatterplot showing the median price of a new car from 1930 to 2005 with the linear model and the residuals plot.

(continued)

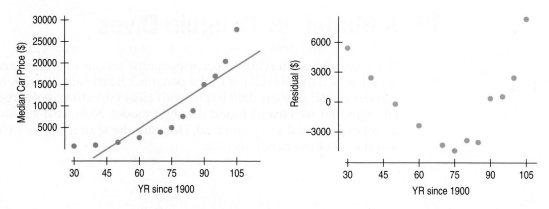

Question: Why isn't this a useful model? What model do you suggest using instead?

The curve in the scatterplot shows up even more dramatically in the residuals plot. A linear model isn't appropriate for describing the increase in car prices. Because the prices may have grown by a similar percentage each year, I'll try an exponential model.

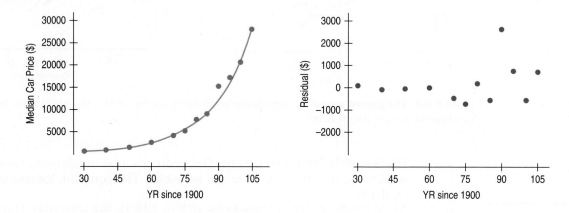

This model seems to fit well. The residuals appear to be randomly scattered, with an unusually large increase in prices in 1990. I think the exponential model $\widehat{Price} = 114.06(1.0535)^{YRsince1900}$ would be useful. My model suggests that since 1930 car prices have increased, on average, about 5.35% each year.

TI Tips — Creating an exponential model

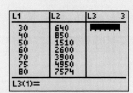

It's easy to create the exponential model for car prices with your calculator. First you'll need to enter the data. The values are shown at the left.

- Put *YRsince1900* in L1.
- Put *Price* in L2.

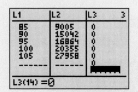

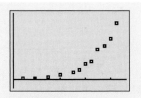

Remember the rules: make a picture!

- Set up STAT PLOT Plot1 as a scatterplot with XList:L1 and YList:L2, then ZoomStat.

The scatterplot clearly shows a curve, so we won't try LinReg.

Notice that a little farther down the STAT CALC menu you can also choose ExpReg. Let's give that a try.

- Enter the command ExpReg L1, L2, Y1.

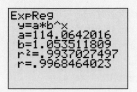

There's the exponential model. The TI shows you the form of the equation and tells you the values of a and b. The model is:

$$\widehat{Price} = 114.06(1.0535)^{YRsince1900}$$

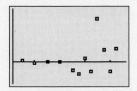

Before you jump into using the model, remember to check the residuals plot to be sure it's appropriate.

- Turn Plot1 Off.
- Set up STAT PLOT Plot2 as a scatterplot with XList:L1 and YList:RESID.
- Hit ZoomStat to create the plot.

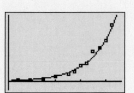

Aside from the outlier (sticker shock in 1990!), the residuals look reasonably random, so let's have a look at the model.

- Turn Plot2 Off.
- Turn Plot1 back on, and ZoomStat again.

Isn't it satisfying watching the curve hit those points?

Want to use the model to make a prediction? That works the same way it did for linear models. Let's see what the average new car might cost in 2020:

- Choose VARS, Y-VARS, Y₁
- Enter the year in parentheses; remember that we're counting from 1900, so for 2020 you enter 120.

Yikes! \$59413? For the *average* car? Well, maybe, but there's hope it won't be that high. After all, this *is* an extrapolation . . . .[1]

---

[1]So, of course, the price could turn out to be even *higher*.

STEP-BY-STEP EXAMPLE   Exponential Models

When you take medication, the active ingredient reaches your bloodstream soon after you swallow the pill. Then your body begins to absorb and use the medicine, reducing the amount remaining in your blood. It's important that doctors know how fast this happens, so that they can prescribe how long you should wait before taking another pill.

Question: What model describes the *Level* (in parts per billion) of the drug in patients' blood over *Time* in hours since they took the first pill?

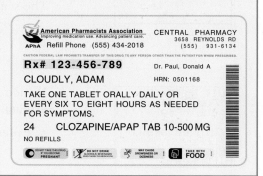

American Pharmacists Association
Improving medication use. Advancing patient care.
APhA Refill Phone (555) 434-2018

CENTRAL PHARMACY
3658 REYNOLDS RD
(555) 931-6134

CAUTION FEDERAL LAW PROHIBITS TRANSFER OF THIS DRUG TO ANY PERSON OTHER THAN THE PATIENT FOR WHOM PRESCRIBED.

Rx# 123-456-789   Dr. Paul, Donald A
CLOUDLY, ADAM   HRN: 0501168

TAKE ONE TABLET ORALLY DAILY OR EVERY SIX TO EIGHT HOURS AS NEEDED FOR SYMPTOMS.
24   CLOZAPINE/APAP TAB 10-500 MG
NO REFILLS

DO NOT TAKE THIS DRUG IF YOU BECOME PREGNANT | DO NOT DRINK ALCOHOLIC BEVERAGES WHEN TAKING THIS MEDICATION | MAY CAUSE DROWSINESS OR DIZZINESS | TAKE WITH FOOD

THINK

Plan State the problem.

Variables Identify the variables.

Check the appropriate assumptions and conditions. As always, we make a picture.

The bend in the scatterplot makes it clear we can't use a linear model.

I want to model the relationship between the time that has passed since a patient took a certain medication and the amount of the active ingredient remaining in the patient's blood. I have data collected in a medical research study.

Let *Level* = amount of the drug in patient's blood (parts per billion)

Let *Time* = hours since the patient took the pill.

✔ **Quantitative Variables Condition:** Both the blood level and time are quantitative.

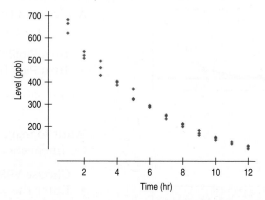

✘ **Straight Enough Condition:** There is an obvious bend in this relationship. I can't use a linear model.

✔ **Outlier Condition:** I don't see any outliers in the scatterplot.

Mechanics Thinking that perhaps the rate at which the body absorbs the medication may depend on the current level in the person's blood, it makes sense to try an exponential model.

I'll create an exponential model.

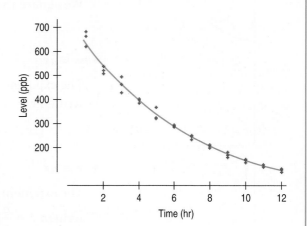

Always check the randomness of the residuals to see if the model is appropriate.

Here's my residuals plot:

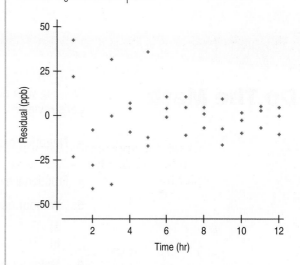

There's no sign of a curve in the residuals, so my exponential model should be useful:

$$\widehat{Level} = 757.3(0.85)^{Time}$$

Give the equation of the model.

Conclusion Interpret your model.

Based on my exponential model, it appears this dosage has an initial level of about 757 parts per billion in the bloodstream, and about 15% of the remaining drug is absorbed each hour.

Plan B: Power Models

Not all curves are exponential. Sometimes we spot a bend in a scatterplot, try an exponential model, and still see curvature in the residuals. When an exponential model doesn't fit, a power model may prove useful. A power model involves exponents again, but now instead of using x as the exponent, we raise the x values to a power.

Equations of **power models** look like this:

$$\hat{y} = a(x^b)$$

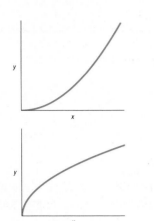

You have seen some common examples in your algebra classes:

- $\hat{y} = ax^2$

    We square the x value (sometimes called a quadratic model).

- $y = ax^{1/2}$

    The exponent of $^1/_2$ means take the square root of the x value (sometimes written $y = a\sqrt{x}$).

- $y = ax^{-1}$

    The exponent of -1 tells us to use the reciprocal of the x value (sometimes written $y = \dfrac{a}{x}$).

Do The Math

Working with power functions involves understanding all kinds of exponents. Let's review:

- Negative exponents represent reciprocals: $4^{-1} = \dfrac{1}{4}$; $3^{-2} = \dfrac{1}{3^2} = \dfrac{1}{9}$

- Fractional exponents represent roots: $25^{\frac{1}{2}} = \sqrt{25} = 5$; $27^{\frac{1}{3}} = \sqrt[3]{27} = 3$

5. Without using your calculator, evaluate:

    a) 5^{-2} c) $16^{-\frac{1}{2}}$ e) $125^{\frac{1}{3}}$
    b) $16^{\frac{1}{2}}$ d) 2^{-5} f) $27^{-\frac{1}{3}}$

6. Make a table of values and sketch the graph of each power function $y = ab^x$ below. Describe the pattern of change of each using one of the following:

    - as x increases, y decreases at a decreasing rate
    - as x increases, y decreases at an increasing rate
    - as x increases, y increases at a decreasing rate
    - as x increases, y increases at an increasing rate

    a) $y = 0.5x^2$, using $x = 0, 1, 2, 3, 4, 5, 6$
    b) $y = 40x^{-1}$, using $x = 1, 2, 4, 5, 8, 10$
    c) $y = 2x^{0.5}$, using $x = 0, 1, 4, 9, 16, 25$
    d) $y = 30x^{-0.5}$, using $x = 1, 4, 9, 16, 25, 36$

    (Check your answers on pages 214–215.)

Power models are useful in many situations. Rates—miles per hour, say—may use a reciprocal model; think minutes per mile. A variable based on area—for instance, the amount of paint needed—may involve squaring or square roots. Similarly, quantities based on volume—such as weight—may involve cubes or cube roots. If you have studied Physics you learned that energy variables like the intensity of light or force of gravity follow inverse square laws; those are power models using the exponent -2.

If you're wondering how you'll know what exponent to try, we have good news for you: your calculator or a computer can figure that out. Your jobs are to think of trying a power model, and then to use the residuals plot to be sure that model is appropriate.

Modeling Fuel Economy

We know from common sense and from physics that heavier cars need more fuel, but exactly how does a car's weight affect its fuel efficiency? Here are the scatterplot of *Weight* (in pounds) and *Fuel Efficiency* (in miles per gallon) for 38 cars, and the residuals plot for a linear model:

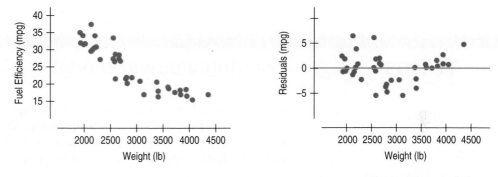

FIGURE 9.5 *Fuel Efficiency* (mpg) vs. *Weight* for 38 cars as reported by *Consumer Reports*. The scatterplot shows a negative direction, roughly linear shape, and strong relationship. However, the residuals from a linear model reveal a bent shape. Looking back at the original scatterplot, you may be able to see the bend.

Hmm . . . . The residuals certainly don't show the random scatter we were hoping for. The shape is clearly bent. Looking back at the first scatterplot, you can probably see the slight bending. Think about the regression line through the points. How heavy would a car have to be to have a predicted gas mileage of 0? It looks like the *Fuel Efficiency* would go negative at about 6000 pounds. A Hummer H2 weighs about 6400 pounds. The H2 is hardly known for fuel efficiency, but it does get more than the *minus* 5 mpg this linear model predicts. Extrapolation is always dangerous, but it's more dangerous the more the model is wrong, because wrong models tend to do even worse the farther you get from the middle of the data.

Okay, a linear model is out. What should we try? Fuel economy is a ratio. We measure it in miles per gallon, but it makes just as much sense to think about gallons per mile—as in, I have to drive 1000 miles, so how much gas will I need? Reciprocals like this can be represented by an exponent of −1, so let's see how a power model works. Here are the scatterplot showing this model and the resulting residuals plot:

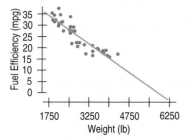

FIGURE 9.6 Extrapolating the regression line gives an absurd answer for vehicles that weigh as little as 6000 pounds.

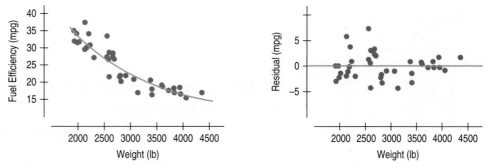

FIGURE 9.7 **A better model.** The power curve seems to fit the bend in the relationship, and if we don't get distracted by a couple of uncharacteristically large residuals, we see that most of the bend is gone.

This residuals plot, while perhaps not ideal, shows we've made a big improvement on the linear model we started with. Our power model should be more useful:

$$\widehat{MPG} = 79157(Weight^{-1.022})$$

How well does this model[2] do? It predicts, for example, that on average cars weighing 3000 pounds should get $79157(3000^{-1.022}) = 22.1$ miles per gallon, a plausible estimate. And what about that Hummer? Well, we're still extrapolating, but our power model estimates that the 6400-pound H2 should get $79157(6400^{-1.022}) = 10.2$ mpg—far more reasonable than what the line predicted, and very close to the reported value of 11.0 miles per gallon. (Of course, *your* mileage may vary . . .)

TI Tips   Creating a power model

Some camera lenses have an adjustable aperture, the hole that lets the light in. The size of the hole is called the f/stop. By making the hole larger, you get to use a faster shutter speed when taking a picture. Here's a table of recommended shutter speeds and f/stops for a digital camera:

Shutter speed:	$\frac{1}{1000}$	$\frac{1}{500}$	$\frac{1}{250}$	$\frac{1}{125}$	$\frac{1}{60}$	$\frac{1}{30}$	$\frac{1}{15}$	$\frac{1}{8}$
f/stop:	2.8	4	5.6	8	11	16	22	32

Let's look at that relationship. First you'll need to enter the data.

- Put *Shutter speed* in L1 (Note that you can enter fractions.)
- Put *f/stop* in L2.

Now create the scatterplot with XList:L1 and YList:L2.

See the curve? We can't use a linear model here. What to try? Well, the size of the hole is an area, and areas are measured in units *squared*. That suggests a power model might work.

- Look in the STAT CALC menu for PwrReg.
- Enter the command PwrReg L1,L2,Y1

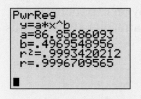

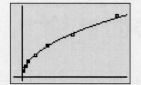

Notice that the exponent $b = 0.497$ is almost ½. That means we're finding a square root, consistent with our suspicion that area is involved.

Hit ZoomStat again to see how that looks.

Not bad! The power model is:

$$\widehat{fstop} = 86.86(Speed^{0.497})$$

---

[2]We tried a power model because we were thinking "reciprocal" (an exponent of -1), and the exponent turned out to be -1.022. Boy, are we good!

STEP-BY-STEP EXAMPLE   A Power Model

Standard (monofilament) fishing line comes in a range of strengths, usually expressed as "test pounds." Five-pound test line, for example, can be expected to withstand a pull of up to five pounds without breaking. When you buy fishing line, the price of a spool doesn't vary with strength. Instead, the length of line on the spool varies. Higher test pound line is thicker, so less fits on the spool. Other spools hold line that is thinner, weaker, and longer. Let's look at the *Length* and *Strength* of spools of monofilament line manufactured by the same company and sold for the same price at one store.

Question: How are the *Length* on the spool and the *Strength* related?

THINK

Plan State the problem.

I want to fit a linear model for the length and strength of monofilament fishing line.

Variables Identify the variables and report the W's.

I have the *length* and "pound test" strength of monofilament fishing line sold by a single vendor at a particular store. Each case is a different strength of line, but all spools of line sell for the same price.

Let *Length* = length (in yards) of fishing line on the spool
Strength = the test strength (in pounds).

Plot Make a scatterplot to look at the form of the relationship.

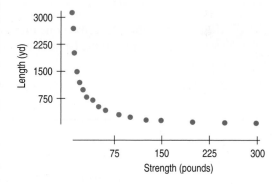

The bend in the scatterplot makes it obvious that a linear model would not be useful.

✔ **Quantitative Variables Condition:** Both the line's *Strength* and *Length* are quantitative.

✘ **Straight Enough Condition:** The relationship is clearly bent. A linear model is not appropriate.

✔ **Outlier Condition.** The plot shows a negative direction and an association that has little scatter and no outliers.

(continued)

Mechanics The length of the line that fits on the spool depends on its thickness, and that cross-sectional area probably determines the strength of the line. With area involved, it makes sense to try a power model.

I'll create a power model.

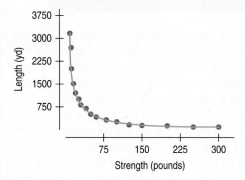

Always check the randomness of the residuals to see if the model is appropriate.

Here's my residuals plot:

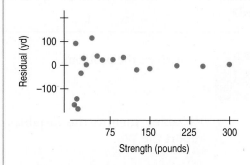

Give the equation of the model.

Although some of the residuals are large, there's no sign of a curve. My power model should be useful:

$$\overgroup{Length} = 31218(Strength^{-1.078})$$

Conclusion Describe your model. If you see any limitations, be sure to mention them.

My power model has an exponent near −1, indicating that the length of fishing line on a spool varies inversely with the line's strength. This model's estimates of length aren't nearly as accurate for the very weak fishing lines.

In Search of the Perfect Model—NOT!

When we try to understand the relationship between two quantitative variables, we start with a scatterplot. If the plot looks Straight Enough, that's good news: we can check out a linear model. If its residuals look nicely scattered with no curvature, that linear model can be useful.

But things get messier when we see curvature, either visible in the scatterplot or not so obvious until we look at the residuals. Then we need a different model. What should we try? As we've seen, sometimes the context provides a hint. Quantities that change based on percentages of current values, like populations or drug absorption, often turn out to be exponential. Power models often work well for quantities involving ratios, areas, or volumes. Sometimes it's hard to guess, though. Whether you think you have a clue or are just winging it, try something. Check the residuals. If you're not happy, try something else.

But beware: the real world is a messy place. Don't expect any model to be perfect. We're looking for something useful. Even when the residuals show a bit of curvature, if they represent generally small errors the model may still be helpful. And keep in mind that you may not find a useful model at all. Some relationships are too complex for our simple linear, exponential, or power models to describe.

Just Checking

After reading each of the following descriptions, tell whether you'd try a linear model, an exponential model, or a power model. If you think none of these might be useful, explain why.

1. You want to model the growth of a college savings fund that Uncle Rich set up when you were born. It has been sitting in the bank earning compound interest ever since.

2. You want to model the relationship between prices for common items like food and clothing in Boston and Tokyo. Your scatterplot shows a generally straight pattern with only moderate amounts of scatter.

3. You want to model the average weekly temperature in Denver over a year.

4. You want to model the relationship between the *Diameter* of rivets used to hold metal beams together and the *Strength* of those rivets (the weight they can hold without breaking).

(Check your answers on page 215.)

🚫 WHAT CAN GO WRONG?

- **Remember the basic rule of data analysis: *Make a picture.*** Before you fit a line, always look at the pattern in the scatterplot. After you fit the line, check for linearity again by plotting the residuals, where a curve will be easier to spot.

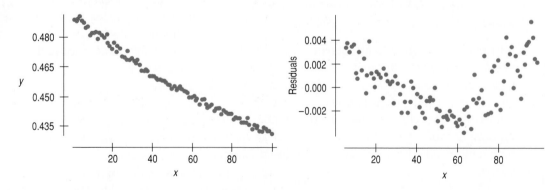

- **Don't expect your model to be perfect.** In Chapter 6 we quoted statistician George Box: "All models are wrong, but some are useful." Be aware that the real world is a messy place and data can be uncooperative. Don't expect to find one elusive model that magically captures every kink in your scatterplot and produces perfect residuals. You aren't looking for the Right Model, because that mythical creature doesn't exist. Find a useful model and use it wisely.

(continued)

■ **Watch out for scatterplots that turn around.** Exponential and power models work for many bent relationships but not those that go up and then down or down and then up. You don't have models for dealing with those.

FIGURE 9.8 Too many bends.
The shape of the scatterplot of *Birth Rates* (births per 100,000 women) in the United States shows a wavy pattern we can't model with exponential or power curves.

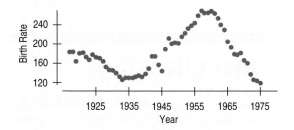

■ **Don't round off too much.** Use several decimal places when you write your model equation. With exponents involved, even minor changes can make big differences in the estimates.

CURVED MODELS **IN YOUR WORLD**

Great Sluggers, or Steroids?

(From "Numbers, Sabermetrics, Joe Jackson, and Steroids"
by Jim Albert, STATS: The Magazine for Students of Statistics, 48:3–6)

Suppose a . . . [baseball] player—such as Barry Bonds—exhibits a large increase in hitting home runs for one season. Does this mean Bonds was on steroids? . . . [A] jump in home run numbers . . . could also be due to the player doing a large amount of training during the off-season or using a new hitting style. It could also just be due to chance variation.

Although it is difficult to say that any player is "juiced" based entirely on . . . a single season, we can look at patterns of hitting over time . . . One useful measure of hitting strength is the *home run rate* defined as the proportion of home runs hit among all the balls a player hits and puts into play.

If we look at the home run rate for a particular player across all the seasons in his career, a typical pattern emerges. A player's home run rate generally increases until the middle of his career—usually when he is around 30 years old—and then diminishes until retirement. For example, the graphs [below] plot the home run rates of four great sluggers—Babe Ruth, Mickey Mantle, Willie Mays, and Mike Schmidt—with best-fitting quadratic curves[3] placed on top.

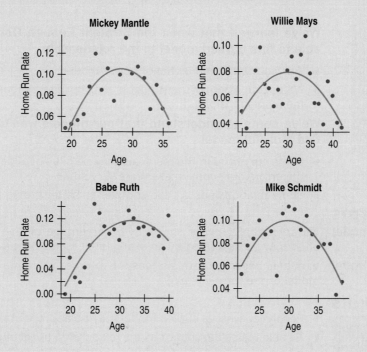

Suppose we look at the home run rates . . . for recent home run leaders McGwire, Bonds, and Sosa, who are suspected of using steroids. The trajectories of McGwire and Bonds show a surprising pattern.

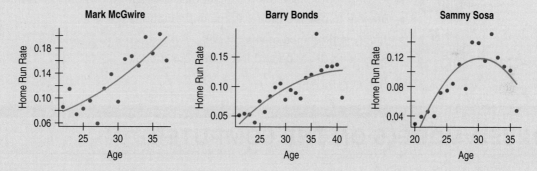

McGwire's home run rate shows a steady increase until his mid-30s, when his home run rate actually exceeded 20% at age 36. Bonds' home run rate also shows an interesting pattern. When Bonds was in his early 30s, his home run rate started dropping off (as would be expected), before jumping back up and increasing until age 40. We also see that Bonds' rate of 19%, when he hit 73 home runs at age 36, appears to be a very unusual event. Sosa shows a more typical career trajectory.

His home run rate peaked in his early 30s and dropped off toward the end of his career.

Have we proven that McGwire and Bonds used steroids when they were playing baseball? No. What these graphs show is that McGwire and Bonds have very unusual career trajectories, in that they peak in hitting home runs at atypically old ages. Few players in baseball history have displayed similar career trajectories. [We don't know] what caused [McGwire and Bonds] to peak at such advanced ages.

---

[3]Quadratic curves are parabolas; you probably met those in an Algebra class. Your TI can fit these models using a procedure named QuadReg.

WHAT HAVE WE LEARNED?

We've learned that when the Straight Enough Condition fails, we may be able to fit a curved model to the relationship.

▶ We can try an exponential or power model.
▶ We decide whether the model is appropriate by looking for random scatter in the residuals plot.

We've come to understand that our models won't be perfect, but we may find a useful model.

▶ Some curvature in the residuals may be okay provided the residuals (and hence the errors our model makes) are very small.
▶ Some relationships are too complex to be described by these simple models.

Terms

Exponential model   Variables that grow or shrink by a percentage of the amount currently present often may be described by an exponential model of the form $\hat{y} = ab^x$.

Power model   Variables based on ratios, areas, or volumes often may be described by a power model of the form $\hat{y} = ax^b$.

Skills

▶ Recognize when a curved model may be appropriate.
▶ Be able to use the context of the relationship to get clues about what model to try.
▶ Recognize when the pattern of the data indicates that no simple model will work.

▶ Know how to fit both exponential and power models using technology.
▶ Know how to check the residuals plot to see if your model is useful.
▶ Be able to use the model to make predictions.

▶ Justify your choice of the model.
▶ Be able to describe and interpret your model.

CURVED MODELS ON THE COMPUTER

Many software packages will also create curved models. Usually you'll get a menu of options to choose from.

You can then see the scatterplot with the curve and the equation of the model.

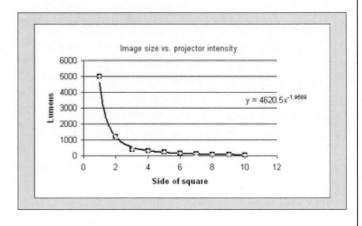

EXERCISES

A

1. Identify, part I. Look at each table of values for two variables. Which relationship is linear, and which is exponential? Explain.

a)

x	y
2	70
4	220
6	370
8	520
10	670

b)

x	y
2	70
4	140
6	280
8	560
10	1120

2. Identify, part II. Look at each table of values for two variables. Which relationship is linear, and which is exponential? Explain.

a)

x	y
10	160
20	240
30	360
40	540
50	810

b)

x	y
10	160
20	360
30	560
40	760
50	960

3. Identify, part III. Look at each table of values for two variables. Which relationship is linear, and which is exponential? Explain.

a)

x	y
10	81
20	54
30	36
40	24
50	16

b)

x	y
10	81
20	66
30	51
40	36
50	21

4. Identify, part IV. Look at each table of values for two variables. Which relationship is linear, and which is exponential? Explain.

a)

x	y
1	80
3	60
5	40
7	20
9	0

b)

x	y
1	80
3	40
5	20
7	10
9	5

5. Predictions. For each of the models listed, predict y when $x = 2$

a) $\hat{y} = 12x^2$
b) $\hat{y} = 12x^{-1}$
c) $\hat{y} = 12(1.5)^x$
d) $\hat{y} = 12(0.5)^x$

6. More predictions. For each of the models listed below, predict y when $x = 10$

a) $\hat{y} = 8x^{-1}$
b) $\hat{y} = 8(0.98)^x$
c) $\hat{y} = 8(1.02)^x$
d) $\hat{y} = 8x^2$

7. Bacteria. An exponential model describing the growth of a laboratory culture of bacteria is $\widehat{Pop} = 500\,(1.06^{Day})$.

a) How many bacteria did the lab culture start with?
b) What's the growth rate?
c) How many bacteria should the lab expect there to be in a week?

8. Homes. An exponential model describing number of homes in a growing suburban community since 1980 is $\widehat{Homes} = 350(1.08^{year})$.

a) How many homes were there in 1980?
b) What's the growth rate?
c) How many homes does the model estimate there to be in 2020?

9. Lead. Through a process called radioactive decay, the isotope polonium-210 changes to lead. The model $\widehat{Pol} = 80(0.966^{Week})$ describes how much polonium (in grams) remains over time in one sample.

a) How many grams of polonium were there initially?
b) At what rate does the polonium decay to lead?
c) How much polonium should remain after a year?

10. Reaction. The model $\widehat{Toxin} = 1200(0.88)^{min}$ describes the number of milligrams of a toxin that remains during a chemical reaction intended to decontaminate some water.

a) How many milligrams of the toxin were in the water initially?
b) At what rate does this reaction reduce the level of toxin?
c) To what level should the toxin be reduced in half an hour?

11. Residuals. Suppose you have fit a model to some data and now take a look at the residuals. For which of the following possible residuals plots would you reject your model? Why?

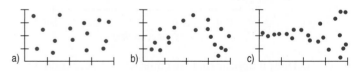

a)   b)   c)

12. Residuals. Suppose you have fit a model to some data and now take a look at the residuals. For which of the following possible residuals plots would you reject your model? Why?

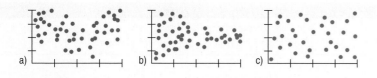

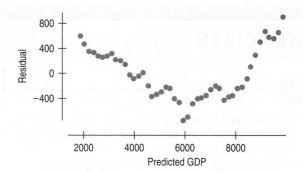

13. Gas mileage. As the example in the chapter indicates, one of the important factors determining a car's *Fuel Efficiency* is its *Weight*. Let's examine this relationship again, for 11 cars.

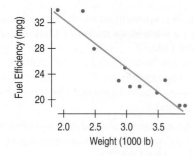

Do you think this linear model is appropriate? Use the residuals plot to explain your decision.

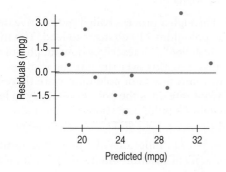

14. GDP. The scatterplot shows the gross domestic product (GDP) of the United States in billions of dollars plotted against years since 1950.

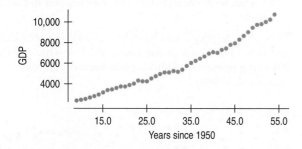

a) Does it appear that we could use a linear model?
b) Here's a scatterplot of the residuals. Now do you think this is a good model for these data? Explain.

15. Gas mileage revisited. Let's try a power model to examine the fuel efficiency of the 11 cars in Exercise 13. Here's the residuals plot:

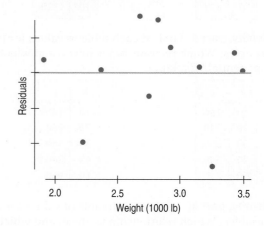

Explain why this model appears to be better than the linear model.

16. Better GDP model? Consider again the post-1950 trend in U.S. GDP we examined in Exercise 14. Here are a regression and residual plot when we use an exponential model. Is this a better model for GDP? Explain.

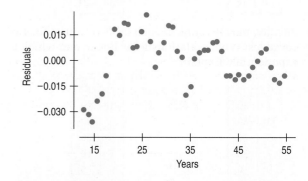

17. Hopkins winds revisited. In Chapter 5, we examined the wind speeds in the Hopkins Memorial Forest over the course of a year. Here's the scatterplot we saw then:

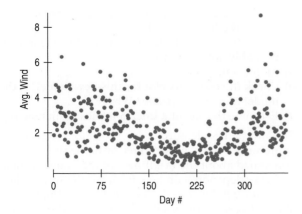

a) Describe the pattern you see here.
b) Do you think either an exponential or power model might be useful? Explain.

18. Treasury bills. The 3-month Treasury bill interest rate is watched by investors and economists. Here's a scatterplot of the 3-month Treasury bill rate since 1950:

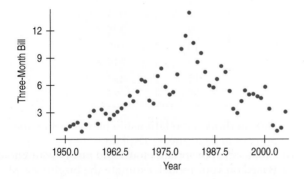

Clearly, the relationship is not linear. Which model, if any, would you suggest? Explain?

B

19. Dying coins. Here's a game about survival. Start with 100 pennies in a large jar. Shake them up and toss them all onto the floor. Consider any penny that lands tails as "dead" and remove it. Put all the "live" pennies (the heads) back in jar. Shake them and toss them out on the floor again. Keep going until all the pennies have died. Here's a table showing what happened in one game:

Toss #	0	1	2	3	4	5	6
Living	100	51	22	11	8	3	1

a) Create a model describing the survivability of these coins. Be sure to look at the residuals to verify that your model is appropriate.
b) There's a 50-50 chance that each coin lands heads. How does that show up in your model?

20. Soup. A student doing a Science Fair experiment put a hot bowl of soup in the refrigerator and checked the temperature of the soup every 2 minutes:

Time (min)	0	2	4	6	8	10	12	14	16	18	20
Temp (°C)	72	54.2	39.5	30.9	22.2	16.3	13.0	9.0	6.4	5.2	4.5

a) Create a model describing how the soup cools. Be sure to look at the residuals to verify that your model is appropriate.
b) Explain what the two values in the equation suggest about the soup.
c) Estimate the temperature of the soup after 3 minutes.
d) Estimate the temperature of the soup after 25 minutes.
e) How much confidence do you place in those estimates. Why?

21. Brakes. The following table shows stopping distances in feet for a car tested 3 times at each of 5 speeds. We hope to create a model that predicts *Stopping Distance* from the *Speed* of the car.

Speed (mph)	Stopping Distances (ft)
20	64, 62, 59
30	114, 118, 105
40	153, 171, 165
50	231, 203, 238
60	317, 321, 276

a) Explain why a linear model is not appropriate.
b) Create an appropriate model.
c) Estimate the stopping distance for a car traveling 55 mph.
d) Estimate the stopping distance for a car traveling 70 mph.
e) How much confidence do you place in these predictions? Why?

22. Planet distances and years. At a meeting of the International Astronomical Union (IAU) in Prague in 2006, Pluto was determined not to be a planet, but rather the largest member of the Kuiper belt of icy objects. Let's examine some facts. Here is a table of the 9 sun-orbiting objects formerly known as planets:

Planet	Position Number	Distance from Sun (million miles)	Length of Year (Earth years)
Mercury	1	36	0.24
Venus	2	67	0.61
Earth	3	93	1.00
Mars	4	142	1.88
Jupiter	5	484	11.86
Saturn	6	887	29.46
Uranus	7	1784	84.07
Neptune	8	2796	164.82
Pluto	9	3707	247.68

a) Plot the *Length* of the year against the *Distance* from the sun. Describe the shape of your plot.
b) Create a model describing the length of a planet's year based on its distance from the sun.
c) Looking at the residuals, comment on how well Pluto fits your model.

23. Planet distances and order. Let's look again at the pattern in the locations of the planets in our solar system seen in the table in Exercise 22.
a) Create a model for the *Distance* from the sun based on the planet's *Position*.
b) Based on this model, would you agree with the International Astronomical Union that Pluto is not a planet? Explain using residuals.

24. Planet Eris? In July 2005, astronomers Mike Brown, Chad Trujillo, and David Rabinowitz announced the discovery of a sun-orbiting object, since named Eris,[4] that is 5% larger than Pluto. Eris orbits the sun once every 560 earth years at an average distance of about 6300 million miles from the sun. Based on these properties, does your model of Exercise 22 suggest Eris might be a planet?

25. Planet Eris revisited. Let's consider the newly discovered body Eris (described in Exercise 24) to be in position 10 from the sun, on average, about 6300 million miles away. Based on the model you created in Exercise 23, do you think Eris should be considered a planet? Explain.

26. How old is that tree? One can determine how old a tree is by counting its rings, but that requires cutting the tree down. Can we estimate the tree's age simply from its diameter? A forester measured 27 trees of the same species that had been cut down, and counted the rings to determine the ages of the trees.

Diameter (in.)	Age (yr)	Diameter (in.)	Age (yr)
1.8	4	10.3	23
1.8	5	14.3	25
2.2	8	13.2	28
4.4	8	9.9	29
6.6	8	13.2	30
4.4	10	15.4	30
7.7	10	17.6	33
10.8	12	14.3	34
7.7	13	15.4	35
5.5	14	11.0	38
9.9	16	15.4	38
10.1	18	16.5	40
12.1	20	16.5	42
12.8	22		

a) Create a scatterplot and describe the association.
b) Create the linear model.
c) Check the residuals. Explain why a linear model is probably not appropriate.
d) Create a new model.
e) Check the residuals plot for this new model. Is this model more appropriate? Why?
f) Estimate the age of a tree 18 inches in diameter.

27. Brightness. People purchasing projectors for use in conference rooms, classrooms, or home theaters are concerned about the brightness of the image, which will vary depending on how far from the screen the projector will be placed. The table shows the relationship between image brightness (measured in candlepower) and distance (in feet) for one projector model.

Distance (ft)	Brightness (cp)
3	3450
4	1910
5	1225
6	850
7	625
8	475
9	375
10	300
12	210
15	135
20	75

a) Do you think the relationship between *Brightness* and *Distance* is linear? Explain.
b) Create an appropriate model and justify your choice.
c) Based on your model, estimate the brightness of the image this projector would produce in a classroom where it would be 18 feet from the screen.
d) What does the model's value of b reveal about this relationship?

28. Growth. Research has found that as boys approach adolescence the level of insulin growth factor (IGF) in their blood increases. Here are the data showing the *Age* (in years) and *IGF* level for 25 of the boys who were studied.

Age	IGF	Age	IGF
0.50	43	9.56	158
1.16	44	10.03	224
1.50	89	10.44	239
2.00	79	10.65	281
2.83	80	11.34	239
4.08	106	11.69	232
5.41	68	12.46	419
5.50	148	12.83	279
6.42	179	13.47	488
7.50	198	13.77	499
8.30	206	14.54	518
8.83	160	14.83	473
9.14	179		

---

[4]Eris is the Greek goddess of warfare and strife who caused a quarrel among the other goddesses that led to the Trojan war. In the astronomical world, Eris stirred up trouble when the question of its proper designation led to the raucous meeting of the IAU in Prague where IAU members voted to demote Pluto and Eris to dwarf-planet status—http://www.gps.caltech.edu/~mbrown/planetlila/#paper.

a) Explain why a linear model isn't appropriate.
b) Find a better model. Justify your choice.
c) Why don't you expect the increasing pattern of this growth factor to continue?

29. Down the drain. Most water tanks have a drain plug so that the tank may be emptied when it's to be moved or repaired. How long it takes a certain size of tank to drain depends on the size of the plug, as shown in the table. Create a useful model.

Plug Dia (in.)	$\frac{3}{8}$	$\frac{1}{2}$	$\frac{3}{4}$	1	$1\frac{1}{4}$	$1\frac{1}{2}$	2
Drain Time (min.)	140	80	35	20	13	10	5

30. Pressure. Scientist Robert Boyle examined the relationship between the volume in which a gas is contained and the pressure in its container. He used a cylindrical container with a moveable top that could be raised or lowered to change the volume. He measured the *Height* in inches by counting equally spaced marks on the cylinder, and measured the *Pressure* in inches of mercury (as in a barometer). Some of his data are listed in the table. Create an appropriate model.

Height	48	44	40	36	32	28
Pressure	29.1	31.9	35.3	39.3	44.2	50.3
Height	24	20	18	16	14	12
Pressure	58.8	70.7	77.9	87.9	100.4	117.6

31. Baseball salaries 2005. Ballplayers have been signing ever larger contracts. The highest salaries (in millions of dollars per season) for some notable players are given in the following table.

Player	Year	Salary (million $)
Nolan Ryan	1980	1.0
George Foster	1982	2.0
Kirby Puckett	1990	3.0
Jose Canseco	1990	4.7
Roger Clemens	1991	5.3
Ken Griffey, Jr.	1996	8.5
Albert Belle	1997	11.0
Pedro Martinez	1998	12.5
Mike Piazza	1999	12.5
Mo Vaughn	1999	13.3
Kevin Brown	1999	15.0
Carlos Delgado	2001	17.0
Alex Rodriguez	2001	22.0
Manny Ramirez	2004	22.5
Alex Rodriguez	2005	26.0

a) Examine a scatterplot of the data. Does it look straight?
b) Find the linear model for *Salary* vs. *Year* (use year since 1980) and plot the residuals. Do they look straight?
c) What model would you use for the trend in salaries?

32. Moore's Law. The transistor is the basic building block of computers; the more transistors the more powerful the machine. In 1965 Gordon Moore predicted that the number of transistors that could be put on an integrated circuit chip would double every 18 months. Three years later, he co-founded a company called Intel, now one of the leading and most innovative manufacturers of computer processors. This table tracks the history of Intel's chips in the 40 years since Moore's prediction, showing the number of transistors (in thousands) and the year that chip was introduced.

Year	Transistors (1000's)	Year	Transistors (1000's)
1971	2.3	1993	3100
1972	3.3	1997	7500
1974	4.5	1999	9500
1979	29	2000	42000
1982	134	2003	77000
1985	275	2004	125000
1989	1200	2006	291000

a) Use years since Moore's prediction (ex: 1971 − 1965 = 6) to create a scatterplot showing the growth in the number of transistors, and explain why you would not try a linear model.
b) Find a better model.
c) What does your model's residuals plot suggest about the 2006 Intel chip?

33. Chips. A start-up company has developed an improved electronic chip for use in laboratory equipment. The company needs to project the manufacturing cost, so it develops a spreadsheet model that takes into account the purchase of production equipment, overhead, raw materials, depreciation, maintenance, and other business costs. The spreadsheet estimates the cost of producing 10,000 to 200,000 chips per year, as seen in the table.

Chips Produced (1000s)	Cost per Chip ($)	Chips Produced (1000s)	Cost per Chip ($)
10	146.10	90	47.22
20	105.80	100	44.31
30	85.75	120	42.88
40	77.02	140	39.05
50	66.10	160	37.47
60	63.92	180	35.09
70	58.80	200	34.04
80	50.91		

a) Why don't you think a linear model is appropriate?
b) Develop a better model to predict *Costs* based on the *level* of production.

34. Pendulum. A student experimenting with a pendulum counted the number of full swings the pendulum made in 20 seconds for various lengths of string. Her data are shown below.

Length (in.)	6.5	9	11.5	14.5	18	21	24	27	30	37.5
Number of Swings	22	20	17	16	14	13	13	12	11	10

a) Explain why a linear model is not appropriate for using the *Length* of a pendulum to predict the *Number of Swings* in 20 seconds.
b) Create an appropriate model.

35. Slower is cheaper? Researchers studying how a car's *Fuel Efficiency* varies with its *Speed* drove a compact car 200 miles at various speeds on a test track. Their data are shown in the table. Why wouldn't you use linear, exponential, or power models?

Speed (mph)	35	40	45	50	55	60	65	70	75
Fuel Eff. (mpg)	25.9	27.7	28.5	29.5	29.2	27.4	26.4	24.2	22.8

36. Tree growth. A 1996 study examined the growth of grapefruit trees in Texas, determining the average trunk *Diameter* (in inches) for trees of varying *Ages:*

Age (yr)	2	4	6	8	10	12	14	16	18	20
Diameter (in.)	2.1	3.9	5.2	6.2	6.9	7.6	8.3	9.1	10.0	11.4

a) Fit a linear model to these data. What concerns do you have about the model?
b) Why would you not use an exponential or power model either?

37. Orange production. The table below shows that as the number of oranges on a tree increases, the fruit tends to get smaller.

Number of Oranges/Tree	Average Weight/Fruit (lb)
50	0.60
100	0.58
150	0.56
200	0.55
250	0.53
300	0.52
350	0.50
400	0.49
450	0.48
500	0.46
600	0.44
700	0.42
800	0.40
900	0.38

a) Create a linear model for this relationship, and express any concerns you may have.
b) Why might you prefer this linear model to a power model?

38. Oranges again. In Exercise 37 you rejected a power model. Try an exponential model instead. What are the advantage and disadvantages of using this model?

Answers

Do the Math

1. a)

x	0	1	2	3	4	5
y	4	12	36	108	324	972

b)

x	0	1	2	3	4	5
y	256	192	144	108	81	60.75

2. a) Linear; *y* increases by adding the same amount (5) each time.
b) Exponential; *y* doubles each time—that's multiplying by the same factor.
c) Exponential; *y* is divided by 3 each time.
d) Linear; *y* decreases by subtracting 3 each time.

3. a) Linear: $y = 30 + 5x$
b) Exponential: $y = 30 \cdot 5^x$
c) Linear: $y = 64 - 7x$
d) Exponential: $y = 64(0.5)^x$
e) Exponential: $y = 5000(1.15^x)$
f) Linear: $y = 5000 + 50x$
g) Exponential: $y = 800(0.95^x)$

4. a) 675   b) −75   c) 1530.75

5. a) $^1/_{25}$ b) 4   c) $^1/_4$
d) $^1/_{32}$ e) 5   f) $^1/_3$

6. a) As x increases, y increases at an increasing rate

x	0	1	2	3	4	5	6
y	0	0.5	2	4.5	8	12.5	18

b) As x increases, y decreases at a decreasing rate

x	1	2	4	5	8	10
y	40	20	10	8	5	4

c) As x increases, y increases at a decreasing rate

x	0	1	4	9	16	25
y	0	2	4	6	8	10

d) As x increases, y decreases at a decreasing rate

x	1	4	9	16	25	36
y	30	15	10	7.5	6	5

Just Checking

1. Exponential (Interest earned each year is a percentage of the amount present.)

2. Linear

3. None of these. The temperature probably starts low in the winter months, goes up during the spring, peaks in the summer, then goes back down again through the fall and into the next winter.

4. Power. (Strength may depend on the cross-sectional area of the rivet.)

Gathering Data

Samples

Y ou've heard about all of these situations. What do they have in common?

- Pollsters predict which candidate will win an upcoming election.
- Traffic researchers report that more people are wearing seat belts.
- Government crash tests produce safety ratings for automobiles.
- A soft drink company asks people to taste-test a new flavor of soda.
- A quality control inspector tests some of a factory's products.

In each case people look at a few individuals—a sample—in hopes of learning something about the larger world—a population. Taking that step from a small sample to the entire population is impossible without understanding Statistics. To make business decisions, to do science, to choose wise investments, or to understand what voters think they'll do in the next election, we need to stretch beyond the data at hand to the world at large.

To make that stretch, we need three ideas. You'll find the first one natural. The second may be more surprising. The third is one of the strange but true facts that often confuse those who don't know Statistics.

Idea 1: Examine a Part of the Whole

The first idea is to draw a sample. We'd like to know about an entire **population,** but examining everyone is usually impractical, if not impossible. So we settle for examining a smaller group—a **sample**—selected from the population.

You do this every day. For example, suppose you wonder how the vegetable soup you're cooking for dinner tonight is going to go over with your friends. To decide whether it meets your standards, you only need to try a small amount. You might taste just a spoonful or two. You certainly don't have to consume the whole pot. You trust that the taste will *represent* the flavor of the entire pot. The idea behind your tasting is that a small sample, if selected properly, can represent the entire population.

It's hard to go a day without hearing about the latest opinion poll. These polls are examples of **sample surveys,** designed to ask questions of a small group of people in the hope of learning something about the entire population. Most likely, you've never been selected to be part of one of these national opinion polls. That's true of most people. So how can the pollsters claim that a sample is representative of the entire population? The answer is that professional pollsters work quite hard to ensure that the "taste"—the sample that they take—represents the population. If not, the sample can give misleading information about the population.

THE W'S AND SAMPLING

The population we are interested in is usually determined by the *Why* of our study. The sample we draw will be the *Who.* *When* and *How* we draw the sample may depend on what is practical.

Bias

Selecting a sample to represent the population fairly is more difficult than it sounds. Polls or surveys most often fail because they use a sampling method that overlooks or favors some part of the population. The method may overlook subgroups that are harder to find (such as the homeless or those who use only cell phones) or favor others (such as Internet users who like to respond to online surveys). Sampling methods that, by their nature, tend to over- or underemphasize some characteristics of the population are said to be **biased.** Bias is the Achilles heel of sampling—the one thing above all to avoid. Conclusions based on samples drawn with biased methods are inherently flawed. There is usually no way to fix bias after the sample is drawn and no way to salvage useful information from it.

Here's a famous example of a really dismal failure. During the 1936 presidential campaign between Alf Landon and Franklin Delano Roosevelt, the magazine *Literary Digest* mailed more than 10 million ballots and got back an astonishing 2.4 million. The results were clear: Alf Landon would be the next president by a landslide, 57% to 43%. You remember President Landon? No? In fact, Landon carried only two states. Roosevelt won, 62% to 37%, and, perhaps coincidentally, the *Digest* went bankrupt soon afterward.

What went wrong? One problem was that the *Digest's* sample wasn't representative. Where would *you* find 10 million names and addresses to sample? The *Digest* used the phone book, as many surveys do.[1] But in 1936, at the

In 1936, a young pollster named George Gallup used a subsample of only 3000 of the 2.4 million responses that the *Literary Digest* received to reproduce the wrong prediction of Landon's victory over Roosevelt. He then used an entirely different sample of 50,000 and predicted that Roosevelt would get 56% of the vote to Landon's 44%. His sample was apparently much more *representative* of the actual voting populace. The Gallup Organization went on to become one of the leading polling companies.

---

[1]Today phone numbers are computer-generated to make sure that unlisted numbers are included. But even now, cell phones and VOIP Internet phones are often not included.

height of the Great Depression, telephones were a real luxury, so they sampled more rich than poor voters. The campaign of 1936 focused on the economy, and those who were less well off were more likely to vote for the Democrat. So the *Digest*'s sample was hopelessly biased.

How do modern polls get their samples to *represent* the entire population? You might think that they'd handpick individuals to sample with care and precision. But in fact, they do something quite different: They select individuals to sample *at random*. The importance of deliberately using randomness is one of the great insights of Statistics.

Idea 2: Randomize

Think back to the soup sample. Suppose you add some salt to the pot. If you sample soup from the top before stirring, you'll get the misleading idea that the whole pot is salty. If you sample from the bottom, you'll get an equally misleading idea that the whole pot is bland. By stirring before you sample, you *randomize* the amount of salt throughout the pot, making each taste more typical of the whole pot.

Not only does randomization protect you against factors that you know are in the data, it can also help protect against factors that you didn't even know were there. Suppose, while you weren't looking, a friend added a handful of peas to the soup. If they're down at the bottom of the pot, and you don't randomize the soup by stirring, your test spoonful won't have any peas. By stirring in the salt, you *also* randomize the peas throughout the pot, making your sample taste more typical of the overall pot *even though you didn't know the peas were there*. So randomizing protects us even in this case.

How do we "stir" populations? We select from them at random. **Randomizing** protects us from the influences of *all* the features of our population by making sure that, *on average*, the sample looks like the rest of the population.

> **Why not match the sample to the population?** Rather than randomizing, we could try to design our sample so that the people we choose are typical in terms of every characteristic we can think of. We might want the income levels of those we sample to match the population. How about age? Political affiliation? Marital status? Having children? Living in the suburbs? We can't possibly think of all the things that might be important. Even if we could, we wouldn't be able to match our sample to the population for all these characteristics.

For Example Is a Random Sample Representative?

Here are summary statistics comparing two samples of 8000 drawn at random from a company's database of 3.5 million customers:

Mean Age (yr)	White (%)	Female (%)	Mean # of Children	Income Bracket (1–7)	Wealth Bracket (1–9)	Homeowner? (% Yes)
61.4	85.12	56.2	1.54	3.91	5.29	71.36
61.2	84.44	56.4	1.51	3.88	5.33	72.30

Question: Do you think these samples are representative of the population? Explain.

The two samples look very similar with respect to these seven variables. It appears that randomizing has automatically matched them pretty closely. We can reasonably assume that since the two samples don't differ too much from each other, they don't differ much from the rest of the population either.

Idea 3: It's the Sample Size

How large a random sample do we need for the sample to be reasonably representative of the population? Most people think that we need a large percentage, or *fraction*, of the population, but it turns out that what matters is the *number* of individuals *in the sample*, not the size of the population. A random sample of 100 students in a college represents the student body just about as well as a random sample of 100 voters represents the entire electorate of the United States. This is the *third* idea and probably the most surprising one in designing surveys.

How can it be that only the size of the sample, and not the population, matters? Well, let's return one last time to that pot of soup. If you're cooking for a banquet rather than just for a few people, your pot will be bigger, but do you need a bigger spoon to decide how the soup tastes? Of course not. The same-size spoonful is probably enough to make a decision about the entire pot, no matter how large the pot. The *fraction* of the population that you've sampled doesn't matter.[2] It's the **sample size** itself that's important.

How big a sample do you need? That depends on what you're estimating. To get an idea of what's really in the soup, you'll need a large enough taste to get a *representative* sample from the pot. For a survey that tries to find the proportion of the population falling into a category, you'll usually need several hundred respondents to say anything precise enough to be useful.[3]

Populations and Samples.
How well can a sample reveal the population's shape, center, and spread? Explore what happens as you change the sample size.

TI-*nspire*

Does a Census Make Sense?

Wouldn't it be better to just include everyone and "sample" the entire population? Such a special sample is called a **census.** Although a census would appear to provide the best possible information about the population, there are a number of reasons why it might not.

First, it can be difficult to complete a census. Some individuals in the population will be hard (and expensive) to locate. Or a census might just be impractical. If you were a taste tester for the Hostess™ Company, you probably wouldn't want to census *all* the Twinkies on the production line. Not only might this be life-endangering, but you wouldn't have any left to sell.

Second, populations rarely stand still. In populations of people, babies are born and folks die or leave the country. In opinion surveys, events may cause a shift in opinion during the survey. A census takes longer to complete and the population changes while you work. A sample surveyed in just a few days may give more accurate information.

Third, taking a census can be more complex than sampling. For example, the U.S. Census records too many college students. Many are counted once with their families and are then counted a second time in a report filed by their schools.

A friend who knows that you are taking Statistics asks your advice on her study. What can you possibly say that will be helpful? Just say, "If you could just get a larger sample, it would probably improve your study." Even though a larger sample might not be worth the cost, it will almost always make the results more precise.

---

[2]Well, that's not exactly true. If the population is small enough and the sample is more than 10% of the whole population, it *can* matter. It doesn't matter whenever, as usual, our sample is a very small fraction of the population.

[3]Chapter 17 gives the details behind this statement and shows how to decide on a sample size for a survey.

> **The undercount.** It's particularly difficult to compile a complete census of a population as large, complex, and spread out as the U.S. population. The U.S. Census is known to miss some residents. On occasion, the undercount has been striking. For example, there have been blocks in inner cities in which the number of residents recorded by the Census was smaller than the number of electric meters for which bills were being paid. What makes the problem particularly important is that some groups have a higher probability of being missed than others—undocumented immigrants, the homeless, the poor. The Census Bureau proposed the use of random sampling to estimate the number of residents missed by the ordinary census. Unfortunately, the resulting debate has become more political than statistical.

Populations and Parameters

Any quantity that we calculate from data could be called a "statistic." But in practice, we usually use a statistic to estimate a population parameter.

Remember: Population model parameters are not just unknown—usually they are *unknowable*. We have to settle for sample statistics.

A study found that teens were less likely to "buckle up." The National Center for Chronic Disease Prevention and Health Promotion reports that 21.7% of U.S. teens never or rarely wear seat belts. We're sure they didn't take a census, so what *does* the 21.7% mean? We can't know what percentage of all teenagers wear seat belts. Reality is just too complex. But we can simplify the question by building a model.

Models use mathematics to represent reality. Parameters are the key numbers in those models. A parameter used in a model for a population is sometimes called (redundantly) a **population parameter.**

But let's not forget about the data. We use summaries of the data to estimate the population parameters. As we know, any summary found from the data is a **statistic.** Sometimes you'll see the term **sample statistic** (also redundant)[4].

We've already met two parameters in Chapter 6: the mean, μ, and the standard deviation, σ. We'll try to keep denoting population model parameters with Greek letters and the corresponding statistics with Latin letters. So the sample mean is $\bar{y}$ and the population mean is μ. The standard deviation of the data is s, and the corresponding parameter is σ (Greek for s).

Get the pattern? Good. But it breaks down. Suppose we want to talk about the proportion of teens who don't wear seat belts. Now we'll use p for the population model parameter and $\hat{p}$ for the proportion from the data (since, like $\hat{y}$ in regression, it's an estimated value).

Here's a table summarizing the notation:

Name	Statistic	Parameter
Mean	$\bar{y}$	μ (mu, pronounced "meeoo," not "moo")
Standard deviation	s	σ (sigma)
Proportion	$\hat{p}$	p

We draw samples because we can't work with the entire population, but we want the statistics we compute from a sample to reflect the corresponding parameters accurately. A sample that does this is said to be **representative.** A biased sampling methodology tends to over- or underestimate the parameter of interest.

---

[4]Where else besides a sample *could* a statistic come from?

Just Checking

1. Various claims are often made for surveys. Why is each of the following claims not correct?

   a) It is always better to take a census than to draw a sample.

   b) Stopping students on their way out of the cafeteria is a good way to sample if we want to know about the quality of the food there.

   c) We drew a sample of 100 from the 3000 students in a school. To get the same level of precision for a town of 30,000 residents, we'll need a sample of 1000.

   d) A poll taken at a statistics support website garnered 12,357 responses. The majority said they enjoy doing statistics homework. With a sample size that large, we can be pretty sure that most Statistics students feel this way, too.

   e) The true percentage of all Statistics students who enjoy the homework is called a "population statistic."

   (Check your answers on page 240.)

Simple Random Samples

Insist that every possible *sample* of the size you plan to draw has an equal chance to be selected. This guarantees that each person has an equal chance of being selected and each *combination* of people has an equal chance of being selected as well. A sample drawn in this way is called a **Simple Random Sample,** usually abbreviated **SRS.** An SRS is the standard against which we measure other sampling methods, and the sampling method on which the theory of working with sampled data is based.

To select a sample at random, we first need to define where the sample will come from. The **sampling frame** is a list of individuals from which the sample is drawn. For example, to draw a random sample of students at a college, we might obtain a list of all registered full-time students and sample from that list. In defining the sampling frame, we must deal with the details of defining the population. Are part-time students included? How about those who are attending school elsewhere and transferring credits back to the college?

Once we have a sampling frame, the easiest way to choose an SRS is to assign a number to each individual in the sampling frame. We then select those whose numbers are randomly generated by a computer or calculator[5] or read from a table.

For Example Using Random Numbers to Get an SRS

There are 80 students enrolled in an introductory Statistics class; you are to select a sample of 5.

Question: How can you select an SRS of 5 students using these random digits found on the Internet: 05166 29305 77482?

First I'll number the students from 00 to 79. Taking the random numbers two digits at a time gives me 05, 16, 62, 93, 05, 77, and 48. I'll ignore 93 because the students were numbered only up to 79. And, so as not to pick the same person twice, I'll skip the repeated number 05. My simple random sample consists of students with the numbers 05, 16, 62, 77, and 48.

---

[5]Chapter 12 will discuss ways of finding and working with random numbers.

Samples drawn at random generally differ one from another. Each draw of random numbers selects *different* people for our sample. These differences lead to different values for the variables we measure. We call these sample-to-sample differences **sampling variability.** Surprisingly, sampling variability isn't a problem; it's an opportunity. In future chapters we'll investigate what the variation in a sample can tell us about its population.

ERROR OKAY, BIAS BAD!

Sampling variability is sometimes referred to as *sampling error*, making it sound like it's some kind of mistake. It's not. We understand that samples will vary, so "sampling error" is to be expected. It's *bias* we must strive to avoid. Bias means our sampling method distorts our view of the population, and that will surely lead to mistakes.

Stratified Sampling

Simple random sampling is not the only fair way to sample. More complicated designs may save time or money or help avoid sampling problems. All statistical sampling designs have in common the idea that chance, rather than human choice, is used to select the sample.

Designs that are used to sample from large populations—especially populations of people residing across large areas—are often more complicated than simple random samples. Sometimes the population is first sliced into groups of similar individuals called **strata,** before the sample is selected. Then simple random sampling is used within each stratum before the results are combined. This common sampling design is called **stratified random sampling.**

Why would we want to complicate things? Here's an example. Suppose we want to learn how students feel about funding for the football team at a large university. The campus is 60% men and 40% women, and we suspect that men and women have different views on the funding. If we use simple random sampling to select 100 people for the survey, we could end up with 70 men and 30 women or 35 men and 65 women. Our resulting estimates of the level of support for the football funding could vary widely. To help reduce this sampling variability, we can decide to force a representative balance, selecting 60 men at random and 40 women at random. This would guarantee that the proportions of men and women within our sample match the proportions in the population, and that should make such samples more accurate in representing population opinion.

You can imagine the importance of stratifying polls by race, income, age, and other characteristics. Samples taken within a stratum vary less, so our estimates can be more precise. This reduced sampling variability is the most important benefit of stratifying.

For Example Stratifying the Sample

Recap: You're trying to find out what freshmen think of the food served on campus. Food Services believes that men and women typically have different opinions about the importance of the salad bar.

Question: You initially considered an SRS, but how should you adjust your sampling strategy to allow for this difference between men and women?

I will stratify my sample by drawing an SRS of men and a separate SRS of women—assuming that the data from the registrar include information about each person's sex.

Cluster and Multistage Sampling

Suppose we wanted to assess the reading level of this textbook based on the length of the sentences. Simple random sampling could be awkward; we'd have to number each sentence, then find, for example, the 576th sentence or the 2482nd sentence, and so on. Doesn't sound like much fun, does it?

It would be much easier to pick a few *pages* at random and count the lengths of the sentences on those pages. That works if we believe that each page is representative of the entire book in terms of reading level. Splitting the population into representative **clusters** can make sampling more practical. Then we could simply select one or a few clusters at random and perform a census within each of them. This sampling design is called **cluster sampling.** If each cluster represents the full population fairly, cluster sampling will be unbiased.

For Example Cluster Sampling

Recap: In trying to find out what freshmen think about the food served on campus, you've considered both an SRS and a stratified sample. Now you have run into a problem: It's simply too difficult and time consuming to track down the individuals whose names were chosen for your sample. Fortunately, freshmen at your school are all housed in 10 freshman dorms.

Questions: How could you use this fact to draw a cluster sample? How might that alleviate the problem? What concerns do you have?

To draw a cluster sample, I would select one or two dorms at random and then try to contact everyone in each selected dorm. I could save time by simply knocking on doors on a given evening and interviewing people. I'd have to assume that freshmen were assigned to dorms pretty much at random and that the people I'm able to contact are representative of everyone in the dorm.

What's the difference between cluster sampling and stratified sampling? We stratify to ensure that our sample represents different groups in the population, then sample randomly within each stratum. Strata are internally similar, but differ from one another. By contrast, clusters are internally varied, each resembling the overall population. We select clusters to make sampling more practical or affordable.

Stratified vs. cluster sampling. Boston cream pie consists of a layer of yellow cake, a layer of pastry creme, another cake layer, and then a chocolate frosting. Suppose you are a professional taster (yes, there really are such people) whose job is to check your company's pies for quality. You'd need to eat small samples of randomly selected pies, tasting all three components: the cake, the creme, and the frosting.

One approach is to cut a thin vertical slice out of the pie. Such a slice will be a lot like the entire pie, so by eating that slice, you'll learn about the whole pie. This vertical slice containing all the different ingredients in the pie would be a *cluster* sample.

Another approach is to sample in *strata:* Select some tastes of the cake at random, some tastes of creme at random, and some bits of frosting at random. You'll end up with a reliable judgment of the pie's quality.

(continued)

Many populations you might want to learn about are like this Boston cream pie. You can think of the subpopulations of interest as horizontal strata, like the layers of pie. Cluster samples slice vertically across the layers to obtain clusters, each of which is representative of the entire population. Stratified samples represent the population by drawing some from each layer, reducing variability in the results that could arise because of the differences among the layers.

STRATA OR CLUSTERS?
We may split a population into strata or clusters. What's the difference? We create strata by dividing the population into groups of similar individuals so that each stratum is different from the others. By contrast, since clusters each represent the entire population, they all look pretty much alike.

Sometimes we use a variety of sampling methods together. In trying to assess the reading level of this book, we might worry that it starts out easy and then gets harder as the concepts become more difficult. If so, we'd want to avoid samples that selected heavily from early or from late chapters. To guarantee a fair mix of chapters, we could randomly choose one chapter from each of the five parts of the book and then randomly select a few pages from each of those chapters. If, altogether, that made too many sentences, we might select a few sentences at random from each of the chosen pages. So, what is our sampling strategy? First we stratify by the part of the book and randomly choose a chapter to represent each stratum. Within each selected chapter, we choose pages as clusters. Finally, we consider an SRS of sentences within each cluster. Sampling schemes that combine several methods are called **multistage samples.** Most surveys conducted by professional polling organizations use some combination of stratified and cluster sampling as well as simple random samples.

For Example Multistage Sampling

Recap: Having learned that freshmen are housed in separate dorms allowed you to sample their attitudes about the campus food by going to dorms chosen at random, but you're still concerned about possible differences in opinions between men and women. It turns out that these freshmen dorms house the sexes on alternate floors.

Question: How can you design a sampling plan that uses this fact to your advantage?

Now I can stratify my sample by sex. I would first choose one or two dorms at random and then select some dorm floors at random from among those that house men and, separately, from among those that house women. I could then treat each floor as a cluster and interview everyone on that floor.

Systematic Samples

Some samples select individuals systematically. For example, you might survey every 10th person on an alphabetical list of students. To make it random, you still must start the systematic selection from a randomly selected individual. When the order of the list is not associated in any way with the responses sought, **systematic sampling** can give a representative sample. Systematic sampling can be much less expensive than true random sampling.

Think about the reading-level sampling example again. Suppose we have chosen a chapter of the book at random, then three pages at random from that chapter, and now we want to select a sample of 10 sentences from the 73 sentences found on those pages. Instead of numbering each sentence so we can pick a simple random sample, it would be easier to sample systematically.

A quick calculation shows 73/10 = 7.3, so we can get our sample by just picking every seventh sentence on the page. But where should you start? At random, of course. We've accounted for 10 × 7 = 70 of the sentences, so we'll throw the extra 3 into the starting group and choose a sentence at random from the first 10. Then we pick every seventh sentence after that and record its length.

Just Checking

2. We need to survey a random sample of the 300 passengers on a flight from San Francisco to Tokyo. Name each sampling method described below.

a) Pick every 10th passenger as people board the plane.

b) From the boarding list, randomly choose 5 people flying first class and 25 of the other passengers.

c) Randomly generate 30 seat numbers and survey the passengers who sit there.

d) Randomly select a seat position (right window, right center, right aisle, etc.) and survey all the passengers sitting in those seats.

(Check your answers on page 240.)

STEP–BY–STEP EXAMPLE   Sampling

An assignment says, "Conduct your own sample survey to find out how many hours per week students at your school spend watching TV during the school year." Let's see how we might do this step by step. (Remember, though—actually collecting the data from your sample can be difficult and time consuming.)

Question: How would you design this survey?

THINK

Plan State what you want to know.

I wanted to design a study to find out how many hours of TV students at my school watch per week.

Population and Parameter Identify the W's of the study. The *Why* determines the population and the associated sampling frame. The *What* identifies the parameter of interest and the variables measured. The *Who* is the sample we actually draw. The *How, When,* and *Where* are given by the sampling plan.

Often, thinking about the *Why* will help us see whether the sampling frame and plan are adequate to learn about the population.

The population studied was students at our school. I obtained a list of all students currently enrolled and used it as the sampling frame. The parameter of interest was the number of TV hours watched per week during the school year, which I attempted to measure by asking students how much TV they watched during the previous week.

(continued)

Sampling Plan Specify the sampling method and the sample size, n. Specify how the sample was actually drawn. What is the sampling frame? How was the randomization performed?

A good description should be complete enough to allow someone to replicate the procedure, drawing another sample from the same population in the same manner.

I decided against stratifying by grade or sex because I didn't think TV watching would differ much between males and females or across grades. I selected a systematic random sample of students from the list by rolling dice to decide where to start and then choosing every 10th student. This method generated a sample of 213 students from the population of 2133 students.

Sampling Practice Specify *When*, *Where*, and *How* the sampling was performed. Specify any other details of your survey, such as how respondents were contacted, what incentives were offered to encourage them to respond, how nonrespondents were treated, and so on.

The question you ask also matters. It's better to be specific ("How many hours did you watch TV last week?") than to ask a general question ("How many hours of TV do you usually watch in a week?").

The survey was taken over the period Oct. 15 to Oct. 25. Surveys were sent to selected students in homeroom, with the request that they return them to the homeroom teacher.

The survey the students received asked the following question: "How many hours did you spend watching television last week?"

Summary and Conclusion This report should include a discussion of the process and results. Show a display of the data, provide and interpret the statistics from the sample, and state the conclusions that you reached about the population.

In addition, it's good practice to discuss any special circumstances. Professional polling organizations report the *When* of their samples but will also note, for example, any important news that might have changed respondents' opinions during the sampling process. In this survey, perhaps, a major news story or sporting event might change students' TV viewing behavior.

Of the 213 students surveyed, 110 responded. It's possible that the nonrespondents differ in the number of TV hours watched from those who responded, but I was unable to follow up on them due to limited time. The 110 respondents reported an average 3.62 hours of TV watching per week. The median was only 2 hours per week. A histogram of the data shows that the distribution is highly right-skewed, indicating that the median might be a more appropriate summary of the typical TV watching of the students.

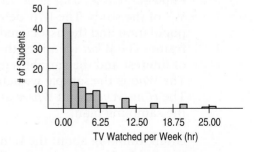

Most of the students (90%) watch between 0 and 10 hours per week, while 30% reported watching less than 1 hour per week. A few watch much more. About 3% reported watching more than 20 hours per week.

Defining the "Who": You Can't Always Get What You Want

Before you start a survey, think first about the population you want to study. You may find that it's not the well-defined group you thought it was. Who, exactly, is a student, for example? Even if the population seems well defined, it may not be a practical group from which to draw a sample. For example, election polls want to sample from all those who will vote in the next election—a population that is impossible to identify before Election Day.

Next, you must specify the sampling frame. (Do you have a list of students to sample from? How about a list of registered voters?) Usually, the sampling frame is not the group you *really* want to know about. (All those registered to vote are not equally likely to show up.) The sampling frame limits what your survey can find out.

Then there's your target sample. These are the individuals for whom you *intend* to measure responses. You're not likely to get responses from all of them. ("I know it's dinnertime, but I'm sure you wouldn't mind answering a few questions. It'll only take 20 minutes or so. Oh, you're busy?") Nonresponse is a problem in many surveys.

Finally, there's your sample—the actual respondents. These are the individuals about whom you *do* get data and can draw conclusions. Unfortunately, they might not be representative of the sampling frame or the population. A careful study should pay attention to how well the sample represents the population of interest.

> The population is determined by the *Why* of the study. Unfortunately, the sample is just those we can reach to obtain responses—the *Who* of the study. This difference could undermine even a well-designed study.

CALVIN AND HOBBES © 1993 Watterson. Reprinted with permission of Universal Press Syndicate. All rights reserved.

WHAT CAN GO WRONG?—OR, HOW TO SAMPLE BADLY

Bad sample designs yield worthless data. Many of the most convenient forms of sampling can be seriously biased. And there is no way to correct for the bias from a bad sample. So it's wise to pay attention to sample design—and to beware of reports based on poor samples.

Sample Badly with Volunteers

One of the most common dangerous sampling methods is a voluntary response sample. In a **voluntary response sample,** a large group of individuals is invited to respond, and all who do respond are counted. This method is used by

(continued)

call-in shows, 900 numbers, Internet polls, and letters written to members of Congress. Voluntary response samples are almost always biased, and so conclusions drawn from them are almost always wrong. Such samples are often biased toward those with strong opinions or those who are strongly motivated. People with very negative opinions tend to respond more often than those with equally strong positive opinions. The sample is not representative, even though every individual in the population may have been offered the chance to respond. The resulting **voluntary response bias** invalidates the survey.

> **If you had it to do over again, would you have children?**
> Ann Landers, the advice columnist, asked parents this question. The overwhelming majority—70% of the more than 10,000 people who wrote in—said no, kids weren't worth it. A more carefully designed survey later showed that about 90% of parents actually are happy with their decision to have children. What accounts for the striking difference in these two results? What parents do you think are most likely to respond to the original question?

For Example Bias in Sampling

Recap: You're trying to find out what freshmen think of the food served on campus, and have thought of a variety of sampling methods, all time consuming. A friend suggests that you set up a "Tell Us What You Think" website and invite freshmen to visit the site to complete a questionnaire.

Question: What's wrong with this idea?

Letting each freshman decide whether to participate makes this a voluntary response survey. Students who were dissatisfied might be more likely to go to the website to record their complaints, and this could give me a biased view of the opinions of all freshmen.

Sample Badly, but Conveniently

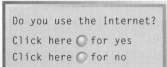

Internet convenience surveys are worthless. As voluntary response surveys, they have no well-defined sampling frame (all those who use the Internet and visit their site?) and thus report no useful information. Do not believe them.

Another sampling method that doesn't work is convenience sampling. As the name suggests, in **convenience sampling** we simply include the individuals who are convenient for us to sample. Unfortunately, this group may not be representative of the population. A recent survey of 437 potential home buyers in Orange County, California, found, among other things, that

> *All but 2 percent of the buyers have at least one computer at home, and 62 percent have two or more. Of those with a computer, 99 percent are connected to the Internet (Jennifer Hieger, "Portrait of Homebuyer Household: 2 Kids and a PC," Orange County Register, 27 July 2001).*

Later in the article, we learn that the survey was conducted via the Internet! That was a convenient way to collect data and surely easier than drawing a simple random sample, but perhaps home builders shouldn't conclude from this study that *every* family has a computer and an Internet connection.

Many surveys conducted at shopping malls suffer from the same problem. People in shopping malls are not necessarily representative of the population of interest. Mall shoppers tend to be more affluent and include a larger percentage of teenagers and retirees than the population at large. To make matters worse, survey interviewers tend to select individuals who look "safe," or easy to interview.

For Example Bias in Sampling

Recap: To try to gauge freshman opinion about the food served on campus, Food Services suggests that you just stand outside a school cafeteria at lunchtime and stop people to ask them questions.

Questions: What's wrong with this sampling strategy?

This would be a convenience sample, and it's likely to be biased. I would miss people who use the cafeteria for dinner, but not for lunch, and I'd never hear from anyone who hates the food so much that they have stopped coming to the school cafeterias.

Undercoverage

Many designs suffer from **undercoverage,** in which some portion of the population is not sampled at all or has a smaller representation in the sample than it has in the population. Undercoverage can arise for a number of reasons, but it's always a potential source of bias because the individuals included may differ from those who are missed. People in prison, homeless people, students, and long-term travelers are all likely to be left out. In telephone surveys, people who have only cell phones or who use VOIP Internet phones are often missing from the sample.

WHAT ELSE CAN GO WRONG?

A SHORT SURVEY

Given the fact that those who understand Statistics are smarter and better looking than those who don't, don't you think it is important to take a course in Statistics?

- **Watch out for nonrespondents.** A common and serious potential source of bias for most surveys is **nonresponse bias.** No survey succeeds in getting responses from everyone. The problem is that those who don't respond may differ from those who do. And they may differ on just the variables we care about. The lack of response will bias the results. Rather than sending out a large number of surveys for which the response rate will be low, it is often better to design a smaller randomized survey for which you have the resources to ensure a high response rate. One of the problems with nonresponse bias is that it's usually impossible to tell what the nonrespondents might have said.

- **Work hard to avoid influencing responses. Response bias**[6] refers to anything in the survey design that influences the responses. Response biases include the tendency of respondents to tailor their responses to try to please the interviewer, the natural unwillingness of respondents to reveal personal facts or admit to illegal or unapproved behavior, and the ways in which the wording of the questions can influence responses.

(continued)

___

[6]Response bias is not the opposite of nonresponse bias. (We don't make these terms up; we just try to explain them.)

A respondent may not understand the question—or may understand the question differently than the researcher intended it. ("Does anyone in your family belong to a union?" Do you mean just me, my spouse, and my children? Or does "family" include my father, my siblings, and my second cousin once removed? A question like "Do you approve of the recent actions of the Secretary of Labor?" is likely not to measure what you want if many respondents don't know who the Secretary of Labor is or what actions he or she recently made.

Respondents may even lie or shade their responses if they feel embarrassed by the question ("Did you have too much to drink last night?"), are intimidated or insulted by the question ("Could you understand our new *Instructions for Dummies* manual, or was it too difficult for you?"), or if they want to avoid offending the interviewer ("Would you hire a man with a tattoo?" asked by a tattooed interviewer).

The wording of the question can affect the responses. In January 2006, the *New York Times* asked half of the 1229 U.S. adults in their sample the following question:

> After 9/11, President Bush authorized government wiretaps on some phone calls in the U.S. without getting court warrants, saying this was necessary to reduce the threat of terrorism. Do you approve or disapprove of this?

They found that 53% of respondents approved. But when they asked the other half of their sample a question with only slightly different phrasing,

> After 9/11, George W. Bush authorized government wiretaps on some phone calls in the U.S. without getting court warrants. Do you approve or disapprove of this?

only 46% approved.

THE WIZARD OF ID   parker and hart

How to Think About Biases

- **Look for biases in any survey you encounter.** If you design one of your own, ask someone else to help look for biases that may not be obvious to you. And do this *before* you collect your data. There's no way to recover from a biased sampling method or a survey that asks biased questions. Sorry, it just can't be done.

  A bigger sample size for a biased study just gives you a bigger useless study. A really big sample gives you a really big useless study. (Think of the 2.4 million *Literary Digest* responses.)

A researcher distributed a survey to an organization before some money-saving changes were to be made. She asked how people felt about a proposed cutback in secretarial and administrative support on a seven-point scale from Very Happy to Very Unhappy.

But virtually all respondents were very unhappy about the cutbacks, so the results weren't particularly useful. If she had pretested the question, she might have chosen a scale that ran from Unhappy to Outraged.

- **Spend your time and resources reducing biases.** No other use of resources is as worthwhile as reducing the biases.
- **If you can, pilot-test your survey.** Administer the survey in the exact form that you intend to use it to a small sample drawn from the population you intend to sample. Look for misunderstandings, misinterpretation, confusion, or other possible biases. Then refine your survey instrument.
- **Always report your sampling methods in detail.** Others may be able to detect biases where you did not expect to find them.

SAMPLING IN YOUR WORLD

Facebook and low grades: a dangerous mix?

The Lantern, Friday April 17, 2009

Facebook users be wary. In a survey conducted on 219 Ohio State students, an OSU doctoral student says there is [an association] between Facebook use and lower grades.

But "the idea of this study was so exploratory and inconclusive and should be further researched," said Aryn Karpinski, co-author of the study, during a phone interview. Karpinski is a doctoral student in the school of educational policy and leadership.

The study found that students who use Facebook have grade point averages between 3.0 to 3.5, while non-users have GPAs between 3.5 to 4.0. In addition, students who use Facebook only study one to five hours per week, while non-users study 11 to 15 hours per week. But the study found that 79 percent of Facebook users do not think that their time spent on the site interferes with their schoolwork.

In order to survey students, Karpinski said she e-mailed professors and asked them if she could pass out the survey in class. The survey was six pages long and asked a series of open-ended questions.

She said she realizes the sample size was small, but the results were statistically significant. Of the 219 students interviewed, 102 were undergraduate students and 117 were graduate students. The study found that 148 of the students had a Facebook account.

What the study finds is that there may be a relationship between Facebook and lower GPAs, but not that it directly causes it. Because of the smaller sample size, critics have quickly jumped in . . . Brennon Slattery of pcworld.com slammed the findings of the survey and called the study "dubious." "To generate any broad-ranging claim based on such a small collection of data is rash," he said on the Web site.

http://media.www.thelantern.com/media/storage/paper333/news/2009/04/15/Campus/Facebook.And.Low.Grades.A.Dangerous.Mix-3709908.shtml

WHAT HAVE WE LEARNED?

We've learned that a representative sample can offer us important insights about populations. It's the size of the sample—and not its fraction of the larger population—that determines the precision of the statistics it yields.

We've learned several ways to draw samples, all based on the power of randomness to make them representative of the population of interest:

► A Simple Random Sample (SRS) is our standard. Every possible group of n individuals has an equal chance of being our sample. That's what makes it *simple*.

► Stratified samples can reduce sampling variability by identifying subgroups of similar individuals and then randomly sampling within each.

► Cluster samples randomly select among subgroups that each resemble the population at large, making our sampling tasks more manageable.

► Systematic samples can work in some situations and are often the least expensive method of sampling. But we still want to start them randomly.

► Multistage samples combine several random sampling methods.

We've learned that bias can destroy our ability to gain insights from our sample:

► Nonresponse bias can arise when sampled individuals will not or cannot respond.

► Response bias arises when respondents' answers might be affected by external influences, such as question wording or interviewer behavior.

We've learned that bias can also arise from poor sampling methods:

► Voluntary response samples are almost always biased and should be avoided and distrusted.

► Convenience samples are likely to be flawed for similar reasons.

► Even with a reasonable design, sampling frames may not be representative. Undercoverage occurs when individuals from a subgroup of the population are selected less often than they should be.

Finally, we've learned to look for biases in any survey we find and to be sure to report our methods whenever we perform a survey so that others can evaluate the fairness and accuracy of our results.

Terms

Population   The entire group of individuals or instances about whom we hope to learn.

Sample   A (representative) subset of a population, examined in hope of learning about the population.

Sample survey   A study that asks questions of a sample drawn from some population in the hope of learning something about the entire population. Polls taken to assess voter preferences are common sample surveys.

Bias   Any systematic failure of a sampling method to represent its population is bias. Biased sampling methods tend to over- or underestimate parameters. It is almost impossible to recover from bias, so efforts to avoid it are well spent. Common errors include

► relying on voluntary response.

► undercoverage of the population.

► nonresponse bias.

► response bias.

Randomization   The best defense against bias is randomization, in which each individual is given a fair, random chance of selection.

Sample size	The number of individuals in a sample. The sample size determines how well the sample represents the population, not the fraction of the population sampled.
Census	A sample that consists of the entire population is called a census.
Population parameter	A numerical characteristic of a model for a population. We rarely expect to know the true value of a population parameter, but we do hope to estimate it from sampled data. For example, the mean income of all employed people in the country is a population parameter.
Statistic, sample statistic	Statistics are values calculated for sampled data and are used to estimate a population parameter. For example, the mean income of all employed people in a representative sample can provide a good estimate of the corresponding population parameter. The term "sample statistic" is sometimes used, usually to parallel the corresponding term "population parameter."
Representative	A sample is said to be representative if the statistics computed from it accurately reflect the corresponding population parameters.
Simple random sample (SRS)	In a simple random sample each subset of the population has an equal chance of selection.
Sampling frame	A list of individuals from whom the sample is drawn.
Sampling variability	The natural tendency of randomly drawn samples to differ, one from another. Sometimes called *sampling error*, sampling variability is no error at all, but just the natural result of random sampling.
Stratified random sample	A sampling design in which the population is divided into several subpopulations, or **strata,** and random samples are then drawn from each stratum. If the strata are internally similar, but are different from each other, a stratified sample may yield more consistent results than an SRS.
Cluster sample	A sampling design in which entire groups, or **clusters,** are chosen at random. Cluster sampling is usually selected as a matter of convenience, practicality, or cost. Each cluster should be representative of the population, so all the clusters should be internally varied and similar to each other.
Multistage sample	Sampling schemes that combine several sampling methods are called multistage samples. For example, a national polling service may stratify the country by geographical regions, select a random sample of cities from each region, and then interview a cluster of residents in each city.
Systematic sample	A sample drawn by selecting individuals systematically from a sampling frame. When there is no relationship between the order of the sampling frame and the variables of interest, a systematic sample can be representative.
Pilot	A small trial run of a survey to check whether questions are clear. A pilot study can reduce errors due to ambiguous questions.
Voluntary response bias	Bias introduced to a sample when individuals can choose on their own whether to participate in the sample. Samples based on voluntary response are always invalid and cannot be recovered, no matter how large the sample size.
Convenience sample	A convenience sample consists of the individuals who are conveniently available. Convenience samples often fail to be representative because every individual in the population is not equally convenient to sample.
Undercoverage	A sampling scheme that biases the sample in a way that gives a part of the population less representation than it has in the population.
Nonresponse bias	Bias introduced when a large fraction of those sampled fails to respond. Those who do respond are likely to not represent the entire population. Voluntary response bias is a form of nonresponse bias, but nonresponse may occur for other reasons. For example, those who are at work during the day won't respond to a telephone survey conducted only during working hours.
Response bias	Anything in a survey design that influences responses falls under the heading of response bias. One typical response bias arises from the wording of questions, which may suggest a favored response. Voters, for example, are more likely to express support of "the president" than support of the particular person holding that office at the moment.

(continued)

Skills

THINK
- ▶ Know the basic concepts and terminology of sampling (see the preceding list).
- ▶ Recognize population parameters in descriptions of populations and samples.
- ▶ Understand the value of randomization as a defense against bias.
- ▶ Understand the value of sampling to estimate population parameters from statistics calculated on representative samples drawn from the population.
- ▶ Understand that the size of the sample (not the fraction of the population) determines the precision of estimates.

SHOW
- ▶ Know that we draw a simple random sample from a master list of a population, using a calculator, a computer, or a table of random numbers.

TELL
- ▶ Know what to report about a sample as part of your statistical analysis.
- ▶ Report possible sources of bias in sampling methods. Recognize voluntary response and nonresponse as sources of bias in a sample survey.

SAMPLING ON THE COMPUTER

Computer-generated pseudorandom numbers are usually good enough for drawing random samples. But there is little reason not to use the truly random values available on the Internet. We'll see more about this in Chapter 12.

Here's a convenient way to draw an SRS of a specified size using a computer-based sampling frame. The sampling frame can be a list of names or of identification numbers arrayed, for example, as a column in a spreadsheet, statistics program, or database:

1. Generate random numbers of enough digits so that each exceeds the size of the sampling frame list by several digits. This makes duplication unlikely.
2. Assign the random numbers arbitrarily to individuals in the sampling frame list. For example, put them in an adjacent column.
3. Sort the list of random numbers, carrying along the sampling frame list.
4. Now the first *n* values in the sorted sampling frame column are an SRS of *n* values from the entire sampling frame.

EXERCISES

A

1. **Roper.** Through their *Roper Reports Worldwide*, GfK Roper conducts a global consumer survey to help multinational companies understand different consumer attitudes throughout the world. Within 30 countries, the researchers interview 1000 people aged 13–65. Their samples are designed so that they get 500 males and 500 females in each country. (www.gfkamerica.com)
   a) Are they using a simple random sample? Explain.
   b) What kind of design do you think they are using?

2. **Student center survey.** For their class project, a group of Statistics students decide to survey the student body to assess opinions about the proposed new student center. Their sample of 200 contained 50 first-year students, 50 sophomores, 50 juniors, and 50 seniors.
   a) Do you think the group was using an SRS? Why?
   b) What sampling design do you think they used?

3. **Emoticons.** The website www.gamefaqs.com asked, as their question of the day to which visitors to the site were invited to respond, *"Do you ever use emoticons*

when you type online?" Of the 87,262 respondents, 27% said that they did not use emoticons. ;-(
a) What kind of sample was this?
b) How much confidence would you place in using 27% as an estimate of the fraction of people who use emoticons?

4. Drug tests. Major League Baseball tests players to see whether they are using performance-enhancing drugs. Officials select a team at random, and a drug-testing crew shows up unannounced to test all 40 players on the team. Each testing day can be considered a study of drug use in Major League Baseball.
a) What kind of sample is this?
b) Is that choice appropriate?

5. Survey questions. Examine each of the following questions for possible bias. If you think the question is biased, indicate how and propose a better question.
a) Should companies that pollute the environment be compelled to pay the costs of cleanup?
b) Given that 18-year-olds are old enough to vote and to serve in the military, is it fair to set the drinking age at 21?

6. More survey questions. Examine each of the following questions for possible bias. If you think the question is biased, indicate how and propose a better question.
a) Do you think high school students should be required to wear uniforms?
b) Given humanity's great tradition of exploration, do you favor continued funding for space flights?

7. Mistaken poll. A local TV station conducted a "Pulse-Poll" about the upcoming mayoral election. Evening news viewers were invited to phone in their votes, with the results to be announced on the late-night news. Based on the phone calls, the station predicted that Amabo would win the election with 52% of the vote. They were wrong: Amabo lost, getting only 46% of the vote. Do you think the station's faulty prediction is more likely to be a result of bias or sampling error? Explain.

8. Another mistaken poll. Prior to the mayoral election discussed in Exercise 7, the newspaper also conducted a poll. The paper surveyed a random sample of registered voters stratified by political party, age, sex, and area of residence. This poll predicted that Amabo would win the election with 52% of the vote. The newspaper was wrong: Amabo lost, getting only 46% of the vote. Do you think the newspaper's faulty prediction is more likely to be a result of bias or sampling error? Explain.

9–16. What did they do? *For the following reports about statistical studies, identify the following items (if possible). If you can't tell, then say so—this often happens when we read about a survey.*
a) The population
b) The population parameter of interest
c) The sampling frame
d) The sample
e) The sampling method, including whether or not randomization was employed

f) Any potential sources of bias you can detect and any problems you see in generalizing to the population of interest

9. Consumers Union asked all subscribers whether they had used alternative medical treatments and, if so, whether they had benefited from them. For almost all of the treatments, approximately 20% of those responding reported cures or substantial improvement in their condition.

10. A question posted on the Lycos website on 18 June 2000 asked visitors to the site to say whether they thought that marijuana should be legally available for medicinal purposes. (www.lycos.com)

11. Researchers waited outside a bar they had randomly selected from a list of such establishments. They stopped every 10th person who came out of the bar and asked whether he or she thought drinking and driving was a serious problem.

12. Hoping to learn what issues may resonate with voters in the coming election, the campaign director for a mayoral candidate selects one block from each of the city's election districts. Staff members go there and interview all the residents they can find.

13. The Environmental Protection Agency took soil samples at 16 locations near a former industrial waste dump and checked each for evidence of toxic chemicals. They found no elevated levels of any harmful substances.

14. State police set up a roadblock to estimate the percentage of cars with up-to-date registration, insurance, and safety inspection stickers. They usually find problems with about 10% of the cars they stop.

15. A company packaging snack foods maintains quality control by randomly selecting 10 cases from each day's production and weighing the bags. Then they open one bag from each case and inspect the contents.

16. Dairy inspectors visit farms unannounced and take samples of the milk to test for contamination. If the milk is found to contain dirt, antibiotics, or other foreign matter, the milk will be destroyed and the farm reinspected until purity is restored.

B

17. Parent opinion, part 1. In a large city school system with 20 elementary schools, the school board is considering the adoption of a new policy that would require elementary students to pass a test in order to be promoted to the next grade. The PTA wants to find out whether parents agree with this plan. Listed below are some of the ideas proposed for gathering data. For each, indicate what kind of sampling strategy is involved and what (if any) biases might result.
a) Put a big ad in the newspaper asking people to log their opinions on the PTA website.
b) Randomly select one of the elementary schools and contact every parent by phone.

c) Send a survey home with every student, and ask parents to fill it out and return it the next day.

d) Randomly select 20 parents from each elementary school. Send them a survey, and follow up with a phone call if they do not return the survey within a week.

18. **Parent opinion, part 2.** Let's revisit the school system described in Exercise 17. Four new sampling strategies have been proposed to help the PTA determine whether parents favor requiring elementary students to pass a test in order to be promoted to the next grade. For each, indicate what kind of sampling strategy is involved and what (if any) biases might result.

a) Run a poll on the local TV news, asking people to dial one of two phone numbers to indicate whether they favor or oppose the plan.

b) Hold a PTA meeting at each of the 20 elementary schools, and tally the opinions expressed by those who attend the meetings.

c) Randomly select one class at each elementary school and contact each of those parents.

d) Go through the district's enrollment records, selecting every 40th parent. PTA volunteers will go to those homes to interview the people chosen.

19. **Churches.** For your political science class, you'd like to take a survey from a sample of all the Catholic Church members in your city. A list of churches shows 17 Catholic churches within the city limits. Rather than try to obtain a list of all members of all these churches, you decide to pick 3 churches at random. For those churches, you'll ask to get a list of all current members and contact 100 members at random.

a) What kind of design have you used?

b) What could go wrong with your design?

20. **Playground.** Some people have been complaining that the children's playground at a municipal park is too small and is in need of repair. Managers of the park decide to survey city residents to see if they believe the playground should be rebuilt. They hand out questionnaires to parents who bring children to the park. Describe possible biases in this sample.

21. **Roller coasters.** An amusement park has opened a new roller coaster. It is so popular that people are waiting for up to 3 hours for a 2-minute ride. Concerned about how patrons (who paid a large amount to enter the park and ride on the rides) feel about this, they survey every 10th person in the line for the roller coaster, starting from a randomly selected individual.

a) What kind of sample is this?

b) What is the sampling frame?

c) Is it likely to be representative?

22. **Playground, act two.** The survey described in Exercise 20 asked,

Many people believe this playground is too small and in need of repair. Do you think the playground

should be repaired and expanded even if that means raising the entrance fee to the park?

Describe two ways this question may lead to response bias.

23. **Wording the survey.** Two members of the PTA committee in Exercises 17 and 18 have proposed different questions to ask in seeking parents' opinions.

Question 1: Should elementary school–age children have to pass high-stakes tests in order to remain with their classmates?
Question 2: Should schools and students be held accountable for meeting yearly learning goals by testing students before they advance to the next grade?

a) Do you think responses to these two questions might differ? How? What kind of bias is this?

b) Propose a question with more neutral wording that might better assess parental opinion.

24. **Banning ephedra.** An online poll at a website asked:

A nationwide ban of the diet supplement ephedra went into effect recently. The herbal stimulant has been linked to 155 deaths and many more heart attacks and strokes. Ephedra manufacturer NVE Pharmaceuticals, claiming that the FDA lacked proof that ephedra is dangerous if used as directed, was denied a temporary restraining order on the ban yesterday by a federal judge. Do you think that ephedra should continue to be banned nationwide?

65% of 17,303 respondents said "yes." Comment on each of the following statements about this poll:

a) With a sample size that large, we can be pretty certain we know the true proportion of Americans who think ephedra should be banned.

b) The wording of the question is clearly very biased.

c) The sampling frame is all Internet users.

d) Results of this voluntary response survey can't be reliably generalized to any population of interest.

25. **Sampling methods.** Consider each of these situations. Do you think the proposed sampling method is appropriate? Explain.

a) We want to know what percentage of local doctors accept Medicaid patients. We call the offices of 50 doctors randomly selected from local Yellow Pages listings.

b) We want to know what percentage of local businesses anticipate hiring additional employees in the upcoming month. We randomly select a page in the Yellow Pages and call every business listed there.

26. **More sampling methods.** Consider each of these situations. Do you think the proposed sampling method is appropriate? Explain.

a) We want to know if there is neighborhood support to turn a vacant lot into a playground. We spend a Saturday afternoon going door-to-door in the neighborhood, asking people to sign a petition.

b) We want to know if students at our college are satisfied with the selection of food available on campus. We go to the largest cafeteria and interview every 10th person in line.

27. Accounting. Between quarterly audits, a company likes to check on its accounting procedures to address any problems before they become serious. The accounting staff processes payments on about 120 orders each day. The next day, the supervisor rechecks 10 of the transactions to be sure they were processed properly.

a) Propose a sampling strategy for the supervisor.

b) How would you modify that strategy if the company makes both wholesale and retail sales, requiring different bookkeeping procedures?

28. Happy workers? A manufacturing company employs 14 project managers, 48 foremen, and 377 laborers. In an effort to keep informed about any possible sources of employee discontent, management wants to conduct job satisfaction interviews with a sample of employees every month.

a) Do you see any potential danger in the company's plan? Explain.

b) Propose a sampling strategy that uses a simple random sample.

c) Why do you think a simple random sample might not provide the representative opinion the company seeks?

d) Propose a better sampling strategy.

C ━━━━━━━━━━━

29. Gallup. At its website (www.gallup.com) the Gallup Poll publishes results of a new survey each day. Scroll down to the end, and you'll find a statement that includes words such as these:

Results are based on telephone interviews with 1,008 national adults, aged 18 and older, conducted April 2–5, 2007. . . . In addition to sampling error, question wording and practical difficulties in conducting surveys can introduce error or bias into the findings of public opinion polls.

a) For this survey, identify the population of interest.

b) Gallup performs its surveys by phoning numbers generated at random by a computer program. What is the sampling frame?

c) What problems, if any, would you be concerned about in matching the sampling frame with the population?

30. Gallup World. At its website (www.gallupworldpoll.com) the Gallup World Poll describes their methods. After one report they explained:

Results are based on face-to-face interviews with randomly selected national samples of approximately 1,000 adults, aged 15 and older, who live permanently in each of the 21 sub-Saharan African nations surveyed. Those countries include Angola (areas where land mines might be expected were excluded), Benin, Botswana, Burkina Faso, Cameroon, Ethiopia, Ghana, Kenya, Madagascar (areas where interviewers had to walk more than 20 kilometers from a road were excluded), Mali, Mozambique, Niger, Nigeria, Senegal, Sierra Leone, South Africa, Tanzania, Togo, Uganda (the area of activity of the Lord's Resistance Army was excluded

from the survey), Zambia, and Zimbabwe. . . . In all countries except Angola, Madagascar, and Uganda, the sample is representative of the entire population.

a) Gallup is interested in sub-Saharan Africa. What kind of survey design are they using?

b) Some of the countries surveyed have large populations. (Nigeria is estimated to have about 130 million people.) Some are quite small. (Togo's population is estimated at 5.4 million.) Nonetheless, Gallup sampled 1000 adults in each country. How does this affect the precision of its estimates for these countries?

31. Phone surveys. Anytime we conduct a survey, we must take care to avoid undercoverage. Suppose we plan to select 500 names from the city phone book, call their homes between noon and 4 P.M., and interview whoever answers, anticipating contacts with at least 200 people.

a) Why is it difficult to use a simple random sample here?

b) Describe a more convenient, but still random, sampling strategy.

c) What kinds of households are likely to be included in the eventual sample of opinion? Excluded?

d) Suppose, instead, that we continue calling each number, perhaps in the morning or evening, until an adult is contacted and interviewed. How does this improve the sampling design?

e) Random-digit dialing machines can generate the phone calls for us. How would this improve our design? Is anyone still excluded?

32. Cell phone survey. What about drawing a random sample only from cell phone exchanges? Discuss the advantages and disadvantages of such a sampling method compared with surveying randomly generated telephone numbers from non–cell phone exchanges. Do you think these advantages and disadvantages have changed over time? How do you expect they'll change in the future?

33. Arm length. How long is your arm compared with your hand size? Put your right thumb at your left shoulder bone, stretch your hand open wide, and extend your hand down your arm. Put your thumb at the place where your little finger is, and extend down the arm again. Repeat this a third time. Now your little finger will probably have reached the back of your left hand. If the fourth hand width goes past the end of your middle finger, turn your hand sideways and count finger widths to get there.

a) How many hand and finger widths is your arm?

b) Suppose you repeat your measurement 10 times and average your results. What parameter would this average estimate? What is the population?

c) Suppose you now collect arm lengths measured in this way from 9 friends and average these 10 measurements. What is the population now? What parameter would this average estimate?

d) Do you think these 10 arm lengths are likely to be representative of the population of arm lengths in your community? In the country? Why or why not?

34. Fuel economy. Occasionally, when I fill my car with gas, I figure out how many miles per gallon my car got. I wrote down those results after 6 fill-ups in the past few months. Overall, it appears my car gets 28.8 miles per gallon.
   a) What statistic have I calculated?
   b) What is the parameter I'm trying to estimate?
   c) How might my results be biased?
   d) When the Environmental Protection Agency (EPA) checks a car like mine to predict its fuel economy, what parameter is it trying to estimate?

Answers

Just Checking

1. a) It can be hard to reach all members of a population, and it can take so long that circumstances change, affecting the responses. A well-designed sample is often a better choice.
   b) This sample is probably biased—students who didn't like the food at the cafeteria might not choose to eat there.
   c) No, only the sample size matters, not the fraction of the overall population.
   d) Students who frequent this website might be more enthusiastic about Statistics than the overall population of Statistics students. A large sample cannot compensate for bias.
   e) It's the population "parameter." "Statistics" describe samples.

2. a) systematic
   b) stratified
   c) simple
   d) cluster

Observational Studies and Experiments

Who gets good grades? And, more importantly, why? Is there something schools and parents could do to help students improve their grades? Some people think they have an answer: music! No, not your iPod, but an instrument. In a study conducted at Mission Viejo High School, in California, researchers compared the academic performances of music students and non-music students. Guess what? The music students had much higher average grades than the non-music students. Not only that: A whopping 16% of the music students had all A's compared with only 5% of the non-music students.

As a result of this study and others, many parent groups and educators pressed for expanded music programs in the nation's schools. They argued that the work ethic, discipline, and feeling of accomplishment fostered by learning to play an instrument also enhance a person's ability to succeed in school. They thought that involving more students in music would raise academic performance. What do you think? Does this study provide solid evidence? Or are there other possible explanations for the difference in grades? Is there any way to really prove such a claim?

Observational Studies

This research tried to show an association between music education and grades. But it wasn't a survey. Nor did it assign students to get music education. Instead, it simply observed students "in the wild," recording the choices they made and the outcome. Such studies are called **observational studies.** In observational studies, researchers don't *assign* choices; they simply observe them. In addition, this was a **retrospective study,** because researchers first identified subjects who studied music and then collected data on their past grades.

What's wrong with concluding that music education causes good grades? Sure, one high school during one academic year may not be representative of the whole United States. That's true, but the real problem is that the claim that music study *caused* higher grades depends on there being *no other differences* between the groups that could account for the differences in grades. Perhaps studying music was not the *only* difference between the two groups of students.

We can think of lots of lurking variables that might cause the groups to perform differently. Students who study music may have better work habits to start with, and this makes them successful in both music and classes. Music students may have more parental support (someone had to pay for all those lessons), and that support may have helped their academic performances, too. Maybe they came from wealthier homes and had other advantages. Or it could be that smarter kids just like to play musical instruments.

For rare illnesses, it's not practical to draw a large enough sample to see many ill respondents, so the only option remaining is to develop retrospective data. For example, researchers can interview those who have become ill. The likely causes of both Legionnaires' disease and HIV were initially identified from such retrospective studies of the small populations who were initially infected. But to confirm the causes, researchers needed laboratory-based experiments.

Observational studies are valuable for discovering trends and possible relationships. They are used widely in public health and marketing. Observational studies that try to discover variables related to rare outcomes, such as specific diseases, are often retrospective. They first identify people with the disease and then look into their history and heritage in search of things that may be related to their condition. But retrospective studies are usually restricted to a small part of the entire population. And because retrospective records are based on historical data, they can have errors. (Do you recall *exactly* what you ate even yesterday? How about last Wednesday?)

A somewhat better approach is to observe individuals over time, recording the variables of interest and ultimately seeing how things turn out. For example, we might start by selecting young students who have not begun music lessons. We could then track their academic performance over several years, comparing those who later choose to study music with those who do not. Identifying subjects in advance and collecting data as events unfold would make this a **prospective study.**

Although an observational study may identify important variables related to the outcome we are interested in, there is no guarantee that we have found the right or the most influential variables. Students who choose to study an instrument might still differ from the others in some important way that we missed. It may be this difference—whether we know what it is or not—rather than music itself that leads to better grades. It's just not possible for observational studies, whether prospective or retrospective, to demonstrate cause-and-effect relationships.

For Example Designing an Observational Study

In early 2007, a larger-than-usual number of cats and dogs developed kidney failure; many died. Initially, researchers didn't know why, so they used an observational study to investigate.

Question: Suppose you were called on to plan a study seeking the cause of this problem. Would your design be retrospective or prospective? Explain why.

I would use a retrospective observational study. Even though the number of pets with this disease was higher than usual, it was still rare. Surveying all pets would have been impractical. Instead, it makes sense to locate some who were sick and ask about their diets, exposure to toxins, and other possible causes.

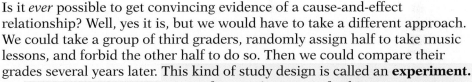

Randomized, Comparative Experiments

Experimental design was advanced in the 19th century by work in psychophysics by Gustav Fechner (1801–1887), the founder of experimental psychology. Fechner designed ingenious experiments that exhibited many of the features of modern designed experiments. Fechner was careful to control for the effects of factors that might affect his results. For example, in his 1860 book *Elemente der Psychophysik* he cautioned readers to group experiment trials together to minimize the possible effects of time of day and fatigue.

Is it *ever* possible to get convincing evidence of a cause-and-effect relationship? Well, yes it is, but we would have to take a different approach. We could take a group of third graders, randomly assign half to take music lessons, and forbid the other half to do so. Then we could compare their grades several years later. This kind of study design is called an **experiment.**

An experiment requires a **random assignment** of subjects to treatments. Only an experiment can justify a claim like "Music lessons cause higher grades." Questions such as "Does taking vitamin C reduce the chance of getting a cold?" and "Does working with computers improve performance in Statistics class?" and "Is this drug a safe and effective treatment for that disease?" require a designed experiment to establish cause and effect.

Experiments study the relationship between two or more variables. An experimenter must identify at least one explanatory variable, called a **factor,** to manipulate and at least one **response variable** to measure. What distinguishes an experiment from other types of investigation is that the experimenter actively and deliberately manipulates the factors to control the details of the possible treatments, and assigns the subjects to those treatments *at random.* The experimenter then observes the response variable and *compares* responses for different groups of subjects who have been treated differently.

For example, imagine an experiment to see whether the amount of sleep and exercise you get affects your performance. The individuals on whom we experiment are known by a variety of terms. Humans are commonly called **subjects** or **participants.** Other individuals (rats, days, petri dishes of bacteria) are commonly referred to by the more generic term **experimental units.** When we recruit subjects for our sleep deprivation experiment by advertising in Statistics class, we'll probably have better luck if we invite them to be participants than if we advertise that we need experimental units.

The specific values that the experimenter chooses for a factor are called the **levels.** We might assign our participants to sleep for 4, 6, or 8 hours. Often there are several factors at a variety of levels. The combination of levels from all the factors that an experimental unit receives is known as its **treatment.**

How should we assign our participants to the treatments? Some students prefer 4 hours of sleep, while others need 8. Should we let the students choose the treatment they'd prefer? No. That would not be a good idea. To have any hope of drawing a fair and meaningful conclusion, we must assign our partici-pants to their treatments *at random*.

AN EXPERIMENT:

Manipulates the factor levels to create treatments. *Randomly assigns* subjects to these treatment levels. *Compares* the responses of the subject groups across treatment levels.

The women's health initiative is a major 15-year research program funded by the National Institutes of Health to address the most common causes of death, disability, and poor quality of life in older women. It consists of both an observational study with more than 93,000 participants and several randomized comparative experiments. The goals of this study include

- giving reliable estimates of the extent to which known risk factors predict heart disease, cancers, and fractures;
- identifying "new" risk factors for these and other diseases in women;
- comparing risk factors with new occurrences of disease during the study; and
- identifying biological indicators of disease, especially those found in blood.

That is, the study seeks to identify possible risk factors and assess how serious they might be. There would be no way to find out these things with an experiment because we could never control risk factors we don't know about yet.

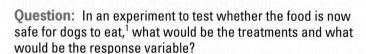

For Example Determining the Treatments and Response Variable

Recap: In 2007, deaths of a large number of pet dogs and cats were ultimately traced to contamination of some brands of pet food. The manufacturer now claims that the food is safe, but before it can be released, it must be tested.

Question: In an experiment to test whether the food is now safe for dogs to eat,[1] what would be the treatments and what would be the response variable?

The treatments would be ordinary-size portions of two dog foods: the new one from the company (the test food) and one that I was certain was safe (perhaps prepared in my kitchen or laboratory). The response would be a veterinarian's assessment of the health of the test animals.

The Three Principles of Experimental Design

1. **Control.** We control sources of variation other than the factors we are testing by making conditions as similar as possible for all treatment groups. For human subjects, we try to treat them alike. Controlling sources of variation makes it easier to detect any differences caused by the treatments.

---

[1]It may disturb you (as it does us) to think of deliberately putting dogs at risk in this experiment, but in fact that is what is done. The risk is borne by a small number of dogs so that the far larger population of dogs can be kept safe.

2. **Randomize.** As in sample surveys, **randomization** allows us to equalize the effects of unknown or uncontrollable sources of variation. It cannot eliminate the effects of these sources, but it should spread them out fairly equally across the treatment levels so that we can see past them. There's an adage that says "control what you can, and randomize the rest."

3. **Replicate.** Two kinds of replication show up in comparative experiments. First, we should apply each treatment to several subjects. That lets us estimate the variability of responses. If we have not assessed the variation, the experiment is not complete.

   A second kind of replication is important because the experimental units are usually not a representative sample from the population of interest. We may believe that what is true of the students in Psych 101 who volunteered for the sleep experiment is true of all humans, but we'll feel more confident if our results for the experiment are *replicated* in another part of the country, with people of different ages, and at different times of the year. **Replication** of an entire experiment with other subjects is an essential step in science.

No drug can be sold in the United States without first showing, in a suitably designed experiment approved by the Food and Drug Administration (FDA), that it's safe and effective. The small print on the booklet that comes with many prescription drugs usually describes the outcomes of that experiment.

For Example Control, Randomize, and Replicate

Recap: We're planning an experiment to see whether the new pet food is safe for dogs to eat. We'll feed some animals the new food and others a food known to be safe, comparing their health after a period of time.

Questions: In this experiment, how will you implement the principles of control, randomization, and replication?

I'd control the portion sizes eaten by the dogs. To reduce possible variability from factors other than the food, I'd standardize other aspects of their environments—housing the dogs in similar pens and ensuring that each got the same amount of water, exercise, play, and sleep time, for example. I might restrict the experiment to a single breed of dog and to adult dogs to further minimize variation.

To equalize traits, pre-existing conditions, and other unknown influences, I would assign dogs to the two feed treatments randomly.

I would replicate by assigning more than one dog to each treatment to allow for variability among individual dogs. If I had the time and funding, I might replicate the entire experiment using, for example, a different breed of dog.

Diagrams

An experiment is carried out over time with specific actions occurring in a specified order. A diagram of the procedure can help in thinking about experiments.[2]

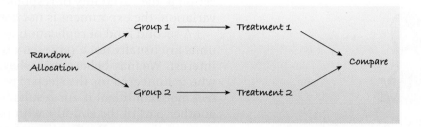

The diagram emphasizes the random allocation of subjects to treatment groups, the separate treatments applied to these groups, and the ultimate comparison of results. It's best to specify the responses that will be compared. A good way to start comparing results for the treatment groups is with boxplots.

STEP-BY-STEP EXAMPLE | Designing an Experiment

An ad for OptiGro plant fertilizer claims that with this product you will grow "juicier, tastier" tomatoes. You'd like to test this claim, and wonder whether you might be able to get by with half the specified dose. How can you set up an experiment to check out the claim?

Of course, you'll have to get some tomatoes, try growing some plants with the product and some without, and see what happens. But you'll need a clearer plan than that. How should you design your experiment?

Let's work through the design, step by step. We'll design the simplest kind of experiment, a **completely randomized experiment in one factor.** Since this is a *design* for an experiment, most of the steps are part of the *Think* stage. The statements in the right column are the kinds of things you would need to say in *proposing* an experiment. You'd need to include them in the "methods" section of a report once the experiment is run.

Question: How would you design an experiment to test OptiGro fertilizer?

A **completely randomized experiment** is the ideal simple design, just as a *simple random sample* is the ideal simple sample—and for many of the same reasons.

 THINK

Plan State what you want to know.

I want to know whether tomato plants grown with OptiGro yield juicier, tastier tomatoes than plants raised in otherwise similar circumstances but without the fertilizer. If so, I wonder if half as much fertilizer would also work.

---

[2] Diagrams of this sort were introduced by David Moore in his textbooks and are widely used.

Response Specify the response variable.

I'll evaluate the juiciness and taste of the tomatoes by asking a panel of judges to rate them on a scale from 1 to 7 in juiciness and in taste.

Treatments Specify the factor levels and the treatments.

The factor is fertilizer, specifically OptiGro fertilizer. I'll grow tomatoes at three different factor levels: some with no fertilizer, some with half the specified amount of OptiGro, and some with the full dose of OptiGro. These are the three treatments.

Experimental Units Specify the experimental units.

I'll buy 24 tomato plants of the same variety from a local garden store.

Experimental Design Observe the principles of design:

▶ **Control** any sources of variability you know of and can control.

I'll locate the garden plots near each other so that the plants get similar amounts of sun and rain and experience similar temperatures. I will weed the plots equally and otherwise treat the plants alike.

▶ **Replicate** results by placing more than one plant in each treatment group.

I'll use 8 plants in each treatment group.

▶ **Randomly assign** experimental units to treatments, to equalize the effects of unknown or uncontrollable sources of variation.

I'll take 24 cards—8 hearts, 8 diamonds, and 8 spades—from a deck and shuffle them thoroughly. Then I'll deal out one card in front of each tomato plant. I'll put the 8 plants with hearts in Group 1, the 8 with diamonds in Group 2, and the remaining 8 with spades in Group 3.

Describe how the randomization will be accomplished.

Make a Picture A diagram of your design can help you think about it clearly.

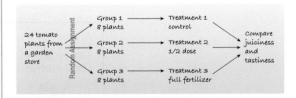

Specify any other experiment details. You must give enough details so that another experimenter could exactly replicate your experiment. It's generally better to include details that might seem irrelevant than to leave out matters that could turn out to make a difference.

I will grow the plants until the tomatoes are mature, as judged by reaching a standard color.

I'll harvest the tomatoes when ripe and store them for evaluation.

Specify how to measure the response.

I'll set up a numerical scale of juiciness and one of tastiness for the taste testers. I'll ask several people to taste slices of tomato and rate them.

(continued)

Once you collect the data, you'll need to display them and compare the results for the three treatment groups.	I will display the results with side-by-side box-plots to compare the tasters' evaluations of the three treatment groups.
TELL To answer the initial question, we ask whether the differences we observe in the means of the three groups are meaningful. Because this is a randomized experiment, we can attribute significant differences to the treatments. To do this properly, we'll need methods from what is called "statistical inference," the subject of the rest of this book.	If the differences in taste and juiciness among the groups are greater than I would expect by knowing the usual variation among tomatoes, I may be able to conclude that these differences can be attributed to the fertilizer.

Does the Difference Make a Difference?

If the differences among the treatment groups are big enough, we'll conclude they result from the treatments, but how can we decide whether the differences are big enough?

Would we expect the group means to be identical? Not really. Even if the treatment made no difference whatever, there would still be some variation. We assigned the tomato plants to treatments at random, and a different random assignment would have led to different results. Even a repeat of the *same* treatment on a different randomly assigned set of plants would lead to different data. The real question is whether the differences we observed are only as big as we might get just by chance alone, or whether they're bigger than that. If we decide that they're bigger, we'll attribute the differences to the treatments. In that case we say the differences are **statistically significant.**

How will we decide if the results are different enough to be considered statistically significant? Later chapters will offer methods to help answer that question, but to get some intuition, think about deciding whether a coin is fair. If we flip a fair coin 100 times, we expect, *on average,* to get 50 heads. Suppose we get 54 heads out of 100. That doesn't seem very surprising. It's well within the bounds of ordinary random fluctuations. What if we'd seen 94 heads? That's clearly outside the bounds. We'd be pretty sure that the coin flips were not random. But what about 74 heads? Is that far enough from 50% to arouse our suspicions? That's the sort of question we need to ask of our experiment results.

In Statistics terminology, 94 heads would be a statistically significant difference from 50, and 54 heads would not. Whether 74 is *statistically*

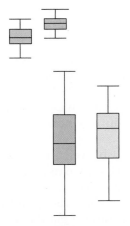

FIGURE 11.1 The boxplots in both pairs have centers the same distance apart, but when the spreads are large, the observed difference may be just from random variation.

significant or not would depend on the chance of getting 74 heads in 100 flips of a fair coin and on our tolerance for believing that rare events can happen to us. A difference is statistically significant if we don't believe that it's likely to have occurred only by chance.

Back at the tomato patch, we ask whether the differences we see among the treatment groups are the kind of differences we'd expect by chance. A good way to get a feeling for that is to look at how much our results vary among plants that get the *same* treatment. Boxplots of our results by treatment group can give us a general idea.

For example, Figure 11.1 shows two pairs of boxplots whose centers differ by exactly the same amount. In the upper set, that difference appears to be larger than we'd expect just by chance. Why? Because the variation is quite small *within* treatment groups, so the larger difference *between* the groups is unlikely to be just from the randomization. In the bottom pair, that same difference between the centers looks less impressive. There the variation *within* each group swamps the difference *between* the two medians. We'd say the difference is statistically significant in the upper pair and not statistically significant in the lower pair.

Just Checking

1. At one time, a method called "gastric freezing" was used to treat people with peptic ulcers. An inflatable bladder was inserted down the esophagus and into the stomach, and then a cold liquid was pumped into the bladder. Now you can find the following notice on the Internet site of a major insurance company:

   [Our company] does not cover gastric freezing (intragastric hypothermia) for chronic peptic ulcer disease. . . .

   Gastric freezing for chronic peptic ulcer disease is a non-surgical treatment which was popular about

   20 years ago but now is seldom performed. It has been abandoned due to a high complication rate, only temporary improvement experienced by patients, and a lack of effectiveness when tested by double-blind, controlled clinical trials.

   What did that "controlled clinical trial" (experiment) probably look like? (Don't worry about "double-blind"; we'll get to that soon.)

   a) What was the factor in this experiment?
   b) What was the response variable?
   c) What were the treatments?
   d) How did researchers decide which subjects received which treatment?
   e) Were the results statistically significant?

   (Check your answers on page 264.)

Experiments vs. Samples

Experiments are rarely performed on random samples from a population. Don't describe the subjects in an experiment as a random sample unless they really are. More likely, the randomization was in assigning subjects to treatments.

Both experiments and sample surveys use randomization to get unbiased data. But they do so in different ways and for different purposes. Sample surveys try to estimate population parameters, so the sample needs to be as representative of the population as possible. By contrast, experiments try to assess the effects of treatments. Experimental units are not always drawn randomly from the population. For example, a medical experiment may deal only with local patients who have the disease under study. The randomization is in the assignment of their therapy. We want a sample to exhibit the diversity and variability of the population, but for an experiment the more alike the subjects the more easily we'll spot differences in the effects of the treatments. Unless the experimental units are chosen from the population at random, you should be cautious about generalizing experiment results to larger populations until the experiment has been repeated under different circumstances.

Control Treatments

Suppose you wanted to test a $300 piece of software designed to shorten download times. You could just try it on several files and record the download times, but you probably want to *compare* the speed with what would happen *without* the software installed. Such a baseline measurement is called a **control treatment**, and the experimental units to whom it is applied are called a **control group.**

This is a different use of the word "control." Previously, we controlled known sources of variation by keeping them constant. Here, we use a control treatment as another *level* of the factor in order to compare the treatment results to a situation in which "nothing happens." That's what we did in the tomato experiment when we used no fertilizer on the 8 tomato plants in Group 1.

Blinding

Humans are notoriously susceptible to errors in judgment.[3] All of us. When we know what treatment was assigned, it's difficult not to let that knowledge influence our assessment of the response, even when we try to be careful.

Suppose you were trying to advise your school on which brand of cola to stock in the school's vending machines. You set up an experiment to see which of the three competing brands students prefer (or whether they can tell the difference at all). But people have brand loyalties. You probably prefer one brand already. So if you knew which brand you were tasting, it might influence your rating. To avoid this problem, it would be better to disguise the brands as much as possible. This strategy is called **blinding** the participants to the treatment.[4]

But it isn't just the subjects who should be blind. Experimenters themselves often subconsciously behave in ways that favor what they believe. Even technicians may treat plants or test animals differently if, for example, they expect them to die. An animal that starts doing a little better than others by showing an increased appetite may get fed a bit more than the experimental protocol specifies.

Blinding is important. People are so good at picking up subtle cues about treatments that it's important to keep *anyone* who could affect the outcome or the measurement of the response from knowing which subjects have been assigned to which treatments. So, not only should your cola-tasting subjects be blinded, but also *you*, as the experimenter, shouldn't know which drink is which, either—at least until you're ready to analyze the results.

There are two main classes of individuals who can affect the outcome of the experiment:

- those who could influence the results (the subjects, treatment administrators, or technicians)
- those who evaluate the results (judges, treating physicians, etc.)

BLINDING BY MISLEADING

Social science experiments can sometimes blind subjects by misleading them about the purpose of a study. One of the authors participated as an undergraduate volunteer in a (now infamous) psychology experiment using such a blinding method. The subjects were told that the experiment was about three-dimensional spatial perception and were assigned to draw a model of a horse. While they were busy drawing, a loud noise and then groaning were heard coming from the room next door. The *real* purpose of the experiment was to see how people reacted to the apparent disaster. The experimenters wanted to see whether the social pressure of being in groups made people react to the disaster differently. Subjects had been randomly assigned to draw either in groups or alone; that was the treatment. The experimenter had no interest in how well the subjects could draw the horse, but the subjects were blinded to the treatment because they were misled.

---

[3]For example, here we are in Chapter 11 and you're still reading the footnotes.
[4]In 1885 C. S. Peirce introduced randomization and also recommended blinding.

When all the individuals in either one of these classes are blinded, an experiment is said to be **single-blind.** When everyone in *both* classes is blinded, we call the experiment **double-blind.**

In our tomato experiment, we certainly don't want the people judging the taste to know which tomatoes got the fertilizer. That makes the experiment single-blind. We might also not want the people caring for the tomatoes to know which ones were being fertilized, in case they might treat them differently in other ways, too. We can accomplish this double-blinding by having some fake fertilizer for them to put on the other plants. Read on.

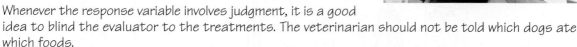

For Example   Blinding

Recap: In our experiment to see if the new pet food is now safe, we're feeding one group of dogs the new food and another group a food we know to be safe. Our response variable is the health of the animals as assessed by a veterinarian.

Questions: Should the vet be blinded? Why or why not? How would you do this? (Extra credit: Can this experiment be double-blind? Would that mean that the test animals wouldn't know what they were eating?)

Whenever the response variable involves judgment, it is a good idea to blind the evaluator to the treatments. The veterinarian should not be told which dogs ate which foods.

Extra credit: There is a need for double-blinding. In this case, the workers who care for and feed the animals should not be aware of which dogs are receiving which food. We'll need to make the "safe" food look as much like the "test" food as possible.

Placebos

Often, simply applying *any* treatment can induce an improvement. Every parent knows the medicinal value of a kiss to make a toddler's scrape or bump stop hurting. Some of the improvement seen with a treatment—even an effective treatment—can be due simply to the act of treating. To separate these two effects, we can use a control treatment that mimics the treatment itself.

A "fake" treatment that looks just like the treatments being tested is called a **placebo.** Placebos are the best way to blind subjects from knowing whether they are receiving the treatment or not. One common version of a placebo in drug testing is a "sugar pill." Especially when psychological attitude can affect the results, control group subjects treated with a placebo may show an improvement.

The fact is that subjects treated with a placebo sometimes improve. It's not unusual for 20% or more of subjects given a placebo to report reduction in pain, improved movement, or greater alertness, or even to demonstrate improved health or performance. This **placebo effect** highlights both the importance of effective blinding and the importance of comparing treatments with a control. Placebo controls are so effective that you should use them as an essential tool for blinding whenever possible.

The placebo effect is stronger when placebo treatments are administered with authority or by a figure who appears to be an authority. "Doctors" in white coats generate a stronger effect than salespeople in polyester suits. But the placebo effect is not reduced much even when subjects know that the effect exists. People often suspect that they've gotten the placebo if nothing at all happens. So, recently, drug manufacturers have gone so far in making placebos realistic that they cause the same side effects as the drug being tested! Such "active placebos" usually induce a stronger placebo effect. When those side effects include loss of appetite or hair, the practice may raise ethical questions.

The best experiments are usually

- randomized.
- comparative.
- double-blind.
- placebo-controlled.

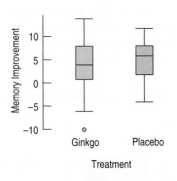

> **Does ginkgo biloba improve memory?** Researchers investigated the purported memory-enhancing effect of ginkgo biloba tree extract (P. R. Solomon, F. Adams, A. Silver, J. Zimmer, R. De Veaux, "Ginkgo for Memory Enhancement. A Randomized Controlled Trial." *JAMA* 288 [2002]: 835–840). In a randomized, comparative, double-blind, placebo-controlled study, they administered treatments to 230 elderly community members. One group received Ginkoba™ according to the manufacturer's instructions. The other received a similar-looking placebo. Thirteen different tests of memory were administered before and after treatment. The placebo group showed greater improvement on 7 of the tests, the treatment group on the other 6. None showed any significant differences. Here are boxplots of one measure.

By permission of John L. Hart FLP and Creators Syndicate, Inc.

Blocking

Suppose that we had hoped to use 24 tomato plants of the same variety for our experiment, but the garden store had only 12 plants left. So we drove down to the nursery and bought their last 6 plants of that variety. We can run the experiment with only 18 plants, but now we worry that the tomato plants from the two stores are different somehow, and, in fact, they don't really look the same.

How can we design the experiment so that the differences between the stores don't mess up our attempts to see differences among fertilizer levels? When there are pre-existing differences between groups of experimental units, it's often a good idea to gather them together into **blocks.** The randomization is introduced when we randomly assign treatments within each block.

Here, we would define the plants from each store to be a block because we realize that the store may have an effect. To isolate the store effect, we block on store by assigning the plants from each store to treatments at random. So we now have six treatment groups, three for each block. Within each block, we'll randomly assign the same number of plants to each of the three treatments. The experiment is still fair because each treatment is still applied (at random) to the same number of plants and to the same proportion from each store: 4 from store A and 2 from store B. Because the randomization occurs only within the blocks (plants from one store cannot be assigned to treatment groups for the other), we call this a **randomized block design.**

In effect, we conduct two parallel experiments, one for tomato plants from each store, and then combine the results. The picture tells the story:

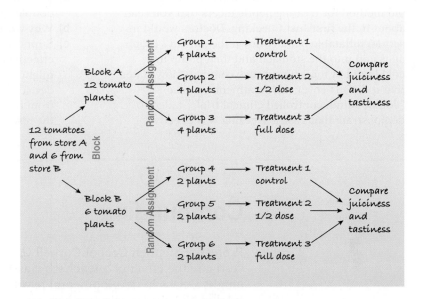

In a retrospective or prospective study, subjects are sometimes paired because they are similar in ways *not* under study. **Matching** subjects in this way can reduce variation in much the same way as blocking. For example, a retrospective study of music education and grades might match each student who studies an instrument with someone of the same sex who is similar in family income but didn't study an instrument. When we compare grades of music students with those of non-music students, the matching would reduce the variation due to income and sex differences.

Blocking is the same idea for experiments as stratifying is for sampling. Both methods group together subjects who are similar and randomize within those groups as a way to remove unwanted variation. (But be careful to keep the terms straight. Don't say that we "stratify" an experiment or "block" a sample.)

For Example Blocking

Recap: In 2007, pet food contamination put cats at risk, as well as dogs. Our experiment should probably test the safety of the new food on both animals.

Questions: Why shouldn't we randomly assign a mix of cats and dogs to the two treatment groups? What would you recommend instead?

Dogs and cats might respond differently to the foods, and that variability could obscure my results. Blocking by species can remove that superfluous variation. I'd randomize cats to the two treatments (test food and safe food) separately from the dogs. I'd measure their responses separately and look at the results afterward.

Just Checking

2. Recall the experiment about gastric freezing, an old method for treating peptic ulcers that you read about in the first Just Checking. Doctors would insert an inflatable bladder down the patient's esophagus and into the stomach and then pump in a cold liquid. A major insurance company now states that it doesn't cover this treatment because "double-blind, controlled clinical trials" failed to demonstrate that gastric freezing was effective.

a) What does it mean that the experiment was double-blind?
b) Why would you recommend a placebo control?
c) Suppose that researchers suspected that the effectiveness of the gastric freezing treatment might depend on whether a patient had recently developed the peptic ulcer or had been suffering from the condition for a long time. How might the researchers have designed the experiment?

(Check your answers on page 264.)

Confounding

Professor Stephen Ceci of Cornell University performed an experiment to investigate the effect of a teacher's classroom style on student evaluations. He taught a class in developmental psychology during two successive terms to a total of 472 students in two very similar classes. He kept everything about his teaching identical (same text, same syllabus, same office hours, etc.) and modified only his style in class. During the fall term, he remained low-key. During the spring term, he lectured with more enthusiasm, varying his vocal pitch and using more hand gestures. He had students fill out a standard evaluation form at the end of each term.

The students in the fall term class rated him only an average teacher. Those in the spring term class rated him an excellent teacher, praising his knowledge and accessibility, and even the quality of the textbook. On the question "How much did you learn in the course?" the average response changed from 2.93 to 4.05 on a 5-point scale.[5]

How much of this difference was due to his classroom behavior, and how much might have been due to the season of the year? Fall term in Ithaca, NY (home of Cornell University), starts out colorful and pleasantly warm but ends cold and bleak. Spring term starts out bitter and snowy and ends with blooming flowers and singing birds. Might students' overall happiness have been affected by the season and reflected in their evaluations?

Unfortunately, there's no way to tell. Nothing in the data enables us to tease apart these two effects, because all the students who experienced the calm and quiet manner did so during the fall term and all who experienced the energetic manner did so during the spring. When the levels of one factor are entangled with the levels of another factor, we say that these two factors are **confounded.**

It's worth noting that the role of blinding in an experiment is to combat a possible source of confounding. There's a risk that knowledge about the treatments could lead the subjects or those interacting with them to behave differently or could influence judgments made by the people evaluating the responses. That means we won't know whether the treatments really do produce different results or if we're being fooled by these confounding influences.

---

[5]But the two classes performed almost identically well on the final exam.

For Example Confounding

Recap: After many dogs and cats suffered health problems caused by contaminated foods, we're trying to find out whether a newly formulated pet food is safe. Our experiment will feed some animals the new food and others a food known to be safe, and a veterinarian will check the response.

Question: Why would it be a bad design to feed the test food to some dogs and the safe food to cats?

This would create confounding. We would not be able to tell whether any differences in animals' health were attributable to the food they had eaten or to differences in how the two species responded.

But Is It Ethical?

Obviously, experiments play a critical role in scientific investigations. And the benefits can be unmistakable. As just one example, consider the rapid advances in medicine over the past hundred years. If you had been born a century earlier, your life expectancy would have been about 50 years. Today people can expect to live well into their 70s and 80s. A hundred years ago people feared diseases like polio and tuberculosis, no treatments were available for cancer and heart disease, and even the most basic drugs like aspirin and penicillin were unknown. Now researchers have developed vaccines that virtually eliminate many dangerous diseases, have made great progress in fighting others, and have formulated many drugs that are highly effective in combating a wide variety of ailments.

In large part, this amazing progress has been possible because of experiments using animals and humans as subjects. And that raises numerous questions about the ethics involved. Let's consider a few.

- Researchers intentionally give rats cancer to explore how tumors grow and what kinds of treatments may slow the progress of or cure the disease. Is it okay to inflict cancer in rats to improve medical care for humans?
- Because their hearts closely resemble ours, pigs are often used in research on heart disease. In fact, many pigs are killed so that their heart valves can be transplanted to humans whose own valves have failed. Do you think this is ethical?
- Our closest genetic relatives are monkeys, making them desirable subjects for many medical experiments. In addition to whatever suffering these animals may experience, often they must be killed and dissected to evaluate the effectiveness of the treatments under investigation. Are these sacrifices justifiable if they help save the life of someone you love?
- Even if you are okay with the use of animal subjects in medical experiments, what about other research? Cosmetics companies test makeup to be sure it's safe for human use. In the Draize test, for example, caustic substances are placed in the eyes of conscious rabbits to see if they cause damage to sensitive eye tissues. Is it ethical to induce pain in animals so that humans can look prettier?
- Many experiments use human subjects. Parkinson's disease involves neural malfunctions deep in the brain. In one possible treatment under investigation, doctors drill a small hole in the patient's skull to insert an instrument

that injects new nerve cells, hoping these may grow and replace the damaged ones. Some subjects got "placebo holes" without any injections, so that the experiment could be double-blind. Is it ethical to just drill a hole in someone's skull, even if not doing so interferes with the researcher's ability to evaluate the effectiveness of the treatment?

Would you volunteer for an experiment not knowing whether the treatment would work—or even be safe—and knowing you might only receive the placebo? What if you were paid, or had no health insurance and were offered free medical care? Suppose you were told that your condition was life-threatening and there were no other options? Do human subjects participate out of desperation, or out of a genuine desire to help others who may suffer from the same condition?

There are lots of difficult questions—we're sure you can think of others— and no easy answers.

WHAT CAN GO WRONG?

- **Don't give up just because you can't run an experiment.** Sometimes we can't run an experiment because we can't identify or control the factors. Sometimes it would simply be unethical to run the experiment. (Consider randomly assigning students to take—and be graded in—a Statistics course deliberately taught to be boring and difficult or one that had an unlimited budget to use multimedia, real-world examples, and field trips to make the subject more interesting.) If we can't perform an experiment, often an observational study is a good choice.

- **Beware of confounding.** Use randomization whenever possible to ensure that the factors not in your experiment are not confounded with your treatment levels. Be alert to confounding that cannot be avoided, and report it along with your results.

EXPERIMENTS **IN YOUR WORLD**

Coffee Kills Pain When You Exercise

April 1, 2009—Java will reduce pain when you exercise . . . That's what Robert Motl, cyclist and professor, says he has found. "We've shown that caffeine reduces pain reliably, consistently during cycling, across different intensities, across different people, different characteristics," says the former competitive cyclist, University of Illinois kinesiology and community health professor.

The study's 25 participants were fit, college-aged males divided into two distinct groups: subjects whose everyday caffeine consumption was extremely low to nonexistent, and those with an average caffeine intake of about 400 milligrams a day, the equivalent of three to four cups of coffee.

After completing an initial exercise test in the lab on an ergometer, or stationary cycle, for determination of . . . aerobic power, subjects returned for two monitored high-intensity, 30-minute exercise sessions. An hour prior to each session, cyclists—who had been instructed not to consume caffeine during the prior 24-hour period—were given a pill. On one occasion, it contained a dose of caffeine measuring 5 milligrams per kilogram of body weight (equivalent to two to three cups of coffee); the other time, they received a placebo. During both exercise periods, subjects' perceptions of quadriceps muscle pain was recorded at regular intervals, along with data on oxygen consumption, heart rate and work rate.

Motl said another research direction might be to determine caffeine's effect on sport performance. "We've shown that caffeine reduces pain reliably, consistently during cycling, across different intensities, across different people, different characteristics. But does that reduction in pain translate into an improvement in sport performance?"

http://www.seniorjournal.com/NEWS/Fitness/2009/20090401-GoodNewsForSeniors.htm

WHAT HAVE WE LEARNED?

We've learned to recognize sample surveys, observational studies, and randomized experiments. We know that these methods collect data in different ways and lead us to different conclusions.

We've learned to identify retrospective and prospective observational studies and understand the advantages and disadvantages of each.

We've learned that only well-designed experiments can allow us to reach cause-and-effect conclusions. We manipulate levels of treatments to see if the factor we have identified produces differences in our response variable.

We've learned the principles of experimental design:

▶ We want to be sure that variation in the response variable can be attributed to our factor, so we identify and control as many other sources of variability as possible.

▶ Because there are many possible sources of variability that we cannot identify, we try to equalize those by randomly assigning experimental units to treatments.

▶ We replicate the experiment on as many subjects as possible.

▶ We consider blocking to reduce variability from sources we recognize but cannot control.

We've learned the value of having a control group and of using blinding and placebo controls.

Finally, we've learned to recognize the problems posed by confounding variables in experiments and lurking variables in observational studies.

Terms

Observational study   A study based on data in which no manipulation of factors has been employed.

Retrospective study   An observational study in which subjects are selected and then their previous conditions or behaviors are determined. Retrospective studies need not be based on random samples and they usually focus on estimating differences between groups or associations between variables.

Prospective study   An observational study in which subjects are followed to observe future outcomes. Because no treatments are deliberately applied, a prospective study is not an experiment.

(continued)

Experiment	An experiment *manipulates* factor levels to create treatments, *randomly assigns* subjects to these treatment levels, and then *compares* the responses of the subject groups across treatment levels.
Random assignment	To be valid, an experiment must assign experimental units to treatment groups at random.
Factor	A variable whose levels are manipulated by the experimenter in an attempt to discover any effects that the factor levels may have on the response variable.
Response	A variable whose values are compared across different treatments. In a randomized experiment, large response differences can be attributed to the effect of differences in treatment level.
Experimental units	Individuals on whom an experiment is performed. Usually called **subjects** or **participants** when they are human.
Level	The specific values that the experimenter chooses for a factor are called the levels of the factor.
Treatment	The process applied to randomly assigned experimental units. Treatments are the different levels of the factor.
Principles of experimental design	▶ **Control** aspects of the experiment that we know may have an effect on the response. ▶ **Randomize** subjects to treatments to even out effects that we cannot control. ▶ **Replicate** over as many subjects as possible. If, as often happens, the subjects of the experiment are not a representative sample from the population of interest, replicate the entire study with a different group of subjects, preferably from a different part of the population.
Statistically significant	When an observed difference is too large for us to believe that it is likely to have occurred by chance, we consider the difference to be statistically significant.
Control group	The experimental units assigned to a baseline treatment level, or a placebo treatment. Their responses provide a basis for comparison.
Blinding	Any individual associated with an experiment who is not aware of how subjects have been allocated to treatment groups is said to be blinded.
Single-blind Double-blind	There are two main classes of individuals who can affect the outcome of an experiment: ▶ those who could *influence the results* (the subjects, treatment administrators, or technicians). ▶ those who *evaluate the results* (judges, treating physicians, etc.). When every individual in *one* of these groups is blinded, an experiment is said to be single-blind. When everyone in *both* groups is blinded, we call the experiment double-blind.
Placebo	A treatment known to have no effect, administered so that all groups experience the same conditions.
Placebo effect	The tendency of many human subjects (often 20% or more of experiment subjects) to show a response even when administered a fake treatment.
Blocking	When there are pre-existing differences among groups of experimental units, it is often a good idea to gather them together into blocks. By blocking, we isolate the differences so that we can see the effects of the treatments more clearly.
Matching	In a retrospective or prospective study, subjects who are similar in ways not under study may be matched and then compared with each other on the variables of interest. Matching, like blocking, reduces unwanted variation.
Designs	In **a completely randomized design,** all experimental units have an equal chance of receiving any treatment. In **a randomized block design,** the randomization occurs only within blocks.
Confounding	When the levels of one factor are associated with the levels of another factor in such a way that their effects cannot be separated, we say that these two factors are confounded.

Skills

▶ Recognize when an observational study would be appropriate.

▶ Be able to identify observational studies as retrospective or prospective, and understand the strengths and weaknesses of each method.

▶ Know the three basic principles of sound experimental design—control, randomize, replicate—and be able to explain each.

▶ Be able to recognize the factors, the treatments, and the response variable in a description of a designed experiment.

▶ Understand the essential importance of randomization in assigning treatments to experimental units.

▶ Understand the importance of replication in generalizing conclusions.

▶ Understand the value of blocking so that variability due to differences in attributes of the subjects can be removed.

▶ Understand the importance of a control group and the need for a placebo treatment in some studies.

▶ Understand the importance of blinding and double-blinding in studies on human subjects, and be able to identify blinding and the need for blinding in experiments.

▶ Understand the value of a placebo in experiments with human participants.

▶ Be able to design a completely randomized experiment to test the effect of a single factor.

▶ Know how to use graphical displays to compare responses for different treatment groups.

▶ Know how to report the results of an observational study. Identify the subjects, how the data were gathered, and any potential biases or flaws you may be aware of. Identify the factors known and those that might have been revealed by the study.

▶ Know how to compare the responses in different treatment groups to assess whether the differences are larger than could be reasonably expected from ordinary sampling variability.

▶ Know how to report the results of an experiment. Tell who the subjects are and how their assignment to treatments was determined. Report how and in what units the response variable was measured.

▶ Understand that your description of an experiment should be sufficient for another researcher to replicate the study with the same methods.

▶ Be able to report on the statistical significance of the result in terms of whether the observed group-to-group differences are larger than could be expected from chance alone.

EXPERIMENTS ON THE COMPUTER

Data from most experiments are analyzed with a statistics package. You should almost always display the results of a comparative experiment with side-by-side boxplots. You may also want to display the means and standard deviations of the treatment groups in a table.

EXERCISES

 A

1. **Standardized test scores.** For his Statistics class experiment, researcher J. Gilbert decided to study how parents' income affects children's performance on standardized tests like the SAT. He proposed to collect information from a random sample of test takers and examine the relationship between parental income and SAT score.
   a) Is this an experiment? If not, what kind of study is it?
   b) If there is a relationship between parental income and SAT score, why can't we conclude that differences in score are caused by differences in parental income?

2. **Heart attacks and height.** Researchers who examined health records of thousands of males found that men who died of myocardial infarction (heart attack) tended to be shorter than men who did not.
   a) Is this an experiment? If not, what kind of study is it?
   b) Is it correct to conclude that shorter men are at higher risk for heart attack? Explain.

3. **MS and vitamin D.** Multiple sclerosis (MS) is an autoimmune disease that strikes more often the farther people live from the equator. Could vitamin D—which most people get from the sun's ultraviolet rays—be a factor? Researchers compared vitamin D levels in blood samples from 150 U.S. military personnel who have developed MS with blood samples of nearly 300 who have not. The samples were taken, on average, five years before the disease was diagnosed. Those with the highest blood vitamin D levels had a 62% lower risk of MS than those with the lowest levels. (The link was only in whites, not in blacks or Hispanics.)
   a) What kind of study was this?
   b) Is that an appropriate choice for investigating this problem? Explain.
   c) Who were the subjects?
   d) What were the variables?

4. **Super Bowl commercials.** When spending large amounts to purchase advertising time, companies want to know what audience they'll reach. In January 2007, a poll asked 1008 American adults whether they planned to watch the upcoming Super Bowl. Men and women were asked separately whether they were looking forward more to the football game or to watching the commercials. Among the men, 16% were planning to watch and were looking forward primarily to the commercials. Among women, 30% were looking forward primarily to the commercials.
   a) Was this a stratified sample or a blocked experiment? Explain.
   b) Was the design of the study appropriate for the advertisers' questions?

5. **Menopause.** Researchers studied the herb black cohosh as a treatment for hot flashes caused by menopause. They randomly assigned 351 women aged 45 to 55 who reported at least two hot flashes a day to one of five groups: (1) black cohosh, (2) a multiherb supplement with black cohosh, (3) the multiherb supplement plus advice to consume more soy foods, (4) estrogen replacement therapy, or (5) a placebo. After a year, only the women given estrogen replacement therapy had symptoms that improved significantly more than those of the placebo group. [*Annals of Internal Medicine* 145:12, 869–897]
   a) What kind of study was this?
   b) Who were the subjects?
   c) Identify the treatment and response variables.
   d) What does "improved significantly" mean in a statistical sense?

6. **Honesty.** Coffee stations in offices often just ask users to leave money in a tray to pay for their coffee, but many people cheat. Researchers at Newcastle University replaced the picture of flowers on the wall behind the coffee station with a picture of staring eyes. They found that the average contribution increased significantly above the well-established standard when people felt they were being watched, even though the eyes were patently not real. (*NY Times* 12/10/06)
   a) Was this a survey, an observational study, or an experiment? How can we tell?
   b) Identify the variables.
   c) What does "increased significantly" mean in a statistical sense?

7–18. **What's the design?** *Read each brief report of statistical research, and identify*
   a) whether it was an observational study or an experiment.

   If it was an observational study, identify (if possible)
   b) whether it was retrospective or prospective.
   c) the subjects studied and how they were selected.
   d) the parameter of interest.
   e) the nature and scope of the conclusion the study can reach.

   If it was an experiment, identify (if possible)
   b) the subjects studied.
   c) the factor in the experiment and the levels.
   d) the number of treatments.
   e) the response variable measured.
   f) the design (completely randomized, blocked, or matched).
   g) whether it was blind (or double-blind).
   h) the nature and scope of the conclusion the experiment can reach.

7. Over a 4-month period, among 30 people with bipolar disorder, patients who were given a high dose (10 g/day) of omega-3 fats from fish oil improved more than those given a placebo. (*Archives of General Psychiatry* 56 [1999]: 407)

8. Among a group of disabled women aged 65 and older who were tracked for several years, those who had a vitamin B_{12} deficiency were twice as likely to suffer severe depression as those who did not. (*American Journal of Psychiatry* 157 [2000]: 715)

9. In a test of roughly 200 men and women, those with moderately high blood pressure (averaging 164/89 mm Hg) did worse on tests of memory and reaction time than those with normal blood pressure. (*Hypertension* 36 [2000]: 1079)

10. Researchers have linked an increase in the incidence of breast cancer in Italy to dioxin released by an industrial accident in 1976. The study identified 981 women who lived near the site of the accident and were under age 40 at the time. Fifteen of the women had developed breast cancer at an unusually young average age of 45. Medical records showed that they had heightened concentrations of dioxin in their blood and that each tenfold increase in dioxin level was associated with a doubling of the risk of breast cancer. (*Science News*, Aug. 3, 2002)

11. In 2002 the journal *Science* reported that a study of women in Finland indicated that having sons shortened the lifespans of mothers by about 34 weeks per son, but that daughters helped to lengthen the mothers' lives. The data came from church records from the period 1640 to 1870.

12. Scientists at a major pharmaceutical firm investigated the effectiveness of an herbal compound to treat the common cold. They exposed each subject to a cold virus, then gave him or her either the herbal compound or a sugar solution known to have no effect on colds. Several days later they assessed the patient's condition, using a cold severity scale ranging from 0 to 5. They found no evidence of benefits associated with the compound.

13. The May 4, 2000, issue of *Science News* reported that, contrary to popular belief, depressed individuals cry no more often in response to sad situations than nondepressed people. Researchers studied 23 men and 48 women with major depression and 9 men and 24 women with no depression. They showed the subjects a sad film about a boy whose father has died, noting whether or not the subjects cried. Women cried more often than men, but there were no significant differences between the depressed and nondepressed groups.

14. Some people who race greyhounds give the dogs large doses of vitamin C in the belief that the dogs will run faster. Investigators at the University of Florida tried three different diets in random order on each of five racing greyhounds. They were surprised to find that when the dogs ate high amounts of vitamin C they ran more slowly. (*Science News*, July 20, 2002)

15. After menopause, some women take supplemental estrogen. There is some concern that if these women also drink alcohol, their estrogen levels will rise too high. Twelve volunteers who were receiving supplemental estrogen were randomly divided into two groups, as were 12 other volunteers not on estrogen. In each case, one group drank an alcoholic beverage, the other a nonalcoholic beverage. An hour later, everyone's estrogen level was checked. Only those on supplemental estrogen who drank alcohol showed a marked increase.

16. A dog food company wants to compare a new lower-calorie food with their standard dog food to see if it's effective in helping inactive dogs maintain a healthy weight. They have found several dog owners willing to participate in the trial. The dogs have been classified as small, medium, or large breeds, and the company will supply some owners of each size of dog with one of the two foods. The owners have agreed not to feed their dogs anything else for a period of 6 months, after which the dogs' weights will be checked.

17. Athletes who had suffered hamstring injuries were randomly assigned to one of two exercise programs. Those who engaged in static stretching returned to sports activity in a mean of 15.2 days faster than those assigned to a program of agility and trunk stabilization exercises. (*Journal of Orthopaedic & Sports Physical Therapy* 34 [March 2004]: 3)

18. Pew Research compared respondents to an ordinary 5-day telephone survey with respondents to a 4-month-long rigorous survey designed to generate the highest possible response rate. They were especially interested in identifying any variables for which those who responded to the ordinary survey were different from those who could be reached only by the rigorous survey.

19. **Omega-3.** Exercise 7 describes an experiment that showed that high doses of omega-3 fats might be of benefit to people with bipolar disorder. The experiment involved a control group of subjects who received a placebo. Why didn't the experimenters just give everyone the omega-3 fats to see if they improved?

20. **Insomnia.** An experiment showed that exercise helped people sleep better. The experiment also involved another group of subjects who didn't exercise. Why didn't the experimenters just have all the subjects exercise and see if their ability to sleep improved?

21. **Omega-3 revisited.** Exercises 7 and 19 describe an experiment investigating a dietary approach to treating bipolar disorder. Researchers randomly assigned 30 subjects to two treatment groups, one group taking a high dose of omega-3 fats and the other a placebo.
    a) Why was it important to randomize in assigning the subjects to the two groups?
    b) What would be the advantages and disadvantages of using 100 subjects instead of 30?

22. **Insomnia revisited.** Exercise 20 describes an experiment investigating the effectiveness of exercise in combating insomnia. Researchers randomly assigned half of the 40 volunteers to an exercise program.

a) Why was it important to randomize in deciding who would exercise?

b) What would be the advantages and disadvantages of using 100 subjects instead of 40?

23. **Omega-3, again.** Exercises 7, 19, and 21 describe an experiment investigating the effectiveness of omega-3 fats in treating bipolar disorder. Suppose some of the 30 subjects were very active people who walked a lot or got vigorous exercise several times a week, while others tended to be more sedentary, working office jobs and watching a lot of TV. Why might researchers choose to block the subjects by activity level before randomly assigning them to the omega-3 and placebo groups?

24. **Insomnia again.** Exercises 20 and 22 describe an experiment investigating the effectiveness of exercise in combating insomnia. Suppose some of the 40 subjects had maintained a healthy weight, but others were quite overweight. Why might researchers choose to block the subjects by weight level before randomly assigning some of each group to the exercise program?

25. **Omega-3, finis.** The experiment described in Exercise 7 showed that high doses of omega-3 helped people with bipolar disorder. The researchers said that the difference between the omega-3 group and the placebo group were "statistically significant." Explain what that means in this context.

26. **Insomnia, finis.** The researchers who conducted the Exercise 20 insomnia experiment reported that the group who exercised got more sleep than the group who didn't, and said that the difference was "statistically significant." Explain what that means in this context.

27. **Shoes.** A running-shoe manufacturer wants to test the effect of its new sprinting shoe on 100-meter dash times. The company sponsors 5 athletes who are running the 100-meter dash in the 2004 Summer Olympic games. To test the shoe, it has all 5 runners run the 100-meter dash with a competitor's shoe and then again with their new shoe. The company uses the difference in times as the response variable.
a) Suggest some improvements to the design.
b) Why might the shoe manufacturer not be able to generalize the results they find to all runners?

28. **Swimsuits.** A swimsuit manufacturer wants to test the speed of its newly designed suit. The company designs an experiment by having 6 randomly selected Olympic swimmers swim as fast as they can with their old swimsuit first and then swim the same event again with the new, expensive swimsuit. The company will use the difference in times as the response variable. Criticize the experiment and point out some of the problems with generalizing the results.

29. **Reading.** Some schools teach reading using phonics (the sounds made by letters) and others using whole language (word recognition). Suppose a school district

wants to know which method works better. Suggest a design for an appropriate experiment.

30. **Healing.** A medical researcher suspects that giving post-surgical patients large doses of vitamin E will speed their recovery times by helping their incisions heal more quickly. Design an experiment to test this conjecture. Be sure to identify the factors, levels, treatments, response variable, and the role of randomization.

31. **Hamstrings.** Exercise 17 discussed an experiment to see if the time it took athletes with hamstring injuries to be able to return to sports was different depending on which of two exercise programs they engaged in.
a) Explain why it was important to assign the athletes to the two different treatments randomly.
b) There was no control group consisting of athletes who did not participate in a special exercise program. Explain the advantage of including such a group.
c) How might blinding have been used?
d) One group returned to sports activity in a mean of 37.4 days ($SD = 27.6$ days) and the other in a mean of 22.2 days ($SD = 8.3$ days). Do you think this difference is statistically significant? Explain.

32. **Diet and blood pressure.** An experiment that showed that subjects fed the DASH diet were able to lower their blood pressure by an average of 6.7 points compared to a group fed a "control diet." All meals were prepared by dieticians.
a) Why were the subjects randomly assigned to the diets instead of letting people pick what they wanted to eat?
b) Why were the meals prepared by dieticians?
c) Why did the researchers need the control group? If the DASH diet group's blood pressure was lower at the end of the experiment than at the beginning, wouldn't that prove the effectiveness of that diet?
d) What additional information would you want to know in order to decide whether an average reduction in blood pressure of 6.7 points was statistically significant?

33. **Dowsing.** Before drilling for water, many rural homeowners hire a dowser (a person who claims to be able to sense the presence of underground water using a forked stick). Suppose we wish to set up an experiment to test one dowser's ability. We get 20 identical containers, fill some with water, and ask him to tell which ones they are.
a) How will we randomize this procedure?
b) The dowser correctly identifies the contents of 12 out of 20 containers. Do you think this level of success is statistically significant? Explain.
c) How many correct identifications (out of 20) would the dowser have to make to convince you that the forked-stick trick works? Explain.

34. Mozart. Will listening to a Mozart piano sonata make you smarter? In a 1995 study published in the journal *Psychological Science,* Rauscher, Shaw, and Ky reported that when students were given a spatial reasoning section of a standard IQ test, those who listened to Mozart for 10 minutes improved their scores more than those who simply sat quietly.

a) These researchers said the differences were statistically significant. Explain what that means in context.

b) Steele, Bass, and Crook tried to replicate the original study. In their study, also published in *Psychological Science* (1999), the subjects were 125 college students who participated in the experiment for course credit. Subjects first took the test. Then they were assigned to one of three groups: listening to a Mozart piano sonata, listening to music by Philip Glass, and sitting for 10 minutes in silence. Three days after the treatments, they were retested. Draw a diagram displaying the design of this experiment.

c) These boxplots show the differences in score before and after treatment for the three groups. Did the Mozart group show greater improvement?

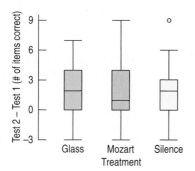

d) Do you think the results prove that listening to Mozart is beneficial? Explain.

35. Wine. A 2001 Danish study published in the *Archives of Internal Medicine* casts significant doubt on suggestions that adults who drink wine have higher levels of "good" cholesterol and fewer heart attacks. These researchers followed a group of individuals born at a Copenhagen hospital between 1959 and 1961 for 40 years. Their study found that in this group the adults who drank wine were richer and better educated than those who did not.

a) What kind of study was this?

b) It is generally true that people with high levels of education and high socioeconomic status are healthier than others. How does this call into question the supposed health benefits of wine?

c) Can studies such as these prove causation (that wine helps prevent heart attacks, that drinking wine makes one richer, that being rich helps prevent heart attacks, etc.)? Explain.

36. Swimming. Recently, a group of adults who swim regularly for exercise were evaluated for depression. It turned out that these swimmers were less likely to be depressed than the general population. The researchers said the difference was statistically significant.

a) What does "statistically significant" mean in this context?

b) Is this an experiment or an observational study? Explain.

c) News reports claimed this study proved that swimming can prevent depression. Explain why this conclusion is not justified by the study. Include an example of a possible lurking variable.

d) But perhaps it is true. We wonder if exercise can ward off depression, and whether anaerobic exercise (like weight training) is as effective as aerobic exercise (like swimming). We find 120 volunteers not currently engaged in a regular program of exercise. Design an appropriate experiment.

37. Weekend deaths. A study published in the *New England Journal of Medicine* (Aug. 2001) suggests that it's dangerous to enter a hospital on a weekend. During a 10-year period, researchers tracked over 4 million emergency admissions to hospitals in Ontario, Canada. Their findings revealed that patients admitted on weekends had a much higher risk of death than those who went on weekdays.

a) The researchers said the difference in death rates was "statistically significant." Explain in this context what that means.

b) What kind of study was this? Explain.

c) If you think you're quite ill on a Saturday, should you wait until Monday to seek medical help? Explain.

d) Suggest some possible explanations for this troubling finding.

38. Full moon. It's a common belief that people behave strangely when there's a full moon and that as a result police and emergency rooms are busier than usual. Design a way you could find out whether there is any merit to this belief.[6] Will you use an observational study or an experiment? Explain.

39. Shingles. A research doctor has discovered a new ointment that she believes will be more effective than the current medication in the treatment of shingles (a painful skin rash). Eight patients have volunteered to participate in the initial trials of this ointment. You are the statistician hired as a consultant to help design a completely randomized experiment.

a) Describe how you will conduct this experiment.

b) Can you make this experiment double-blind? How?

c) The initial experiment revealed that males and females may respond differently to the ointment. Further testing of the drug's effectiveness is now planned, and many patients have volunteered. What changes in your first design, if any, would you make for this second stage of testing?

---

[6]Lots of people have done such studies, finding no evidence of any "full moon effect."

40. **SAT prep.** Can special study courses actually help raise SAT scores? One organization says that the 30 students they tutored achieved an average gain of 60 points when they retook the test.
    a) Explain why this does not necessarily prove that the special course caused the scores to go up.
    b) Propose a design for an experiment that could test the effectiveness of the tutorial course.
    c) Suppose you suspect that the tutorial course might be more helpful for students whose initial scores were particularly low. How would this affect your proposed design?

Answers
Just Checking

1. a) The factor was type of treatment for peptic ulcer.
    b) The response variable could be a measure of relief from gastric ulcer pain or an evaluation by a physician of the state of the disease.
    c) Treatments would be gastric freezing and some alternative control treatment.
    d) Treatments should be assigned randomly.
    e) No. The website reports "lack of effectiveness," indicating that no large differences in patient healing were noted.

2. a) Neither the patients who received the treatment nor the doctor who evaluated them afterward knew what treatment they had received.
    b) The placebo is needed to accomplish blinding. The best alternative would be using body-temperature liquid rather than the freezing liquid.
    c) The researchers should block the subjects by the length of time they had had the ulcer, then randomly assign subjects in each block to the freezing and placebo groups.

Using Randomness

We all know what it means for something to be random. Or do we? Many children's games rely on chance outcomes. Rolling dice, spinning spinners, and shuffling cards are all random. Adult games use randomness as well, from card games to lotteries to bingo. What's the most important aspect of the randomness in these games? They must be fair.

What makes random selection seem fair? It's really two things. First, nobody can guess the outcome before it happens. Second, when we want things to be fair, all outcomes should be equally likely.

Randomness is an essential tool of Statistics. Without deliberately applying randomness, we couldn't do most of Statistics, and this book would stop right about here.[1] As you already know, we must choose samples at random. In experiments, we must assign our subjects to treatments at random. We can also use randomness to simulate real situations, to see what might happen and how often. Knowing how likely certain outcomes are is an essential part of judging what's statistically significant. To do all of these, we rely on random numbers.

But truly random values are surprisingly hard to get. Just to see how fair humans are at selecting, pick a number at random from the top of the next page. Go ahead. Turn the page, look at the numbers quickly, and pick a number at random.

Ready?

Go.

"The most decisive conceptual event of twentieth century physics has been the discovery that the world is not deterministic. . . . A space was cleared for chance."

— Ian Hocking,
The Taming of Chance

---

[1]Don't get your hopes up.

1 2 3 4

It's Not Easy Being Random

"The generation of random numbers is too important to be left to chance."

—Robert R. Coveyou,
Oak Ridge National Laboratory

An ordinary deck of playing cards, like the ones used in bridge and many other card games, consists of 52 cards. There are numbered cards (2 through 10), and face cards (Jack, Queen, King, Ace) whose value depends on the game you are playing. Each card is also marked by one of four suits (clubs, diamonds, hearts, or spades) whose significance is also game-specific.

Did you pick 3? If so, you've got company. Almost 75% of all people pick the number 3. About 20% pick either 2 or 4. If you picked 1, well, consider yourself a little different. Only about 5% choose 1. Psychologists have proposed reasons for this phenomenon, but for us, it simply serves as a lesson that we've got to find a better way to choose things at random.

So how should we generate **random numbers?** It's surprisingly difficult to get random values even when they're equally likely. Computers have become a popular way to generate random numbers, although technically these are only *pseudorandom*. Computers follow programs. Start a computer from the same place, and it will always follow exactly the same path. So pseudorandom values are generated in a fixed sequence. Fortunately, pseudorandom values are good enough for most purposes because they are virtually indistinguishable from truly random numbers.

There *are* ways to generate random numbers so that they are both equally likely and truly random. There are books with lists of carefully generated random numbers.[2] Several Internet sites use methods like timing the decay of a radioactive element or even the random changes of lava lamps to generate truly random digits.[3] In either case, a string of random digits might look like this:

```
2217726304387410092537086270581997622725849795907032825001108963
3217535822643800292254644943760642389043766557204107354186024508
8906427308645681412198226653885873285801699027843110380420067664
8740522639824530519902027044464984322000946238678577902639002954
8887003319933147508331265192321413908608674496383528968974910533
6944182713168919406022181281304751019321546303870481407676636740
6070204916508913632855535136136104379429342848690946288143179336O
7706356513310563210508993624272872250535395513645991015328128202
```

---

[2] You'll find a table of random digits of this kind in the back of this book.
[3] For example, www.random.org or www.randomnumbers.info.

Not exactly thrilling reading, is it? Let's discuss ways to use such random digits to apply randomness to real situations.

> **Aren't you done shuffling yet?** Even something as common as card shuffling may not be as random as you might think. If you shuffle cards by the usual method in which you split the deck in half and try to let cards fall roughly alternately from each half, you're doing a "riffle shuffle."
>
> How many times should you shuffle cards to make the deck random? A surprising fact was discovered by statisticians Persi Diaconis, Ronald Graham, and W. M. Kantor. It takes seven riffle shuffles. Fewer than seven leaves order in the deck, but after that, more shuffling does little good. Most people, though, don't shuffle that many times.
>
> When computers were first used to generate hands in bridge tournaments, some professional bridge players complained that the computer was making too many "weird" hands—hands with 10 cards of one suit, for example. Suddenly these hands were appearing more often than players were used to when cards were shuffled by hand. The players assumed that the computer was doing something wrong. But it turns out that it's humans who hadn't been shuffling enough to make the decks really random and have those "weird" hands appear as often as they should.

Random Selection

In order to give us an unbiased look at a population, polls and surveys must be based on random samples. Indeed, any time we want a fair snapshot of a large group based on a few of its members, we need to select the individuals at random. Consider the problem of quality control. In a factory manufacturing candy, it's someone's job to make sure the process continues to work correctly. This inspector will pick some candies at random, check to see that the wrapper is printed and sealed correctly, put each on a scale to be sure its net weight is correct, and, yes, even taste test the candy. Suppose this hour's candy wrappers carry production codes numbered 205 through 680 and we want to test 4 of them. The inspector could choose her sample using our list of random numbers that starts with 221772630438741009253708627 . . . It's pretty easy:

1. Since the wrappers carry 3-digit codes, divide the random digits into groups of three: 221 772 630 438 741 009 253 708 627
2. Pick candy 221.
3. Skip 772, because no wrapper had that code.
4. Pick candies 630 and 438.
5. Ignore 741 and 009, then pick 253 as the fourth one.

Note that in addition to ignoring numbers that didn't appear on any of the wrappers, the inspector would also ignore any repeated numbers—she won't taste the same candy twice!

Unlike these candies, most populations are not already numbered, but that's no problem. As a first step, we simply assign numbers to everyone, then get out a list of random digits and start selecting the individuals for our sample.

For Example   Using Random Numbers to Get an SRS[4]

There are 80 students enrolled in an introductory Statistics class; you are to select a sample of 5.

Question: How can you select an SRS of 5 students using these random digits found on the Internet?

05166 29305 77482

First I'll number the students from 00 to 79. Taking the random numbers two digits at a time gives me 05, 16, 62, 93, 05, 77, and 48. I'll ignore 93 because the students were numbered only up to 79. And, so as not to pick the same person twice, I'll skip the repeated number 05. My simple random sample consists of students with the numbers 05, 16, 62, 77, and 48.

Just Checking

1. You are in charge of choosing the band for the Senior Prom. This is a big decision, so you want to get input from the seniors. You worry that only some of the seniors would voluntarily fill out and return a survey, and that their opinions might not represent the interests of the entire class. Describe how you can use a list of the 454 seniors and a table of random digits (or a computer or calculator) to choose a simple random sample of 40 students for your survey.

(Check your answers on page 281.)

Random Assignment

When we're running an experiment, we usually don't choose our subjects at random. Instead, we need to randomly assign each individual to a treatment group. But don't let that fool you: the process is quite similar. Suppose we have 40 elementary school children to be assigned to Class A (a traditional math approach emphasizing drill and practice) or Class B (a new approach relying on investigation and discovery). Here are two ways we might create these classes at random.

- Number the students 00–39, then use a list of random digits (2 digits at a time) to select 20 kids for Class A. Put the other 20 in Class B.
- Assign each student a random digit from the list, one at a time. For example, you might get Sally = 2, José = 2, Billy = 1, Keniesha = 7, etc. Put all the students with odd numbers in Class A and all the students with even numbers in Class B. (Because you don't end up ignoring any of the digits, this method may be much easier, but note that the classes may not end up exactly the same size.)

---

[4]Deja vu? Yes, we looked at this example back in Chapter 10 when we first discussed sampling. We promised then we'd show you more details about working with random numbers later on. Like now.

For Example

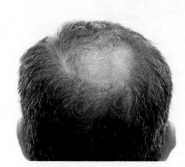

An experiment will test the effectiveness of a new lotion that claims to prevent hair loss in men who are going bald. Our volunteers are 150 balding men who have agreed to try this product. We need to divide them into three groups. The subjects in Group A will apply the lotion twice a day, those in Group B will apply it once a day, and we'll give the men in Group C a placebo (a lotion that looks the same but is known to have no effect on hair loss).

Question: How would you use random numbers from a computer, calculator, or random number table to randomly assigning each participant to Group A, B, or C?

Use a list of random digits, and skipping all the O's, assign each man a number between 1 and 9. Put all the men with numbers 1, 2, or 3 in Group A, the men with a 4, 5, or 6 in Group B, and the others in Group C. This should result in three groups of about 50 men each.

Let's Pretend . . .

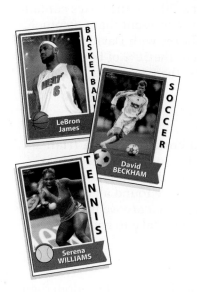

Suppose a cereal manufacturer puts pictures of famous athletes on cards in boxes of cereal in the hope of boosting sales. The manufacturer announces that 20% of the boxes contain a picture of LeBron James, 30% a picture of David Beckham, and the rest a picture of Serena Williams. You want all three pictures. How many boxes of cereal do you expect you'll have to buy in order to get the complete set?

How can we answer questions like this? Well, one way is to buy hundreds of boxes of cereal to see what might happen. But let's not. Instead, we'll consider using a random model. Why random? When we pick a box of cereal off the shelf, we don't know what picture is inside. We'll assume that the pictures are randomly placed in the boxes and that the boxes are distributed randomly to stores around the country. Why a model? Because we won't actually buy the cereal boxes. We can't afford all those boxes and we don't want to waste food. So we need an imitation of the real process that we can manipulate and control. In short, we're going to **simulate** reality.

A Simulation

The question we've asked is how many boxes do you expect to buy to get a complete card collection. But we can't answer that by completing a card collection just once. We want to understand the *typical* number of boxes to open, how that number varies, and, often, the shape of the distribution. So we'll have to do this over and over. We call each time we obtain a simulated answer to our question a **trial**.

For the sports cards, a trial's outcome is the number of boxes. We'll need at least 3 boxes to get one of each card, but with really bad luck, you could empty the shelves of several supermarkets before finding the card you need to get all 3. So, the possible outcomes of a trial are 3, 4, 5, or lots more. But we can't simply pick one of those numbers at random, because they're not equally likely. We'd be surprised if we only needed 3 boxes to get all the cards,

but we'd probably be even more surprised to find that it took exactly 7,359 boxes. In fact, the reason we're doing the simulation is that it's hard to guess how many boxes we'd expect to open.

Building a Simulation

We know how to find equally likely random digits. How can we get from there to simulating the trial outcomes? We know the relative frequencies of the cards: 20% LeBron, 30% Beckham, and 50% Serena. So, we could interpret the digits 0 and 1 as finding LeBron; 2, 3, and 4 as finding Beckham; and 5 through 9 as finding Serena to simulate opening one box. Opening one box is the basic building block, called a **component** of our simulation. But the component's outcome isn't the result we want. We need to open a sequence of boxes until our card collection is complete. The *trial's* outcome is called the **response variable**; for this simulation that's the *number* of components (boxes) in the sequence.

Let's look at the steps for making a simulation:

Specify how to model a component outcome using equally likely random digits:

1. **Identify the component to be repeated.** In this case, our component is the opening of a box of cereal.
2. **Explain how you will model the component's outcome.** The digits from 0 to 9 are equally likely to occur. Because 20% of the boxes contain LeBron's picture, we'll use 2 of the 10 digits to represent that outcome. Three of the 10 digits can model the 30% of boxes with David Beckham cards, and the remaining 5 digits can represent the 50% of boxes with Serena. One possible assignment of the digits, then, is

$$0, 1 = \text{LeBron} \quad 2, 3, 4 = \text{Beckham} \quad 5, 6, 7, 8, 9 = \text{Serena}.$$

Specify how to simulate trials:

3. **Explain how you will combine the components to model a trial.** We pretend to open boxes (repeat components) until our collection is complete. We do this by looking at each random digit and indicating what picture it represents. We continue until we've found all three.
4. **State clearly what the response variable is.** What are we interested in? We want to find out the number of boxes it might take to get all three pictures.

Put it all together to run the simulation:

5. **Run several trials.** For example, consider the third line of random digits shown earlier (p. 266):

   8906427308645681412198226653885587328580169902784311038042006766 4.

   Let's see what "happens."
   The first random digit, 8, means you get Serena's picture. The second digit, 9, means Serena's picture is also in the next box. Continuing to interpret the random digits, we get LeBron's picture (0) in the third, Serena's (6) again in the fourth, and finally Beckham (4) on the fifth box. Since we've now found all three pictures, we've finished one trial of our simulation. This trial's outcome is 5 boxes.
   Now we keep going, running more trials by looking at the rest of our line of random digits:

   89064 2730 8645681 41219 822665388587328580 169902 78431 1038 042006 7664.

It's best to create a chart to keep track of what happens:

Trial Number	Component Outcomes	Trial Outcomes: y = Number of Boxes
1	89064 = Serena, Serena, LeBron, Serena, Beckham	5
2	2730 = Beckham, Serena, Beckham, LeBron	4
3	8645681 = Serena, Serena, Beckham, ..., LeBron	7
4	41219 = Beckham, LeBron, Beckham, LeBron, Serena	5
5	822665388587328580 = Serena, Beckham, ..., LeBron	18
6	169902 = LeBron, Serena, Serena, Serena, LeBron, Beckham	6
7	78431 = Serena, Serena, Beckham, Beckham, LeBron	5
8	1038 = LeBron, LeBron, Beckham, Serena	4
9	042006 = LeBron, Beckham, Beckham, LeBron, LeBron, Serena	6
10	7664... = Serena, Serena, Serena, Beckham...	?

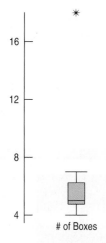

Analyze the response variable:

6. **Collect and summarize the results of all the trials.** You know how to summarize and display a response variable. You'll certainly want to report the shape, center, and spread, and depending on the question asked, you may want to include more.
7. **State your conclusion,** as always, in the context of the question you wanted to answer. Based on this simulation, we estimate that customers hoping to complete their card collection will need to open a median of 5 boxes, but it could take a lot more.

If you fear that these may not be accurate estimates because we ran only nine trials, you are absolutely correct. The more trials the better, and nine is woefully inadequate. Twenty trials is probably a reasonable minimum if you are doing this by hand. Even better, use a computer and run a few hundred trials.

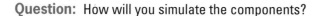

For Example Simulating a Dice Game

The game of 21 can be played with an ordinary 6-sided die. Competitors each roll the die repeatedly, trying to get the highest total less than or equal to 21. If your total exceeds 21, you lose.

Suppose your opponent has rolled an 18. Your task is to try to beat him by getting more than 18 points without going over 21. How many rolls do you expect to make, and what are your chances of winning?

Question: How will you simulate the components?

A component is one roll of the die. I'll simulate each roll by looking at a random digit from a table or my calculator. The digits 1 through 6 will represent the results on the die; I'll ignore digits 7–9 and 0.

Question: How will you combine components to model a trial? What's the response variable?

I'll add components (rolls) until my total is greater than 18, counting the number of rolls. If my total is greater than 21, I lose; if not, I win. There are two response variables. I'll count the number of times I roll the die, and I'll keep track of whether I win or lose.

(continued)

Question: How would you use these random digits to run trials? Show your method clearly for two trials.

91129 58757 69274 92380 82464 33089

I've marked the discarded digits in color.

Trial #1:	9	1	1	2	9	5	8	7	5	7	6	
Total:		1	2	4		9			14		20	Outcomes: 6 rolls, won
Trial #2:	9	2	7	4	9	2	3	8	0	8	2	4	6
Total:		2		6		8	11				13	17	23	Outcomes: 7 rolls, lost

Question: Suppose you run 30 trials, getting the outcomes tallied here. What is your conclusion?

Based on my simulation, when competing against an opponent who has a score of 18, I expect my turn to usually last 5 or 6 rolls, and I should win about 70% of the time.

Number of rolls			Result	
4	///		Won	༔ 𝄂 𝄂 𝄂 𝄂 /
5	𝄂 𝄂		Lost	𝄂 ////
6	𝄂 𝄂 /			
7	𝄂			
8	/			

Just Checking

2. The baseball World Series consists of up to seven games. The first team to win four games wins the series. The first two are played at one team's home ballpark, the next three at the other team's park, and the final two (if needed) are played back at the first park. Records over the past century show that there is a home field advantage; the home team has about a 55% chance of winning any game. Does the current system of alternating ballparks even out the home field advantage? How often will the team that begins at home win the series?

   Let's set up the simulation:

   a) What is the component to be repeated?
   b) How will you model each component from equally likely random digits?
   c) How will you model a trial by combining components?
   d) What is the response variable?
   e) How will you analyze the response variable?

   (Check your answers on page 281.)

How Likely Is That?

Sometimes we want to know what the chances are that some event happens. Take the cereal boxes again, for example. Instead of continuing to buy cereal until we get pictures of LeBron, Beckham, and Serena, suppose we decide to buy only 5 boxes and hope that's enough. How likely are we to end up with at least one photo of each athlete?

Let's answer this new question with a simulation. We start out the same as before, using random digits 0–9 for each cereal box. Because there's a LeBron James picture in 20% of all boxes, we'll let digits 0 and 1 represent LeBron. Beckham's picture is in 30% of the boxes, so we'll use 2, 3, and 4 for him, and that leaves digits 5–9 to represent the 50% of all boxes containing Serena's picture.

Now there's a difference, though. This time we'll use a group of 5 digits to represent the 5 boxes of cereal we buy, and see whose pictures we get. If our first set of digits is 96299, that represents pictures of Serena, Serena, Beckham, Serena, and Serena. Our question was "Did we get all three pictures?," so for this trial the answer (our response variable) is "No."

As always, to begin to understand what might happen we'll need to repeat this several times. Here are 10 trials, based on the first row of random digits in the table in the back of the book:

Trial #	Random Digits	Outcome	Response
1	96299	Serena, Serena, Beckham, Serena, Serena	No
2	07196	LeBron, Serena, LeBron, Serena, Serena	No
3	98642	Serena, Serena, Serena, Beckham, Beckham	No
4	20639	Beckham, LeBron, Serena, Beckham, Serena	Yes
5	23185	Beckham, Beckham, LeBron, Serena, Serena	Yes
6	56282	Serena, Serena, Beckham, Serena, Beckham	No
7	69929	Serena, Serena, Serena, Beckham, Serena	No
8	14125	LeBron, Beckham, LeBron, Beckham, Serena	Yes
9	38872	Beckham, Serena, Serena, Serena, Beckham	No
10	94168	Serena, Beckham, LeBron, Serena, Serena	Yes

We found all three pictures 4 times in these 10 trials. Based on this simulation we can estimate that there's about a 40% chance that we'd get a complete set of the three pictures by buying 5 boxes of cereal.

Assessing the likelihood of certain outcomes in this way can help us decide whether the results of a study are *statistically significant*. Remember, that means that what we've seen happen is so unusual that we don't think it happened just by chance. If we actually were to buy 5 boxes of cereal and managed to score all three athletes' photos, that wouldn't be unusual. After all, it happened 40% of the time in our simulation. But opening all five boxes and never finding a Serena picture should make us suspicious, because our simulation suggests that's very unlikely to happen by chance alone. We'd wonder if the boxes are being distributed randomly, or if Serena's picture isn't really in 50% of them, as advertised.

For Example

Foresters know that historically about 10% of trees in a certain national forest have been infested with bark beetles. This spring, though, they found signs of the beetles in 8 of 50 randomly chosen trees.

Question: Is this apparent increase in beetle infestation statistically significant? (In other words, is finding 8 of 50 trees infested so unlikely that the foresters should be concerned that the true rate is now higher than 10%?)

I'll use a simulation to see how many trees out of 50 might have bark beetles if in fact 10% of all trees are infested.

- A component is one tree. For each tree I'll look at a random digit from 0 to 9. I'll use 0 to indicate bark beetles and 1–9 to indicate that the tree is not infested.

(continued)

- I'll use a line of 50 digits from the random number table to represent a sample of 50 trees.
- I want to see how likely it is to find beetles in 8 or more of the 50 trees.

Trial #1:

96299 0̲7196 98642 2̲0̲639 23185 56282 66929 14125 38872 94168

ooooo Xoooo ooooo oXooo ooooo ooooo ooooo ooooo ooooo ooooo

2 infested, 48 OK

Trial #2:

71622 3594̲0̲ 818̲0̲7 59225 18192 0̲871̲0̲ 8̲0̲777 67395 69563 8628̲0̲

ooooo ooooX oooXo ooooo ooooo XoooX oXooo ooooo ooooo ooooX

6 infested, 44 OK

Here's a table summarizing my results for 10 trials:

Trial	1	2	3	4	5	6	7	8	9	10
Infested	2	6	5	4	5	9	5	3	7	5
OK	48	44	45	46	45	41	45	47	43	45

In my simulation I found 8 or more infested trees once in 10 trials, so maybe the beetle problem really is no worse and the outcome the foresters observed happened just by chance. However, only once in 10 trials seems somewhat unusual, so the foresters should probably check more trees (and I should run a simulation with more trials!).

STEP-BY-STEP EXAMPLE   Simulations and Significance

Fifty-seven students participated in a lottery for a particularly desirable dorm room—a triple with a fireplace and private bath in the tower. Twenty of the participants were members of the same varsity team. When all three winners were members of the team, the other students cried foul.

Question: Could an all-team outcome reasonably be expected to happen if everyone had a fair shot at the room?

Plan State the problem. Identify the important parts of your simulation.	I'll use a simulation to investigate whether it's likely that three varsity athletes would get the great room in the dorm if the lottery were fair.
Components Identify the components.	A component is the selection of a student.
Outcomes State how you will model each component using equally likely random digits. You can't just use the digits from 0 to 9 because the outcomes you are simulating are not multiples of 10%.	I'll look at two-digit random numbers. Let 00–19 represent the 20 varsity applicants. Let 20–56 represent the other 37 applicants.

There are 20 and 37 students in the two groups. This time you must use *pairs* of random digits (and ignore some of them) to represent the 57 students.

Skip 57–99. If I get a number in this range, I'll throw it away and go back for another two-digit random number.

Trial Explain how you will combine the components to simulate a trial. In each of these trials, you can't choose the same student twice, so you'll need to ignore a random number if it comes up a second or third time. Be sure to mention this in describing your simulation.

I'll look at pairs of digits, identifying them as V (varsity) or N (nonvarsity) until 3 people are chosen. I'll ignore out-of-range numbers and also repeated numbers (X)—I can't put the same person in the room twice.

Response Variable Define your response variable.

The response variable is whether or not all three selected students are on the varsity team.

Mechanics Run several trials. Carefully record the random numbers, indicating

1) the corresponding component outcomes (here, Varsity, Nonvarsity, or ignored number) and
2) the value of the response variable.

Trial Number	Component Outcomes	All Varsity?
1	74 02 94 39 02 77 55 X V X N X X N	No
2	18 63 33 25 V X N N	No
3	05 45 88 91 56 V N X X N	No
4	39 09 07 N V V	No
5	65 39 45 95 43 X N N X N	No
6	98 95 11 68 77 12 17 X X V X X V V	Yes
7	26 19 89 93 77 27 N V X X X N	No
8	23 52 37 N N N	No
9	16 50 83 44 V N X N	No
10	74 17 46 85 09 X V N X V	No

Analyze Summarize the results across all trials to answer the initial question.

"All varsity" occurred once, or 10% of the time.

Conclusion Describe what the simulation shows, and interpret your results in the context of the real world.

In my simulation of "fair" room draws, the three people chosen were all varsity team members only 10% of the time. While this result could happen by chance, it is not particularly likely. I'm suspicious, but I'd need many more trials and a smaller frequency of the all-varsity outcome before I would make an accusation of unfairness.

TI Tips

Generating random numbers

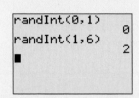

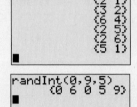

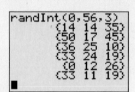

Instead of using coins, dice, cards, or tables of random numbers, you may decide to use your calculator for simulations. There are several random number generators offered in the MATH PRB menu.

5:randInt(is of particular importance. This command will produce any number of random integers in a specified range.

Here are some examples showing how to use randInt for simulations:

- randInt(0,1) randomly chooses a 0 or a 1. This is an effective simulation of a coin toss. You could let 0 represent tails and 1 represent heads.
- randInt(1,6) produces a random integer from 1 to 6, a good way to simulate rolling a die.

- randInt(1,6,2) simulates rolling *two* dice. To do several rolls in a row, just hit ENTER repeatedly.

- randInt(0,9,5) produces five random integers that might represent the pictures in the cereal boxes. Our trial gave us two LeBrons (0, 1), no Beckhams (2, 3, 4), and three Serenas (5–9).

- randInt(0,56,3) produces three random integers between 0 and 56, a nice way to simulate the dorm room lottery. The window shows 6 trials, but we would skip the first one because one student was chosen twice. In none of the remaining 5 trials did three athletes (0–19) win.

⊘ WHAT CAN GO WRONG?

- **Don't overstate your case.** Let's face it: In some sense, a simulation is *always* wrong. After all, it's not the real thing. We didn't buy any cereal, inspect any candy bars or trees, or actually run a room draw. So beware of confusing what *really* happens with what a simulation suggests *might* happen. Never forget that future results will not match your simulated results exactly.

- **Model outcome chances accurately.** A common mistake is to use a strategy that may appear to produce the right kind of results, but that does not accurately model the situation. For example, in our room draw, we could have gotten 0, 1, 2, or 3 team members. Why not just see how often these digits occur in random digits from 0 to 9, ignoring the digits 4 and up?

3 2 1 7 9 0 0 5 9 7 3 7 3 7 9 2 5 2 4 1 3 8

3 2 1 x x 0 0 x x x 3 x x 2 x 2 x 1 3 x

This "simulation" makes it seem fairly likely that three team members would be chosen. There's a big problem with this approach, though: The digits 0, 1, 2, and 3 occur with equal frequency among random digits, making each outcome appear to happen 25% of the time. In fact, the selection of 0, 1, 2, or all 3 team members are not all equally likely outcomes. In our correct simulation, we estimated that all 3 would be chosen only about 10% of the time. If your simulation overlooks important aspects of the real situation, your model will not be accurate.

TI-*nspire*

Simulations. Improve your predictions by running thousands of trials.

- **Run enough trials.** Simulation is cheap and fairly easy to do. Don't try to draw conclusions based on 5 or 10 trials (even though we did for illustration purposes here). Always use a large numbers of trials.

SIMULATIONS **IN YOUR WORLD**

Is Global Warming Real?

Signs that the earth is warming are recorded all over the globe. Although we can't look at thermometers going back thousands of years, we do have some records that help us figure out what temperatures and concentrations were like in the distant past.

Each year, trees grow thicker and form new rings. In warmer and wetter years, the rings are thicker. Old trees and wood can tell us about conditions hundreds or even several thousands of years ago.

Keys to the past are also buried under lakes and oceans. Pollen, creatures and particles fall to the bottom of oceans and lakes each year, forming sediments. . . which contain a wealth of information about what was in the air and water when they fell. Scientists reveal this record by inserting hollow tubes into the mud to collect sediment layers going back millions of years.

For a direct look at the atmosphere of the past, scientists drill cores through the earth's polar ice sheets. Tiny gas bubbles trapped in the ice are actually pieces of the earth's past atmosphere, frozen in time. That's how we know that the concentrations of greenhouse gases since the industrial revolution are higher than they've been for hundreds of thousands of years.

Computer models help scientists to understand the Earth's climate, or long-term weather patterns. Models also allow scientists to make predictions about the future climate. Basically, models simulate how the

Going, Going . . . : Peru's Quelccaya ice cap is the largest in the tropics. If it continues to melt at its current rate—contracting more than 600 feet (182.8 meters) a year in some places—it will be gone by 2100, leaving thousands who rely on its water for drinking and electricity high, dry, and in the dark.

atmosphere and oceans absorb energy from the sun and transport it around the globe. Factors that affect the amount of the sun's energy reaching Earth's surface are what drive the climate in these models, as in real life. These include things like greenhouse gases, particles in the atmosphere (such as from volcanoes), and changes in energy coming from the sun itself.

http://environment.nationalgeographic.com/environment/global-warming/gw-real.html

WHAT HAVE WE LEARNED?

We've learned to harness the power of randomness. We've seen how to select random samples and to randomly assign subjects to treatments. We've learned that a simulation model can help us investigate a question for which many outcomes are possible, we can't (or don't want to) collect data, and a mathematical answer is hard to calculate. Like all models, simulations can provide us with useful insights about the real world, including judgments about statistical significance. We've learned how to use random values to achieve all of these goals.

Terms

Random   An outcome is random if we know the possible values it can have, but not which particular value it takes.

Generating random numbers   Random numbers are hard to generate. Nevertheless, several Internet sites offer an unlimited supply of equally likely random values.

Simulation   A simulation models a real-world situation by using random-digit outcomes to mimic the uncertainty of a response variable of interest.

Simulation component   A component uses equally likely random digits to model simple random occurrences whose outcomes may not be equally likely.

Trial   The sequence of several components representing events that we are pretending will take place.

Response variable   Values of the response variable record the results of each trial with respect to what we were interested in.

Statistical significance   Outcomes are said to be statistically significant if they are unlikely to occur simply by chance.

Skills

THINK
- ▶ Be able to recognize random outcomes in a real-world situation.
- ▶ Be able to recognize when a simulation might usefully model random behavior in the real world.

SHOW
- ▶ Know how to use random numbers or some other source of random values, such as dice, a spinner, or a table of random numbers.
- ▶ Know how to choose a sample at random.
- ▶ Know how to randomly assign subjects to experimental treatments.
- ▶ Know how to simulate real-world outcomes.

TELL
- ▶ Be able to describe a randomization process or a simulation so that others can repeat it.
- ▶ Be able to discuss the results of a simulation study and draw conclusions about the question being investigated.

SIMULATION ON THE COMPUTER

Simulations are best done with the help of technology simply because more trials makes a better simulation, and computers are fast. There are special computer programs designed for simulation, and most statistics packages and calculators can at least generate random numbers to support a simulation.

EXERCISES

1. **Coin toss.** Is a coin flip random? Why or why not?

2. **Casino.** A casino claims that its electronic "video roulette" machine is truly random. What should that claim mean?

3. **The lottery.** Many states run lotteries, giving away millions of dollars if you match a certain set of winning numbers. How are those numbers determined? Do you think this method guarantees randomness? Explain.

4. **Games.** Many kinds of games people play rely on randomness. Cite three different methods commonly used in the attempt to achieve this randomness, and discuss the effectiveness of each.

5. **Birth defects.** The American College of Obstetricians and Gynecologists says that out of every 100 babies born in the United States, 3 have some kind of major birth defect. How would you assign random numbers to conduct a simulation based on this statistic?

6. **Colorblind.** By some estimates, about 10% of all males have some color perception defect, most commonly red–green colorblindness. How would you assign random numbers to conduct a simulation based on this statistic?

7. **Geography.** An elementary school teacher with 25 students plans to have each of them make a poster about two different states. The teacher first numbers the states (in alphabetical order, from 1-Alabama to 50-Wyoming), then uses a random number table to decide which states each kid gets. Here are the random digits:

   45921 01710 22892 37076

   a) Which two state numbers does the first student get?
   b) Which two state numbers go to the second student?

8. **Get rich.** Your state's BigBucks Lottery prize has reached $100,000,000, and you decide to play. You have to pick five numbers between 1 and 60, and you'll win if your numbers match those drawn by the state. You decide to pick your "lucky" numbers using a random number table. Which numbers do you play, based on these random digits?

   43680 98750 13092 76561 58712

9. **Class project.** Here is a list of the 20 students in an English class.

Boys		Girls	
Campbell	Lallas	Buhrman	Hatch
Frechet	Lobkovsky	Burbank	Inglis
Grubb	Osgood	Colongeli	Liu
Kahn	Shiraishi	Crumley	Quach
Kong		Gifford	Widmann
		Harris	

The teacher will assign a group of four students to prepare a class presentation on American poets. Using these random numbers below, choose any four students for this project:

40151 15312 11732 40184 10075 10494 14657 70989 02899

10. **Customers.** A building supplies company filled 25 orders today, 17 for do-it-yourself retail customers and 8 for contractor customers.

Retail		Contractors
Baldwin	Molina	A-1 Homes
Berg	Morse	Best
Borbat	Sharkness	Davis & Davis
Dennis	Snyder	Home4U
Garner	Sternstein	Morton
Gupta	Stratakos	Nash Builders
Hart	Washington	Sunrise Homes
Karandeyev	Yuen	VIP Structures
McPherson		

In order to improve customer service, the company plans to survey three of the customers to see whether they were fully satisfied with their transactions. Use these random numbers to select any three of the 25 customers for the survey:

03634 27522 99160 78671 17410 35600

11. **Another project.** The teacher of the English class in Exercise 9 wants to randomly select students to present a scene from a Shakespeare play in class. There are four roles to fill, two for males and two for females. Describe a process that would use random numbers to select the students.

12. **Better survey.** The building supply store wants to improve the survey described in Exercise 10 to be sure they get a sample that includes both types of customers. They plan to call three retail customers and two contractors. Describe a process that would use random numbers to select the people to contact.

13. **Bad simulations.** Explain why each of the following simulations fails to model the real situation properly:
   a) Use a random integer from 0 through 9 to represent the number of heads when 9 coins are tossed.
   b) A basketball player takes a foul shot. Look at a random digit, using an odd digit to represent a good shot and an even digit to represent a miss.
   c) Use random numbers from 01 through 13 to represent the denominations of the cards in a five-card poker hand.

14. **More bad simulations.** Explain why each of the following simulations fails to model the real situation:
   a) Use random numbers 2 through 12 to represent the sum of the faces when two dice are rolled.

b) Use a random integer from 0 through 5 to represent the number of boys in a family of 5 children.

c) Simulate a baseball player's performance at bat by letting 0 = an out, 1 = a single, 2 = a double, 3 = a triple, and 4 = a home run.

15. Wrong conclusion. A Statistics student properly simulated the length of checkout lines in a grocery store and then reported, "The average length of the line will be 3.2 people." What's wrong with this conclusion?

16. Another wrong conclusion. After simulating the spread of a disease, a researcher wrote, "24% of the people contracted the disease." What should the correct conclusion be?

B

17. Cereal. In the chapter's example, 20% of the cereal boxes contained a picture of LeBron James, 30% David Beckham, and the rest Serena Williams. Suppose you buy four boxes of cereal. Create a simulation to estimate the probability that you end up with a complete set of the pictures. Your simulation should have at least 20 runs.

18. Cereal again. Suppose you really want the LeBron James picture. How many boxes of cereal do you need to buy to be pretty sure of getting at least one? Investigate using a simulation with at least 10 trials.

19. Multiple choice. You take a quiz with 6 multiple choice questions. After you studied, you estimated that you would have about an 80% chance of getting any individual question right.

a) Each question is a component. Explain how you will use random digits to represent the 80% chance that you get the right answer.

b) Explain how you will simulate your results for the 6 questions on the test.

c) What are your chances of getting all 6 right? Base your answer on at least 20 trials.

20. Lucky guessing? A friend of yours who took the multiple choice quiz in Exercise 19 got all 6 questions right, but now claims to have guessed blindly on every question. Do you believe her?

a) Each question had 4 possible answers. Explain how you will use random digits to determine whether someone guesses the right answer on any one question.

b) Explain how you will simulate the results of guessing at all 6 questions on the test.

c) Run at least 10 trials, and use the results to comment on your friend's claim that she got them all right just by guessing.

21. Beat the lottery. Many states run lotteries to raise money. A website advertises that it knows "how to increase YOUR chances of Winning the Lottery." They offer several systems and criticize others as foolish. One system is called *Lucky Numbers*. People who play the *Lucky Numbers* system just pick a "lucky" number to play, but maybe some numbers are luckier than others. Let's use a simulation to see how well this system works.

To make the situation manageable, simulate a simple lottery in which a single digit from 0 to 9 is selected as the winning number. Pick a single value to bet, such as 1, and keep playing it over and over. You'll want to run at least 100 trials. (If you can program the simulations on a computer, run several hundred. Or generalize the questions to a lottery that chooses two- or three-digit numbers—for which you'll need thousands of trials.)

a) What proportion of the time do you expect to win?

b) Would you expect better results if you picked a "luckier" number, such as 7? (Try it if you don't know.) Explain.

22. Random is as random does. The "beat the lottery" website discussed in Exercise 21 suggests that because lottery numbers are random, it is better to select your bet randomly. For the same simple lottery in Exercise 21 (random values from 0 to 9), generate each bet by choosing a separate random value between 0 and 9. Play many games. What proportion of the time do you win?

23. It evens out in the end. The "beat the lottery" website of Exercise 21 notes that in the long run we expect each value to turn up about the same number of times. That leads to their recommended strategy. First, watch the lottery for a while, recording the winners. Then bet the value that has turned up the least, because it will need to turn up more often to even things out. If there is more than one "rarest" value, just take the lowest one (since it doesn't matter). Simulating the simplified lottery described in Exercise 21, play many games with this system. What proportion of the time do you win?

24. Play the winner? Another strategy for beating the lottery is the reverse of the system described in Exercise 23. Simulate the simplified lottery described in Exercise 21. Each time, bet the number that just turned up. The website suggests that this method should do worse. Does it? Play many games and see.

25. The family. Many married couples want to have both a boy and a girl. If they decide to continue to have children until they have one child of each sex, what would we expect the average family size to be? Assume that boys and girls are equally likely.

26. A bigger family. Suppose a married couple will continue having children until they have at least two children of each sex (two boys *and* two girls). How many children might they expect to have?

27. Basketball strategy. Late in a basketball game, the team that is behind often fouls someone in an attempt to get the ball back. Usually the opposing player will get to shoot foul shots "one and one," meaning he gets a shot, and then a second shot only if he makes the first one. Suppose the opposing player has made 72% of his foul shots this season.

a) Each foul shot is a component. Explain how you will use random digits to represent whether the player hits or misses the shot.

b) Explain how you will model the one-and-one situation.

c) After running at least 10 trials, estimate the average number of points the shooter will score.

28. Blood donors. A person with type O-positive blood can receive blood only from other type O donors. About 44% of the U.S. population has type O blood. At a blood drive, how many potential donors do you expect to examine in order to get three units of type O blood?
a) Each blood donor is a component. Explain how you will use random digits to signify whether the donor has type O blood.
b) Explain how you will model a blood drive looking for three type O donors.
c) After running at least 10 trials, estimate the number of donors required in order to get three units of type O blood.

29. Dice game. You are playing a children's game in which the number of spaces you get to move is determined by the rolling of a die. You must land exactly on the final space in order to win. If you are 10 spaces away, how many turns might it take you to win?

30. Parcheesi. You are three spaces from a win in Parcheesi. On each turn, you will roll two dice. To win, you must roll a total of 3 or roll a 3 on one of the dice. How many turns might you expect this to take?

31. Free groceries. To attract shoppers, a supermarket runs a weekly contest that involves "scratch-off" cards. With each purchase, customers get a card with a black spot obscuring a message. When the spot is scratched away, most of the cards simply say, "Sorry—please try again." But during the week, 100 customers will get cards that make them eligible for a drawing for free groceries. Ten of the cards say they may be worth $200, 10 others say $100, 20 may be worth $50, and the rest could be worth $20. To register those cards, customers write their names on them and put them in a barrel at the front of the store. At the end of the week the store manager draws cards at random, awarding the lucky customers free groceries in the amount specified on their card. The drawings continue until the store has given away at least $500 of free groceries. Estimate the average number of winners each week.

32. Find the ace. A new electronics store holds a contest to attract shoppers. Once an hour someone in the store is chosen at random to play the Music Game. Here's how it works: An ace and four other cards are shuffled and placed face down on a table. The customer gets to turn cards over one at a time, looking for the ace. The person wins $100 worth of free CDs or DVDs if the ace is the first card, $50 if it is the second card, and $20, $10, or $5 if it is the third, fourth, or fifth card chosen. What is the average dollar amount of music we expect the store will give away?

33. The hot hand. A basketball player with a 65% shooting percentage has just made 6 shots in a row. The announcer says this player "is hot tonight! She's in the zone!" Assume the player takes about 20 shots per game. Is it unusual for her to make 6 or more shots in a row during a game?

34. Teammates. Four couples at a dinner party played a board game after the meal. They decided to play as teams of two and to select the teams randomly. All eight people wrote their names on slips of paper. The slips were thoroughly mixed, then drawn two at a time. It turned out that every person was teamed with someone other than the person he or she came to the party with. Is that outcome unusual?

35. Job discrimination? A company with a large sales staff announces openings for three positions as regional managers. Twenty-two of the current salespersons apply, 12 men and 10 women. After the interviews, when the company announces the newly appointed managers, all three positions go to women. The men complain of job discrimination. Do they have a case? Simulate a random selection of three people from the applicant pool, and make a decision about the likelihood that a fair process would result in hiring all women.

36. Cell phones. A proud legislator claims that your state's new law against talking on a cell phone while driving has reduced cell phone use to less than 12% of all drivers. While waiting for your bus the next morning, you notice that 4 of the 10 people who drive by are using their cell phones. Does this cast doubt on the legislator's figure of 12%? Use a simulation to estimate the likelihood of seeing at least 4 of 10 randomly selected drivers talking on their cell phones if the actual rate of usage is 12%. Explain your conclusion clearly.

Answers
Just Checking

1. Number the listed students from 000 to 453. Divide a list of random numbers into groups of three digits. Choose the students whose numbers appear, ignoring values 454–999 and skipping any repeats, until you have a sample of 40 seniors.

2. a) The component is one game.
   b) I'll generate random numbers and assign numbers from 00 to 54 to represent the home team winning and from 55 to 99 to represent the visitors winning.
   c) I'll generate components until one team wins 4 games. I'll record which team wins the series.
   d) The response is who wins the series.
   e) I'll calculate the proportion of wins by the team that starts at home.

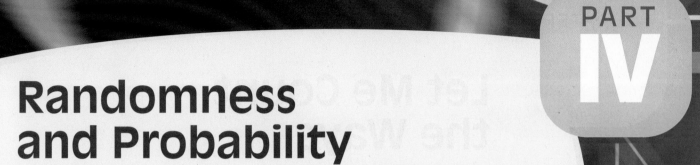

Randomness and Probability

Let Me Count the Ways

Early humans saw a world filled with random events. To help them make sense of the chaos around them, they sought out seers, consulted oracles, and read tea leaves. As science developed, we learned to recognize some events as predictable. We can now forecast the change of seasons, tell when eclipses will occur precisely, and even make a reasonably good guess at how warm it will be tomorrow. But many other events are still essentially random. Will the stock market go up or down today? When will the next car pass this corner? And we now know from quantum mechanics that the universe is in some sense random at the most fundamental levels of subatomic particles.

But we have also learned to understand randomness. The surprising fact is that in the long run, even truly random phenomena settle down in a way that's consistent and predictable. It's this property of random phenomena that makes the next steps we're about to take in Statistics possible.

Dealing with Random Phenomena

Every day you drive through the intersection at College and Main. Even though it may seem that the light is never green when you get there, you know this can't really be true. In fact, if you try really hard, you can recall just sailing through the green light once in a while.

What's random here? The light itself is governed by a timer. Its pattern isn't haphazard. In fact, the light may even be red at precisely the same times each day. It's the pattern of *your driving* that is random. No, we're certainly not insinuating that you can't keep the car on the road. At the precision level of the 30 seconds or so that the light spends being red or green, what time you arrive at the light *is random*. Even if you try to leave your house at exactly the same time every day, whether the light is red or green as *you* reach the intersection is a **random phenomenon.**[1]

Is the color of the light completely unpredictable? When you stop to think about it, it's clear that you do expect some kind of *regularity* in your long-run experience. Some *fraction* of the time, the light will be green as you get to the intersection. How can you figure out what that fraction is?

Suppose you record what happens at the intersection each day. The first day, it was green. Then on the next day it was red, the day after that green again, then green, red, and red. If you plot the percentage of green lights against days, the graph would start at 100% (because the first time, the light was green, so 1 out of 1, for 100%). Then the next day it was red, so the accumulated percentage dropped to 50% (1 out of 2). The third day it was green again (2 out of 3, or 67% green), then green (3 out of 4, or 75%), then red twice in a row (3 out of 5, for 60% green, and then 3 out of 6, for 50%), and so on. A graph of the *accumulated percentage* of green lights looks like this:

Day	Light	% Green so far
1	Green	100
2	Red	50
3	Green	66.7
4	Green	75
5	Red	60
6	Red	50
⋮	⋮	⋮

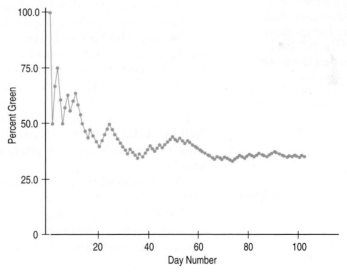

FIGURE 13.1 Green lights over time. The overall percentage of times the light is green settles down as you see more outcomes.

As you collect a new data value for each day, each new outcome becomes a smaller and smaller fraction of the accumulated experience, so, in the long run, the graph settles down. As it settles down, you can see that, in fact, the light is green about 35% of the time.

---

[1]If you somehow managed to leave your house at *precisely* the same time every day and there was *no* variation in the time it took you to get to the light, then there wouldn't be any randomness, but that's not very realistic.

The Law of Large Numbers

What's the *probability* of a green light at College and Main? Based on the graph, it looks like the relative frequency of green lights settles down to about 35%, so saying that the probability is about 0.35 seems like a reasonable answer.

This illustrates a principle called the **Law of Large Numbers** (LLN) for events that are **independent.** (Informally, independent means that the outcome of one trial doesn't affect the outcomes of the others.) The LLN says that as the number of independent trials increases, the long-run *relative frequency* of repeated events gets closer and closer to some value.

The LLN guarantees that relative frequencies settle down in the long run, and we call the value that they approach the **probability** of the event. If the relative frequency of green lights at that intersection settles down to 35% in the long run, we say that the probability of encountering a green light is 0.35, and we write $P(\text{green}) = 0.35$.

> **PROBABILITY**
>
> For any event **A**,
>
> $$P(\mathbf{A}) = \frac{\#\text{ times } \mathbf{A} \text{ occurs}}{\text{total }\#\text{ of trials}}$$
>
> in the long run.

The Nonexistent Law of Averages

Even though the LLN seems natural, it is often misunderstood because the idea of the *long run* is hard to grasp. Many people believe, for example, that an outcome of a random event that hasn't occurred in many trials is "due" to occur. Many gamblers bet on numbers that haven't been seen for a while, mistakenly believing that they're likely to come up sooner. A common term for this is the "Law of Averages." After all, we know that in the long run, the relative frequency will settle down to the probability of that outcome, so now we have some "catching up" to do, right?

Wrong. The Law of Large Numbers says nothing about short-run behavior. Relative frequencies even out *only in the long run*. And, according to the LLN, the long run is *really* long (*infinitely* long, in fact).

The so-called Law of Averages doesn't exist at all. But you'll hear people talk about it as if it does. Is a good hitter in baseball who has struck out the last six times *due* for a hit his next time up? If you've been doing particularly well in weekly quizzes in Statistics class, are you *due* for a bad grade? No. This isn't the way random phenomena work. There is *no* Law of Averages for short runs.

The lesson of the LLN is that sequences of random events don't compensate in the *short* run and don't need to do so to get back to the right long-run probability. If the probability of an outcome doesn't change and the events are independent, the probability of any outcome in another trial is *always* what it was, no matter what has happened in other trials.

Don't let yourself think that there's a Law of Averages that promises short-term compensation for recent deviations from expected behavior. A belief in such a "Law" can lead to money lost in gambling and to poor business decisions.

"Slump? I ain't in no slump. I just ain't hittin'."

—Yogi Berra

> **Coins, keno, and the law of averages** You've just flipped a fair coin and seen six heads in a row. Does the coin "owe" you some tails? Suppose you spend that coin and your friend gets it in change. When she starts flipping the coin, should she expect a run of tails? Of course not. Each flip is a new event. The coin can't "remember" what it did in the past, so it can't "owe" any particular outcomes in the future.

TI-*nspire*

The Law of Large Numbers. Watch the relative frequency of a random event approach the true probability *in the long run*.

Just to see how this works in practice, we ran a computer simulation of 100,000 flips of a fair coin. We collected 100,000 random numbers, letting the numbers 0 to 4 represent heads and the numbers 5 to 9 represent tails. In our 100,000 "flips," there were 2981 streaks of at least 5 heads. The "Law of Averages" suggests that the next flip after a run of 5 heads should be tails more often to even things out. Actually, the next flip was heads more often than tails: 1550 times to 1431 times. That's 51.9% heads, hardly what the "Law of Averages" suggests (and very close to the theoretical 50%).

Keno is a simple casino game in which numbers from 1 to 80 are chosen. The numbers, as in most lottery games, are supposed to be equally likely. Payoffs depend on how many of those numbers you match on your card. A group of graduate students from a Statistics department decided to take a field trip to Reno. They (*very* discreetly) wrote down the outcomes of the games for a couple of days, then drove back to test whether the numbers were, in fact, equally likely. It turned out that some numbers were *more likely* to come up than others. Rather than bet on the Law of Averages and put their money on the numbers that were "due," the students put their faith in the LLN—and all their (and their friends') money on the numbers that had come up more often before. After they pocketed more than $50,000, they were escorted off the premises and invited never to show their faces in that casino again.

Just Checking

1. One common proposal for beating the lottery is to note which numbers have come up lately, eliminate those from consideration, and bet on numbers that have not come up for a long time. Proponents of this method argue that in the long run, every number should be selected equally often, so those that haven't come up are due. Explain why this is faulty reasoning.

(Check your answers on page 304.)

Modeling Probability

Maybe you don't feel like waiting around for an infinite number of trials to find out a probability? Thankfully, there are other ways. Probability was first studied extensively by a group of French mathematicians who were interested in games of chance.[2] Rather than *experiment* with the games (and risk losing their money), they developed mathematical models of **theoretical probability.** To make things simple (as we usually do when we build models), they started by looking at games in which the different outcomes were equally likely. Fortunately, many games of chance are like that. Any of 52 cards is equally likely to be the next one dealt from a well-shuffled deck. Each face of a die is equally likely to land up (or at least it *should be*).

---

[2]Ok, gambling.

A phenomenon consists of **trials.** Each trial has an **outcome.** A **sample space** is a list of all possible outcomes. Outcomes combine to make **events.**

Let's define some terms. In general, each time we observe a random phenomenon is called a **trial.** We call that result the trial's **outcome.** (If this language reminds you of simulations, that's *not* unintentional.) Rolling a die is a trial. Seeing it land a 5 is an outcome.

We sometimes talk about the collection of *all possible outcomes* and call that event the **sample space.**[3] We'll denote the sample space **S**. For the die, **S** = {1, 2, 3, 4, 5, 6}.

Often we're more interested in a combination of outcomes rather than in the individual ones. Suppose you roll a die, needing a 5 or 6 to win a game. When we combine outcomes like that, the resulting combination is an **event.**[4]

It's easy to find probabilities for events that are made up of several *equally likely* outcomes. We just count all the outcomes that the event contains. The probability of the event is the number of outcomes in the event divided by the total number of possible outcomes. We can write

$$P(\mathbf{A}) = \frac{\# \text{ outcomes in } \mathbf{A}}{\# \text{ of possible equally likely outcomes}}.$$

We often use capital letters—and usually from the beginning of the alphabet—to denote events. We *always* use P to denote probability. So,

$$P(\mathbf{A}) = 0.35$$

means "the probability of the event **A** is 0.35."

For example, the probability of drawing a face card (JQK) from a deck is

$$P(\text{face card}) = \frac{\# \text{ face cards}}{\# \text{ cards}} = \frac{12}{52} = \frac{3}{13}.$$

This may seem like a pretty straightforward way to find a probability, but there's a critical issue lurking here: *the outcomes you count must be equally likely.* Consider: either there's an earthquake today where you live or there isn't. One out of the two possible outcomes is "Earthquake". That doesn't mean the probability you experience an earthquake today is 50%! Before you count outcomes to find a probability, be sure to check the

Equally Likely Condition: The outcomes being counted are all equally likely to occur.

For Example Sample Spaces and Probability

In lots of games you roll two dice, then do something based on the total. It's easy to determine the probabilities of various outcomes.

Question: What's the probability of rolling a total of 7?

First I'll make a sample space showing all the possible outcomes when rolling two dice. I could list the totals— {2,3,4,5,6,7,8,9,10,11,12}—but they're not all equally likely. After all, there's only one way to get a total of 2, but lots of ways to get a total of 7.

---

[3]Mathematicians like to use the term "space" as a fancy name for a set. Sort of like referring to that closet colleges call a dorm room as "living space." But remember that it's really just the set of all outcomes.

[4]Each individual outcome is also an event.

It'll be more helpful to list all the pairs of numbers the two dice could show—as in {(1,1), (1,2), (1,3), and so on, through (6,6)}—but that's a pain. This table is a much easier way to see the sample space.

✔ **Equally Likely Condition:** Assuming the dice are fair, all 36 possible outcomes are equally likely.

There are 6 ways to roll a total of 7, so $P(7) = \dfrac{6}{36} = \dfrac{1}{6}$.

Dice Total	Second Die					
	1	**2**	**3**	**4**	**5**	**6**
1	2	3	4	5	6	7
2	3	4	5	6	7	8
3	4	5	6	7	8	9
4	5	6	7	8	9	10
5	6	7	8	9	10	11
6	7	8	9	10	11	12

First Die (row label)

Just Checking

2. Suppose a family has two children. In order of birth, they may be a boy and a boy, a boy and then a girl, a girl and then a boy, or a girl and a girl. Let's represent those outcomes as the sample space **S** = {BB, BG, GB, GG}.
   a) What are we assuming in thinking that these four outcomes are "equally likely"?
   b) What's the probability a 2-child family has two girls?
   c) What's the probability there's at least one girl?
   d) What's the probability both children are the same sex?

3. Now let's think about families with three children.
   a) Create a sample space of equally likely outcomes.
   b) What's the probability a 3-child family has at least one girl?
   c) What's the probability that there are both boys and girls in the family?

(Check your answers on page 304.)

How to Count (Cleverly!)

Finding the probability of any event when the outcomes are equally likely is straightforward, but not necessarily easy. Sure, if you need to roll a 5 or 6 on a die to win a game, $P(\text{win}) = \frac{2}{6}$ because the die has 6 equally likely faces and 2 of them are winners. Simple enough. It gets hard though when the number of outcomes in the event (and in the sample space) gets big. Think about flipping two coins. The sample space is **S** = {HH, HT, TH, TT} and each outcome is equally likely. So, what's the probability of getting *exactly* one head and one tail? Let's call that event **A**. Well, there are two outcomes in the event **A** = {HT, TH} out of the 4 possible equally likely ones in **S**, so $P(\mathbf{A}) = \frac{2}{4}$, or $\frac{1}{2}$.

OK, now flip 100 coins. What's the probability of exactly 67 heads? Well, first, how many outcomes are in the sample space? **S** = {HHHHHHHHHHH . . . H, HH . . .T, . . .} Hmm. A lot. In fact, there are 1,267,650, 600,228,229, 401,496, 703,205,376 different outcomes possible when flipping 100 coins.

To answer the question, we'd still have to figure out how many ways there are to get 67 heads!

Yikes! If you're thinking (hoping?) there has to be an easier way, you're right. Let's not go to the trouble of listing the outcomes in the sample space and then counting in the usual 1, 2, 3, . . . way. There are other ways of counting that are quite clever, and far more efficient.

Suppose you stop at a local café and find that the Lunch Special menu lists 4 salads and 5 sandwiches. For $5 you can have any salad or sandwich for lunch. We don't need to list all the options one by one to know that you have 4 + 5 = 9 choices. The key word here was "or"; that's what told us we could add the number of salads and the number of sandwiches to see how many outcomes were possible. The basic rule is "*or* means *add*," but math folks like to make these things look way more complicated:

Fundamental Counting Principle (Part 1: OR)

If event **A** has m outcomes and event **B** has n different outcomes, then the number of outcomes in event **A *or* B** is $m + n$.

For bigger eaters this café offers the Hungry Special: any salad *and* any sandwich for $8. How many different lunches are possible now? Imagine starting to make the list: spinach salad & roast beef sandwich, spinach salad & BLT, and so on. There are 5 different lunches that start with the spinach salad, and then another 5 with the fruit salad paired with each sandwich, and so on. Because we'll need to include every salad/sandwich combo, we'll end up listing each of the 4 salads with each of the 5 sandwiches; 4 groups of 5 means we'd have $4 \cdot 5 = 20$ different possibilities.

The key word this time was "and"; that's what led us to multiply the number of salads times the number of sandwiches to see how many outcomes were possible. Here the basic rule is "*and* means *multiply*." We know you just can't wait to see the formal statement, so here you are:

Fundamental Counting Principle (Part 2: AND)

If event **A** has m outcomes and independent event **B** has n outcomes, then the number of outcomes in event **A *and* B** is mn.

For Example   Counting License Plates

Colorado license plates like the one shown consist of three digits and then three letters.

Question: Using this pattern of letters and numbers, how many different license plates are possible?

The first space could be any of 10 different digits (0, 1, 2, 3, . . . , 9) and the second space any of 10, and the third, too. The fourth space and the fifth space and the sixth could each be any of 26 letters. Since "and" means multiply, the total number of possibilities is

$$10 \cdot 10 \cdot 10 \cdot 26 \cdot 26 \cdot 26 = 17,576,000 \text{ different license plates.}$$

Just Checking

4. Chances are you use a password of some kind every day, whether you're getting cash from an ATM, logging in to get your email, or checking up on your friends on Facebook. One factor that helps make these codes secure is the large number of possibilities. In each of the situations below, how many different passwords are there?

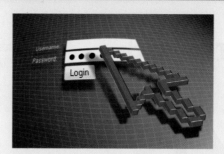

   a) The access code for an ATM is 4 digits (example: 3551).
   b) The login code for an email site is 5 characters, alternating letter-digit-letter-digit-letter (example: a7r3w).
   c) The email login code can alternate letters and digits with either a letter or a digit first (examples: a7r3w or 8m8k3).
   d) That email login code can be any mixture of letters and digits in any order, and is case-sensitive (example: mp27g isn't the same as mP27g).

(Check your answers on page 304.)

On Winning the Lottery

States run lotteries to make money. They get people to pay (or "play," as the ads like to say) by offering bazillion dollar payouts. "All" you have to do[5] is pick a few numbers. Hit the right ones, and you're a millionaire. It's easy, right—just "a dollar and a dream"? But what are your chances, really? To find out, we'll need to do some even cleverer counting.

Step 1: Permutations: Permu—what? Don't panic, "permutations" is just a big word for "orders" or "arrangements." Let's say there are 8 runners in a race to determine who advances to the State Championships. We wonder, in how many different orders can they cross the finish line? Our Fundamental Counting Principle can answer that question. Any one of the 8 runners could be first, *and* then any one of the other 7 second, *and* any one of the remaining 6 third, *and* so on. Remember, *and* means multiply, so there are $8 \cdot 7 \cdot 6 \cdot 5 \cdot 4 \cdot 3 \cdot 2 \cdot 1 = 5040$ different orders in which the runners can finish.

A product like this, starting with some number n and multiplying by each smaller number down to 1, is called **n factorial.** Factorials are used so often for counting that there's a special notation, too: $n!$. Yes, that's an exclamation point—as in $8! = 5040$. For some practice with factorials, visit the Math Box on the next page. (Then come back!)

At a track meet we're usually not interested in the entire order of finish for a race, just who comes in first, second, and third. To count those possibilities, we start out just like a factorial, but stop after just 3 factors: $8 \cdot 7 \cdot 6 = 336$. That's called "permutations of 8 things taken 3 at a time," and yes, there's special math notation for it: $_8P_3$. Our factorial notation works so well that we can't resist using it here, too. If we start with $8!$ we need to chop off the $5 \cdot 4 \cdot 3 \cdot 2 \cdot 1$ part. Dividing $8!$ by $5!$ cancels out the factors we don't want, leaving just what we were after:

$$_8P_3 = \frac{8!}{5!} = \frac{8 \cdot 7 \cdot 6 \cdot \cancel{5} \cdot \cancel{4} \cdot \cancel{3} \cdot \cancel{2} \cdot \cancel{1}}{\cancel{5} \cdot \cancel{4} \cdot \cancel{3} \cdot \cancel{2} \cdot \cancel{1}} = 8 \cdot 7 \cdot 6 = 336.$$

NOTATION ALERT

An exclamation point always means "factorial." For example, $3! = 3 \cdot 2 \cdot 1 = 6$

---

[5]They make it sound so easy.

Do The Math

You'll often rely on counting methods to find probabilities, and that means you'll be using factorials a lot. For practice, try evaluating these:

1. $4!$
2. $6!$
3. $\dfrac{5!}{3!}$
4. $\dfrac{10!}{9!}$ (Hint: Don't multiply; lots of factors cancel out.)
5. $\dfrac{20!}{18!2!}$
6. $\dfrac{10!}{4!6!}$
7. $\dfrac{100!}{100}$ (Express the answer as a factorial.)
8. Look again at problem 7. What happens when you divide the factorial of a number by the number itself? As a factorial, what's $\dfrac{51!}{51}$? $\dfrac{6!}{6}$?
9. Based on your answer to problem 8, what's $\dfrac{1!}{1}$ as a factorial, and as a number?

(Check your answers on page 304.)

NOTATION ALERT

$_nP_r$ is pronounced "permutations of n objects r at a time" and represents the number of different orders in which we could *arrange* any r items out of a group of n different items.

We call the number of ways to *arrange* any r items from a group of n different items **permutations of *n* objects *r* at a time,** and calculate $_nP_r = \dfrac{n!}{(n-r)!}$.

Evaluating permutations is usually easier than it looks, because with factorials in both numerator and denominator, lots of numbers cancel out. There's just one thing you may find a little strange at first. In our first look at the race we cared about the order of finish for all 8 runners. Think of that as permutations of 8 people all 8 at a time, and apply our new formula:

$$_8P_8 = \frac{8!}{(8-8)!} = \frac{8!}{0!}.$$

Hmmm . . . what's $0!$ mean? Look at it this way: there's only one way to arrange no objects: don't! Therefore $0! = 1$, and $_8P_8 = \dfrac{8!}{0!} = \dfrac{8!}{1} = 8!$, as we found earlier.

Just Checking

5. You just downloaded 5 new albums to your iPod, and are eager to listen to them. How many different orders can you play them in?

6. You take your iPod to the beach, and select a playlist of your 8 favorite albums to be played in random order. You'll be there long enough to hear 5 of them. How many different orders of albums could you get to listen to?

(Check your answers on page 304.)

Step 2: Combinations. Sometimes we don't care what *order* things are in; all that matters is which ones are selected. Take that qualifying race with 8 runners, for example. If the fastest 3 competitors will move on to the State Championships, it really doesn't matter whether someone finishes first, second, or third—a person just wants to be in the top 3.

We figured out that there could be $_8P_3 = 336$ possible orders of the first three finishers. Let's say the winners were Anne, Bonita, and Celeste.[6] Egos aside, the orders ABC, ACB, BAC, BCA, CAB, and CBA are all the same: these 3 girls advance to the next meet. In fact, every possible group of three qualifiers is counted $3! = 3 \cdot 2 \cdot 1 = 6$ times. By dividing the total number of permutations by this number of repetitions we learn that there are $\frac{_8P_3}{3!} = \frac{336}{6} = 56$ different possible selections.

In general, we call the number of ways to *select* any r items from a group of n different items **combinations of n objects chosen r at a time,** and calculate $_nC_r = \dfrac{n!}{r!(n-r)!}$.

Again, lots of numbers in these calculations cancel out because of the factorials. For our track example, it works like this:

$$_8C_3 = \frac{8!}{3!5!} = \frac{8 \cdot 7 \cdot 6 \cdot \cancel{5} \cdot \cancel{4} \cdot \cancel{3} \cdot \cancel{2} \cdot \cancel{1}}{3 \cdot 2 \cdot 1 \cdot \cancel{5} \cdot \cancel{4} \cdot \cancel{3} \cdot \cancel{2} \cdot \cancel{1}} = \frac{8 \cdot 7 \cdot 6}{3 \cdot 2 \cdot 1} = 56$$

And now we're finally ready to answer the question we started with: What are your chances of winning the lottery? (Here's a hint: don't go spending that jackpot yet!)

Wait! Should I use permutations or combinations? To count correctly, you need to be able to make that decision. Here are the key questions to ask yourself:

- *Do I care what **order** things happen in?*
  If the arrangement matters, use permutations.
- *Or do I just care which ones are **chosen**?*
  If it's only the selection that matters, use combinations.

For Example   Lotteries (and lightning)

The betting slips for New Jersey's Pick-6 Lotto game offer a field of numbers from 1 to 49. Each bettor chooses any 6 of them. You win the grand prize if your six numbers match those randomly chosen on TV by the NJ Lottery.

Question: How many different selections are possible?

To win, it only matters that I *select* the right numbers, not what *order* they're in, so I'll use combinations. The number of ways to choose any 6 numbers from the 49 offered is

$$_{49}C_6 = \frac{49!}{6!43!} = \frac{49 \cdot 48 \cdot 47 \cdot 46 \cdot 45 \cdot 44}{6 \cdot 5 \cdot 4 \cdot 3 \cdot 2 \cdot 1} = 13,983,816$$

That means I have about 1 chance in 14 million of winning the Pick-6 Lotto!

(continued)

---

[6]Isn't it strange how people in math books always have names starting with A, B, and C?

(For comparison, the National Weather Service reports that in any year about 400 of the 300 million Americans—about 1 in every 700,000—are struck by lightning. That makes you about 20 times as likely to be zapped as to win the lottery! The ads say you can't win if you don't buy a ticket. That's true, of course, but the ads don't say that buying a ticket improves your chance of winning by only a ridiculously tiny amount.)

Just Checking

7. Don't do any arithmetic—just tell whether you'd count these outcomes using permutations or combinations.

   a) How many 13-card bridge hands could you be dealt from a deck of cards?

   b) How many 3-turn combinations could a padlock have if there are 60 numbers on the dial and combinations can't have repeated numbers?

   c) How many batting orders could a baseball manager create using the 9 players in his starting lineup?

   d) How many 6-player volleyball squads could a coach put on the floor if there are 9 women on the team?

8. A county legislature consists of 13 elected representatives, 8 Democrats and 5 Republicans.

They're setting up a 4-person committee to study the proposal to build a new library. How many different committees could be formed if the group will consist of:

   a) 4 Republicans?

   b) 4 Democrats?

   c) 2 Democrats and 2 Republicans? (Remember, "and" means multiply.)

   d) 3 Democrats and 1 Republican?

   e) 2 or 3 Democrats (and the rest Republicans).

(Check your answers on page 304.)

Combinations and Probability

In our quest to create models for probability we've developed two important insights:

- When a sample space consists of equally likely outcomes, the probability of an event is the number of ways the event can happen divided by the total number of possible outcomes.
- We don't have to actually list all the outcomes to find out how many there are. It may be easier to just count them using permutations, combinations, or the Fundamental Counting Principle.

Together these provide us with some clever ways to determine probabilities.

In the last chapter we asked whether there was something fishy about a college dorm lottery. A total of 57 students, including 20 people on the same varsity team, applied for just 3 very desirable rooms. When all 3 winners turned out to be team members, the other students cried foul. Is it reasonable to believe this could have happened by chance? Or would it be really unusual if the process were fair?[7] Earlier we used a simulation to help us think about this issue. Now let's find the probability that all the rooms would go to the teammates if everything were done fairly.

---

[7]Read that "Is it statistically significant?"

The dorm lottery sample space would consist of all the ways to choose any 3 students from the 57 applicants. There are $_{57}C_3 = \dfrac{57!}{3!54!} = 29{,}260$ possible outcomes,[8] all equally likely if the selection were truly random.

Of those outcomes, how many consist of team members only? That event includes all the ways to pick any 3 athletes from the 20 varsity applicants: $_{20}C_3 = \dfrac{20!}{3!17!} = 1{,}140.$ And therefore:

$$P(3 \text{ team winners}) = \dfrac{\text{\# of ways to choose 3 team members}}{\text{total \# of possible choices}} = \dfrac{_{20}C_3}{_{57}C_3}$$

$$= \dfrac{1140}{29260} \approx 0.039$$

So, was the lottery fair? Well, there's only about a 4% chance that all 3 rooms would have gone to team members by chance. This makes the outcome look very suspicious, but the probability is high enough that we can't prove for sure that someone cheated.

STEP-BY-STEP EXAMPLE   Using Combinations to Find a Probability

A video rental store is running a "DVD Grab Bag" special. For $20 you can buy a bag containing 6 used DVDs randomly selected from 15 recently popular movies. The possible titles listed in the store's ads included 4 comedies, 8 dramas, and 3 animated features.

Question: If you buy one of the random Grab Bags, what's the probability you get two comedies, three dramas, and one animated flick?

THINK

Plan Organize the information given and what is asked for.

I am being asked to find the probability that my Grab Bag contains

2 of the 4 comedies
3 of the 8 dramas, and
1 of the 3 animations, for a total of
6 of the 15 DVDs.

Decide which counting method you'll use. Remember:
    permutations = arrangements,
    combinations = choices.

I don't care what order these DVDs come in, just which ones were selected for my Grab Bag. I'll use combinations to count the number of possibilities.

Check the conditions.

✔ **Equally Likely Condition.** *The movies are being chosen at random, so all the outcomes are equally likely.*

(continued)

---

[8]Aren't you glad we don't have to list them all?

Mechanics Show your work. The probability of an event **A** is:

$$P(A) = \frac{\text{\# of outcomes in A}}{\text{\# of possible outcomes}}.$$

Remember: "and" means multiply.

If the messy arithmetic annoys you, relax. You'll love the TI Tips, coming next.

I am asked to find a probability so I'll write my answer as a fraction. The denominator is the total number of ways to choose any 6 of the 15 movies. That's $_{15}C_{16}$.

The numerator is the number of selections that include any 2 of 4 comedies, and 4 of 8 dramas, and 1 of the 3 animated movies:

$$_4C_2 \cdot {_8}C_3 \cdot {_3}C_1$$

The probability is:

$$P(A) = \frac{_4C_2 \cdot {_8}C_3 \cdot {_3}C_1}{_{15}C_6}$$

$$= \frac{6 \cdot 56 \cdot 3}{5005}$$

$$\approx 0.201$$

Conclusion Interpret your results in the proper context.

There's about a 20% chance that my Grab Bag will have two comedies, three dramas, and one animated movie.

TI Tips   Counting

Tired of writing down 4, 3, 2, 1 over and over, then canceling and multiplying? You'll be happy to know your TI can do the work for you. Let's try it out.

Factorials.
The qualifying race at the track meet had 8 runners who could finish in 8! orders.

```
8!
           40320
15!
   1.307674368E12
■
```

- Type the 8.
- Hit the MATH button and go all the way to the right to the PRB menu. (You can "wrap" there using the left arrow.)
- Choose 4:!, then hit ENTER.

Note that factorials can get quite large. With only 15 runners the number of arrangements is so large the answer requires scientific notation! Try it.

Permutations.
The number of different orders for the top 3 runners could finish in is $_8P_3$. To evaluate that:

```
8 nPr 3
            336
■
```

- Type the 8.
- Go to the MATH PRB menu again. This time choose 2:nPr.
- Type the 3, then hit ENTER.

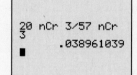

Combinations.

The number of possible groups of 3 qualifying runners (order of finish doesn't matter) is $_8C_3$. Evaluate that the same way you did $_8P_3$, only this time choose 3:nCr from the MATH PRB menu.

Probability.

You can even put a bunch of these calculations in the same expression. Try calculating the probability that those 3 dorm rooms all went to varsity teammates by entering $\frac{_{20}C_3}{_{57}C_3}$.

Pretty nice, eh? Take these buttons out for a test drive with the following *Just Checking* questions.

Just Checking

9. The DVD Grab Bag described in the Step-By-Step Example is to contain 6 movies chosen at random from 15 recently popular titles, including

4 comedies, 8 dramas, and 3 animated features. Find the probability you get:

a) nothing but dramas.
b) 3 of the comedies and 3 of the dramas.
c) all 3 of the animated flicks and any 3 other movies.

(Check your answers on page 304.)

⊘ WHAT CAN GO WRONG?

- **Don't mistake the Law of Large Numbers for the so-called "Law of Averages."** The LLN tells us that *in the long run* the frequency with which things happen will approach the true probability. But the "long run" means forever, not today! People who reason by the "Law of Averages" think that things have to even out right away. Seeing 6 heads in a row doesn't mean a coin is more likely to land tails on the next toss. If you were to flip the coin 94 more times and get around 47 heads and 47 tails, you'd end up with a total of 53 heads and 47 tails in 100 tosses. There's nothing strange about that. The coin doesn't compensate for the 6 extra heads. Rather, whatever happens in the short run gets drowned out in the long run. *There's no such thing as the "Law of Averages."*

- **Don't think that random events are always equally likely.** The chance of winning a lottery—especially lotteries with very large payoffs—is small. Regardless, people continue to buy tickets. In an attempt to understand why, an interviewer asked someone who had just purchased a lottery ticket, "What do you think your chances are of winning the lottery?" The reply was, "Oh, about 50–50." The shocked interviewer asked, "How do you get that?" to which the response was, "Well, the way I figure it, either I win or I don't!" The moral of this story is that you must be sure the events in your sample space are equally likely before thinking about probability.

(continued)

■ **Don't mix up permutations and combinations.** These two counting methods seem pretty similar, but the difference is critical. If you need to know how many *orders* something can happen in, or how many different ways you can *arrange* things, those are permutations. When you don't care what order things are in, combinations count the number of possible *choices* or *selections*.

■ **Don't think a "personal probability" is mathematically valid.** What's the probability that your grade in this Statistics course will be an **A**? You may be able to offer a number that seems reasonable. How did you come up with this probability? Do you plan on taking the course over and over (and over . . .), calculating the proportion of times you get an **A**? Or do you assume the outcomes are equally likely?

People use the language of probability in everyday speech to express a degree of uncertainty about an outcome. We call this kind of probability a subjective or **personal probability.** Although personal probabilities may be based on experience, they're not based either on long-run relative frequencies or on equally likely events. So they don't display the kind of consistency that we'll need probabilities to have. For that reason, we'll stick to formally defined probabilities. You should be alert to the difference.

COUNTING **IN YOUR WORLD**

Massachusetts Needs New License Plates

There is only one state in the union with older license plates on the street than Massachusetts . . . If you don't retire the old [green] plates, and restart a new series of plates, you exhaust many millions of available letter-number combinations.

Consider the math. The first series of license plates had three letters and three numbers, and the first digit was never a 0. The letters I, O, and Q were never used. For math geeks, these rules create 12,167 different letter combinations. Each three digit number can generate 900 plates (numbers 100-999), so the 12,167 letter combinations generate 10,950,300 possible plates. When Massachusetts exhausted the three-letter three-number combinations, they simply reversed the order, generating another 10,950,300 possibilities.

In 1987 the Registry started manufacturing the red "Spirit of Massachusetts" plates . . . Soon after the red plates began to appear, the NNNLLL[9] combinations were exhausted, and the state went to a series of two letter-four number plates. Given the same rules, four number-two letter plates generate 4,761,000 combinations. Reversing the order generates another 4,761,000 plates.

---

[9]number-number-number-letter-letter-letter

Those combinations were exhausted, and the state then placed the two letters in the middle of the number combination. That generates another 4,761,000 plates.

This numbering scheme is now exhausted, and we appear to be going to a scheme of three digits, two letters, followed by another number. So, here we are in a state of 6.4 million people, and we have exhausted more than 37 million possible plate combinations.

JUN ○ *Massachusetts* ○ 06

59E C16

○ *The Spirit of America* ○

http://sacredcod.blogspot.com/2008/09/massachusetts-needs-new-license-plates.html

WHAT HAVE WE LEARNED?

We've learned that probability is based on long-run relative frequencies. We've thought about the Law of Large Numbers and noted that it speaks only of long-run behavior. Because the long run is a very long time, we need to be careful not to misinterpret the Law of Large Numbers. Even when we've observed a string of heads, we shouldn't expect extra tails in subsequent coin flips.

Also, we've learned some basic tools for modeling probabilities of outcomes and more complex events. These include

▶ sample spaces,
▶ the Fundamental Counting Principle,
▶ permutations, and
▶ combinations.

Terms

Random phenomenon	A phenomenon is random if we know what outcomes could happen, but not which particular values will happen.
Law of Large Numbers	The Law of Large Numbers states that the long-run *relative frequency* of repeated independent events gets closer and closer to the *true* relative frequency as the number of trials increases.
Probability	The probability of an event is a number between 0 and 1 that identifies the likelihood that the event happens. We write $P(\mathbf{A})$ for the probability of the event **A**.
Independence (informally)	Two events are *independent* if learning that one event occurs does not change the probability that the other event occurs.
Trial	A single attempt or realization of a random phenomenon.
Outcome	The outcome of a trial is the observed result.
Event	A collection of outcomes. Usually, we identify events so that we can attach probabilities to them. We denote events with bold capital letters such as **A, B,** or **C.**
Sample Space	The collection of all possible outcomes.
Fundamental Counting Principle	If event **A** has m outcomes and event **B** has n outcomes, then
	▶ the number of outcomes in event **A** *or* **B** is $m + n$;
	▶ the number of outcomes in event **A** *and* **B** is mn.
Permutations	We call the number of ways to arrange any r items from a group of n different items permutations of n objects r at a time, calculated $_nP_r = \dfrac{n!}{(n-r)!}$.

(continued)

Combinations We call the number of ways to select any r items from a group of n different items combinations of n objects chosen r at a time, calculated

$$_nC_r = \frac{n!}{r!(n-r)!}.$$

Equally Likely Condition The outcomes in a sample space (whether listed or just counted) are all equally likely to occur. If this condition is satisfied, the probability of an event **A** is

$$P(A) = \frac{\text{number of outcomes in } \mathbf{A}}{\text{number of possible outcomes}}$$

Skills

THINK

▶ Understand that random phenomena are unpredictable in the short term but show long-run regularity.

▶ Be able to recognize random outcomes in a real-world situation.

▶ Know that the relative frequency of a random event settles down to a value called the probability. Know that this is guaranteed for independent events by the Law of Large Numbers.

▶ Recognize when outcomes in a sample space are equally likely.

▶ Recognize the difference between permutations (arrangements) and combinations (selections).

SHOW

▶ Be able to create a sample space for a random phenomenon.

▶ Be able to count how many outcomes an event may have using the Fundamental Counting Principle, permutations, or combinations.

▶ Be able to use lists or counts of equally likely outcomes to find the probability of an event.

TELL

▶ Know and be able to use the terms "sample space," "outcomes," and "events" correctly.

▶ Be able to use statements about probability in describing a random phenomenon.

▶ Be able to use probability as evidence that an observed outcome may be unusual (statistically significant).

EXERCISES

A

1. **Sample spaces.** For each of the following, list the sample space and tell whether you think the events are equally likely:
   a) Draw a card from a well-shuffled deck; record the suit.
   b) Roll two dice; record the larger number.
   c) Toss 2 coins; record the order of heads and tails.
   d) Flip a coin until you get a head or 3 consecutive tails; record each flip.

2. **Sample spaces.** For each of the following, list the sample space and tell whether you think the events are equally likely:
   a) Shake a coin out of a piggy bank containing a penny, a nickel, a dime, and a quarter; record the value.

   b) Shake two coins out of the piggy bank in part a; record the total value.
   c) A family has 3 children; record each child's sex in order of birth.
   d) Toss a coin 10 times; record the length of the longest run of heads.

3. **Roulette.** A casino claims that its roulette wheel is truly random. What should that claim mean?

4. **Ping pong.** Many state lotteries determine winners by using a machine that rapidly mixes numbered ping-pong balls, eventually allowing some to escape from the container. What assumptions make this method of selecting the winning numbers truly random?

5. **Survival.** A doctor tells a patient just diagnosed with a serious disease that there's a 60% chance she'll live at least 5 years. Where do you think that probability comes from?

6. **Rain.** The weather reporter on TV makes predictions such as a 25% chance of rain. What do you think is the meaning of such a phrase?

7. **Winter.** Comment on the following quotation:

"What I think is our best determination is it will be a colder than normal winter," said Pamela Naber Knox, a Wisconsin state climatologist. "I'm basing that on a couple of different things. First, in looking at the past few winters, there has been a lack of really cold weather. Even though we are not supposed to use the law of averages, we are due." (Associated Press, fall 1992, quoted by Schaeffer et al.)

8. **Snow.** After an unusually dry autumn, a radio announcer is heard to say, "Watch out! We'll pay for these sunny days later on this winter." Explain what he's trying to say, and comment on the validity of his reasoning.

9. **Cold streak.** A batter who had failed to get a hit in seven consecutive times at bat then hits a game-winning home run. When talking to reporters afterward, he says he was very confident that last time at bat because he knew he was "due for a hit." Comment on his reasoning.

10. **Fouls.** Near the end of a basketball game the team that's behind often commits a foul in order to get the ball back quickly. Of course, they hope the other team's player won't make his free throws. Whom should they foul? Here are three possible strategies:
    - Foul the player who has missed the highest percentage of his foul shots tonight, because he's a cold shooter.
    - Foul the player who has made the highest percentage of his foul shots tonight, because he's due to miss.
    - Foul the player with the lowest season-long foul-shooting percentage, because he's the one most likely to miss.

    Based on your understanding of probability, which is the best strategy, and why?

11. **Crash.** Commercial airplanes have an excellent safety record. Nevertheless, there are crashes occasionally, with the loss of many lives. In the weeks following a crash, airlines often report a drop in the number of passengers, probably because people are afraid to risk flying. A travel agent suggests that since the law of averages makes it highly unlikely to have two plane crashes within a few weeks of each other, flying soon after a crash is the safest time. What do you think? Explain.

12. **Crash, part II.** If the airline industry proudly announces that it has set a new record for the longest period of safe flights, would you be reluctant to fly? Are the airlines due to have a crash? Explain.

13. **Literature.** Your American Lit class will read 4 novels this year, chosen by class vote from a list of 12 possible books offered by the teacher.
    a) How many different ways could the course unfold, given that it probably matters what order you read the books in?
    b) How many different choices of books could the class make?

14. **Relay team.** A swim coach is considering 7 swimmers as possible members of a 4-person relay team.
    a) How many different choices of 4 swimmers does the coach have?
    b) Given that the relay team's performance may depend on what order the athletes swim in, how many different teams could the coach create?

15. **Experiment.** A researcher needs to split 12 plants into two treatment groups of 6 plants each. How many different groupings can be formed?

16. **Customer satisfaction.** A furniture store routinely calls some customers to inquire about their buying experience in the showroom, the delivery and setup, and their overall satisfaction with the purchase. Today the calls will go to 4 homes randomly selected from last week's 21 customers. How many different groups of customers could the store call?

17. **Dice product.** You are playing a game that involves rolling two dice, but instead of using the sum, you'll get paid in tokens equal to the *product* of the numbers you roll. (For example, if you roll a 3 and a 4, you win 12 tokens.)
    a) Create a sample space for the number of tokens you may win. Be careful to do this in a way that makes all the outcomes you list equally likely.
    b) What's the probability you win an odd number of tokens?

18. **Telemarketing.** Some companies who make those annoying phone calls to try to sell you something generate the phone numbers randomly.
    a) Given that seven-digit phone numbers NNN-NNNN can't have a 0 or a 1 as the first digit, how many different phone numbers are possible (within one area code)?
    b) One company is working your area code tonight, 10 people each making an average of 20 calls per hour. If it takes your family a half hour to eat supper, what's the probability you get a call during mealtime?

19. **Poker.** You're playing poker with some friends, and are about to be dealt 5 cards from a well-shuffled standard deck.
    a) How many different hands are possible?
    b) How many hands contain only hearts?
    c) What's the probability you'll be dealt a flush (5 cards all the same suit)?

20. **Hearts.** You're playing hearts with some friends, and are about to be dealt 7 cards from a well-shuffled deck.
    a) How many different hands are possible?
    b) How many of those hands don't contain any hearts at all?
    c) What's the probability that there will be some hearts in your hand?

B

21. **Fire insurance.** Insurance companies collect annual payments from homeowners in exchange for paying to rebuild houses that burn down.

a) Why should you be reluctant to accept a $300 payment from your neighbor to replace his house should it burn down during the coming year?

b) Why can the insurance company make that offer?

22. Jackpot. On January 20, 2000, the International Gaming Technology company issued a press release:

(LAS VEGAS, Nev.)—Cynthia Jay was smiling ear to ear as she walked into the news conference at The Desert Inn Resort in Las Vegas today, and well she should. Last night, the 37-year-old cocktail waitress won the world's largest slot jackpot—$34,959,458—on a Megabucks machine. She said she had played $27 in the machine when the jackpot hit. Nevada Megabucks has produced 49 major winners in its 14-year history. The top jackpot builds from a base amount of $7 million and can be won with a 3-coin ($3) bet.

a) How can the Desert Inn afford to give away millions of dollars on a $3 bet?

b) Why did the company issue a press release? Wouldn't most businesses want to keep such a huge loss quiet?

23. Blood. There are four basic human blood types, in order of frequency: O, A, B, and AB.

a) Make a sample space showing the possible pairings of blood types for a mother and father.

b) Why can't you use this sample space to find the probability that a baby has parents whose blood types match?

24. Cereal. Recall the cereal company that was putting photos of athletes in boxes of cereal? It puts pictures of Serena Williams in half the boxes and pictures of either LeBron James or David Beckham in the others. Hoping to get a complete set, you buy three boxes of cereal.

a) Make a sample space showing the possible outcomes as you open the boxes one at a time.

b) Why can't you use this sample space to find the probability that you get one picture of each athlete?

25. Family. A family has 3 children.

a) Create a sample space for the number of boys the family could have.

b) Explain why counting outcomes in that sample space can't tell you the probability that there are 2 boys.

c) Use a different sample space to find the probability that 2 of the 3 children are boys.

d) What assumptions did you make in using this sample space to find the probability?

26. Coins. You're going to toss 4 coins.

a) Create a sample space for the number of tails you could get.

b) Explain why counting outcomes in that sample space can't tell you the probability that you'll get exactly 2 tails.

c) Use a different sample space to find the probability that you'll toss 2 heads and 2 tails.

27. Bigger die. In the game you're playing, you'll roll 2 dice. The number of spaces you get to move will be the larger of the two numbers you roll.

a) Create a sample space showing the number of spaces you may get to move.

b) You need to move 5 spaces in order to win the game. Why can't you count outcomes in this sample space to find the probability of winning?

c) Use a different sample space to find the probability that you win the game.

28. Three dice. You are playing a game that involves rolling a die 3 times in a row.

a) The sample space of all possible outcomes is a long list, so don't bother creating it—but how many outcomes are there?

b) You'll win the game if each roll is a bigger number than the roll before. (For example, 2-3-6 is a winner, but 1-5-3 and 3-4-4 are not.) List the outcomes that are winners.

c) What's the probability that you win?

29. Lottery. A scratch-off lottery ticket has 12 concealed spaces among which are 4 symbols saying "Win!". The person who bought the ticket scratches the black yucky stuff off 3 spaces, winning an instant $10 if all three are winners. What's the probability of winning this game?

30. Second chance lottery. On the scratch-off lottery ticket described in Exercise 29, 5 of the spaces conceal symbols saying "Free Play." Buyers who reveal 3 of those get another ticket for free. What's the probability this happens?

31. Audit. As a check against possible accounting errors, each day a building supplies company audits 5 recently active accounts. Yesterday they made sales to 18 retail customers and 8 contractors.

a) If they select the 5 accounts to be audited as a simple random sample from all accounts, what's the probability that they don't check any contractor accounts?

b) If they want to audit 3 retail accounts and 2 contractor accounts, what's a better sampling design?

c) How many ways are there to select the 3 retail accounts and 2 contractor accounts?

d) What's the probability that an SRS would produce this more representative sample by chance?

32. Border inspections. Three buses, 9 cars, and 5 trucks are currently waiting in line to enter the United States at a border checkpoint. Customs agents will select 5 vehicles at random for a more thorough inspection.

a) What's the probability that a simple random sample of 5 vehicles doesn't include any of the buses?

b) If agents want to be sure to inspect one bus, 2 cars, and 2 trucks, what's a better sampling design?

c) How many different ways are there to create a sample of one bus, 2 cars, and 2 trucks?

d) What's the probability that an SRS would produce this more representative sample by chance?

33. Poker again. What's the probability that a 5-card poker hand contains exactly 3 or 4 queens?

34. Another poker hand. What's the probability that a 5-card poker hand contains two pairs, one of them a pair of aces?

35. **Legislature.** In a state legislature the elected representatives include 17 Democrats, 13 Republicans, and 4 Independents. What's the probability that a random selection of 6 legislators would include 2 of each?

36. **Team.** On a high school varsity baseball team, 5 players are sophomores, 11 are juniors, and 10 are seniors. The team selects 4 honorary co-captains by drawing names from a hat. What's the probability they pick 2 sophomores and 2 seniors?

37. **Legislature revisited.** What's the probability that the randomly selected group of legislators described in Exercise 35 would end up being 6 members of the same party?

38. **Second team.** What's the probability that as a result of the process described in Exercise 36 all 4 co-captains come from the same class?

39. **Motorcycles.** A state's motorcycle license plates are a random combination of three letters followed by two numbers (such as BLE-12). Because of possible confusion with numbers 0 and 1, they never use the letters I, O, or Q.
    a) How many different motorcycle plates are possible?
    b) What's the probability that a plate chosen at random contains all vowels? (Include Y as a vowel.)

40. **Trailers.** License plates for trailers in one state are a random combination of two letters followed by four numbers (such as PN-5423). Because of possible confusion with numbers 0 and 1, they never use the letters I, O, or Q.
    a) How many different license plates are possible?
    b) What is the probability that a plate chosen at random contains all even digits?

C

41. **Lotto winners.** To win the New Jersey Pick-6 Lotto jackpot (described in the chapter's For Example) you have to successfully match 6 numbers chosen from the numbers 1 through 49. The lottery also pays smaller prizes if you match only 3, 4, or 5 of the numbers.
    a) What's the probability that your 6 choices match 3 of the 6 winning numbers?
    b) What's the probability your ticket wins some prize?

42. **Quality control.** Each hour a shipping inspector at an Internet retailer spot-checks 5 outgoing packages to be sure they are completely sealed. While some occasional problems are to be expected, he will shut down the process to look for serious trouble if more than one of the packages he checks is improperly sealed. He's unaware of the fact that the machine stapling the packages shut has begun to misfire sporadically, leaving 7 of the 50 packages that were prepared this hour unsatisfactorily sealed.
    a) What's the probability that 2 of the defective packages show up in his sample?
    b) What's the probability he detects the problem?

43. **Job opening.** A city recently held interviews for 4 jobs available. Even though there were 12 Republicans among the 22 applicants, none of them was hired. Noting that the mayor is a Democrat, the local Republican Party chairperson has challenged the fairness of what was supposed to be a non-political process. What do you think? Assuming that the qualifications of the applicants don't depend on their political affiliation, is it reasonable to believe that this outcome could simply have happened by chance? Support your conclusion with a statistical argument.

44. **Kittens.** Are people equally likely to abandon male kittens as female kittens? If so, you'd think that among 50 kittens recently turned in at a local animal shelter there should be half males and half females. However, a friend of yours who works there just told you that the staff fed the kittens today, all 7 of the ones she picked up (essentially at random) were female. Does this make you suspect that female kittens are more likely to be turned in, or could this have happened purely by chance? Support your conclusion with a statistical argument.

45. **More about permutations.** In the chapter you learned that there are 6! ways to arrange any 6 different objects—the letters in the word MEDIAN, say. But what if some of the letters are the same? How many ways can we arrange the letters in MAXIMUM?

    Well, there are 7 letters, so that would be 7! if they were all different. But that counts MAXIMUM as a new arrangement (we switched the M's around when you weren't looking). To avoid recounting these "arrangements," note that every ordering appears 3! = 6 times because of the matching Ms. We can eliminate all those duplicates from consideration by simply dividing by 3!, revealing that there are actually $\frac{7!}{3!} = 840$ distinct permutations of the letters in MAXIMUM. Similarly, there would be $\frac{7!}{2!2!} = 1260$ different ways to arrange the letters in DOTPLOT: 7! if they were all different, reduced by a factor of 2! for the matching Os and 2! for the matching Ts.

    How many ways could the letters in these words be arranged?
    a) SKEWED
    b) EXPERIMENT
    c) PROBABILITY
    d) MINIMUM
    e) STATISTICS
    f) ASSOCIATION

46. **States.** Use the counting method described in Exercise 45 to find out how many ways you could arrange the letters in the names of each of these states.
    a) OHIO
    b) ALASKA
    c) HAWAII
    d) ARKANSAS
    e) INDIANA
    f) MISSISSIPPI

Answers

Do The Math

1. 24

2. 720

3. 20

4. 10

5. 190

6. 210

7. 99!

8. You get the next lower number, factorial:
   $\dfrac{n!}{n} = (n-1)!. \ \dfrac{51!}{51} = 50!; \ \dfrac{6!}{6} = 5!$

9. $\dfrac{1!}{1} = 0!$, and $\dfrac{1!}{1} = 1$, so $0! = 1$.

Just Checking

1. The LLN works only in the short run. The random methods for selecting lottery numbers have no memory of previous picks, so there is no change in the probability that a certain number will come up.

2. a) We assume that babies are 50-50 boy-girl, and that sexes of babies born in the same family are independent.
   b) $\dfrac{1}{4}$ c) $\dfrac{3}{4}$ d) $\dfrac{1}{2}$

3. a) S = {BBB, BBG, BGB, GBB, BGG, GBG, GGB, GGG}
   b) $\dfrac{7}{8}$ c) $\dfrac{3}{4}$

4. a) $10 \cdot 10 \cdot 10 \cdot 10 = 10,000$
   b) $26 \cdot 10 \cdot 26 \cdot 10 \cdot 26 = 1,757,600$
   c) $1,757,600 + 10 \cdot 26 \cdot 10 \cdot 26 \cdot 10 = 2,433,600$
   d) 26 capitals + 26 lowercase letters + 10 digits = 62 possible characters in each of the 5 positions; $62^5 = 916,132,832$

5. $_5P_5 = \dfrac{5!}{0!} = 5 \cdot 4 \cdot 3 \cdot 3 \cdot 1 = 120$

6. $_8P_5 = \dfrac{8!}{3!} = 8 \cdot 7 \cdot 6 \cdot 5 \cdot 4 = 6,720$

7. a) combinations
   b) permutations
   c) permutations
   d) combinations

8. a) $_5C_4 = \dfrac{5!}{4!1!} = \dfrac{5 \cdot 4 \cdot 3 \cdot 2}{4 \cdot 3 \cdot 2 \cdot 1} = 5$
   b) $_8C_4 = \dfrac{8!}{4!4!} = \dfrac{8 \cdot 7 \cdot 6 \cdot 5}{4 \cdot 3 \cdot 2 \cdot 1} = 70$
   c) $_8C_2 \cdot {}_5C_2 = \dfrac{8!}{2!6!} \cdot \dfrac{5!}{2!3!} = \dfrac{8 \cdot 7}{2 \cdot 1} \cdot \dfrac{5 \cdot 4}{2 \cdot 1} = 280$
   d) $_8C_3 \cdot {}_5C_1 = \dfrac{8!}{3!5!} \cdot \dfrac{5!}{1!4!} = 280$
   e) $_8C_2 \cdot {}_5C_2 + {}_8C_3 \cdot {}_5C_1 = 280 + 280 = 560$

9. a) $\dfrac{{}_8C_6}{{}_{15}C_6} \approx 0.0056$
   b) $\dfrac{{}_4C_3 \cdot {}_8C_3}{{}_{15}C_6} \approx 0.045$
   c) $\dfrac{{}_3C_3 \cdot {}_{12}C_3}{{}_{15}C_6} \approx 0.044$

What Are the Chances?

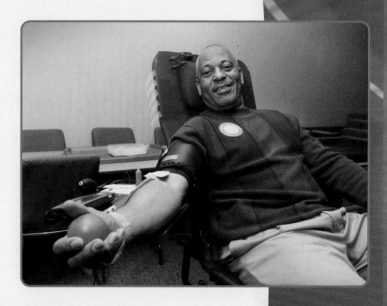

The Red Cross reminds us that every minute of every day, someone needs blood. People who have been in accidents, have serious illnesses, or are undergoing surgery all require blood transfusions. The blood comes from volunteer donors, often through Red Cross blood drives. But it's not as simple as just collecting blood from someone and giving it to someone else. There are several different types of blood: O, A, B, and AB, each sub-classified as positive or negative depending on whether a certain antigen is present. Severe reactions can occur if people are given the wrong type of blood, and the options are more restrictive for some types than for others. People who are Type AB+ are fortunate; they're "universal recipients," meaning they can receive blood of any type. At the other extreme are folks with O− blood, who can receive blood only from other O−'s—a scant 6.6% of the population. Interestingly, these same O− people are "universal donors," because their blood can be transfused to anyone of any blood type.

Worldwide the frequencies of the various blood types vary by geographic region and ethnicity. This table shows the distribution in the United States population. Because Type O blood (both + and −) is the only blood type that can be transfused to patients with other blood types, the Red Cross is always eager to find Type O donors. And that leads us to some probability questions. A few are easy to answer, but others aren't.

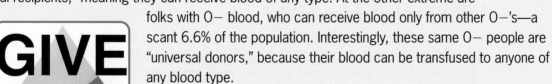

Human Blood Types		
Frequency of occurrence in the United States		
Type	Freq	Can donate to:
O+	37.4%	O+, A+, B+, AB+
A+	35.7%	A+, AB+
B+	8.5%	B+, AB+
O−	6.6%	Anyone!
A−	6.3%	A+, A−, AB+, AB−
AB+	3.4%	AB+
B−	1.5%	B+, B−, AB+, AB−
AB−	0.6%	AB+, AB−

- *What's the probability a randomly selected donor is Type O−?* The table indicates that the population-wide proportion of people with O− blood is 6.6%. Since this is akin to a "long run" frequency, we're comfortable saying P(O−) = 0.066.
- *What's the probability a randomly selected donor is Type O, either + or −?* Be careful here. Our sample space O+, O−, A+, etc. lists 8 types and two of them are O's, but the probability isn't $\frac{2}{8}$. That's because the blood types aren't equally likely outcomes. The last chapter's counting approach won't work here, but if you think finding this probability is as easy as adding the 37.4% O+'s to the 6.6% O−'s, you're absolutely right: *P*(O) = 0.44. (More about that soon.)
- *What's the probability that the Red Cross won't find a Type O donor until they get to the 5th volunteer?* This is a much harder question. The sample space isn't any help, and it's not so clear how to use the frequencies in the table either. Stay tuned, though. You'll be able to answer this question by the end of the chapter.
- *What's the probability the Red Cross finds at least 50 Type O donors among the 150 volunteers who show up for a blood drive?* While this may be the kind of question of greatest interest to the Red Cross (and all the people needing blood today), it's not one we can answer yet.[1] Eventually you'll know how, but that requires understanding ways to work with probabilities of outcomes even if they're not equally likely. Let's get started.

The First Three Rules for Working with Probability

1. Make a picture.
2. Make a picture.
3. Make a picture.

We're dealing with probabilities now, not data, but the three rules don't change. The most common kind of picture to make is called a Venn diagram. We'll use Venn diagrams throughout the rest of this chapter. Even experienced statisticians make Venn diagrams to help them think about probabilities of compound and overlapping events. You should, too.

John Venn (1834–1923) created the Venn diagram. His book on probability, *The Logic of Chance*, was "strikingly original and considerably influenced the development of the theory of Statistics," according to John Maynard Keynes, one of the luminaries of Economics.

Formal Probability

In order to answer the Blood Drive questions and attack other interesting probability issues, we'll need to understand some formal rules[2] about how probability works.

1. If the probability is 0, the event can't occur. If it has probability 1, it *always* occurs. Even if you think an event is very unlikely, its probability can't be

---

[1]Coming soon to a textbook near you. Chapter 16, in fact.

[2]Actually, in mathematical terms, these are axioms—statements that we assume to be true of probability. We'll derive other rules from these in the next chapter.

negative, and even if you're really sure it will happen, its probability can't be greater than 1. So:

A probability is a number between 0 and 1.

For any event A, $0 \leq P(A) \leq 1$.

The probability a person has red blood is 1; the probability it's green is 0.[3]

2. We need to distribute all the probability among the various outcomes a trial can have. How can we do that so that it makes sense? For example, consider what you're doing as you read this book. The possible outcomes might be

A. You read to the end of this chapter before stopping.
B. You finish this section but stop reading before the end of the chapter.
C. You bail out before the end of this section.[4]

When we assign probabilities to these outcomes, the first thing to be sure of is that we distribute *all* of the available probability. One of them has to occur, so the probability of the entire sample space is 1.

Making this more formal gives the **Probability Assignment Rule.**

The set of all possible outcomes of a trial must have probability 1.

$$P(\mathbf{S}) = 1$$

Check out the Blood Types table on p. 306. You'll see that the frequencies total 100% (as, logically, they must).

3. Suppose there's an 80% chance that you get to class on time. What are the chances that you don't? Yes, the other 20%. The set of outcomes that are *not* in the event **A** is called the **complement** of **A**, written **A**^{C}. This leads to the **Complement Rule:**

The probability an event doesn't occur is 1 minus the probability that it does.

$$P(\mathbf{A}^C) = 1 - P(\mathbf{A})$$

Suppose we want to know the probability that a person volunteering to give blood is not a universal donor (O−). One way to find out would be to add up the percentages of all the other blood types. But if you're lazy (smart!), you'll take the shortcut:

$$P(not\ \text{O}-) = P(\text{O}-^C) = 1 - P(\text{O}-) = 1 - 0.066 = 0.934$$

In other words, 93.4% of people are *not* universal donors.

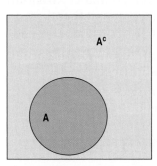

The set **A** and its complement **A**^{C}. Together, they make up the entire sample space **S**.

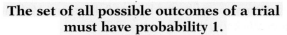

For Example Applying the Complement Rule

Recap: We opened the last chapter by looking at the traffic light at the corner of College and Main, observing that when we arrive at that intersection, the light is green about 35% of the time.

Question: If $P(\text{green}) = 0.35$, what's the probability the light isn't green when you get to College and Main?

"Not green" is the complement of "green," so $P(\text{not green}) = 1 - P(\text{green})$
$$= 1 - 0.35 = 0.65$$

There's a 65% chance I won't have a green light.

[3]Not counting Mr. Spock.
[4]Please don't stop right here!

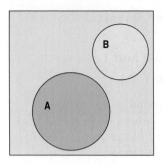

Two disjoint sets, **A** and **B**.

4. Based on our blood types table the probability that (**A**) a randomly selected blood donor is O⁺ is 0.374, and the probability that (**B**) he or she is O⁻ is 0.066. What is the probability that the donor is Type O—*either* O⁺ *or* O⁻, written $P(\mathbf{A} \cup \mathbf{B})$? If you guessed 0.374 + 0.066 = 0.44, you've come up with the **Addition Rule,** which says that you can add the probabilities of events that are disjoint. Here that's true: no one is both O⁺ and O⁻. **Disjoint** (or **mutually exclusive**) events have no outcomes in common (in other words, they don't overlap). The **Addition Rule** states,

For two disjoint events A and B, the probability that one or the other occurs is the sum of the probabilities.

$$P(A \text{ } or \text{ } B) = P(A \cup B) = P(A) + P(B), \text{ provided that}$$
$$A \text{ and } B \text{ are disjoint.}$$

For Example   Applying the Addition Rule

Recap: When you get to the light at College and Main, it's either red, green, or yellow. We know that $P(\text{green}) = 0.35$.

Question: Suppose we find out that $P(\text{yellow})$ is about 0.04. What's the probability the light is red?

To find the probability that the light is green or yellow, I can use the Addition Rule because these are disjoint events: The light can't be both green and yellow at the same time.

$$P(\text{green} \cup \text{yellow}) = 0.35 + 0.04 = 0.39$$

Red is the only remaining alternative, and the probabilities must add up to 1, so

$$P(\text{red}) = P(\text{not }(\text{green} \cup \text{yellow}))$$
$$= 1 - P(\text{green} \cup \text{yellow})$$
$$= 1 - 0.39 = 0.61$$

"Baseball is 90% mental. The other half is physical."

—Yogi Berra

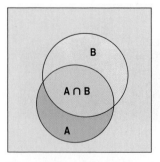

Two sets **A** and **B** that are not disjoint. The event (**A** ∩ **B**) is their intersection.

Because the outcomes listed in a sample space are all disjoint, we have an easy way to check whether the probabilities we've assigned them are **legitimate.** The Probability Assignment Rule tells us that the sum of the probabilities must be exactly 1. No more, no less. For example, if we were told that the probabilities of the four major blood types (O, A, B, AB) were 0.40, 0.40, 0.10, and 0.04, respectively, we would know that something was wrong. These "probabilities" add up to only 0.94, so this is not a legitimate probability assignment. Either a value is wrong, or we missed some previously unknown blood type that soaks up the remaining 0.06. Similarly, a claim that the probabilities were 0.45, 0.45, 0.10, and 0.05 would be wrong because these "probabilities" total more than 1.

But be careful: The Addition Rule doesn't work for events that aren't disjoint. If the probability of owning an MP3 player is 0.80 and the probability of owning a computer is 0.90, the probability of owning either an MP3 player or a computer may be pretty high, but it is *not* 1.70! Why can't you add probabilities like this? Because these events are not disjoint. You *can* own both. In the next chapter, we'll see how to add probabilities for events like these, but we'll need a somewhat different rule.

5. We know that 44% of people have Type O blood, but what are the chances that two strangers waiting in line to donate will *both* be Type O? That's the same as asking for the probability that the first donor is Type O *and* the second donor is, too. For independent events, the answer is very simple. Remember, independence means that the probability of one event doesn't influence the probability of another. Unless the two people are close relatives, there's no reason to believe the second person's blood type is in any way related to the first person's. Since we said these two donors are strangers, it's reasonable to assume their blood types are independent. We expect that 44% of the time the first person will be Type O, and 44% of that 44% the next person will be, too. Since 44% of 44% = $0.44 \times 0.44 = 0.1936$, it appears there's about a 19% chance that two consecutive unrelated blood donors will both be Type O.

The **Multiplication Rule** says that for independent events, to find the probability that both events occur, we just multiply the probabilities together. Formally,

> **For two independent events A and B, the probability that both A *and* B occur is the product of the probabilities of the two events.**
>
> $$P(A \text{ } and \text{ } B) = P(A \cap B) = P(A) \times P(B), \text{ provided that } A \text{ and } B \text{ are independent.}$$

This rule can be extended to more than two independent events. What's the chance of finding four unrelated Type O donors in a row? We can multiply the probabilities for each person:

$$0.44 \times 0.44 \times 0.44 \times 0.44 = 0.0375$$

or just under 4%. Of course, to calculate this probability, we have assumed the four events are independent.

Many Statistics methods require an **Independence Assumption,** but *assuming* independence won't make it true. Always *Think* about whether that assumption is reasonable before using the Multiplication Rule.

For Example   Applying the Multiplication Rule (and others)

Recap: The probability that we encounter a green light at the corner of College and Main is 0.35, a yellow light 0.04, and a red light 0.61. Let's think about your morning commute for the week ahead.

Question: What's the probability you find the light red both Monday and Tuesday?

Because the color of the light I see on Monday doesn't influence the color I'll see on Tuesday, these are independent events; I can use the Multiplication Rule:

$$P(\text{red Monday} \cap \text{red Tuesday}) = P(\text{red}) \times P(\text{red})$$
$$= (0.61)(0.61)$$
$$= 0.3721$$

There's about a 37% chance I'll hit red lights both Monday and Tuesday mornings.

Question: What's the probability you don't hit a red light until Wednesday?

For that to happen, I'd have to see green or yellow on Monday, green or yellow on Tuesday, and then red on Wednesday. I can simplify this by thinking of it as not red on Monday and not red Tuesday and then red on Wednesday.

(continued)

$$P(\text{not red}) = 1 - P(\text{red}) = 1 - 0.61 = 0.39, \text{so}$$

$$P(\text{not red Monday} \cap \text{not red Tuesday} \cap \text{red Wednesday}) = P(\text{not red}) \times P(\text{not red}) \times P(\text{red})$$

$$= (0.39)(0.39)(0.61)$$

$$= 0.092781$$

There's about a 9% chance that this week I'll hit my first red light there on Wednesday morning.

Question: What's the probability that you'll have to stop *at least once* during the week?

Having to stop at least once means that I have to stop for the light either 1, 2, 3, 4, or 5 times next week. It's easier to think about the complement: never having to stop at a red light. Having to stop at least once means that I didn't make it through the week with no red lights.

P(having to stop at the light at least once in 5 days)

$$= 1 - P(\text{no red lights for 5 days in a row})$$

$$= 1 - P(\text{not red} \cap \text{not red} \cap \text{not red} \cap \text{not red} \cap \text{not red})$$

$$= 1 - (0.39)(0.39)(0.39)(0.39)(0.39)$$

$$= 1 - 0.0090$$

$$= 0.991$$

There's over a 99% chance that light will be red at least once this week.

Note that the phrase "at least" is often a tip-off to think about the complement. Something that happens *at least once* <u>does</u> happen. Happening at least once is the complement of not happening at all, and that's easier to find.

> In informal English, you may see "some" used to mean "at least one." "What's the probability that some of the eggs in that carton are broken?" means at least one.

STEP-BY-STEP EXAMPLE   Probability

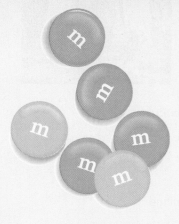

The five rules we've seen can be used many ways to answer a surprising number of questions. Let's see how we might go about it.

In 2001, Masterfoods, the manufacturers of M&M's® milk chocolate candies, decided to add another color to the standard brown, yellow, red, orange, blue, and green. To decide which color to add, they surveyed people in nearly every country of the world and asked them to vote among purple, pink, and teal. The global winner was purple!

In the United States, 42% of those who voted said purple, 37% said teal, and only 19% said pink. But in Japan the percentages were 38% pink, 36% teal, and only 16% purple. Let's use Japan's percentages to ask some questions:

1. What's the probability that a Japanese M&M's survey respondent selected at random preferred either pink or teal?
2. If we pick two respondents at random, what's the probability that they both selected purple?
3. If we pick three respondents at random, what's the probability that *at least one* preferred purple?

The probability of an event is its long-term relative frequency. We can find it several ways: by looking at many trials, by counting equally likely events, or by using some other information. Here, we are told the relative frequencies of the three responses.

Make sure the probabilities are legitimate. Here, they're not. Either there was a mistake, or the other voters must have given a response other than the three colors given.

The M&M's website reports the proportions of Japanese votes by color. These give the probability of selecting a voter who preferred each:

$$P(\text{pink}) = 0.38$$
$$P(\text{teal}) = 0.36$$
$$P(\text{purple}) = 0.16$$

Each is between 0 and 1, but they don't all add up to 1. The remaining 10% of the voters must have not expressed a preference or written in another color. I'll put them together into "other" and add $P(\text{other}) = 0.10$.

Question 1. What's the probability that a Japanese M&M's survey respondent selected at random preferred either pink or teal?

Plan Decide which rules to use and check the conditions they require.

The events "Pink" and "Teal" are individual outcomes (a person can't choose both), so they are disjoint. I can apply the Addition Rule.

Mechanics Show your work.

$$P(\text{pink} \cup \text{teal}) = P(\text{pink}) + P(\text{teal})$$
$$= 0.38 + 0.36 = 0.74$$

Conclusion Interpret your results in the proper context.

The probability that a person said pink or teal is 0.74.

Question 2. If we pick two respondents at random, what's the probability that they both said purple?

Plan The word "both" suggests we want $P(A \text{ and } B)$, which calls for the Multiplication Rule. Think about the assumption.

✔ **Independence Assumption:** It's unlikely that the choice made by one randomly chosen person affected the choice of the other, so the events seem to be independent. I can use the Multiplication Rule.

Mechanics Show your work. For both respondents to pick purple, each one has to pick purple.

$P(\text{both purple})$
$$= P(\text{first respondent picks purple} \cap \text{second respondent picks purple})$$
$$= P(\text{purple}) \times P(\text{purple})$$
$$= 0.16 \times 0.16 = 0.0256$$

Conclusion Interpret your results in the proper context.

The probability that two people both pick purple is 0.0256.

(continued)

Question 3. If we pick three respondents at random, what's the probability that at least one preferred purple?

Plan The phrase "at least . . ." often flags a question best answered by looking at the complement, and that's the easiest approach here. The complement of "At least one preferred purple" is "None of them preferred purple."

Think about the assumption.

$$P(\text{at least one picked purple})$$
$$= P(\{\text{none picked purple}\}^C)$$
$$= 1 - P(\text{none picked purple}).$$
$$= 1 - P(\text{not purple} \cap \text{not purple} \cap \text{not purple}).$$

✔ **Independence Assumption:** These are independent events because they are choices by three random by chosen people. I can use the Multiplication Rule.

Mechanics First we find $P(\text{not purple})$ with the Complement Rule.

Next we calculate $P(\text{none picked purple})$ by using the Multiplication Rule.

Then we can use the Complement Rule to get the probability we want.

$$P(\text{not purple}) = 1 - P(\text{purple})$$
$$= 1 - 0.16 = 0.84$$

$$P(\text{at least one picked purple})$$
$$= 1 - P(\text{none picked purple})$$
$$= 1 - P(\text{not purple} \cap \text{not purple} \cap \text{not purple})$$
$$= 1 - (0.84)(0.84)(0.84)$$
$$= 1 - 0.5927$$
$$= 0.4073$$

Conclusion Interpret your results in the proper context.

There's about a 40.7% chance that at least one of three people picked purple.

Just Checking

Opinion polling organizations contact their respondents by telephone. Random telephone numbers are generated, and interviewers try to contact those households. According to the Pew Research Center for the People and the Press, the contact success rate is about 76%. We can reasonably assume each household's response to be independent of the others. What's the probability that . . .

1. the interviewer successfully contacts the next household on her list?

2. the interviewer successfully contacts both of the next two households on her list?

3. the interviewer's first successful contact is the third household on the list?

4. the interviewer makes at least one successful contact among the next five households on the list?

(Check your answers on page 318.)

 WHAT CAN GO WRONG?

- **Beware of probabilities that don't add up to 1.** To be a legitimate probability assignment, the sum of the probabilities for all possible outcomes must be 1. If the sum is less than 1, you may need to add another category ("other") and assign the remaining probability to that outcome. If the sum is more than 1, check that the outcomes are disjoint. If they're not, you need to correct your sample space.

- **Don't add probabilities of events if they're not disjoint.** Events must be disjoint to use the Addition Rule. The probability of being under 80 *or* a female is not the probability of being under 80 *plus* the probability of being female. That sum is surely more than 1.

- **Don't multiply probabilities of events if they're not independent.** The probability of selecting a student at random who is over 6'10" tall *and* on the basketball team is *not* the probability the student is over 6'10" tall *times* the probability he's on the basketball team. Knowing that the student is over 6'10" makes him more likely to be on the basketball team, so these events aren't independent. Don't multiply these probabilities. Multiplying probabilities of events that are not independent is one of the most common errors people make.

- **Don't confuse disjoint and independent.** Disjoint events *can't* be independent. If **A** = {you get an **A** in this class} and **B** = {you get a **B** in this class}, **A** and **B** are disjoint. Are they independent? If you find out that **A** is true, does that change the probability of **B**? You bet it does! So they can't be independent. We'll return to this issue in the next chapter.

PROBABILITY **IN YOUR WORLD**

Holy Craps! How a Gambling Grandma Broke the Record

It sounds like a homework problem out of a high school math book: What is the probability of rolling a pair of dice 154 times continuously at a craps table, without throwing a seven?

The answer is roughly 1 in 1.56 trillion, and on May 23, Patricia Demauro, a New Jersey grandmother, beat those odds at Atlantic City's Borgata Hotel Casino and Spa. Demauro's 154-roll lucky streak, which lasted four hours and 18 minutes, broke the world records for the longest craps roll and the most successive dice rolls without "sevening out." According to Stanford University statistics professor Thomas Cover, the chances of that happening are smaller than getting struck by lightning (one in a million) being hit by an errant ball at a baseball game (one in 1.5 million) or winning the lottery (one in 100 million, depending on the game).

A craps turn begins with an initial or "come out" roll, in which the player tries to establish a "point number"—that is, when the dice add up to four, five, six, eight, nine or 10. Once that happens, the player must roll the point again before throwing a seven, which is statistically the most likely outcome on a pair of dice. If the player rolls a seven before the point, the turn ends. . . The average number of dice rolls before sevening out? Eight.

http://www.time.com/time/nation/article/0,8599,1901663,00.html

WHAT HAVE WE LEARNED?

We've learned some basic rules for combining probabilities of outcomes to find probabilities of more complex events. These include

▶ the Probability Assignment Rule,
▶ the Complement Rule,
▶ the Addition Rule for disjoint events, and
▶ the Multiplication Rule for independent events.

Terms

The Probability Assignment Rule The probability of the entire sample space must be 1. $P(\mathbf{S}) = 1$.

Complement Rule The probability an event doesn't occur is 1 minus the probability that it does.

$$P(\mathbf{A}^{\mathbf{C}}) = 1 - P(\mathbf{A})$$

Disjoint (Mutually exclusive) Two events are disjoint if they never can happen at the same time. If **A** and **B** are disjoint, then knowing that **A** occurs tells us that **B** cannot. Disjoint events are also called "mutually exclusive."

Addition Rule If **A** and **B** are disjoint events, then the probability of **A** *or* **B** is

$$P(\mathbf{A} \cup \mathbf{B}) = P(\mathbf{A}) + P(\mathbf{B}).$$

Legitimate probability assignment An assignment of probabilities to outcomes is legitimate if
▶ each probability is between 0 and 1 (inclusive).
▶ the sum of the probabilities is 1.

Multiplication Rule If **A** and **B** are independent events, then the probability of **A** and **B** is

$$P(\mathbf{A} \cap \mathbf{B}) = P(\mathbf{A}) \times P(\mathbf{B})$$

Independence Assumption We often require events to be independent. Two events are *independent* if learning that one event occurs does not change the probability that the other event occurs. (You should think about whether this assumption is reasonable.)

Skills

▶ Know the rules of probability.
▶ Recognize when events are disjoint and when events are independent. Understand the difference and that disjoint events cannot be independent.

▶ Be able to use the facts about probability to determine whether an assignment of probabilities is legitimate. Each probability must be a number between 0 and 1, and the sum of the probabilities assigned to all possible outcomes must be 1.

▶ Know how and when to apply the Addition Rule. Know that events must be disjoint for the Addition Rule to work.

▶ Know how and when to apply the Multiplication Rule. Know that events must be independent for the Multiplication Rule to work.

▶ Know how to use the Complement Rule to make calculating probabilities simpler. Recognize that probabilities of "at least . . ." are likely to be easier to find this way.

▶ Be able to use statements about probability in describing a random phenomenon.
▶ Know and be able to use the terms "disjoint events" and "independent events" correctly.

EXERCISES

A

1. **Spinner.** The plastic arrow on a spinner for a child's game stops rotating to point at a color that will determine what happens next. Which of the following probability assignments are possible? Explain what's wrong with the others.

	Probabilities of . . .			
	Red	**Yellow**	**Green**	**Blue**
a)	0.25	0.25	0.25	0.25
b)	0.10	0.20	0.30	0.40
c)	0.20	0.30	0.40	0.50
d)	0	0	1.00	0
e)	0.10	0.20	1.20	−1.50

2. **Scratch off.** Many stores run "secret sales": Shoppers receive cards that determine how large a discount they get, but the percentage is revealed by scratching off that black stuff (what *is* that?) only after the purchase has been totaled at the cash register. The store is required to reveal (in the fine print) the distribution of discounts available. Which of these probability assignments are legitimate? Explain what's wrong with the others.

	Probabilities of . . .			
	10% off	**20% off**	**30% off**	**50% off**
a)	0.20	0.20	0.20	0.20
b)	0.50	0.30	0.20	0.10
c)	0.80	0.10	0.05	0.05
d)	0.75	0.25	0.25	−0.25
e)	1.00	0	0	0

3. **Environment.** In 2009, a Gallup poll found that 47% of Americans favored protection of the environment at the risk of limiting energy supplies, while 46% favored the development of U.S. energy supplies even if that risked harming the environment. Another 3% felt that energy and the environment were equally important.
   a) Based on this poll, what's the probability a person had "No Opinion"?
   b) What's the probability that a respondent rated protecting the environment as equally important or more important than developing energy sources?

4. **Failing fathers?** A Pew Research poll asked U.S. adults whether fathers today were doing as good a job of fathering as fathers of 20–30 years ago. Here's how they responded:

Response	**Percentage**
Better	21%
Same	28%
Worse	47%

If we select a respondent at random from this sample,
   a) what is the probability that the selected person responded "No Opinion"?
   b) what is the probability that the person responded the "Same" or "Better"?

5. **Environment again.** Consider again the poll on energy and the environment described in Exercise 3. Suppose we select two of the respondents at random.
   a) What is the probability that both favored the environment?
   b) What's the probability that neither favored the environment?
   c) What assumption did you make in computing these probabilities?
   d) Why do you think that assumption is reasonable?

6. **Fathers revisited.** Consider again the results of the poll about fathering discussed in Exercise 4. If we select two people at random from this sample,
   a) what is the probability that both think fathers are better today?
   b) what is the probability that neither thinks fathers are better today?
   c) What assumption did you make in computing these probabilities?
   d) Explain why you think that assumption is reasonable.

7. **Car repairs.** A consumer organization estimates that over a 1-year period 17% of cars will need to be repaired once, 7% will need repairs twice, and 4% will require three or more repairs. What is the probability that a car chosen at random will need
   a) no repairs?
   b) no more than one repair?
   c) some repairs?

8. **Stats projects.** In a large Introductory Statistics lecture hall, the professor reports that 55% of the students enrolled have never taken a Calculus course, 32% have taken only one semester of Calculus, and the rest have taken two or more semesters of Calculus. The professor randomly assigns students to groups of three to work on a project for the course. What is the probability that the first groupmate you meet has studied
   a) two or more semesters of Calculus?
   b) some Calculus?
   c) no more than one semester of Calculus?

9. More repairs. Consider again the auto repair rates described in Exercise 7. If you own two cars, what is the probability that
   a) neither will need repair?
   b) both will need repair?
   c) at least one car will need repair?

10. Another project. You are assigned to be part of a group of three students from the Intro Stats class described in Exercise 8. What is the probability that of your other two groupmates,
   a) neither has studied Calculus?
   b) both have studied at least one semester of Calculus?
   c) at least one has had more than one semester of Calculus?

11. Repairs again. You used the Multiplication Rule to calculate repair probabilities for your cars in Exercise 9.
   a) What must be true about your cars in order to make that approach valid?
   b) Do you think this assumption is reasonable? Explain.

12. Final project. You used the Multiplication Rule to calculate probabilities about the Calculus background of your Statistics groupmates in Exercise 10.
   a) What must be true about the groups in order to make that approach valid?
   b) Do you think this assumption is reasonable? Explain.

13. Vehicles. Suppose that 46% of families living in a certain county own a car and 18% own an SUV. The Addition Rule might suggest, then, that 64% of families own either a car or an SUV. What's wrong with that reasoning?

14. Homes. Funding for many schools comes from taxes based on assessed values of local properties. People's homes are assessed higher if they have extra features such as garages and swimming pools. Assessment records in a certain school district indicate that 37% of the homes have garages and 3% have swimming pools. The Addition Rule might suggest, then, that 40% of residences have a garage or a pool. What's wrong with that reasoning?

15. Speeders. Traffic checks on a certain section of highway suggest that 60% of drivers are speeding there. Since $0.6 \times 0.6 = 0.36$, the Multiplication Rule might suggest that there's a 36% chance that two vehicles in a row are both speeding. What's wrong with that reasoning?

16. Lefties. Although it's hard to be definitive in classifying people as right- or left-handed, some studies suggest that about 14% of people are left-handed. Since $0.14 \times 0.14 = 0.0196$, the Multiplication Rule might suggest that there's about a 2% chance that a brother and a sister are both lefties. What's wrong with that reasoning?

17. M&M's. The Masterfoods company says that before the introduction of purple, yellow candies made up 20% of their plain M&M's, red another 20%, and orange, blue, and green each made up 10%. The rest were brown. If you pick an M&M at random, what is the probability that
   a) it is brown?
   b) it is yellow or orange?
   c) it is not green?
   d) it is striped?

18. Blood. The American Red Cross says that about 44% of the U.S. population has Type O blood, 42% Type A, 10% Type B, and the rest Type AB. Someone volunteers to give blood. What is the probability that this donor
   a) has Type AB blood?
   b) has Type A or Type B?
   c) is not Type A?

B

19. More M&M's. The probabilities of getting an M&M of the various colors are described in Exercise 17. If you pick three M&M's in a row, what is the probability that
   a) they are all brown?
   b) the third one is the first one that's red?
   c) none are yellow?
   d) at least one is green?

20. More blood. The probabilities an American has each of the major blood types are given in Exercise 18. Among four potential donors, what is the probability that
   a) all are Type O?
   b) no one is Type AB?
   c) they are not all Type A?
   d) at least one person is Type B?

21. Disjoint or independent? In Exercises 17 and 19 you calculated probabilities of getting various M&M's. Some of your answers depended on the assumption that the outcomes described were *disjoint;* that is, they could not both happen at the same time. Other answers depended on the assumption that the events were *independent;* that is, the occurrence of one of them doesn't affect the probability of the other. Think about the difference between disjoint and independent: If you draw one M&M, are the events of getting a red one and getting an orange one disjoint, independent, or neither?

22. Disjoint or independent? In Exercises 18 and 20 you calculated probabilities involving various blood types. Some of your answers depended on the assumption that the outcomes described were *disjoint;* that is, they could not both happen at the same time. Other answers depended on the assumption that the events were *independent;* that is, the occurrence of one of them doesn't affect the probability of the other. Think about the difference between disjoint and independent: If you examine one person, are the events that the person is Type A and that the person is Type B disjoint, independent, or neither?

23. Disjoint vs. independent again. Think once more about the difference between disjoint and independent events. If you draw two M&M's one after the other, are the events of getting a red on the first and a red on the second disjoint, independent, or neither?

24. Disjoint vs. independent again. Think once more about the difference between disjoint and independent events. If you examine two people, are the events that the first is Type A and the second Type B disjoint, independent, or neither?

25. Dice. You roll a fair die three times. What is the probability that
a) you roll all 6's?
b) you roll all odd numbers?
c) none of your rolls gets a number divisible by 3?
d) you roll at least one 5?
e) the numbers you roll are not all 5's?

26. Slot machine. A slot machine has three wheels that spin independently. Each has 10 equally likely symbols: 4 bars, 3 lemons, 2 cherries, and a bell. If you play, what is the probability that
a) you get 3 lemons?
b) you get no fruit symbols?
c) you get 3 bells (the jackpot)?
d) you get no bells?
e) you get at least one bar (an automatic loser)?

27. Champion bowler. A certain bowler can bowl a strike 70% of the time. What's the probability that she
a) goes three consecutive frames without a strike?
b) makes her first strike in the third frame?
c) has at least one strike in the first three frames?
d) bowls a perfect game (12 consecutive strikes)?

28. The train. To get to work, a commuter must cross train tracks. The time the train arrives varies slightly from day to day, but the commuter estimates he'll get stopped on about 15% of work days. During a certain 5-day work week, what is the probability that he
a) gets stopped on Monday and again on Tuesday?
b) gets stopped for the first time on Thursday?
c) gets stopped every day?
d) gets stopped at least once during the week?

29. Voters. Suppose that in your city 37% of the voters are registered as Democrats, 29% as Republicans, and 11% as members of other parties (Liberal, Right to Life, Green, etc.). Voters not aligned with any official party are termed "Independent." You are conducting a poll by calling registered voters at random. In your first three calls, what is the probability you talk to
a) all Republicans?
b) no Democrats?
c) at least one Independent?

30. Religion. Census reports for a city indicate that 62% of residents classify themselves as Christian, 12% as Jewish, and 16% as members of other religions (Muslims, Buddhists, etc.). The remaining residents classify themselves as nonreligious. A polling organization seeking information about public opinions wants to be sure to talk with people holding a variety of religious views, and makes random phone calls. Among the first four people they call, what is the probability they reach
a) all Christians?
b) no Jews?
c) at least one person who is nonreligious?

31. Tires. You bought a new set of four tires from a manufacturer who just announced a recall because 2% of those tires are defective. What is the probability that at least one of yours is defective?

32. Pepsi. For a sales promotion, the manufacturer places winning symbols under the caps of 10% of all Pepsi bottles. You buy a six-pack. What is the probability that you win something?

C

33. College admissions. For high school students graduating in 2007, college admissions to the nation's most selective schools were the most competitive in memory. (*The New York Times*, "A Great Year for Ivy League Schools, but Not So Good for Applicants to Them," April 4, 2007). Harvard accepted about 9% of its applicants, Stanford 10%, and Penn 16%. Jorge has applied to all three. Assuming that he's a typical applicant, he figures that his chances of getting into both Harvard and Stanford must be about 0.9%.
a) How has he arrived at this conclusion?
b) What additional assumption is he making?
c) Do you agree with his conclusion?

34. College admissions II. In Exercise 33, we saw that in 2007 Harvard accepted about 9% of its applicants, Stanford 10%, and Penn 16%. Jorge has applied to all three. He figures that his chances of getting into at least one of the three must be about 35%.
a) How has he arrived at this conclusion?
b) What assumption is he making?
c) Do you agree with his conclusion?

35. Polling. As mentioned in the chapter, opinion-polling organizations contact their respondents by sampling random telephone numbers. Although interviewers now can reach about 76% of U.S. households, the percentage of those contacted who agree to cooperate with the survey has fallen from 58% in 1997 to only 38% in 2003 (Pew Research Center for the People and the Press). Each household, of course, is independent of the others.
a) What is the probability that the next household on the list will be contacted but will refuse to cooperate?
b) What is the probability (in 2003) of failing to contact a household or of contacting the household but not getting them to agree to the interview?
c) Show another way to calculate the probability in part b.

36. Polling, part II. According to Pew Research, the contact rate (probability of contacting a selected household) was 69% in 1997 and 76% in 2003. However, the cooperation rate (probability of someone at the contacted household agreeing to be interviewed) was 58% in 1997 and dropped to 38% in 2003.
a) What is the probability (in 2003) of obtaining an interview with the next household on the sample list? (To obtain an interview, an interviewer must both contact the household and then get agreement for the interview.)

b) Was it more likely to obtain an interview from a randomly selected household in 1997 or in 2003?

37. Recalls. An automaker has discovered defects in some cars that were manufactured recently. The company has sent recall notices to 12% of the owners telling them to bring their cars in to get the trunk latch replaced, and another notice to 9% of the owners offering to replace the brake system warning light.
   a) Explain what it would mean if the two defects were disjoint.
   b) The company says that the two defects are independent. Explain what that means in this context.
   c) Can having these defects be both disjoint and independent? Explain.

38. Activities. At Springfield High School, 22% of the students are involved in the sports program and 17% in the music program.
   a) Explain what it would mean if being involved in sports and music were disjoint events.
   b) Explain what it would mean if being involved in sports and music were independent events.
   c) Can involvement in sports and music be both independent and disjoint? Explain.

39. 9/11? On September 11, 2002, the first anniversary of the terrorist attack on the World Trade Center, the New York State Lottery's daily number came up 9–1–1. An interesting coincidence or a cosmic sign? (The lottery's ping-pong ball machine creates a winning number between 000 and 999 at random.)

a) What is the probability that the winning three numbers match the date on any given day?
b) What is the probability that a whole year passes without this happening?
c) What is the probability that the date and winning lottery number match at least once during any year?
d) If every one of the 50 states has a three-digit lottery, what is the probability that at least one of them will come up 9–1–1 on September 11?

40. Red cards. You shuffle a deck of cards and then start turning them over one at a time. The first one is red. So is the second. And the third. In fact, you are surprised to get 10 red cards in a row. You start thinking, "The next one is due to be black!"
   a) Are you correct in thinking that there's a higher probability that the next card will be black than red? Explain.
   b) Is this an example of the Law of Large Numbers? Explain.

Answers

Just Checking

1. 0.76

2. $0.76(0.76) = 0.5776$

3. $(1 - 0.76)^2(0.76) = 0.043776$

4. $1 - (1 - 0.76)^5 = 0.9992$

Probability Rules!

L et's think about money—specifically, all the kinds of bills currently in circulation in the United States. The sample space is
S = {$1 bill, $2 bill, $5 bill, $10 bill,$20 bill, $50 bill, $100 bill}.[1] Now pick one of them at random. What are your chances of getting the $100 bill? That's easy: since we're choosing from this collection of 7 bills and all outcomes are equally likely, $P(100) = \frac{1}{7}$.

We can combine the outcomes lots of ways to make many different events. For example, let event **A** be selecting an odd dollar value. Then **A** = {$1, $5}, and $P(A) = \frac{2}{7}$. How about the probability of event **B** = {a bill with a building on the back}? Well, the back of a $1 has the word ONE in the center, and the $2 bill shows the signing of the Declaration of Independence. All the rest have images of famous buildings, so **B** = {$5, $10, $20, $50, $100}, and $P(B) = \frac{5}{7}$.

Let's get to the point.[2] What's the probability that a bill selected at random has an odd number value *or* a building on the back? Yes, "or" means add, but if you follow our Addition Rule you'll get the wrong answer: $\frac{2}{7} + \frac{5}{7} = 1$. That probability of 1 should mean that *every* bill must either be odd or have a building. Not the $2 bill, though: even value and no building.

---

[1] Are you surprised that those are *all* the possible outcomes? In spite of what you may have seen in bank robbery movies, there are no $500 or $1,000 bills.

[2] Yes, we do have a point to make.

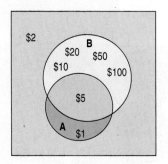

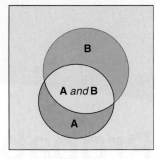

What went wrong? As we pointed out in the last chapter, that Addition Rule requires that the events be *disjoint*. Our Venn diagram showing that had two circles that don't overlap. Now, though, they do, because the $5 bill is in both event **A** (odd) and **B** (building). So what can we do? We need to revise our probability rule.

Events **A** and **B** and their intersection.

The General Addition Rule

Here's the Venn diagram for our sample space of bills. Notice that the $2 bill, being neither odd nor having a building, sits outside both circles. Because the $5 bill is both odd *and* has a building, it's in the *intersection* of **A** and **B**. The events are not disjoint, so we can't use the Addition Rule. If we simply add the probabilities of **A** and **B**, we'll count the $5 bill twice.

There's an easy fix. We can add the two probabilities, then compensate by *subtracting* out the probability of that $5 bill (so we only count it once):

$$P(A \cup B) = P(\text{odd number value } or \text{ building})$$
$$= P(\text{odd}) + P(\text{building}) - P(\text{odd } and \text{ building})$$
$$= P(\$1, \$5) + P(\$5, \$10, \$20, \$50, \$100) - P(\$5)$$
$$= \frac{2}{7} + \frac{5}{7} - \frac{1}{7}$$
$$= \frac{6}{7}$$

Denominations of bills that are odd (**A**) or that have a building on the reverse side (**B**). The two sets both include the $5 bill, and both exclude the $2 bill.

It worked! That's the right answer: 6 out of 7 bills (all but the $2) have an odd value or show a building.

This method works in general. We add the probabilities of two events and then subtract out the probability of their intersection. This approach gives us the **General Addition Rule,** which does not require disjoint events:

$$P(\mathbf{A} \cup \mathbf{B}) = P(\mathbf{A}) + P(\mathbf{B}) - P(\mathbf{A} \cap \mathbf{B}).$$

For Example Using the General Addition Rule

A survey of college students found that 56% live in a campus residence hall, 62% participate in a campus meal program, and 42% do both.

Question: What's the probability that a randomly selected student either lives or eats on campus?

Let **L** = {student lives on campus} and **M** = {student has a campus meal plan}.

$P(\text{a student either lives or eats on campus}) = P(\mathbf{L} \cup \mathbf{M})$
$$= P(\mathbf{L}) + P(\mathbf{M}) - P(\mathbf{L} \cap \mathbf{M})$$
$$= 0.56 + 0.62 - 0.42$$
$$= 0.76$$

There's a 76% chance that a randomly selected college student either lives or eats on campus.

Sometimes rules like this are helpful; other times trying to fit a rule to the question can be the Hard Way. For one thing, natural language can be ambiguous. Would you like pie or ice cream? Does that mean you must choose one or the other, or can you have both? Sometimes the word "or" means "one or the other or both," yet at other times it excludes the possibility of both, as in, "Would you like the steak *or* the vegetarian dinner?" In addition, we'll encounter still other wordings, like "pie but no ice cream," "neither pie nor ice cream," and so on.

What to do? *Make a picture.* It's almost always easier to think about these situations by making a Venn diagram.

For Example   Using Venn Diagrams

Recap: We return to our survey of college students: 56% live on campus, 62% have a campus meal program, and 42% do both.

Questions: Based on a Venn diagram, what is the probability that a randomly selected student

a) lives off campus and doesn't have a meal program?
b) lives in a residence hall but doesn't have a meal program?

Let $L = \{$student lives on campus$\}$ and $M = \{$student has a campus meal plan$\}$. In the Venn diagram, the intersection of the circles is $P(L \cap M) = 0.42$. Since $P(L) = 0.56$, $P(L \cap M^C) = 0.56 - 0.42 = 0.14$. Also, $P(L^C \cap M) = 0.62 - 0.42 = 0.20$. So far, $0.14 + 0.42 + 0.20 = 0.76$, leaving $1 - 0.76 = 0.24$ for the region outside both circles.

Now: $P(\text{off campus and no meal program}) = P(L^C \cap M^C) = 0.24$
$P(\text{on campus and no meal program}) = P(L \cap M^C) = 0.14$

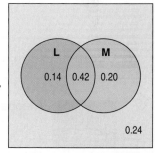

STEP-BY-STEP EXAMPLE   Using the General Addition Rule

Police report that 78% of drivers stopped on suspicion of drunk driving are given a breath test, 36% a blood test, and 22% both tests.

Question: What is the probability that a randomly selected DWI (DUI) suspect is given

1. a test?
2. a blood test or a breath test, but not both?
3. neither test?

THINK

Plan Define the events we're interested in. There are no conditions to check; the General Addition Rule works for any events!

Let $A = \{$suspect is given a breath test$\}$.

Let $B = \{$suspect is given a blood test$\}$.

(continued)

Plot Make a picture, and use the given probabilities to find the probability for each region.

The blue region represents **A** but not **B**. The green intersection region represents **A** *and* **B**. Note that since $P(\mathbf{A}) = 0.78$ and $P(\mathbf{A} \cap \mathbf{B}) = 0.22$, the probability of **A** but not **B** must be $0.78 - 0.22 = 0.56$.

The yellow region is **B** but not **A**.

The gray region outside both circles represents the outcome neither **A** nor **B**. All the probabilities must total 1, so you can determine the probability of that region by subtraction.

Now, figure out what you want to know. The probabilities can come from the diagram or a formula. Sometimes translating the words to equations is the trickiest step.

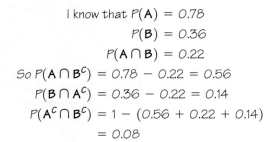

I know that $P(\mathbf{A}) = 0.78$
$P(\mathbf{B}) = 0.36$
$P(\mathbf{A} \cap \mathbf{B}) = 0.22$
So $P(\mathbf{A} \cap \mathbf{B}^C) = 0.78 - 0.22 = 0.56$
$P(\mathbf{B} \cap \mathbf{A}^C) = 0.36 - 0.22 = 0.14$
$P(\mathbf{A}^C \cap \mathbf{B}^C) = 1 - (0.56 + 0.22 + 0.14)$
$= 0.08$

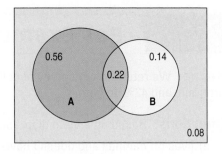

Question 1. What is the probability that the suspect is given a test?

Mechanics The probability the suspect is given a test is $P(\mathbf{A} \cup \mathbf{B})$. We can use the General Addition Rule, or we can add the probabilities seen in the diagram.

$$P(\mathbf{A} \cup \mathbf{B}) = P(\mathbf{A}) + P(\mathbf{B}) - P(\mathbf{A} \cap \mathbf{B})$$
$$= 0.78 + 0.36 - 0.22$$
$$= 0.92$$

OR

$$P(\mathbf{A} \cup \mathbf{B}) = 0.56 + 0.22 + 0.14 = 0.92$$

Conclusion Don't forget to interpret your result in context.

92% of all suspects are given a test.

Question 2. What is the probability that the suspect gets either a blood test or a breath test but NOT both?

Mechanics We can add the appropriate probabilities seen in the Venn diagram. We want to include the two regions that are inside just one (but not both) of the circles.

$$P(\mathbf{A} \text{ or } \mathbf{B} \text{ but NOT both}) = P(\mathbf{A} \cap \mathbf{B}^C) + P(\mathbf{B} \cap \mathbf{A}^C)$$
$$= 0.56 + 0.14$$
$$= 0.70$$

Conclusion Interpret your result in context.

70% of the suspects get exactly one of the tests.

Question 3. What is the probability that the suspect gets neither test?		
	Mechanics Getting neither test is the complement of getting one or the other. Use the Complement Rule or just notice that "neither test" is represented by the region outside both circles.	$$P(\text{neither test}) = 1 - P(\text{either test})$$ $$= 1 - P(A \cup B)$$ $$= 1 - 0.92 = 0.08$$ *OR* $$P(A^C \cap B^C) = 0.08$$
	Conclusion Interpret your result in context.	*Only 8% of the suspects get no test.*

Just Checking

1. Back in Chapter 1 we suggested that you sample some pages of this book at random to see whether they held a graph or other data display. We actually did just that. We drew a representative sample and found the following:

   48% of pages had some kind of data display,

   27% of pages had an equation, and

   7% of pages had both a data display and an equation.

   a) Display these results in a Venn diagram.
   b) What is the probability that a randomly selected sample page had neither a data display nor an equation?
   c) What is the probability that a randomly selected sample page had a data display but no equation?

   (Check your answers on page 341.)

Conditional Probability: It Depends . . .

Two psychologists surveyed 478 children in grades 4, 5, and 6 in elementary schools in Michigan. They stratified their sample, drawing roughly 1/3 from rural, 1/3 from suburban, and 1/3 from urban schools. Among other questions, they asked the students whether their main goal was to get good grades, to be popular, or to be good at sports. We wonder whether boys and girls at this age had similar goals.

Here's a *contingency table* giving counts of the students by their goals and sex:

		Goals			
		Grades	**Popular**	**Sports**	**Total**
Sex	**Boy**	117	50	60	**227**
	Girl	130	91	30	**251**
	Total	**247**	**141**	**90**	**478**

TABLE 15.1 **The distributions of goals for boys and girls.**

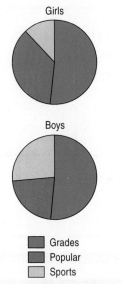

Girls

Boys

Grades
Popular
Sports

FIGURE 15.1 The distributions of goals for boys and girls.

Using these 478 students as our sample space, some probabilities aren't hard to find:

- What's the probability that a person selected at random from these students is a girl? There are 251 girls among the 478 students, so
$$P(\text{girl}) = \frac{251}{478} = 0.525.$$

- The probability that a randomly selected student's goal is to excel in sports is $P(\text{sports}) = \frac{90}{478} = 0.188.$

- What's the probability that a randomly selected student is a girl who hopes to be good at sports? Well, 30 girls named sports as their goal, so the probability is $P(\text{girl} \cap \text{sports}) = \frac{30}{478} = 0.063.$

But how do the goals of boys and girls compare?

We looked at contingency tables and graphed *conditional distributions* back in Chapter 3. These pie charts show the *relative frequencies* with which boys and girls named the three goals.

We see that girls are much less likely to say their goal is to excel at sports than are boys. When we restrict our focus to girls, we look only at the girls' row of the table. Of the 251 girls, only 30 of them said their goal was to excel at sports.

We write the probability that a selected student wants to excel at sports *given that we have selected a girl* as

$$P(\text{sports} \mid \text{girl}) = 30/251 = 0.120$$

What about boys? Look at the top row of the table. There, of the 227 boys, 60 said their goal was to excel at sports. So, $P(\text{sports} \mid \text{boy}) = 60/227 = 0.264$, more than twice the girls' probability.

In general, when we want the probability of an event from a *conditional* distribution, we write $P(\mathbf{B} \mid \mathbf{A})$ and pronounce it "the probability of $\mathbf{B}$ *given* $\mathbf{A}$." A probability that takes into account a given *condition* is called a **conditional probability.**

Let's look at what we did. We worked with the counts, but we could work with percents or probabilities just as well. There were 30 students who both were girls and had sports as their goal, and there were 251 girls. So we found the probability to be 30/251. To find the probability of the event $\mathbf{B}$ *given* the event $\mathbf{A}$, we restrict our attention to the outcomes in $\mathbf{A}$. We then find in what fraction of *those* outcomes $\mathbf{B}$ also occurred. Formally, we write:

$$P(\mathbf{B}|\mathbf{A}) = \frac{P(\mathbf{A} \cap \mathbf{B})}{P(\mathbf{A})}.$$

Thinking this through, we can see that it's just what we've been doing. The result is the same whether we use counts or probabilities because the total number in the sample cancels out:

$$P(\text{sports} \mid \text{girl}) = \frac{P(\text{sports} \cap \text{girl})}{P(\text{girl})} = \frac{30/478}{251/478} = \frac{30}{251}.$$

For Example   Finding a Conditional Probability

Recap: Our survey found that 56% of college students live on campus, 62% have a campus meal program, and 42% do both.

Question: While dining in a campus facility open only to students with meal plans, you meet someone interesting. What is the probability that your new acquaintance lives on campus?

Let L = {student lives on campus} and
    M = {student has a campus meal plan}.

$P(\text{student lives on campus given that the student has a meal plan}) = P(L\,|\,M)$

$$= \frac{P(L \cap M)}{P(M)}$$

$$= \frac{0.42}{0.62}$$

$$\approx 0.677$$

There's a probability of about 0.677 that a student with a meal plan lives on campus.

Be careful, though. When we work with conditional probabilities, we must be sure to pay attention to which event is given. We've just seen that the probability a girl hopes to excel in sports is $\frac{30}{251}$, or about 0.12. That's $P(\text{sports}\,|\,\text{girl})$. What about $P(\text{girl}\,|\,\text{sports})$? First, lets think about what that means. $P(\text{girl}\,|\,\text{sports})$ is the probability that a student with a goal of being good at sports is a girl. Now we restrict our thinking to the 90 students who said they wanted to excel in sports. Among them, 30 are girls, so $P(\text{girl}\,|\,\text{sports}) = \frac{30}{90} = 0.33$, a much different result. Using our conditional probability formula, we'd find

$$P(\text{girl}\,|\,\text{sports}) = \frac{P(\text{girl} \cap \text{sports})}{P(\text{sports})} = \frac{30/478}{90/478} = \frac{30}{90} \approx 0.33$$

So remember, $P(\mathbf{A}\,|\,\mathbf{B})$ is not the same as $P(\mathbf{B}\,|\,\mathbf{A})$. Whenever you're finding a conditional probability, always *Think* about which event is given.

Just Checking

2. A company's Office of Human Resources reports a breakdown of employees by job type and sex, as seen in the table. Suppose we select one employee at random.

   a) What's the probability the employee is a male supervisor?
   b) What's the probability a male employee is a supervisor?
      (*Hint:* Restate the question using the word "given.")
   c) What's the probability that a supervisor is male?

		Sex	
		Male	**Female**
Job Type	**Management**	7	6
	Supervision	8	12
	Production	45	72

(Check your answers on page 341.)

If we had to pick one idea in this chapter that you should understand and remember, it's the definition and meaning of independence. We'll need this idea in every one of the chapters that follow.

Independence

It's time to return to the question of just what it means for events to be independent. We've said informally that what we mean by independence is that the outcome of one event does not influence the probability of the other. With our new notation for conditional probabilities, we can write a formal definition: Events **A** and **B** are **independent** whenever

$$P(\mathbf{B} \mid \mathbf{A}) = P(\mathbf{B}).$$

In other words, the probability that **B** happens is the same whether we know about **A** or not.

Let's look again at the study about the goals of 4th, 5th, and 6th grade children. Is wanting to excel in sports independent of a student's sex? We've already looked at the two probabilities we need:

- $P(\text{sports}) = \dfrac{90}{478} = 0.188$
- $P(\text{sports} \mid \text{girl}) = \dfrac{30}{251} = 0.120$

While nearly 19% of all students had being good in sports as their goal, only about 12% of the girls did. Apparently, girls attach less importance to sports. Because these probabilities aren't equal, choosing success in sports as a goal is *not* independent of the student's sex.

What about grades? Is the probability of having good grades as a goal independent of the sex of the responding student? We need to check whether

$$P(\text{grades} \mid \text{girl}) = P(\text{grades})$$

$$\frac{130}{251} = 0.52 \overset{?}{=} \frac{247}{478} = 0.52$$

To two decimal place accuracy, it looks like we can consider choosing good grades as a goal to be independent of sex.

	Goals			
	Grades	Popular	Sports	Total
Boy	117	50	60	**227**
Girl	130	91	30	**251**
Total	**247**	**141**	**90**	**478**

TABLE 15.2 The distributions of goals for boys and girls.

For Example Checking for Independence

Recap: Our survey told us that 56% of college students live on campus, 62% have a campus meal program, and 42% do both.

Question: Are living on campus and having a meal plan independent? Are they disjoint?

Let L = {**student lives on campus**} and M = {**student has a campus meal plan**}. If these events are independent, then knowing that a student lives on campus doesn't affect the probability that he or she has a meal plan. I'll check to see if $P(M \mid L) = P(M)$:

$$P(M \mid L) = \frac{P(L \cap M)}{P(L)}$$

$$= \frac{0.42}{0.56}$$

$$= 0.75, \quad \text{but } P(M) = 0.62.$$

Because 0.75 ≠ 0.62, the events are not independent; students who live on campus are more likely to have meal plans. Living on campus and having a meal plan are not disjoint either; in fact, 42% of college students do both.

Independent ≠ Disjoint

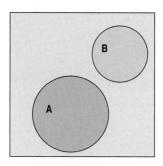

FIGURE 15.2 Because these events are mutually exclusive, learning that **A** happened tells us that **B** didn't. The probability of **B** has changed from whatever it was to zero. So the disjoint events **A** and **B** are not independent.

Are disjoint events independent? These concepts seem to have similar ideas of separation and distinctness about them, but in fact disjoint events *cannot* be independent. Let's see why. Consider the two disjoint events {you get an A in this course} and {you get a B in this course}. They're disjoint because they have no outcomes in common. Suppose you learn that you *did* get an A in the course. Now what is the probability that you got a B? You can't get both grades, so it must be 0.

Think about what that means. Knowing that the first event (getting an A) occurred changed your probability for the second event (down to 0). So these events aren't independent.

Disjoint events can't be independent. They have no outcomes in common, so if one occurs, the other doesn't. A common error is to treat disjoint (mutually exclusive) events as if they were independent and apply the Multiplication Rule for independent events. Don't make that mistake.

Let's summarize:

- Two events could be either independent or disjoint, *but not both*.
- And they could be *neither* disjoint nor independent.

One spring day your high school baseball team will be playing for the league championship right after school. You and your friends are planning to go, but when you look outside at lunchtime it's raining. Consider the events **R**: rain at lunchtime and **B**: baseball after school. They're not disjoint: even though it's raining now, the game may still go on as scheduled. And they're not independent either: you thought you were going to the game, but now that you see the rain, you're not so sure anymore. Knowing **R** changes the probability of **B**.

Just Checking

3. Here's that company's report on job type and sex for their employees.

   a) Are being male and having a supervision job disjoint events?
   b) Is having a supervisor's job independent of the sex of the employee?

		Sex	
		Male	**Female**
Job Type	**Management**	7	6
	Supervision	8	12
	Production	45	72

(Check your answers on page 341.)

Tables vs. Venn Diagrams

One of the easiest ways to think about conditional probabilities is with contingency tables. But sometimes we're given probabilities without a table.

For instance, in the drunk driving example, we were told that 78% of suspect drivers get a breath test, 36% a blood test, and 22% both. Let's use what we know to start a table:

	Breath Test		
	Yes	**No**	**Total**
Yes **No**	0.22		**0.36**
Total	**0.78**		**1.00**

Blood Test

Notice that the 0.78 and 0.36 are *marginal* probabilities and so they go in as totals. The 0.22 is the probability of getting both tests—a breath test *and* a blood test—so it goes in the interior of the table.

By paying attention to the marginal totals filling in the rest of the table is quick:

	Breath Test		
	Yes	**No**	**Total**
Yes	0.22	0.14	**0.36**
No	0.56	0.08	**0.64**
Total	**0.78**	**0.22**	**1.00**

Blood Test

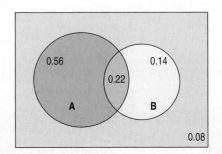

Compare this with the Venn diagram. Notice which entries in the table match up with the regions in this diagram. Whether a Venn diagram or a table is better to use will depend on what you are given and the questions you're being asked. When in doubt, try both.

STEP-BY-STEP EXAMPLE   Are the Events Disjoint? Independent?

Let's take another look at the drunk driving situation. Police report that 78% of drivers are given a breath test, 36% a blood test, and 22% both tests.

Questions: 1. Are giving a DWI (DUI) suspect a blood test and a breath test mutually exclusive?
2. Are giving the two tests independent?

THINK

Plan Define the events we're interested in.

State the given probabilities.

Let **A** = {suspect is given a breath test}

Let **B** = {suspect is given a blood test}.

I know that $P(A) = 0.78$
$P(B) = 0.36$
$P(A \cap B) = 0.22$

Question 1. Are giving a DWI (DUI) suspect a blood test and a breath test mutually exclusive?

Mechanics Disjoint events cannot *both* happen at the same time, so check to see if $P(\mathbf{A} \cap \mathbf{B}) = 0$.

$P(A \cap B) = 0.22$. Since some suspects are given both tests, $P(A \cap B) \neq 0$. The events are not mutually exclusive.

Conclusion State your conclusion in context.

22% of all suspects get both tests, so a breath test and a blood test are not disjoint events.

Question 2. Are the two tests independent?

Plan Make a table.

Always check to be sure the total probability is 1.

		Breath Test		
		Yes	No	Total
Blood Test	Yes	0.22	0.14	**0.36**
	No	0.56	0.08	**0.64**
	Total	**0.78**	**0.22**	**1.00**

Mechanics Does getting a breath test change the probability of getting a blood test? That is, does $P(\mathbf{B} \mid \mathbf{A}) = P(\mathbf{B})$?

Because the two probabilities are *not* the same, the events are not independent.

$$P(B \mid A) = \frac{P(A \cap B)}{P(A)} = \frac{0.22}{0.78} \approx 0.28$$

$$P(B) = 0.36$$

$$P(B \mid A) \neq P(B)$$

Conclusion Interpret your results in context.

Overall, 36% of the drivers get blood tests, but only 28% of those who get a breath test do. Since suspects who get a breath test are less likely to have a blood test, the two events are not independent.

Just Checking

4. Remember our sample of pages in this book from the first Just Checking?

48% of pages had a data display.

27% of pages had an equation, and

7% of pages had both a data display and an equation.

a) Make a contingency table for the variables *Display* (Yes/No) and *Equation* (Yes/No).

b) What is the probability that a randomly selected sample page with an equation also had a data display? (*Hint:* Rewrite this question using the word "given.")

c) Are having an equation and having a data display disjoint events? Explain.

d) Are having an equation and having a data display independent events? Explain.

(Check your answers on page 341.)

The General Multiplication Rule

Remember the Multiplication Rule for the probability of **A** *and* **B**? It said

$$P(\mathbf{A} \cap \mathbf{B}) = P(\mathbf{A}) \times P(\mathbf{B}) \text{ when } \mathbf{A} \text{ and } \mathbf{B} \text{ are independent.}$$

Now we can write a more general rule that doesn't require independence. In fact, we've *already* written it down. We just need to rearrange the equation a bit.

The equation in the definition for conditional probability contains the probability of **A** *and* **B**. Rewriting the equation gives

$$P(\mathbf{A} \cap \mathbf{B}) = P(\mathbf{A}) \times P(\mathbf{B} \mid \mathbf{A}).$$

This is a **General Multiplication Rule** for compound events that does not require the events to be independent. Better than that, it even makes sense. The probability that two events, **A** and **B**, *both* occur is the probability that event **A** occurs multiplied by the probability that event **B** *also* occurs—that is, by the probability that event **B** occurs *given* that event **A** occurs.

> We know:
> $$P(\mathbf{B}|\mathbf{A}) = \frac{P(\mathbf{A} \cap \mathbf{B})}{P(\mathbf{A})}$$
> Multiply both sides of the equation by $P(\mathbf{A})$ to get:
> $$P(\mathbf{A} \cap \mathbf{B}) = P(\mathbf{A}) \cdot P(\mathbf{B}|\mathbf{A})$$

For Example Using the General Multiplication Rule

A factory produces two types of batteries, regular and rechargeable. Quality inspection tests show that 2% of the regular batteries come off the manufacturing line with a defect while only 1% of the rechargeable batteries have a defect. Rechargeable batteries make up 25% of the company's production.

Questions: What's the probability that if we choose one of the company's batteries at random we get

    a) a defective rechargeable battery?
    b) a regular battery and it's not defective?

Let **R** = rechargeable and **B** = a regular battery. It's given that $P(\mathbf{R}) = 0.25$, so $P(\mathbf{B}) = 0.75$.

Let **D** = defective. It's given that $P(\mathbf{D} \mid \mathbf{B}) = 0.02$ and $P(\mathbf{D} \mid \mathbf{R}) = 0.01$.

Now: $P(\mathbf{R} \cap \mathbf{D}) = P(\mathbf{R}) \times P(\mathbf{D} \mid \mathbf{R}) = (0.25)(0.01) = 0.0025$

If 2% of the regular batteries are defective, then the other 98% aren't; in other words: $P(\mathbf{D}^C \mid \mathbf{B}) = 0.98$. So:

$$P(\mathbf{B} \cap \mathbf{D}^C) = P(\mathbf{B}) \times P(\mathbf{D}^C \mid \mathbf{B}) = (0.75)(0.98) = 0.735$$

Only $\frac{1}{4}$ of 1% of the company's batteries are rechargeable and defective, while 73.5% are non-defective regular batteries.

Drawing Without Replacement

Room draw is a process for assigning dormitory rooms to students who live on a college campus. When it's time for you and your friend to draw, there are 12 rooms left. Three are in Gold Hall, a very desirable dorm. You get to draw first, and then your friend will draw. Naturally, you would both like to score rooms in Gold. What are your chances? In particular, what's the probability that you *both* can get rooms in Gold?

When you go first, the chance that *you* will draw one of the Gold rooms is 3/12. Suppose you do. Now, with you clutching your prized room assignment, what chance does your friend have? At this point there are only 11 rooms left and just 2 left in Gold, so your friend's chance is now 2/11.

Using our notation, we write

$$P(\text{friend draws Gold} \mid \text{you draw Gold}) = 2/11.$$

The reason the denominator changes is that we draw these rooms *without replacement.* That is, once one is drawn, it doesn't go back into the pool.

What are the chances that *both* of you will luck out? Well, now we've calculated the two probabilities we need for the General Multiplication Rule, so we can write:

$$P(\text{you draw Gold} \cap \text{friend draws Gold})$$
$$= P(\text{you draw Gold}) \times P(\text{friend draws Gold} \mid \text{you draw Gold})$$
$$= 3/12 \times 2/11 = 1/22 = 0.045\ ^3$$

It doesn't matter who went first, or even if the rooms were drawn simultaneously. Even if the room draw was accomplished by shuffling cards containing the names of the dormitories and then dealing them out to 12 applicants (rather than by each student drawing a room in turn), we can still *think* of the calculation as having taken place in two steps:

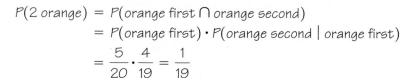

Drawing conditional probabilities this way leads to a more general way of helping us think with pictures: tree diagrams.

For Example Drawing Without Replacement

You just bought a small bag of Skittles. Not that you could know this, but inside are 20 candies: 7 green, 5 orange, 4 red, 3 yellow, and only 1 brown. You tear open one corner of the package and begin eating them by shaking out one at a time.[4]

Questions: What's the probability that your first 2 Skittles are both orange? That none of your first 3 candies is green?

Getting two orange candies in a row means I draw an orange one first *and* then another one second, with one orange candy already missing from the bag:

$$P(2\ \text{orange}) = P(\text{orange first} \cap \text{orange second})$$
$$= P(\text{orange first}) \cdot P(\text{orange second} \mid \text{orange first})$$
$$= \frac{5}{20} \cdot \frac{4}{19} = \frac{1}{19}$$

There's a 1 in 19 chance (just over 5%) that I'd shake out 2 orange Skittles in a row.

(continued)

---

[3]In Chapter 13 you learned another way to find this: $\dfrac{_3C_2}{_{12}C_2}$. Think the answer is the same?

[4]Wow—what self-control!

Not getting any green ones means all of the first 3 Skittles were among the colors (13 candies):

$$P(\text{3 non-greens}) = P(\text{green}^C \cap \text{green}^C \cap \text{green}^C)$$
$$= \frac{13}{20} \cdot \frac{12}{19} \cdot \frac{11}{18} \approx 0.25$$

There's about a 25% chance I won't get any green Skittles among the first 3 I shake out of the bag.

Just Checking

5. Think some more about that bag of Skittles described in the previous *For Example* (7 green, 5 orange, 4 red, 3 yellow, 1 brown). Write out the fractions you'd multiply together to find the probabilities of these outcomes. (Don't bother multiplying them together—unless you're curious.)

   a) The first two are both red.
   b) The first three are all green.
   c) You eat four without seeing a yellow one.
   d) The fifth candy out of the bag is the brown one.

   (Check your answers on page 341.)

Tree Diagrams

A recent Maryland highway safety study found that in 77% of all accidents the driver was wearing a seat belt. Accident reports indicated that 92% of those drivers escaped serious injury (defined as hospitalization or death), but only 63% of the non-belted drivers were so fortunate. Overall, what's the probability that a driver involved in an accident was seriously injured?

The best way to organize information like this is—you guessed it—to make a picture.

Here we'll use a **tree diagram,** because it shows sequences of events, like those we had in the room draw, as paths that look like branches of a tree. It is a good idea to make a tree diagram almost any time you have conditional probabilities and plan to use the General Multiplication Rule. The number of different paths we can take can get large, so we usually draw the tree starting from the left and growing vine-like across the page.

The first branch of our tree separates drivers who had accidents according to whether they wore a seat belt. We label each branch of the tree with a possible outcome and its corresponding probability.

"Why," said the Dodo, "the best way to explain it is to do it."

—Lewis Carroll

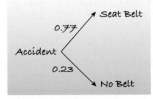

FIGURE 15.3 The first branches. We can diagram the two seat belt possibilities and indicate their respective probabilities with a simple tree diagram.

Notice that because we cover all possible outcomes with the branches, the probabilities add up to one. But we're also interested in injuries. The probability of being seriously injured *depends* on one's seat belt behavior.

Because the probabilities are *conditional,* we draw those outcomes separately on each branch of the tree:

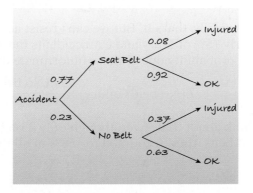

FIGURE 15.4 The second branches. Extending the tree diagram, we can show both seat belt and injury out-comes. The injury probabilities are conditional on the seat belt outcomes, and they change depending on which branch we follow.

On each of the second set of branches, we write the possible outcomes associated with having a car accident (being seriously injured or not) and the associated probability. These probabilities are different because they are *conditional* depending on the driver's seat belt behavior. (It shouldn't be too surprising that those who don't wear their seat belts have a higher probability of serious injury or death.) Each set of probabilities add up to one, because given the outcome on the first branch, these outcomes cover all the possibilities.

Looking back at the General Multiplication Rule, we can see how the tree helps with the calculation.

To find the probability that in a randomly selected accident the driver was wearing a seat belt and was seriously injured, we follow the top branches, multiplying as we go:

$$P(\text{seat belt} \cap \text{injury}) = P(\text{seat belt}) \times P(\text{injury} \mid \text{seat belt})$$
$$= (0.77)(0.08)$$
$$= 0.0616$$

And we do the same for each combination of outcomes:

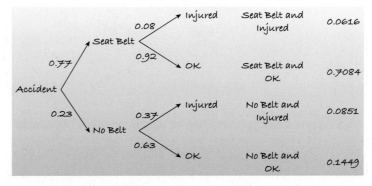

FIGURE 15.5 The completed tree. We can find the probabilities of compound events by multiplying the probabilities along the branch of the tree that leads to the event, just the way the General Multiplication Rule specifies.

Here's a hint to help you check your work creating a tree diagram. All the outcomes at the far right are disjoint and they are *all* the possibilities, so the final probabilities must add up to one. Always check!

And now (at last!) we can answer our original question: Overall, what's the probability that a driver involved in an accident was seriously injured? We simply find all the branches that lead to a serious injury; there are two (the first and the third). Adding the probabilities of those disjoint outcomes we find that

$$P(\text{injury}) = 0.0616 + 0.0851 = 0.1467.$$

Technically, that's it. But we can't resist adding another observation. Overall, in nearly 15% of Maryland car accidents the driver was seriously injured or died. The probability one of those unfortunate drivers wasn't wearing a seat belt is

$$P(\text{No belt} \mid \text{injury}) = \frac{P(\text{No belt} \cap \text{injury})}{P(\text{injury})} = \frac{0.0851}{0.1467} = 0.58$$

Think about that. Even though only 23% of all drivers weren't wearing their seat belts, they accounted for 58% of all the deaths and serious injuries![5]

For Example Tree Diagrams

Recap: Let's revisit the battery factory. Remember, it produced both regular and rechargeable batteries, with 25% of them rechargeable. Past history indicates that 2% of the regular batteries and 1% of the rechargeable batteries have some kind of defect.

Question: What's the probability that a battery chosen at random from a shipment of this factory's products turns out to be defective?

First, I'll create a tree diagram. Batteries are either regular (**B**) or rechargeable (**R**), and each type may be defective (**D**) or **OK**.

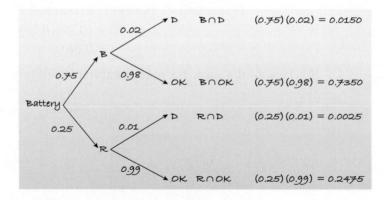

Check to see if the probabilities of all the possible outcomes add up to 1:

$$0.0150 + 0.7350 + 0.0025 + 0.2475 = 1.0000 \text{ (Hooray!)}$$

And now, the probability that a randomly chosen battery is defective is:

$$P(\textbf{D}) = P(\textbf{B} \cap \textbf{D}) + P(\textbf{R} \cap \textbf{D}) = 0.0150 + 0.0025 = 0.0175$$

Overall, 1.75% of the batteries produced at this factory are defective.

---

[5]Just some advice from your friends, the authors: *Please buckle up*. We want you to finish this course.

⃠ WHAT CAN GO WRONG?

- **Don't use a simple probability rule where a general rule is needed.** Don't assume independence without reason to believe it. Don't assume that outcomes are disjoint without checking that they are. Remember that the general rules always apply, even when outcomes are in fact independent or disjoint.

- **Don't reverse conditioning naively.** As we have seen, the probability of **A** *given* **B** is *not* the same as the probability of **B** *given* **A**. Always think carefully about which event is given.

- **Don't confuse "disjoint" with "independent."** Disjoint events *cannot* happen at the same time. When one happens, you know the other did not, so $P(\mathbf{B} \mid \mathbf{A}) = 0$. Independent events *must* be able to happen at the same time. When one happens, you know it has no effect on the other, so $P(\mathbf{B} \mid \mathbf{A}) = P(\mathbf{B})$.

- **Don't find probabilities for samples drawn without replacement as if they had been drawn with replacement.** Remember to adjust the denominator of your probabilities. This warning applies only when we draw from small populations or draw a large fraction of a finite population. When the population is very large relative to the sample size, the adjustments make very little difference, and we ignore them.

INDEPENDENCE **IN YOUR WORLD**

Smoking During Pregnancy—DON'T!!!

Smoking during pregnancy can harm the health of both a woman and her unborn baby. Currently, at least 10 percent of women in the United States smoke during pregnancy.

According to the U.S. Public Health Service, if all pregnant women in this country stopped smoking, there would be an estimated 11 percent reduction in stillbirths [and a] 5 percent reduction in newborn deaths.

Smoking nearly doubles a woman's risk of having a low-birthweight baby. In 2004, 11.9 percent of babies born to smokers in the United States were of low birthweight (less than 5½ pounds), compared to 7.2 percent of babies of nonsmokers.

Smoking also increases the risk of preterm delivery (before 37 weeks of gestation). Premature and low-birthweight babies face an increased risk of serious health problems during the newborn period,

chronic lifelong disabilities (such as cerebral palsy, mental retardation and learning problems), and even death.

A recent study suggests that women who smoke anytime during the month before pregnancy to the end of the

first trimester are more likely to have a baby with birth defects, particularly congenital heart defects.

Smoking is associated with a number of pregnancy complications. Smoking cigarettes doubles a woman's risk of developing . . . bleeding during delivery that can endanger mother and baby.

Babies whose mothers smoked during pregnancy are up to three times as likely to die from sudden infant death syndrome (SIDS) as babies of nonsmokers.

<http://www.marchofdimes.com/professionals/14332_1171.asp>

WHAT HAVE WE LEARNED?

The last chapter's basic rules of probability are important, but they work only in special cases—when events are disjoint or independent.

Now we've learned the more versatile General Addition Rule and General Multiplication Rule.

We've also learned about conditional probabilities, and seen how to use them when sampling without replacement.

We've learned the value of Venn diagrams, tables, and tree diagrams to help organize our thinking about probabilities.

Most important, we've learned to think clearly about independence. We've seen how to use conditional probability to determine whether two events are independent and to work with events that are not independent. A sound understanding of independence will be important throughout the rest of this book.

Terms

General Addition Rule   For any two events, **A** and **B**, the probability of **A** *or* **B** is

$$P(\mathbf{A} \cup \mathbf{B}) = P(\mathbf{A}) + P(\mathbf{B}) - P(\mathbf{A} \cap \mathbf{B}).$$

Conditional probability $P(\mathbf{B} \mid \mathbf{A}) = \dfrac{P(\mathbf{A} \cap \mathbf{B})}{P(\mathbf{A})}$

$P(\mathbf{B} \mid \mathbf{A})$ is read "the probability of **B** *given* **A**."

Independence   Events **A** and **B** are independent when $P(\mathbf{B} \mid \mathbf{A}) = P(\mathbf{B})$.

General Multiplication Rule   For any two events, **A** and **B**, the probability of **A** and **B** is

$$P(\mathbf{A} \cap \mathbf{B}) = P(\mathbf{A}) \times P(\mathbf{B} \mid \mathbf{A}).$$

Tree diagram   A display of conditional events or probabilities that is helpful in organizing our thinking.

Skills

▶ Understand the concept of conditional probability as redefining the *Who* to consider only the event that is *given*.

▶ Understand the concept of independence.

▶ Know how and when to apply the General Addition Rule.

▶ Know how to find probabilities for using a Venn diagram, a two-way table, or a tree diagram.

▶ Know how to determine whether events are independent.

▶ Know how and when to apply the General Multiplication Rule.

▶ Be able to make a clear statement about a conditional probability that shows how the condition affects the probability.

▶ Avoid making statements that assume independence of events when there is no clear evidence that they are in fact independent.

EXERCISES

A

1. **Homes.** Real estate ads suggest that 64% of homes for sale have garages, 21% have swimming pools, and 17% have both features. What is the probability that a home for sale has a pool or a garage?

2. **Travel.** Suppose the probability that a U.S. resident has traveled to Canada is 0.18, to Mexico is 0.09, and to both countries is 0.04. What's the probability that an American chosen at random has traveled to either Canada or Mexico?

3. **Second home.** Look again at the description of real estate ads in Exercise 1.
   a) Create a Venn diagram displaying the percentages of homes for sale that have pools and/or garages.
   b) What's the probability a home has
      i) neither a pool nor a garage?
      ii) a pool but no garage?

4. **More travel.** Look again at the description of travel by Americans in Exercise 2.
   a) Create a Venn diagram displaying the percentages who have traveled to Mexico and/or Canada.
   b) What's the probability an Americans has
      i) traveled to Canada but not Mexico?
      ii) not traveled to either country?

5. **For sale.** For real-estate ads described in Exercise 1, if a home for sale has a garage, what's the probability that it has a pool too?

6. **On the road again.** According to the travel information in Exercise 2, what's the probability that someone who has traveled to Mexico has visited Canada too?

7. **Another home.** Based on your answer to Exercise 5 and the information given in Exercise 1, are having a pool and having a garage independent events for these homes? Explain.

8. **Another trip.** Based on your answer to Exercise 6 and the information given in Exercise 2, are visiting Canada and Mexico independent events? Explain.

9. **Home at last.** For the homes described in Exercise 1, are having a pool and having a garage disjoint events? Explain.

10. **Last journey.** Based on the travel experiences of Americans described in Exercise 2, are having visited Canada and Mexico disjoint events? Explain.

11. **Cards.** You draw a card at random from a standard deck of 52 cards. Find each of the following conditional probabilities:
    a) The card is a heart, given that it is red.
    b) The card is red, given that it is a heart.
    c) The card is an ace, given that it is red.
    d) The card is a queen, given that it is a face card.

12. **Pets.** In its monthly report, the local animal shelter states that it currently has 24 dogs and 18 cats available for adoption. Eight of the dogs and 6 of the cats are male. Find each of the following conditional probabilities if an animal is selected at random:
    a) The pet is male, given that it is a cat.
    b) The pet is a cat, given that it is female.
    c) The pet is female, given that it is a dog.

13. **Two cards.** If you draw 2 cards from a well-shuffled deck, what's the probability that they are both hearts?

14. **Two pets.** If the animal shelter described in Exercise 12 picks two pets at random to show in a TV ad, what's the probability they are both male?

15. **Sick kids.** Seventy percent of kids who visit a doctor have a fever, and 30% of kids with a fever have sore throats. What's the probability that a kid who goes to the doctor has a fever and a sore throat?

16. **Sick cars.** Twenty percent of cars that are inspected have faulty pollution control systems. The cost of repairing a pollution control system exceeds $100 about 40% of the time. When a driver takes her car in for inspection, what's the probability that she will end up paying more than $100 to repair the pollution control system?

17. **Health.** The probabilities that an adult American man has high blood pressure and/or high cholesterol are shown in the table.

		Blood Pressure	
		High	**OK**
Cholesterol	**High**	0.11	0.21
	OK	0.16	0.52

What's the probability that a man has
a) both conditions?
b) high blood pressure?
c) at least one of these conditions?

18. **Death penalty.** The table shows the political affiliations of American voters and their positions on the death?penalty.

		Death Penalty	
		Favor	**Oppose**
Party	**Republican**	0.26	0.04
	Democrat	0.12	0.24
	Other	0.24	0.10

a) What's the probability that
   i) a randomly chosen voter favors the death penalty?
   ii) is a Republican and opposes the death penalty?
b) A candidate thinks she has a good chance of gaining the votes of anyone who is a Republican or in favor of the death penalty. What portion of the voters is that?

19. Health revisited. Based on the information summarized in the table in Exercise 17, what's the probability that
a) a man with high blood pressure has high cholesterol?
b) a man has high blood pressure if it's known that he has high cholesterol?

20. Death penalty revisited. Based on the information summarized in the table in Exercise 18, what's the probability that
a) a Republican favors the death penalty?
b) a voter who favors the death penalty is a Democrat?

21. Luggage. Leah is flying from Boston to Denver with a connection in Chicago. The probability her first flight leaves on time is 0.15. If the flight is on time, the probability that her luggage will make the connecting flight in Chicago is 0.95, but if the first flight is delayed, the probability that the luggage will make it is only 0.65. Are the first flight leaving on time and the luggage making the connection independent events? Explain.

22. Graduation. A private college report contains these statistics:

70% of incoming freshmen attended public schools.
75% of public school students who enroll as freshmen eventually graduate.
90% of other freshmen eventually graduate.

Is there any evidence that a freshman's chances to graduate may depend upon what kind of high school the student attended? Explain.

23. More luggage. Based on the probabilities given in Exercise 21, what's the probability that Leah's luggage arrives in Denver with her? (*Hint:* organize the information in a tree diagram.)

24. Graduation revisited. Based on the probabilities given in Exercise 22, what percent of freshmen eventually graduate? (*Hint:* organize the information in a tree diagram.)

B

25. Amenities. A check of dorm rooms on a large college campus revealed that 38% had refrigerators, 52% had TVs, and 21% had both a TV and a refrigerator. What's the probability that a randomly selected dorm room has
a) a TV but no refrigerator?
b) a TV or a refrigerator, but not both?
c) neither a TV nor a refrigerator?

26. Workers. Employment data at a large company reveal that 72% of the workers are married, that 44% are college graduates, and that half of the college grads are married. What's the probability that a randomly chosen worker
a) is neither married nor a college graduate?
b) is married but not a college graduate?
c) is married or a college graduate?

27. Eligibility. A university requires its Biology majors to take a course called BioResearch. The prerequisite for this course is that students must have taken either a Statistics course or a computer course. By the time they are juniors, 52% of the Biology majors have taken Statistics, 23% have had a computer course, and 7% have done both.
a) What percent of the junior Biology majors are ineligible for BioResearch?
b) What's the probability that a junior Biology major who has taken Statistics has also taken a computer course?
c) Are taking these two courses disjoint events? Explain.
d) Are taking these two courses independent events? Explain.

28. Benefits. Fifty-six percent of all American workers have a workplace retirement plan, 68% have health insurance, and 49% have both benefits. We select a worker at random.
a) What's the probability he has neither employer-sponsored health insurance nor a retirement plan?
b) What's the probability he has health insurance if he has a retirement plan?
c) Are having health insurance and a retirement plan independent events? Explain.
d) Are having these two benefits mutually exclusive? Explain.

29. Cards. If you draw a card at random from a well-shuffled deck, is getting an ace independent of the suit? Explain.

30. Pets again. The local animal shelter in Exercise 12 reported that it currently has 24 dogs and 18 cats available for adoption; 8 of the dogs and 6 of the cats are male. Are the species and sex of the animals independent? Explain.

31. Men's health again. Given the table of probabilities from Exercise 17, are high blood pressure and high cholesterol independent? Explain.

		Blood Pressure	
		High	**OK**
Cholesterol	**High**	0.11	0.21
	OK	0.16	0.52

32. Politics. Given the table of probabilities from Exercise 18, are party affiliation and position on the death penalty independent? Explain.

	Death Penalty	
	Favor	**Oppose**
Republican	0.26	0.04
Democrat	0.12	0.24
Other	0.24	0.10

(Party on left axis)

33. Phone service. According to estimates from the federal government's 2003 National Health Interview Survey, based on face-to-face interviews in 16,677 households, approximately 58.2% of U.S. adults have both a landline in their residence and a cell phone, 2.8% have only cell phone service but no landline, and 1.6% have no telephone service at all.
 a) Polling agencies won't phone cell phone numbers because customers object to paying for such calls. What proportion of U.S. households can be reached by a landline call?
 b) Are having a cell phone and having a landline independent? Explain.

34. Snoring. After surveying 995 adults, 81.5% of whom were over 30, the National Sleep Foundation reported that 36.8% of all the adults snored. 32% of the respondents were snorers over the age of 30.
 a) What percent of the respondents were under 30 and did not snore?
 b) Is snoring independent of age? Explain.

35. Montana. A 1992 poll conducted by the University of Montana classified respondents by sex and political party, as shown in the table. Is party affiliation independent of the respondents' sex? Explain.

	Democrat	Republican	Independent
Male	36	45	24
Female	48	33	16

36. Cars. A random survey of autos parked in student and staff lots at a large university classified the brands by country of origin, as seen in the table. Is country of origin independent of type of driver?

	Driver	
	Student	**Staff**
American	107	105
European	33	12
Asian	55	47

(Origin on left axis)

37. Cards. You are dealt a hand of three cards, one at a time. Find the probability of each of the following.
 a) The first heart you get is the third card dealt.
 b) Your cards are all red (that is, all diamonds or hearts).
 c) You get no spades.
 d) You have at least one ace.

38. Another hand. You pick three cards at random from a deck. Find the probability of each event described below.
 a) You get no aces.
 b) You get all hearts.
 c) The third card is your first red card.
 d) You have at least one diamond.

39. Batteries. A junk box in your room contains a dozen old batteries, five of which are totally dead. You start picking batteries one at a time and testing them. Find the probability of each outcome.
 a) The first two you choose are both good.
 b) At least one of the first three works.
 c) The first four you pick all work.
 d) You have to pick 5 batteries to find one that works.

40. Shirts. The soccer team's shirts have arrived in a big box, and people just start grabbing them, looking for the right size. The box contains 4 medium, 10 large, and 6 extra-large shirts. You want a medium for you and one for your sister. Find the probability of each event described.
 a) The first two you grab are the wrong sizes.
 b) The first medium shirt you find is the third one you check.
 c) The first four shirts you pick are all extra-large.
 d) At least one of the first four shirts you check is a medium.

41. Absenteeism. A company's records indicate that on any given day about 1% of their day-shift employees and 2% of the night-shift employees will miss work. Sixty percent of the employees work the day shift.
 a) Is absenteeism independent of shift worked? Explain.
 b) What percent of employees are absent on any given day?

42. Lungs and smoke. Suppose that 23% of adults smoke cigarettes. It's known that 57% of smokers and 13% of nonsmokers develop a certain lung condition by age 60.
 a) Explain how these statistics indicate that lung condition and smoking are not independent.
 b) What's the probability that a randomly selected 60-year-old has this lung condition?

C

43. Global survey. The marketing research organization GfK Custom Research North America conducts a yearly survey on consumer attitudes worldwide. They collect demographic information on the roughly 1500 respondents from each country that they survey. Here is a table showing the number of people with various levels of education in five countries:

Educational Level by Country						
	Post-graduate	College	Some high school	Primary or less	No answer	Total
China	7	315	671	506	3	**1502**
France	69	388	766	309	7	**1539**
India	161	514	622	227	11	**1535**
U.K.	58	207	1240	32	20	**1557**
USA	84	486	896	87	4	**1557**
Total	**379**	**1910**	**4195**	**1161**	**45**	**7690**

If we select someone at random from this survey,
a) what is the probability that the person is from the United States?
b) what is the probability that the person completed his or her education before college?
c) what is the probability that the person is from France *or* did some post-graduate study?
d) what is the probability that the person is from France *and* finished only primary school or less?

44. Birth order. A survey of students in a large Introductory Statistics class asked about their birth order (1 = oldest or only child) and which college of the university they were enrolled in. Here are the results:

	Birth Order		
College	**1 or only**	**2 or more**	**Total**
Arts & Sciences	34	23	**57**
Agriculture	52	41	**93**
Human Ecology	15	28	**43**
Other	12	18	**30**
Total	**113**	**110**	**223**

Suppose we select a student at random from this class. What is the probability that the person is
a) a Human Ecology student?
b) a firstborn student?
c) firstborn *and* a Human Ecology student?
d) firstborn *or* a Human Ecology student?

45. Global survey, take 2. Look again at the table summarizing the Roper survey in Exercise 43.
a) If we select a respondent at random, what's the probability we choose a person from the United States who has done post-graduate study?
b) Among the respondents who have done post-graduate study, what's the probability the person is from the United States?
c) What's the probability that a respondent from the United States has done post-graduate study?

d) What's the probability that a respondent from China has only a primary-level education?
e) What's the probability that a respondent with only a primary-level education is from China?

46. Birth order, take 2. Look again at the data about birth order of Intro Stats students and their choices of colleges shown in Exercise 44.
a) If we select a student at random, what's the probability the person is an Arts and Sciences student who is a second child (or more)?
b) Among the Arts and Sciences students, what's the probability a student was a second child (or more)?
c) Among second children (or more), what's the probability the student is enrolled in Arts and Sciences?
d) What's the probability that a first or only child is enrolled in the Agriculture College?
e) What is the probability that an Agriculture student is a first or only child?

47. Unsafe food. Early in 2007 *Consumer Reports* published the results of an extensive investigation of broiler chickens purchased from food stores in 23 states. Tests for bacteria in the meat showed that 81% of the chickens were contaminated with campylobacter, 15% with salmonella, and 13% with both.
a) What's the probability that a tested chicken was not contaminated with either kind of bacteria?
b) Are contamination with the two kinds of bacteria disjoint? Explain.
c) Are contamination with the two kinds of bacteria independent? Explain.

48. Birth order, finis. In Exercises 44 and 46 we looked at the birth orders and college choices of some Intro Stats students. For these students:
a) Are enrolling in Agriculture and Human Ecology disjoint? Explain.
b) Are enrolling in Agriculture and Human Ecology independent? Explain.
c) Are being firstborn and enrolling in Human Ecology disjoint? Explain.
d) Are being firstborn and enrolling in Human Ecology independent? Explain.

49. Drunks. Police often set up sobriety checkpoints—roadblocks where drivers are asked a few brief questions to allow the officer to judge whether or not the person may have been drinking. If the officer does not suspect a problem, drivers are released to go on their way. Otherwise, drivers are detained for a Breathalyzer test that will determine whether or not they will be arrested. The police say that based on the brief initial stop, trained officers can make the right decision 80% of the time. Suppose the police operate a sobriety checkpoint after 9 p.m. on a Saturday night, a time when national traffic safety experts suspect that about 12% of drivers have been drinking.
a) You are stopped at the checkpoint and, of course, have not been drinking. What's the probability that you are detained for further testing?
b) What's the probability that any given driver will be detained?

50. No-shows. An airline offers discounted "advance-purchase" fares to customers who buy tickets more than 30 days before travel and charges "regular" fares for tickets purchased during those last 30 days. The company has noticed that 60% of its customers take advantage of the advance-purchase fares. The "no-show" rate among people who paid regular fares is 30%, but only 5% of customers with advance-purchase tickets are no-shows.
a) What percent of all ticket holders are no-shows?
b) Is being a no-show independent of the type of ticket a passenger holds? Explain.

51. Dishwashers. Dan's Diner employs three dishwashers. Al washes 40% of the dishes and breaks only 1% of those he handles. Betty and Chuck each wash 30% of the dishes, and Betty breaks only 1% of hers, but Chuck breaks 3% of the dishes he washes. (He, of course, will need a new job soon. . . .) You go to Dan's for supper one night. What's the probability that you hear a dish break at the sink?

52. Parts. A company manufacturing electronic components for home entertainment systems buys electrical connectors from three suppliers. The company prefers to use supplier A because only 1% of those connectors prove to be defective, but supplier A can deliver only 70% of the connectors needed. The company must also purchase connectors from two other suppliers, 20% from supplier B and the rest from supplier C. The rates of defective connectors from B and C are 2% and 4%, respectively. You buy one of these components. What's the probability that you find that the connector is defective?

53. HIV testing. In July 2005 the journal *Annals of Internal Medicine* published a report on the reliability of HIV testing. Results of a large study suggested that among people with HIV, 99.7% of tests conducted were (correctly) positive, while for people without HIV 98.5% of the tests were (correctly) negative. A clinic serving an at-risk population offers free HIV testing, believing that 15% of the patients may actually carry HIV.
a) What's the overall probability a patient tests negative?
b) What percentage of patients testing negative are truly free of HIV?

54. Polygraphs. Lie detectors are controversial instruments, barred from use as evidence in many courts. Nonetheless, many employers use lie detector screening as part of their hiring process in the hope that they can avoid hiring people who might be dishonest. There has been some research, but no agreement, about the reliability of polygraph tests. Based on this research, suppose that a polygraph can detect 65% of lies, but incorrectly identifies 15% of true statements as lies.

A certain company believes that 95% of its job applicants are trustworthy. The company gives everyone a polygraph test, asking, "Have you ever stolen anything from your place of work?" Naturally, all the applicants answer "No," but the polygraph identifies some of those answers as lies, making the person ineligible for a job.

a) Overall, what percentage of applicants will fail the lie detector test?
b) What percentage of job applicants rejected under suspicion of dishonesty were actually trustworthy?

Answers
Just Checking

1. a)
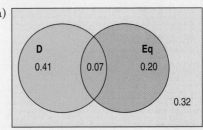

b) 0.32
c) 0.41

2. a) $8/150 = 0.053$
b) $8/60 = 0.133$
c) $8/20 = 0.4$

3. a) No. Some employees are male and supervisors.
b) $P(\text{male}) = 60/150 = 0.4$ and $P(\text{male} \mid \text{supervisor}) = 8/20 = 0.4$. Yes, 40% of employees are males, and 40% of the supervisors are males.

4. a)

		Equation		
		Yes	**No**	**Total**
Display	**Yes**	0.07	0.41	**0.48**
	No	0.20	0.32	**0.52**
	Total	**0.27**	**0.73**	**1.00**

b) $P(\mathbf{D} \mid \mathbf{Eq}) = P(\mathbf{D} \text{ and } \mathbf{Eq})/P(\mathbf{Eq})$
$= 0.07/0.27 = 0.259$
c) No, pages can (and 7% do) have both.
d) To be independent, we'd need $P(\mathbf{D} \mid \mathbf{Eq}) = P(\mathbf{D})$.
$P(\mathbf{D} \mid \mathbf{Eq}) = 0.259$, but $P(\mathbf{D}) = 0.48$. Overall, 48% of pages have data displays, but only about 26% of pages with equations do. They do not appear to be independent.

5. a) $\dfrac{4}{20} \cdot \dfrac{3}{19}$
b) $\dfrac{7}{20} \cdot \dfrac{6}{19} \cdot \dfrac{5}{18}$
c) $\dfrac{17}{20} \cdot \dfrac{16}{19} \cdot \dfrac{15}{18} \cdot \dfrac{14}{17}$
d) $\dfrac{19}{20} \cdot \dfrac{18}{19} \cdot \dfrac{17}{18} \cdot \dfrac{16}{17} \cdot \dfrac{1}{16}$

Probability Models

Insurance companies make bets. They bet that you're going to live a long life. You bet that you're going to die sooner[1]. Both you and the insurance company want the company to stay in business, so it's important to find a "fair price" for your bet. Of course, the right price for *you* depends on many factors, and nobody can predict exactly how long you'll live. But when the company averages over enough customers, it can make reasonably accurate estimates of the amount it can expect to collect on a policy before it has to pay its benefit.

Here's a simple example. An insurance company offers a "death and disability" policy that pays $10,000 when you die or $5000 if you are permanently disabled. It charges a premium of only $50 a year for this benefit. Is the company likely to make a profit selling such a plan? To answer this question, the company needs to know the *probability* that its clients will die or be disabled in any year. From actuarial information like this, the company can calculate the expected value of this policy.

---

[1]And hope you don't "win"!

Random Variables

NOTATION ALERT

The most common letters for random variables are X, Y, and Z. But be cautious: If you see any capital letter, it just might denote a random variable.

We'll want to build a probability model in order to answer the questions about the insurance company's risk. First we need to define a few terms. The amount the company pays out on an individual policy is called a **random variable** because its numeric value is based on the outcome of a random event. We use a capital letter, like X, to denote a random variable. We'll denote any particular value that it can have by x (lowercase). For the insurance company, X = the payout; x can be \$10,000 (if you die that year), \$5000 (if you are disabled), or \$0 (if neither occurs). The collection of all the possible values and the probabilities that they occur is called the **probability model** for the random variable.

Suppose, for example, that the death rate in any year is 1 out of every 1000 people, and that another 2 out of 1000 suffer some kind of disability. Then we can display the probability model for this insurance policy in a table like this:

Policyholder Outcome	Payout x	Probability $P(X = x)$
Death	10,000	$\dfrac{1}{1000}$
Disability	5000	$\dfrac{2}{1000}$
Neither	0	$\dfrac{997}{1000}$

Expected Value (Center)

NOTATION ALERT

The expected value (or mean) of a random variable is written $E(X)$ or μ.

To see what the insurance company can expect to earn from such a policy, imagine that it insures exactly 1000 people. Further imagine that, in a perfect "probability world," 1 of the policyholders dies, 2 are disabled, and the remaining 997 survive the year unscathed. The company would pay \$10,000 to one client and \$5000 to each of 2 clients, with no payout to the other 997 policy holders. That's a total of \$20,000, or an average of 20000/1000 = \$20 per policy. Since it is charging people \$50 for the policy, the company expects to make a profit of \$30 per customer. Not bad!

We can't predict what *will* happen during any given year, but we can say what we *expect* to happen. To do this, we (or, rather, the insurance company) need the probability model. The expected value of a policy is a parameter of this model. In fact, it's the mean. We'll signify this with the notation μ (for population mean) or $E(X)$ for expected value.

How did we come up with \$20 as the expected value of a policy payout? We imagined that we had exactly 1000 clients. Of those, we imagined exactly 1 died and 2 were disabled, corresponding to the probabilities. Our average payout is:

$$\mu = E(X) = \frac{10{,}000(1) + 5000(2) + 0(997)}{1000} = \$20 \text{ per policy.}$$

Instead of writing the expected value as one big fraction, we can rewrite it as separate terms with a common denominator of 1000.

$$\mu = E(X)$$
$$= \$10{,}000\left(\frac{1}{1000}\right) + \$5000\left(\frac{2}{1000}\right) + \$0\left(\frac{997}{1000}\right)$$
$$= \$20.$$

How convenient! See the probabilities? For each policy, there's a 1/1000 chance that we'll have to pay \$10,000 for a death and a 2/1000 chance that we'll have to pay \$5000 for a disability. Of course, there's a 997/1000 chance that we won't have to pay anything.

Take a good look at the expression now. It's easy to calculate the **expected value** of a (discrete) random variable—just multiply each possible value by the probability that it occurs, and find the sum:

$$\mu = E(X) = \sum xP(x).$$

Be sure that every possible outcome is included in the sum. And verify that you have a valid probability model to start with—the probabilities should each be between 0 and 1 and should sum to one.

For Example Love and Expected Values

On Valentine's Day the *Quiet Nook* restaurant offers a *Lucky Lovers Special* that could save couples money on their romantic dinners. When the waiter brings the check, he'll also bring the four aces from a deck of cards. He'll shuffle them and lay them out face down on the table. The couple will then get to turn one card over. If it's a black ace, they'll owe the full amount, but if it's the ace of hearts, the waiter will give them a \$20 Lucky Lovers discount. If they first turn over the ace of diamonds (hey—at least it's red!), they'll then get to turn over one of the remaining cards, earning a \$10 discount for finding the ace of hearts this time.

Question: Based on a probability model for the size of the Lucky Lovers discounts the restaurant will award, what's the expected discount for a couple?

Let $X =$ the Lucky Lovers discount. The probabilities of the three outcomes are:

$$P(X = 20) = P(A\heartsuit) = \frac{1}{4}$$
$$P(X = 10) = P(A\diamondsuit, \text{ then } A\heartsuit) = P(A\diamondsuit) \cdot P(A\heartsuit | A\diamondsuit)$$
$$= \frac{1}{4} \cdot \frac{1}{3} = \frac{1}{12}$$
$$P(X = 0) = P(X \neq 20 \text{ or } 10) = 1 - \left(\frac{1}{4} + \frac{1}{12}\right) = \frac{2}{3}.$$

My probability model is:

Outcome	A♥	A♦, then A♥	Black Ace
x	20	10	0
$P(X=x)$	$\frac{1}{4}$	$\frac{1}{12}$	$\frac{2}{3}$

$$E(X) = 20 \cdot \frac{1}{4} + 10 \cdot \frac{1}{12} + 0 \cdot \frac{2}{3} = \frac{70}{12} \approx 5.83$$

Couples dining at the Quiet Nook can expect an average discount of \$5.83.

Just Checking

1. One of the authors took his minivan in for repair recently because the air conditioner was cutting out intermittently. The mechanic identified the problem as dirt in a control unit. He said that in about 75% of such cases, drawing down and then recharging the coolant a couple of times cleans up the problem—and costs only $60. If that fails, then the control unit must be replaced at an additional cost of $100 for parts and $40 for labor.

   a) Define the random variable and construct the probability model.
   b) What is the expected value of the cost of this repair?
   c) What does that mean in this context?

(Oh—in case you were wondering—the $60 fix worked!)

(Check your answers on page 367.)

First Center, Now Spread . . .

Of course, this expected value (or mean) is not what actually happens to any *particular* policyholder. No individual policy actually costs the company $20. We are dealing with random events, so some policyholders receive big payouts, others nothing. Because the insurance company must anticipate this variability, it needs to know the *standard deviation* of the random variable.

For data, we calculated the **standard deviation** by first computing the deviation from the mean and squaring it. We do that with (discrete) random variables as well. First, we find the deviation of each payout from the mean (expected value):

Policyholder Outcome	Payout x	Probability $P(X = x)$	Deviation $(x - \mu)$
Death	10,000	$\dfrac{1}{1000}$	$(10{,}000 - 20) = 9980$
Disability	5000	$\dfrac{2}{1000}$	$(5000 - 20) = 4980$
Neither	0	$\dfrac{997}{1000}$	$(0 - 20) = -20$

Next, we square each deviation. The **variance** is the expected value of those squared deviations, so we multiply each by the appropriate probability and sum those products. That gives us the variance of X. Here's what it looks like:

$$Var(X) = 9980^2\left(\frac{1}{1000}\right) + 4980^2\left(\frac{2}{1000}\right) + (-20)^2\left(\frac{997}{1000}\right) = 149{,}600.$$

Finally, we take the square root to get the standard deviation:

$$SD(X) = \sqrt{149{,}600} \approx \$386.78.$$

The insurance company can expect an average payout of $20 per policy, with a standard deviation of $386.78.

Think about that. The company charges $50 for each policy and expects to pay out $20 per policy. Sounds like an easy way to make $30. In fact, most of the time (probability 997/1000) the company pockets the entire $50. But would you consider selling your neighbor such a policy? The problem is that occasionally the company loses big. With probability 1/1000, it will pay out $10,000, and with probability 2/1000, it will pay out $5000. That may be more risk than you're willing to take on. The standard deviation of $386.78 gives an indication that it's no sure thing. That's a pretty big spread (and risk) for an average profit of $30.

Here are the formulas for what we just did. Because these are parameters of our probability model, the variance and standard deviation can also be written as σ^2 and σ. You should recognize both kinds of notation.

$$\sigma^2 = Var(X) = \sum (x - \mu)^2 P(x)$$
$$\sigma = SD(X) = \sqrt{Var(X)}$$

For Example Finding the Standard Deviation

Recap: Here's the probability model for the Lucky Lovers restaurant discount.

Outcome	A♥	A♦, then A♥	Black Ace
x	20	10	0
$P(X = x)$	$\dfrac{1}{4}$	$\dfrac{1}{12}$	$\dfrac{2}{3}$

We found that couples can expect an average discount of $\mu = \$5.83$.

Question: What's the standard deviation of the discounts?

First find the variance: $Var(X) = \sum (x - \mu)^2 \cdot P(x)$

$$= (20 - 5.83)^2 \cdot \frac{1}{4} + (10 - 5.83)^2 \cdot \frac{1}{12} + (0 - 5.83)^2 \cdot \frac{2}{3}$$

$$\approx 74.306.$$

So, $SD(X) = \sqrt{74.306} \approx \$8.62$

Couples can expect the Lucky Lovers discounts to average $5.83, with a standard deviation of $8.62.

| STEP-BY-STEP EXAMPLE | **Expected Values and Standard Deviations for Discrete Random Variables** |

As the head of inventory for Knowway computer company, you were thrilled that you had managed to ship 2 computers to your biggest client the day the order arrived. You are horrified, though, to find out that someone had restocked refurbished computers in with the new computers in your storeroom. The shipped computers were selected randomly from the 15 computers in stock, but 4 of those were actually refurbished.

If your client gets 2 new computers, things are fine. If the client gets one refurbished computer, it will be sent back at your expense—$100—and you can replace it. However, if both computers are refurbished, the client will cancel the order this month and you'll lose a total of $1000.

Question: What's the expected value and the standard deviation of the company's loss?

THINK

Plan State the problem.

I want to find the company's expected loss for shipping refurbished computers and the standard deviation.

Variable Define the random variable.

Let X = amount of loss.

Plot Make a picture. This is another job for tree diagrams.

If you prefer calculation to drawing, find $P(\mathbf{NN})$ and $P(\mathbf{RR})$, then use the Complement Rule to find $P(\mathbf{NR}\ or\ \mathbf{RN})$. (The letters stand for **N**ew and **R**efurbished.)

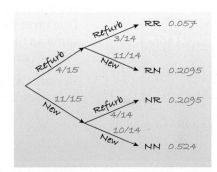

Model List the possible values of the random variable, and determine the probability model.

Outcome	x	$P(X = x)$
Two refurbs	1000	$P(RR) = 0.057$
One refurb	100	$P(NR \cup RN) = 0.2095 +$
		$0.2095 = 0.419$
New/new	0	$P(NN) = 0.524$

SHOW

Mechanics Find the expected value.

$E(X) = 0(0.524) + 100(0.419) + 1000(0.057)$
$\quad\quad = \$98.90$

Find the variance.

$Var(X) = (0 - 98.90)^2(0.524)$
$\quad\quad\quad + (100 - 98.90)^2(0.419)$
$\quad\quad\quad + (1000 - 98.90)^2(0.057)$
$\quad\quad\quad = 51,408.79$

Find the standard deviation.

$SD(X) = \sqrt{51,408.79} = \$226.735$

(continued)

Conclusion Interpret your results in context.		I expect such mistakes to cost the firm an average of $98.90, with a standard deviation of $226.74. The large standard deviation reflects the fact that there's a pretty large range of possible losses.

TI Tips — Finding the mean and SD of a random variable

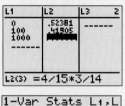

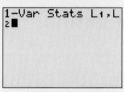

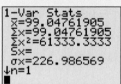

You can easily calculate means and standard deviations for a random variable with your TI. Let's do the Knowway computer example.

- Enter the values of the variable in a list, say, L1: 0, 100, 1000.
- Enter the probability model in another list, say, L2. Notice that you can enter the probabilities as fractions. For example, multiplying along the top branches of the tree gives the probability of a $1000 loss to be $\frac{4}{15} \cdot \frac{3}{14}$. When you enter that, the TI will automatically calculate the probability as a decimal!
- Under the **STAT CALC** menu, ask for 1–Var Stats L1,L2.

Now you see the mean and standard deviation (along with some other things). Don't fret that the calculator's mean and standard deviation aren't precisely the same as the ones we found; it's just being more accurate. Minor differences can arise whenever we round off probabilities to do the work by hand.

Beware: Although the calculator knows enough to call the standard deviation σ, it uses $\bar{x}$ where it should say μ. Make sure you don't make that mistake!

Just Checking

2. A large electronics supply store has sent out sale flyers proclaiming **"Everyone WINS!."** The flyers contain Secret Discount scratch-off coupons, good toward the purchase of a big-screen TV. This weekend only, when you **"Buy a New HDTV,"** you'll also **"Get a FREE Blu-ray Player!!!"** plus an extra cash discount of **"Up to $500 OFF!!!"** to be revealed at checkout time. Other prizes include discounts of $250 or $100, or $10 off the purchase of any movie of your choice. In the fine print at the bottom of the flyer the company reveals:

We have distributed 10,000 Secret Discount coupons. Of these, 5 award $500 discounts, 50 award $250 discounts, and 200 award $100 discounts off the purchase price of a qualifying HDTV. All other coupons have a

value of $10 when applied to the purchase of a Blu-ray movie currently in stock, when purchased in conjunction with a qualifying HDTV. Coupons are good 10 a.m. Friday through 6 p.m. Sunday. Store employees and their families are not eligible to participate.

a) Create a probability model for the random variable X = coupon value.
b) Use technology to find the expected value and the standard deviation.
c) Explain what the expected value and the standard deviation mean in this context.

(Check your answers on page 367.)

Useful Probability Models

Suppose a cereal manufacturer puts pictures of famous athletes on cards in boxes of cereal, in the hope of increasing sales. The manufacturer announces that 20% of the boxes contain a picture of LeBron James, 30% a picture of David Beckham, and the rest a picture of Serena Williams.

Sound familiar? In Chapter 12 we used simulation to find the number of boxes we'd need to open to get one of each card. That's a fairly complex question and one well suited for simulation. But many other questions can be answered more directly by using simple probability models.

Where do these models come from? Well, some are do-it-yourself models, created to fit the situation at hand. We just did that for the insurance company, the Lucky Lovers, and the HDTV coupons, for example. Other times, though, we don't need to custom-build new models. Often we can rely on some very powerful probability models that are useful in a wide variety of situations. You've already learned about one of them, back in Chapter 6: the Normal model. More about that model soon. Right now, let's see how another frequently useful model—the Binomial—can help us think about that LeBron James card.

Searching for LeBron: Bernoulli Trials

Daniel Bernoulli (1700–1782) was the nephew of Jacob, whom you saw in Chapter 13. He was the first to work out the mathematics for what we now call Bernoulli trials.

Suppose you're a huge LeBron James fan. You don't care about completing the whole sports card collection, but you've just *got* to have LeBron's picture. If you buy 5 boxes of cereal, what's the probability you'll get one? Or two? There's a probability model for questions like these.

We'll keep the assumption that pictures are distributed at random and we'll trust the manufacturer's claim that 20% of the cards are LeBron. So, when you open the box, the probability that you succeed in finding LeBron is 0.20. Now we'll call the act of opening *each* box a trial, and note that:

- There are only two possible outcomes (called *success* and *failure*) on each trial. Either you get LeBron's picture (success), or you don't (failure).
- In advance, the probability of success, denoted p, is the same on every trial. Here $p = 0.20$ for each box.
- As we proceed, the trials are independent. Finding LeBron in the first box does not change what might happen when you reach for the next box.

Situations like this occur often, and are called **Bernoulli trials.** Common examples of Bernoulli trials include tossing a coin, looking for defective products rolling off an assembly line, or even shooting free throws in a basketball game. We can use Bernoulli trials to build a useful probability model.

For Example   Spam and Bernoulli Trials

Postini is a global company specializing in communications security. The company monitors over 1 billion Internet messages per day and recently reported that 91% of e-mails are spam!

Let's assume that your e-mail is typical—91% spam. We'll also assume you aren't using a spam filter, so every message gets dumped in your inbox.

Question: Overnight your inbox collects e-mail. When you first check your e-mail in the morning, are your messages Bernoulli trials?

When I check my emails one-by-one:

✔ There are two possible outcomes each time: a real message (success) or spam (failure).
✔ Since 91% of all emails are spam, the probability of success is

$$p = 1 - 0.91 = 0.09.$$

✔ My messages arrive in random order from many different sources, so I think they are independent.

Yes, I can consider my email messages to be Bernoulli trials.

Just Checking

3. Think about each of these situations. Are these random variables based on Bernoulli trials? If you don't think so, explain why not.

a) The waitstaff at a small restaurant consists of 5 males and 8 females. They write their names on slips of paper and the boss chooses 4 people at random to work overtime on a holiday weekend. We count the number of females who are chosen.

b) In the United States about 1 in every 90 pregnant women gives birth to twins. We count the number of twins born to a group of pregnant women who work in the same office.

c) We count the number of times a woman who has been pregnant 3 times gave birth to twins.

d) We pick 40 M&M's at random from a large bag, counting how many of each color we get.

e) The last census found that 26% of all businesses in the United States were owned by women. Suppose that's true in your town. You call 15 businesses randomly chosen from the Yellow Pages, counting the number owned by women.

(Check your answers on page 367.)

Binomial Probabilties

NOTATION ALERT

Now we have two more reserved letters. Whenever we deal with Bernoulli trials, *p* represents the probability of success, and *q* the probability of failure. (Of course, $q = 1 - p$.)

Ok, back to LeBron James. Suppose you buy 5 boxes of cereal. What's the probability you get *exactly* 2 pictures of LeBron? We're talking about Bernoulli trials, and we're asking about the *number of successes* in the 5 trials. We'll let random variable X = number of successes. We want to find $P(X = 2)$. This is an example of a **Binomial probability.** It takes two parameters to define this **Binomial model:** the number of trials, n, and the probability of success, p. We denote this model Binom(n, p). Here, $n = 5$ trials, and $p = 0.2$, the probability of finding a LeBron James card in any trial.

In Binomial models we use q to represent the probability of failure. Here $q = 1 - 0.2 = 0.8$. Exactly 2 successes in 5 trials means 2 successes and 3 failures. It seems logical that the probability should be $(0.2)^2(0.8)^3$. Too bad! It's not that easy. That calculation would give you the probability of finding LeBron in the first 2 boxes and not in the next 3—*in that order*. But you could

COMBINATIONS
Remember? The number of ways to choose r objects from a group of n is
$$_nC_r = \frac{n!}{r!(n-r)!}$$

find LeBron in the third and fifth boxes and still have 2 successes. The probability of those outcomes in that particular order is $(0.8)(0.8)(0.2)(0.8)(0.2)$. That's also $(0.2)^2(0.8)^3$. In fact, the probability will always be $(0.2)^2(0.8)^3$, no matter what order the successes and failures occur in. We only need to count how many different ways we could get 2 successes in 5 trials. The good news is that we learned just the right counting method back in Chapter 13: combinations. The number of different ways 2 cereal boxes out of 5 could have LeBron pictures is:

$$_5C_2 = \frac{5!}{2!(5-2)!} = \frac{5 \times 4 \times 3 \times 2 \times 1}{2 \times 1 \times 3 \times 2 \times 1} = \frac{5 \times 4}{2 \times 1} = 10.$$

So there are 10 ways to get 2 successes in 5 trials, and the probability of each is $(0.2)^2(0.8)^3$. Now we can find what we wanted:

$$P(\#\text{success} = 2) = 10(0.2)^2(0.8)^3 = 0.2048$$

In general, the probability of exactly k successes in n trials is $_nC_k\, p^k q^{n-k}$.

For Example   Spam and Binomial Probability

Recap: The communications monitoring company *Postini* has reported that 91% of e-mail messages are spam. Suppose your unfiltered inbox contains 25 messages.

Questions: What's the probability that you'll find only 1 or 2 real messages?

I assume that messages arrive independently and at random, with the probability of success (a real message) $p = 1 - 0.91 = 0.09$. Let $X =$ the number of real messages among 25. I can use the model Binom(25, 0.09).

$$P(X = 1 \text{ or } 2) = P(X = 1) + P(X = 2)$$
$$= {}_{25}C_1(0.09)^1(0.91)^{24} + {}_{25}C_2(0.09)^2(0.91)^{23}$$
$$= 0.2340 + 0.2777$$
$$= 0.5117$$

There's just over a 50% chance that 1 or 2 of my 25 e-mails will be real messages.

The Binomial Model

Okay, we can find binomial probabilities. To complete our understanding of the Binomial model, we just need to know the mean and standard deviation. What's the expected value? You already know intuitively. If we have 5 boxes, and LeBron's picture is in 20% of them, then we would expect to have $5(0.2) = 1$ success. If we had 100 trials with probability of success 0.2, how many successes would you expect? Can you think of any reason not to say 20? It seems so simple that most people wouldn't even stop to think about it. You just multiply the probability of success by n. In other words, $E(X) = np$.

The standard deviation is less obvious; you can't just rely on your intuition. Fortunately, the formula for the standard deviation also boils down to something simple: $SD(X) = \sqrt{npq}$. In 100 boxes of cereal, we expect to find

20 LeBron James cards, with a standard deviation of $\sqrt{100 \times 0.8 \times 0.2} = 4$ pictures.

It's time to summarize. A Binomial probability model describes the number of successes in a specified number of Bernoulli trials. It takes two parameters to specify this model: the number of trials n and the probability of success p.

BINOMIAL PROBABILITY MODEL FOR BERNOULLI TRIALS: Binom(n, p)

n = number of trials
p = probability of success (and $q = 1 - p$ = probability of failure)
x = number of successes in n trials

$$P(x) = {}_nC_x\, p^x q^{n-x}, where\ {}_nC_x = \frac{n!}{x!(n-x)!}$$

Mean: $\mu = np$
Standard Deviation: $\sigma = \sqrt{npq}$

For Example Spam and the Binomial Model

Recap: The communications monitoring company *Postini* has reported that 91% of e-mail messages are spam. Your inbox contains 25 messages.

Questions: What are the mean and standard deviation of the number of real messages you should expect to find in your inbox?

I assume that messages arrive independently and at random, with the probability of success (a real message) $p = 1 - 0.91 = 0.09$. Let $X =$ the number of real messages among 25. I can use the model Binom(25, 0.09).

$$E(X) = np = 25(0.09) = 2.25$$
$$SD(X) = \sqrt{npq} = \sqrt{25(0.09)(0.91)} = 1.43$$

Among 25 e-mail messages, I expect to find an average of 2.25 that aren't spam, with a standard deviation of 1.43 messages.

Independence

One of the important requirements for Bernoulli trials is that the trials be independent. Sometimes that's a reasonable assumption—when tossing a coin or rolling a die, for example. But that becomes a problem when (often!) we're looking at situations involving samples chosen without replacement. We said that whether we find a LeBron James card in one box has no effect on the probabilities in other boxes. This is *almost* true. Technically, if exactly 20% of the boxes have LeBron James cards, then when you find one, you've reduced the number of remaining LeBron James cards. If you knew there were 2

LeBron James cards hiding in the 10 boxes of cereal on the market shelf, then finding one in the first box you try would clearly change your chances of finding LeBron in the next box. With a few million boxes of cereal, though, the difference is hardly worth mentioning.

Selecting without replacement causes the probabilities to change, making the trials not independent. Obviously, taking 2 out of 10 boxes changes the probability. Taking even a few hundred out of millions, though, makes very little difference. Fortunately, we have a rule of thumb for the in-between cases. It turns out that if we look at less than 10% of the population, we can pretend that the trials are independent and calculate binomial probabilities that are quite close to the true probabilities.

The 10% Condition: Bernoulli trials must be independent. If that assumption is violated, it is still okay to proceed as long as the sample is smaller than 10% of the population.

STEP-BY-STEP EXAMPLE   **Working with a Binomial Model**

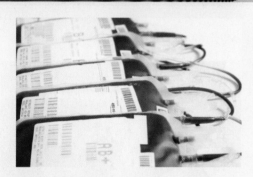

People with O-negative blood are called "universal donors" because O-negative blood can be given to anyone else, regardless of the recipient's blood type. Only about 6% of people have O-negative blood.

Questions:
1. What are the mean and standard deviation of the number of universal donors among them?
2. What is the probability that there are 2 or 3 universal donors?

THINK **Plan** State the question.	I want to know the mean and standard deviation of the number of universal donors among 20 people, and the probability that there are 2 or 3 of them.

Check to see that these are Bernoulli trials.	✔ There are two outcomes:
	    success = O-negative
	    failure = other blood types
	
	✔ $p = 0.06$, because people have lined up at random.
	
	✔ **10% Condition:** Trials are not independent, because the population is finite, but fewer than 10% of all possible donors are lined up.

Variable Define the random variable.	Let X = number of O-negative donors among $n = 20$ people.
Model Specify the model.	I can model X with Binom(20, 0.06).

(continued)

SHOW	**Mechanics** Find the expected value and standard deviation. Calculate the probabilities of 2 or 3 successes using the Binomial formula: $$P(x) = {}_nC_x\, p^x q^{n-x}$$	$E(X) = np = 20(0.06) = 1.2$ $SD(X) = \sqrt{npq} = \sqrt{20(0.06)(0.94)} \approx 1.06$ $P(X = 2 \text{ or } 3) = P(X = 2) + P(X = 3)$ $= {}_{20}C_2(0.06)^2(0.94)^{18}$ $\quad + {}_{20}C_3(0.06)^3(0.94)^{17}$ $\approx 0.2246 + 0.0860$ $= 0.3106$
TELL	**Conclusion** Interpret your results in context.	In groups of 20 randomly selected blood donors, I expect to find an average of 1.2 universal donors, with a standard deviation of 1.06. About 31% of the time, I'd find 2 or 3 universal donors among the 20 people.

Just Checking

4. The Pew Research Center reports that they are only able to contact 76% of randomly selected households drawn for telephone surveys. Suppose a pollster has a list of 12 calls to make. Let X = the number of people successfully contacted.

   a) Specify the Binomial model Binom(n, p).
   b) What's q, the probability of failure?
   c) Find the expected number of successful calls.
   d) Find the standard deviation of the number of successful calls.
   e) Find the probability that exactly 9 of the calls are successful.

(Check your answers on page 367.)

TI Tips   Finding binomial probabilities

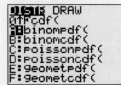

Your TI knows the Binomial model. Just as you saw back in Chapter 6 with the Normal model, commands to calculate probability distributions are found in the **2ⁿᵈ DISTR** menu. Have a look. After many other options (don't drop the course yet!) you'll see two Binomial probability functions.

- **binompdf(**

  The "pdf" stands for "probability density function." That's a fancy way of telling you to use this when you want to find the probability of an *individual* outcome. You need to define the Binomial model by specifying n and p, and then indicate the desired number of successes, x. The format is **binompdf(n,p,x)**.

`binompdf(5,.2,2)`

`                .2048`

For example, recall that LeBron James' picture is in 20% of the cereal boxes. Suppose that we want to know the probability of finding LeBron exactly twice among 5 boxes of cereal. We use $n = 5, p = 0.2$, and $x = 2$, entering the command `binompdf(5,.2,2)`. There's about a 20% chance of getting two pictures of LeBron James in 5 boxes of cereal.

- **binomcdf(**

  Need to add several Binomial probabilities? The "cdf" stands for "cumulative density function," meaning that it finds the sum of the probabilities of a sequence of possible outcomes. To find the total probability of getting x or fewer successes among the n trials use the cumulative Binomial density function `binomcdf(n,p,x)`.

`binomcdf(10,.2,4`
`)`
`     .9672065025`

For example, suppose we have 10 boxes of cereal and wonder about the probability of finding up to 4 pictures of LeBron. That's the probability of 0, 1, 2, 3, or 4 successes, so we specify the command `binomcdf(10,.2,4)`. Pretty likely!

`1-binomcdf(10,.2`
`,3)`
`     .1208738816`

Of course "up to 4" allows for the possibility that we end up with none. What's the probability we get at least 4 pictures of LeBron in 10 boxes? Well, "at least 4" means "not 3 or fewer." That's the complement of 0, 1, 2, or 3 successes. Have your TI evaluate `1-binomcdf(10,.2,3)`. There's about a 12% chance we'll find at least 4 pictures of LeBron in 10 boxes of cereal.

Just Checking

5. Back to The Pew Research Center again. Remember that they say they are able to make phone contact with 76% of randomly selected households. Our pollster has a list of 12 calls to make.

Use TI calculator commands to find the probability that:

a) exactly 9 calls are successful.
b) at most 9 calls are successful.
c) at least 9 calls are successful.

(Check your answers on page 367.)

The Normal Model to the Rescue!

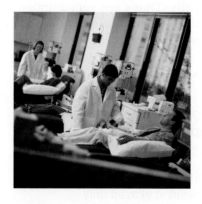

The Tennessee Red Cross anticipates the need for at least 1850 units of O-negative blood this year. It estimates that it will collect blood from 32,000 donors. Recalling that only 6% of people have Type O blood, how great is the risk that the Tennessee Red Cross will fall short of meeting its need? We've just learned how to calculate such probabilities. The Good News: We can use the Binomial model with $n = 32,000$ and $p = 0.06$. The probability of getting *exactly* 1850 units of O-negative blood from 32,000 donors is $_{3200}C_{1850} \times 0.06^{1850} \times 0.94^{30150}$. The Bad News: No calculator on earth can calculate that first factor (it has more than 100,000 digits).[2] And that's just the beginning. The problem said *at least* 1850, so we have to do it again for 1851, for 1852, and all the way up to 32,000. No thanks.

---

[2]If your calculator *can* find Binom(32000,0.06), then it's smart enough to use an approximation. Read on to see how you can, too.

When we're dealing with a large number of trials like this, making direct calculations of the probabilities becomes nasty (or outright impossible). Here an old friend—the Normal model—comes to the rescue.

The Binomial model has mean $np = 1920$ and standard deviation $\sqrt{npq} \approx 42.48$. We could try approximating its distribution with a Normal model, using the same mean and standard deviation. Remarkably enough, that turns out to be a very good approximation. We can estimate the *probability*:

$$P(X < 1850) = P\left(z < \frac{1850 - 1920}{42.48}\right) \approx P(z < -1.65) \approx 0.05$$

There seems to be about a 5% chance that this Red Cross chapter will run short of O-negative blood.

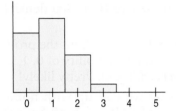

Can we always use a Normal model to make estimates of Binomial probabilities? No. Consider the LeBron James situation—pictures in 20% of the cereal boxes. If we buy five boxes, the actual Binomial probabilities that we get 0, 1, 2, 3, 4, or 5 pictures of LeBron are 33%, 41%, 20%, 5%, 1%, and 0.03%, respectively. The first histogram shows that this probability model is skewed. That makes it clear that we should not try to estimate these probabilities by using a Normal model.

Now suppose we open 50 boxes of this cereal and count the number of LeBron James pictures we find. The second histogram shows this probability model. It is centered at $np = 50(0.2) = 10$ pictures, as expected, and it appears to be fairly symmetric around that center. Let's have a closer look.

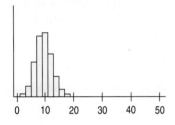

The third histogram again shows Binom(50, 0.2), this time magnified somewhat and centered at the expected value of 10 pictures of LeBron. It looks close to Normal, for sure. With this larger sample size, it appears that a Normal model might be a useful approximation.

A Normal model, then, is a close enough approximation only for a large enough number of trials. And what we mean by "large enough" depends on the probability of success. We'd need a larger sample if the probability of success were very low (or very high). It turns out that a Normal model works pretty well if we expect to see at least 10 successes and 10 failures. That is, we check the **Success/Failure Condition.**

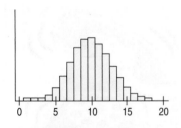

The Success/Failure Condition: A Binomial model is approximately Normal if we expect at least 10 successes and 10 failures:

$$np \geq 10 \text{ and } nq \geq 10.$$

Why 10? Remember how Normal models work. The problem is that a Normal model extends infinitely in both directions. But a Binomial model must have between 0 and n successes, so if we use a Normal to approximate a Binomial, we have to cut off its tails. That's not very important if what we cut off is the tiny area more than three standard deviations from the mean. It turns out that's what happens when n is large enough that np and nq are at least 10.

How close to Normal? How well does a Normal curve fit a binomial model? Check out the Success/Failure Condition for yourself.

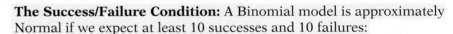

For Example   **Spam and the Normal Approximation to the Binomial**

Recap: The communications monitoring company *Postini* has reported that 91% of e-mail messages are spam. Recently, you installed a spam filter. You observe that over the past week it okayed only 151 of 1422 e-mails you received, classifying the rest as junk. Should you worry that the filtering is too aggressive?

Question: What's the probability that no more than 151 of 1422 e-mails is a real message?

I assume that messages arrive randomly and independently, with a probability of success (a real message) $p = 0.09$. The model Binom(1422, 0.09) applies, but will be hard to work with. Checking conditions for the Normal approximation, I see that:

✔ These messages represent less than 10% of all e-mail traffic.
✔ I expect $np = (1422)(0.09) = 127.98$ real messages and $nq = (1422)(0.91) = 1294.02$ spam messages, both far greater than 10.

It's okay to approximate this binomial probability by using a Normal model.

$$\mu = np = 1422(0.09) = 127.98$$
$$\sigma = \sqrt{npq} = \sqrt{1422(0.09)(0.91)} \approx 10.79$$
$$P(X \le 151) = P\left(z \le \frac{151 - 127.98}{10.79}\right)$$
$$= P(z \le 2.13)$$
$$= 0.9834$$

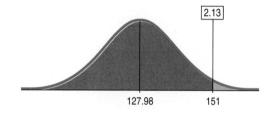

Among my 1422 e-mails, there's over a 98% chance that no more than 151 of them were real messages, so the filter may be working properly.

Just Checking

6. Let's think about the Pew Research pollsters one more time. They tell us they are successful in contacting 76% of the households randomly selected for telephone surveys. When surveying public opinion, they hope to poll at least 1000 adults. Suppose Pew has compiled a list of 1300 phone numbers to call. What's the probability that they'll reach enough people?

   a) Despite the fact that pollsters sample people without replacement, can we think of these calls as independent trials? Check the 10% Condition.
   b) What are the parameters (n and p) for this Binomial model?
   c) Find the mean and standard deviation of $X =$ the number of adults Pew may successfully contact.
   d) We want to find $P(X \ge 1000)$. Can we use a Normal model to approximate this Binomial probability? Check the Success/Failure Condition.
   e) Find the z-score that represents 1000 successful contacts in 1300 phone calls.
   f) Use your z-score to approximate the probability Pew is able to contact at least 1000 voters on their list.

(Check your answers on page 367.)

Should I Be Surprised? A First Look at Statistical Significance

You watch a friend toss a coin 100 times and get 67 heads. That's more heads than you'd expect, but is it enough more that you should think she might be cheating somehow? You probably wouldn't consider 52 or 53 heads instead of a "perfect" 50 to be unusual, but if she tossed heads 90 times out of 100 you'd be really suspicious. How about 67? After all, random outcomes will vary, sometimes ending up higher or lower than expected. Is 67 heads too strange to be explained away as just random chance?

Back in Chapter 11 we started thinking about *statistical significance*. The results of an experiment or a sample are said to be **statistically significant** if it's not reasonable to believe they occurred just by chance. If a new pain reliever cures headaches for 66% of the subjects in an experiment, but the old one worked for 63%, the apparent increase may just have been chance variation among this particular group of subjects. It would take a larger increase in the success rate in order to convince us that the new medication really is better. But how much larger? That's a question we can now begin to answer.

Let's think about your friend's 67 heads in 100 tosses. Coin tosses are Bernoulli trials; here there are $n = 100$ trials with probability of success $p = 0.5$. We can model the random variable $X =$ number of heads with Binom(100,0.5). For our model, the mean is $np = 100(0.5) = 50$. OK, on average we expect 50 heads (duh!), but we know it won't be *exactly* 50 every time. The standard deviation is $\sqrt{npq} = \sqrt{100(0.5)(0.5)} = 5$ heads, and that's our clue about how much variation is reasonable.

Add one more key insight and we're ready to go: since we expect more than 10 successes (50) and more than 10 failures (nq is also 50), a Normal model is useful here. Her 67 heads is 17 more than we expected. Because the SD = 5, we know her results are over 3 standard deviations above the mean. (To be exact, $z = \dfrac{67 - 50}{5} = 3.4$) Remember the 68-95-99.7 Rule?[3] More than 99.7% of the time, the result should be within 3 standard deviations of the mean, but hers isn't. If her coin-tossing method is fair, this would be an exceedingly rare outcome. Such an unusual result is statistically significant—friend or not, we should be very suspicious.

This is a real breakthrough! (Drumroll, please!) For the first time we've been able to decide whether what we've observed is just a chance occurrence or is strong evidence that something unusual is afoot. We'll explore this kind of reasoning in greater detail in the chapters ahead. For now it's enough to recognize that when a Normal model is useful,[4] outcomes more than 2 standard deviations from the expected value should be considered surprising.

STEP-BY-STEP EXAMPLE  Looking for Statistical Significance

Before a blood drive, a local Red Cross agency puts out a plea for universal donors, hoping that they'll get more than the usual 6% among the donors who show up. That day they collected 202 units of blood, and among them 17 units were Type O-negative.

Question: Does this suggest that making a public plea is an effective way to get more O-negative donors to come to blood drives?

---

[3]No? Well, it was a long time ago. Take a quick peek at page 120.
[4]*Always* check the conditions to be sure!

Plan State the question.

I expect 6% of all blood donors to be O-negative. I want to decide whether getting 17 O-negative donors among 202 people is statistically significant evidence that the Red Cross's public plea may have worked.

Variable Define the random variable.

X = number of O-negative donors

Check the conditions We've already confirmed that these are Bernoulli trials (p. 353), but it's critical to be sure that a Normal model applies.

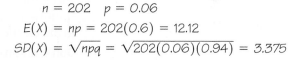

✔ **10% Condition:** 202 < 10% of all possible donors.

✔ **Success/Failure Condition:** Among 202 donors with $P = 0.06$ I expect:
$$np = (202)(0.06) = 12.12 \text{ successes,}$$
$$\text{and}$$
$$nq = (202)(0.94) = 189.88 \text{ failures.}$$
Both are at least 10.

Model Name your model.

OK to use a Normal model.

Mechanics Find the mean and standard deviation.

$$n = 202 \quad p = 0.06$$
$$E(X) = np = 202(0.6) = 12.12$$
$$SD(X) = \sqrt{npq} = \sqrt{202(0.06)(0.94)} = 3.375$$

Find the z-score for the observed result.

$$z = \frac{17 - 12.12}{3.375} = 1.45$$

Use the 68-95-99.7 Rule to think about whether that z-score seems unusual. We shouldn't be surprised unless the outcome is more than 2 standard deviations above or below the mean.

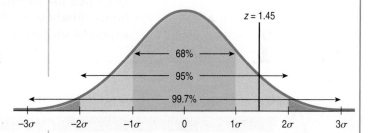

This doesn't look unusual; it's within 2 standard deviations of the mean.

Conclusion Explain (in context, of course) whether or not you consider the outcome to be statistically significant.

Although it was a good turnout, getting 17 Type O-negative donors among 202 people is only about 1.5 standard deviations more than expected. This could have been just random chance, so it's not strong evidence that the Red Cross's public plea raised the number of universal donors who came to the blood drive.

WHAT CAN GO WRONG?

- **Probability models are still just models.** Models can be useful, but they are not reality. Think about the assumptions behind your models. Are your dice really perfectly fair? (They are probably pretty close.) But when you hear that the probability of a nuclear accident is 1/10,000,000 per year, is that likely to be a precise value? Question probabilities as you would data.

- **If the model is wrong, so is everything else.** Before you try to find the mean or standard deviation of a random variable, check to make sure the probability model is reasonable. As a start, the probabilities in your model should add up to 1. If not, you may have calculated a probability incorrectly or left out a value of the random variable. For instance, in the insurance example, the description mentions only death and disability. Good health is by far the most likely outcome, not to mention the best for both you and the insurance company (who gets to keep your money). Don't overlook that.

- **Be sure you have Bernoulli trials before using a Binomial model.** Always check the requirements first: two possible outcomes per trial ("success" and "failure"), a constant probability of success, and independence. Remember to check the 10% Condition when sampling without replacement.

- **Don't assume everything's Normal.** Just because you happen to know a mean and standard deviation doesn't mean that a Normal model will be useful. You must *Think* about whether the **Normality Assumption** is justified. Using a Normal model when it really does not apply will lead to wrong answers and misleading conclusions.

- **Don't use the Normal approximation with small *n*.** To use a Normal approximation in place of a Binomial model, there must be at least 10 expected successes and 10 expected failures.

PROBABILITY AND SIGNIFICANCE **IN YOUR WORLD**

The Devil Is in the Digits

By Bernd Beber and Alexandra Scacco
(Washington Post Saturday, June 20, 2009)

Since the declaration of Mahmoud Ahmadinejad's landslide victory in Iran's presidential election, accusations of fraud have swelled. ... [E]xperts ... speculate that the election results released by Iran's Ministry of the Interior had been altered behind closed doors. We can use statistics ... systematically to show that this is likely what happened. Here's how.

We'll concentrate on vote counts—the number of votes received by different candidates in different provinces—and in particular the last and second-to-last digits of these numbers. For example, if a candidate received 14,579 votes in a province, we'll focus on digits 7 and 9.

This may seem strange, because these digits usually don't change who wins. In fact, last digits in a fair election don't tell us anything about the candidates, the make-up of the electorate or the context of the election. They are random noise in the sense that a fair vote count is as likely to end in 1 as it is to end in 2, 3, 4, or any other numeral. But that's exactly why they can serve as a litmus test for election fraud. For example, an election in which a majority of provincial vote counts ended in 5 would surely raise red flags.

Why would fraudulent numbers look any different? The reason is that humans are bad at making up numbers. Cognitive psychologists have found that study participants in lab experiments asked to write sequences of random digits will tend to select some digits more frequently than others.

So what can we make of Iran's election results? The numbers look suspicious. We find too many 7s and not enough 5s in the last digit. We expect each digit (0, 1, 2, and so on) to appear at the end of 10 percent of the vote counts. But in Iran's provincial results, the digit 7 appears 17 percent of the time, and only 4 percent of the results end in the number 5. Two

such departures from the average—a spike of 17 percent or more in one digit and a drop to 4 percent or less in another—are extremely unlikely. Fewer than four in a hundred non-fraudulent elections would produce such numbers.

But that's not all. Psychologists have also found that humans have trouble generating non-adjacent digits (such as 64 or 17, as opposed to 23) as frequently as one would expect in a sequence of random numbers. To check for deviations of this type, we examined the pairs of last and second-to-last digits in Iran's vote counts. On average, if the results had not been manipulated, 70 percent of these pairs should consist of distinct, non-adjacent digits.

Not so in the data from Iran: Only 62 percent of the pairs contain non-adjacent digits. This may not sound so different from 70 percent, but the probability that a fair election would produce a difference this large is less than 4.2 percent.

Each of these two tests provides strong evidence that the numbers released by Iran's Ministry of the Interior were manipulated. But taken together, they leave very little room for reasonable doubt. The probability that a fair election would produce both too few non-adjacent digits and the suspicious deviations in last-digit frequencies described earlier is less than 0.005. In other words, a bet that the numbers are clean is a one in two-hundred long shot.

http://www.washingtonpost.com/wp-dyn/content/article/2009/06/20/AR2009062000004.html

WHAT HAVE WE LEARNED?

We've learned to work with random variables. We can use the probability model for a random variable to find its expected value and its standard deviation.

▶ We've learned that Bernoulli trials show up in lots of places.

▶ We've learned how to use a Binomial model when we're interested in probabilities for the number of successes in a certain number of Bernoulli trials.

▶ We've learned how to use a Normal model to approximate a Binomial model when we expect at least 10 successes and 10 failures.

(continued)

Terms

Random variable
A random variable assumes any of several different numeric values as a result of some random event. Random variables are denoted by a capital letter such as X.

Probability model
The probability model is a function that associates a probability P with each value of a random variable X.

Expected value
The expected value of a random variable is its theoretical long-run average value, the center of its model. Denoted μ or $E(X)$, it is found by summing the products of variable values and probabilities:
$$\mu = E(X) = \sum xP(x).$$

Variance
The variance of a random variable is the expected value of the squared deviation from the mean:
$$\sigma^2 = Var(X) = \sum (x - \mu)^2 P(x).$$

Standard deviation
The standard deviation of a random variable describes the spread in the model, and is the square root of the variance:
$$\sigma = SD(X) = \sqrt{Var(X)}.$$

Bernoulli trials, if . . .
1. there are two possible outcomes.
2. the probability of success is constant.
3. the trials are independent.

Binomial probability model
A Binomial model is appropriate for a random variable that counts the number of successes in a fixed number of Bernoulli trials. If the probability of success is p, then the probability of x successes in n trials is
$$p(x) = {}_nC_x(p)^x(q)^{n-x}$$

10% Condition
When sampling without replacement, trials are not independent. It's still okay to proceed as long as the sample is smaller than 10% of the population.

Success/Failure Condition
For a Normal model to be a good approximation of a Binomial model, we must expect at least 10 successes and 10 failures. That is, $np \geq 10$ and $nq \geq 10$.

Skills

▶ Be able to recognize random variables.

▶ Be able to find the probability model for a random variable.

▶ Know how to find the mean (expected value) and the variance of a random variable.

▶ Be able to interpret the meaning of the expected value and standard deviation of a random variable in the proper context.

▶ Always use the proper notation for these population parameters: μ or $E(X)$ for the mean, and σ or $SD(X)$ when discussing variability.

▶ Know how to tell if a situation involves Bernoulli trials.

▶ Be able to choose when to use a Binomial model for a random variable involving Bernoulli trials.

▶ Know how to find the mean and standard deviation of a Binomial model.

▶ Be able to calculate Binomial probabilities, perhaps approximating with a Normal model.

THE BINOMIAL MODEL ON THE COMPUTER

Most statistics packages offer functions that compute Binomial probabilities. Some technology solutions automatically use the Normal approximation for the Binomial when the exact calculations become unmanageable.

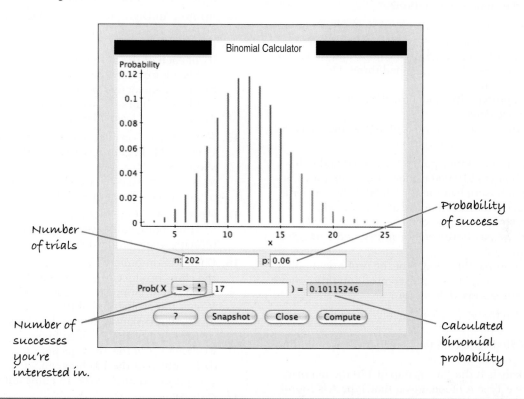

Number of trials

Probability of success

Number of successes you're interested in.

Calculated binomial probability

EXERCISES

1. Expected value. Find the expected value of each random variable:

a)

x	10	20	30
P(X = x)	0.3	0.5	0.2

b)

x	2	4	6	8
P(X = x)	0.3	0.4	0.2	0.1

2. Expected value. Find the expected value of each random variable:

a)

x	0	1	2
P(X = x)	0.2	0.4	0.4

b)

x	100	200	300	400
P(X = x)	0.1	0.2	0.5	0.2

3. Variation 1. Find the standard deviations of the random variables in Exercise 1.

4. Variation 2. Find the standard deviations of the random variables in Exercise 2.

5. Repairs. The following probability model describes the number of repair calls that an appliance repair shop may receive during an hour.

Repair Calls	0	1	2	3
Probability	0.1	0.3	0.4	0.2

a) How many calls should the shop expect per hour?
b) What is the standard deviation?

6. Red lights. A commuter must pass through five traffic lights on her way to work and will have to stop at each one that is red. She estimates the probability model for the number of red lights she hits, as shown below.

X = # of red	0	1	2	3	4	5
P(X = x)	0.05	0.25	0.35	0.15	0.15	0.05

a) How many red lights should she expect to hit each day?
b) What's the standard deviation?

7. **Defects.** A consumer organization inspecting new cars found that many had appearance defects (dents, scratches, paint chips, etc.). While none had more than three of these defects, 7% had three, 11% two, and 21% one defect.
a) Create a probability model for the number of appearance defects.
b) What's the expected number of appearance defects?
c) What's the standard deviation?

8. **Insurance.** An insurance policy costs $100 and will pay policyholders $10,000 if they suffer a major injury (resulting in hospitalization) or $3000 if they suffer a minor injury (resulting in lost time from work). The company estimates that each year 1 in every 2000 policyholders may have a major injury, and 1 in 500 a minor injury only.
a) Create a probability model for the profit on a policy.
b) What's the company's expected profit on this policy?
c) What's the standard deviation?

9. **Bernoulli.** Do these situations involve Bernoulli trials? Explain.
a) We roll 50 dice to find the distribution of the number of spots on the faces.
b) How likely is it that in a group of 120 the majority may have Type A blood, given that Type A is found in 43% of the population?
c) We deal 7 cards from a deck and get all hearts. How likely is that?
d) We wish to predict the outcome of a vote on the school budget, and poll 500 of the 3000 likely voters to see how many favor the proposed budget.
e) A company realizes that about 10% of its packages are not being sealed properly. In a case of 24, is it likely that more than 3 are unsealed?

10. **Bernoulli 2.** Do these situations involve Bernoulli trials? Explain.
a) You are rolling 5 dice and need to get at least two 6's to win the game.
b) We record the distribution of eye colors found in a group of 500 people.
c) A manufacturer recalls a doll because about 3% have buttons that are not properly attached. Customers return 37 of these dolls to the local toy store. Is the manufacturer likely to find any dangerous buttons?
d) A city council of 11 Republicans and 8 Democrats picks a committee of 4 at random. What's the probability they choose all Democrats?
e) A 2002 Rutgers University study found that 74% of high school students have cheated on a test at least once. Your local high school principal conducts a survey in homerooms and gets responses that admit to cheating from 322 of the 481 students.

11. **Hoops.** A basketball player who has made 70% of his foul shots during the season gets to take 5 shots in the first playoff game. Assuming the shots are independent, what's the probability he makes
a) exactly 3 of the 5 shots?
b) exactly 4?
c) all 5?
d) at least 3 of his 5 shots?

12. **Arrows.** An Olympic archer is able to hit the bull's-eye 80% of the time. Assume each shot is independent of the others. If she shoots 6 arrows, what's the probability of each of the following results?
a) She gets exactly 4 bull's-eyes.
b) She gets exactly 5 bull's-eyes.
c) She gets bull's-eyes on all 6 shots.
d) She gets at least 4 bull's-eyes.

13. **More hoops.** Consider our basketball player from Exercise 11. In games where he gets to take 5 foul shots. . .
a) what's the expected number of shots he should make?
b) and what's the standard deviation?

14. **More arrows.** Consider our archer from Exercise 12.
a) How many bull's-eyes do you expect her to get?
b) With what standard deviation?

15. **Chips.** A computer chip manufacturer rejects 2% of the chips produced because they fail to work properly when tested. Suppose the company tests 10 chips. What's the probability of these results?
a) All 10 are okay. (None of the chips fails the test.)
b) Exactly 1 of the 10 chips fails.
c) Exactly 2 of the 10 chips fail.
d) Exactly 3 of the 10 chips fail.
e) No more than 3 of the 10 chips fail the test.

16. **Colorblindness.** About 8% of males are colorblind. A researcher has a list of 12 men who have volunteered to be tested. What's the probability of these results?
a) None of the men is colorblind.
b) Exactly 1 of the 12 men is colorblind.
c) Exactly 2 of the 12 men are colorblind.
d) Exactly 3 of the 12 men are colorblind.
e) At most 3 of the 12 men are colorblind.

17. **Many chips.** The manufacturer in Exercise 15 produces 1000 computer chips a day. What are the mean and standard deviation of the number that fail?

18. **Many men.** Hoping to find many colorblind men for her study, the researcher in Exercise 16 tests 450 volunteers. What are the mean and standard deviation of the number who are colorblind?

B

19. **Pick a card, any card.** You draw a card from a deck. If you get a red card, you win nothing. If you get a spade, you win $5. For any club, you win $10 plus an extra $20 for the ace of clubs.
a) Create a probability model for the amount you win.
b) Find the expected amount you'll win.
c) What would you be willing to pay to play this game?

20. **You bet!** You roll a die. If it comes up a 6, you win $100. If not, you get to roll again. If you get a 6 the second time, you win $50. If not, you lose.

a) Create a probability model for the amount you win.
b) Find the expected amount you'll win.
c) What would you be willing to pay to play this game?

21. Pick another card. Find the standard deviation of the amount you might win drawing a card in Exercise 19.

22. The die. Find the standard deviation of the amount you might win rolling a die in Exercise 20.

23. Cancelled flights. Mary is deciding whether to book the cheaper flight home from college after her final exams, but she's unsure when her last exam will be. She thinks there is only a 20% chance that the exam will be scheduled after the last day she can get a seat on the cheaper flight. If it is and she has to cancel the flight, she will lose $150. If she can take the cheaper flight, she will save $100.
a) If she books the cheaper flight, what can she expect to gain, on average?
b) What is the standard deviation?

24. Day trading. An option to buy a stock is priced at $200. If the stock closes above 30 on May 15, the option will be worth $1000. If it closes below 20, the option will be worth nothing, and if it closes between 20 and 30 (inclusively), the option will be worth $200. A trader thinks there is a 50% chance that the stock will close in the 20–30 range, a 20% chance that it will close above 30, and a 30% chance that it will fall below 20 on May 15.
a) How much does she expect to gain by buying the stock option?
b) What is the standard deviation of her gain?

25. Batteries. In a group of 10 batteries, 3 are dead. You choose 2 batteries at random.
a) Create a probability model for the number of good batteries you get.
b) What's the expected number of good ones you get?
c) What's the standard deviation?

26. Kittens. In a litter of seven kittens, three are female. You pick two kittens at random.
a) Create a probability model for the number of male kittens you get.
b) What's the expected number of males?
c) What's the standard deviation?

27. On time. A Department of Transportation report about air travel found that, nationwide, 76% of all flights are on time. Suppose you are at the airport and your flight is one of 50 scheduled to take off in the next two hours. Can you consider these departures to be Bernoulli trials? Explain.

28. Lost luggage. A Department of Transportation report about air travel found that airlines misplace about 5 bags per 1000 passengers. Suppose you are traveling with a group of people who have checked 22 pieces of luggage on your flight. Can you consider the fate of these bags to be Bernoulli trials? Explain.

29. Still more hoops. Our basketball player from Exercise 11 is able to hit 70% of his foul shots. What's the probability of each of these?
a) He makes 8 shots in a row.
b) He makes exactly 6 of 8 shots.
c) He makes no more than 6 of 8 shots.

d) He makes exactly 10 of 20 shots.
e) He makes at least 10 of 20 shots.

30. Still more arrows. Our archer from Exercise 12 hits bulls-eyes with 80% of her arrows. What's the probability that
a) she never misses in 10 shots?
b) there are exactly 8 bull's-eyes in 10 shots?
c) there are no more than 8 bull's-eyes in 10 shots?
d) she gets exactly 12 bull's-eyes in 20 shots?
e) she gets at least 12 bull's-eyes 20 shots?

31. Tennis, anyone? A certain tennis player makes a successful first serve 70% of the time. Assume that each serve is independent of the others. If she serves 6 times, what's the probability she gets
a) all 6 serves in?
b) exactly 4 serves in?
c) at least 4 serves in?
d) no more than 4 serves in?

32. Frogs. A wildlife biologist examines frogs for a genetic trait he suspects may be linked to sensitivity to industrial toxins in the environment. Previous research had established that this trait is usually found in 1 of every 8 frogs. He collects and examines a dozen frogs. If the frequency of the trait has not changed, what's the probability he finds the trait in
a) none of the 12 frogs?
b) at least 2 frogs?
c) 3 or 4 frogs?
d) no more than 4 frogs?

33. And more tennis. Suppose the tennis player in Exercise 31 serves 80 times in a match.
a) What are the mean and standard deviation of the number of good first serves expected?
b) Verify that you can use a Normal model to approximate the distribution of the number of good first serves.
c) What's the probability she makes at least 65 first serves?

34. More arrows. The archer in Exercise 30 will be shooting 200 arrows in a large competition.
a) What are the mean and standard deviation of the number of bull's-eyes she might get?
b) Is a Normal model appropriate here? Explain.
c) What's the probability she makes over 150 bull's-eyes?

35. Seat belts. Police estimate that 80% of drivers now wear their seat belts. They set up a safety roadblock, stopping 120 cars to check for seat belt use.
a) Find the mean and standard deviation of the number of drivers expected to be wearing seat belts.
b) What's the probability they find at least 20 drivers not wearing their seat belts?

36. Vitamin D. Vitamin D is essential for strong, healthy bones. Recent research indicated that about 20% of British children are deficient in vitamin D. Suppose doctors test a group of 320 elementary school children.
a) Find the mean and standard deviation of the number who may be deficient in vitamin D.
b) What's the probability that no more than 50 of them have the vitamin deficiency?

37. Apples. An orchard owner knows that he'll have to use about 6% of the apples he harvests for cider because they will have bruises or blemishes. He expects a tree to produce about 300 apples.
a) Find the mean and standard deviation for the number of cider apples that may come from that tree. Justify your model.
b) Verify that he can use a Normal model to approximate the distribution of the number of cider apples.
c) Should he be surprised to get more than 50 cider apples? Explain.

38. Frogs, part II. Based on concerns raised by his preliminary research, the biologist in Exercise 32 decides to collect and examine 150 frogs.
a) Assuming the frequency of the trait is still 1 in 8, determine the mean and standard deviation of the number of frogs with the trait he should expect to find in his sample.
b) Verify that he can use a Normal model to approximate the distribution of the number of frogs with the trait.
c) He found the trait in 22 of his frogs. Do you think this proves that the trait has become more common? Explain.

39. Annoying phone calls. A newly hired telemarketer is told he will probably make a sale on about 12% of his phone calls. The first week he called 200 people, but only made 10 sales. Should he suspect he was misled about the true success rate? Explain.

40. The euro. Shortly after the introduction of the euro coin in Belgium, newspapers around the world published articles claiming the coin is biased. The stories were based on reports that someone had spun the coin 250 times and gotten 140 heads—that's 56% heads. Do you think this is evidence that spinning a euro is unfair? Explain.

C ────────────────

41. Kids. A couple plans to have children until they get a girl, but they agree that they will not have more than three children even if all are boys. (Assume boys and girls are equally likely.)
a) Create a probability model for the number of children they might have.
b) Find the expected number of children.
c) Find the standard deviation.

42. Carnival. A carnival game offers a $100 cash prize for anyone who can break a balloon by throwing a dart at it. It costs $5 to play, and you're willing to spend up to $20 trying to win. You estimate that you have about a 10% chance of hitting the balloon on any throw.
a) Create a probability model for the amount you could win.
b) Find your expected winnings.
c) Find the standard deviation.

43. Contest. You play two games against the same opponent. The probability you win the first game is 0.4. If you win the first game, the probability you also win the second is 0.2. If you lose the first game, the probability that you win the second is 0.3.
a) Are the two games independent? Explain.
b) What's the probability you lose both games?
c) What's the probability you win both games?
d) Let random variable X be the number of games you win. Find the probability model for X.
e) What are the expected value and standard deviation?

44. Contracts. Your company bids for two contracts. You believe the probability you get contract #1 is 0.8. If you get contract #1, the probability you also get contract #2 will be 0.2, and if you do not get #1, the probability you get #2 will be 0.3.
a) Are the two contracts independent? Explain.
b) Find the probability you get both contracts.
c) Find the probability you get no contract.
d) Let X be the number of contracts you get. Find the probability model for X.
e) Find the expected value and standard deviation.

45. LeBron again. Let's take one last look at the LeBron James picture search. You know his picture is in 20% of the cereal boxes. You buy five boxes to see how many pictures of LeBron you might get.
a) Describe how you would simulate the number of pictures of LeBron you might find in five boxes of cereal.
b) Run at least 30 trials.
c) Based on your simulation, estimate the probabilities that you get no pictures of LeBron, 1 picture, 2 pictures, etc.
d) Find the actual probability model.
e) Compare the distribution of outcomes in your simulation to the probability model.

46. Seat belts again. Suppose 75% of all drivers always wear their seat belts. Let's investigate how many of the drivers might be belted among five cars waiting at a traffic light.
a) Describe how you would simulate the number of seatbelt-wearing drivers among the five cars.
b) Run at least 30 trials.
c) Based on your simulation, estimate the probabilities there are no belted drivers, exactly one, two, etc.
d) Find the actual probability model.
e) Compare the distribution of outcomes in your simulation to the probability model.

47. Lefties again. A lecture hall has 200 seats with folding arm tablets, 30 of which are designed for left-handers. The typical size of classes that meet there is 188, and we can assume that about 13% of students are left-handed. What's the probability that a right-handed student in one of these classes is forced to use a lefty arm tablet?

48. No-shows. An airline, believing that 5% of passengers fail to show up for flights, overbooks (sells more tickets than there are seats). Suppose a plane will hold 265 passengers, and the airline sells 275 tickets. What's the probability the airline will not have enough seats, so someone gets bumped?

49. Tennis model. Our tennis player in Exercise 31 has a 70% success rate on first serves, and might serve 80 times in a competitive match.
   a) Verify that a Normal model approximates the distribution of successful first serves.
   b) Use the 68–95–99.7 Rule to sketch this model.
   c) Write a few sentences describing what the model says about her serves.

50. Colorblindness model. Our researcher in Exercise 18 is seeking colorblind men for a study. Knowing that only 8% of all men are colorblind, he tests 450 volunteers.
   a) Verify that a Normal model approximates the distribution of the number of these men who might be colorblind.
   b) Use the 68–95–99.7 Rule to sketch this model.
   c) Write a few sentences describing what the model says about the number of colorblind men the researcher might find in this group.

51. Chips model. Historically 2% of the chips produced by our manufacturer in Exercise 17 have proven to be defective. The factory produces 1000 chips a day.
   a) Verify that a Normal model approximates the distribution of the number of defective chips.
   b) Use the 68–95–99.7 Rule to sketch this model.
   c) Write a few sentences describing what the manufacturer might expect based on this model.

52. Frog model. Our biologist in Exercise 32 knows that about 1 of every 8 frogs has a certain genetic trait, and plans to collect and test another group of 120 frogs.
   a) Verify that a Normal model approximates the distribution the number of frogs that should display this trait.
   b) Use the 68–95–99.7 Rule to sketch the model.
   c) Write a few sentences describing what the model says about what the biologist might expect to find.

53. ESP. Scientists wish to test the mind-reading ability of a person who claims to "have ESP." They use five cards with different and distinctive symbols (square, circle, triangle, line, squiggle). Someone picks a card at random and thinks about the symbol. The "mind reader" must correctly identify which symbol was on the card. If the test consists of 100 trials, how many would this person need to get right in order to convince you that ESP may actually exist? Explain.

54. True-False. A true-false test consists of 50 questions. How many does a student have to get right to convince you that he is not merely guessing? Explain.

55. Hot hand. A basketball player who ordinarily makes about 55% of his free throw shots has new sneakers, which he thinks improve his game. Over his past 40 shots, he's made 32—much better than usual. Do you think his chances of making a shot really increased? In other words, is making at least 32 of 40 shots really unusual for him? (Do you think it's his sneakers?)

56. New bow. Our archer in Exercise 12 purchases a new bow, hoping that it will improve her success rate to more than 80% bull's-eyes. She is delighted when she first tests her new bow and hits 45 bull's-eyes in

50 shots. Do you think this is compelling evidence that the new bow is better? In other words, is a streak like this unusual for her? Explain.

Answers
Just Checking

1. a)

Outcome	X = cost	Probability
Recharging works	$60	0.75
Replace control unit	$200	0.25

   b) $60(0.75) + 200(0.25) = \$95$
   c) Car owners with this problem will spend an average of $95 to get it fixed.

2. a)

X = value	$500	$250	$100	$10
$P(X)$	0.0005	0.0050	0.0200	0.9745

   b) $E(X) = \$13.245$; $SD(X) = \$23.654$
   c) The store expects to give customers discounts averaging about $15.25, with a standard deviation of just over $33.

3. a) No; the probability of choosing a female changes with each name drawn.
   b) Yes.
   c) No; women who have had twins are more likely to have them again.
   d) No; there are more than two possible outcomes (colors).
   e) Yes.

4. a) Binom(12, 0.76)
   b) $q = 0.24$
   c) $\mu = np = 12(0.76) = 9.12$
   d) $\sigma = \sqrt{npq} = \sqrt{12(0.76)(0.24)} \approx 1.48$
   e) $_{12}C_9(0.76)^9(0.24)^3 \approx 0.26$

5. a) 0.26   b) 0.58   c) 0.68

6. a) $1300 < 10\%$ of all households.
   b) $n = 1300$ and $p = 0.76$
   c) $\mu = 1300(0.76) = 988$;
       $\sigma = \sqrt{1300(0.76)(0.24)} = 15.4$
   d) Yes; $np = 1300(0.76) = 988$ and $nq = 1300(0.24) = 312$ are both at least 10.
   e) $z = \dfrac{1000 - 988}{15.4} = 0.78$
   f) $P(z > 0.78) = 0.22$

From the Data at Hand to the World at Large

Confidence Intervals for a Proportion

W hen the Harris Poll asked 889 U.S. adults, "Do you believe in ghosts?", 40% said they did. At almost the same time, CBS News polled 808 U.S. adults and asked the same question. 48% of their respondents professed a belief in ghosts. Why don't the polls agree? How can surveys conducted at essentially the same time and asking the same questions get different results?

The answer is at the heart of Statistics. The proportions vary from sample to sample because the samples are composed of different people. The surprise is that it's actually pretty easy to predict how much a proportion can vary under circumstances like this. Understanding the variability of our estimates will enable us to better understand the world.

WHO	U.S. adults
WHAT	Belief in ghosts
WHEN	November 2005
WHERE	United States
WHY	Public attitudes

Variability in Sample Proportions

Imagine

We see only the sample that we actually drew, but by simulating or modeling, we can *imagine* what we might have seen had we drawn other possible random samples.

TI-*nspire*

Sample Proportions. Generate sample after sample to see how the proportions vary.

NOTATION ALERT

The letter p is our choice for the *parameter* of the model for proportions. It violates our "Greek letters for parameters" rule, but if we stuck to that, our natural choice would be π, and then we'd have to write statements like $\pi = 0.46$. That just seems a bit weird to us. (After all, $\pi = 3.1415926 \ldots$ is a hard habit to break.)

So, we'll use p for the model parameter (the probability of a success) and $\hat{p}$ for the observed proportion in a sample.

But be careful. We've already used capital P for a general probability. And we'll soon see another use of P in the next chapter! There are a lot of p's in this course; you'll need to think clearly about the context to keep them straight.

We've talked about *Think, Show,* and *Tell.* Now we add *Imagine.* In order to understand the CBS poll, imagine the results from all the random samples of size 808 that CBS News could have used. What would the histogram of all the sample proportions look like?

The center of that histogram will be at the true proportion of all Americans who believe in ghosts, and we call that p. (See the Notation Alert.) Of course, we don't *know* that value (and probably never will). For the sake of discussion here, let's suppose that 45% of all American adults believe in ghosts, so we'll use $p = 0.45$.

How about the *shape* of the histogram? We don't have to just imagine that. We can simulate a bunch of random samples. Here's a histogram of the proportions saying they believe in ghosts for 2000 simulated samples of 808 adults.

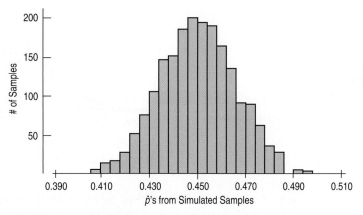

FIGURE 17.1 A histogram of sample proportions for 2000 simulated samples of 808 adults drawn from a population with $p = 0.45$. The sample proportions vary, but their distribution is centered at the true proportion, p.

Look closely. You've seen lots of histograms before, but never one like this. Until now, our histograms have always displayed the distribution of data values in our sample. Not this one. Here you see the distribution of a **sample statistic**, the sample proportion $\hat{p}$. Each $\hat{p}$ comes from a different sample. We have collected those statistics from 2000 different samples, and now examine them as though they were data. The histogram above is a simulation of what we'd get if we could see *all the proportions from all possible samples.* That distribution has a special name. It is called the **sampling distribution** of the proportions.

It should be no surprise that we don't get the same proportion for each sample we draw. Does it surprise you that the histogram is unimodal? Symmetric? That it is centered at p? You probably don't find any of this shocking. Does the shape remind you of any model that we've discussed? It's an amazing and fortunate fact that a Normal model is just the right one for the histogram of sample proportions.

Modeling how sample proportions vary from sample to sample is one of the most powerful ideas we'll see in this course. A **sampling distribution model** for how a sample proportion varies from sample to sample allows us to quantify that variation and to talk about how likely it is that we'd observe a sample proportion in any particular interval.

To use a Normal model, we need to know two parameters: its mean and standard deviation. It's natural to put μ, the mean of the Normal, at p, the center of the histogram.

What about the standard deviation? Once we know the mean, p, we automatically also know the standard deviation. We saw in the last chapter that for a Binomial model the standard deviation of the *number* of successes is $\sqrt{npq}$. Now we want the standard deviation of the *proportion* of successes, $\hat{p}$. The sample proportion $\hat{p}$ is the number of successes divided by the number of trials, n, so the standard deviation is also divided by n:

$$\sigma(\hat{p}) = SD(\hat{p}) = SD\left(\frac{x}{n}\right) = \frac{\sqrt{npq}}{n} = \sqrt{\frac{pq}{n}}.$$

Assuming the true proportion of adults who believe in ghosts to be $p = 0.45$, the standard deviation for the CBS poll is

$$SD(\hat{p}) = \sqrt{\frac{pq}{n}} = \sqrt{\frac{(0.45)(0.55)}{808}} = 0.0175, \text{ or } 1.75\%.$$

Our Normal model for the sample proportions is $N(45\%, 1.75\%)$. We can use the 68–95–99.7 Rule to see what might happen in polls about ghosts:

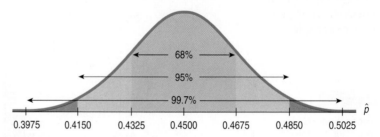

FIGURE 17.2 The sampling model. Using 0.45 for p gives this Normal model for this histogram of the sample proportions of adults believing in ghosts ($n = 808$).

We see that different polls could report varying levels of belief, near the true level of 45%, but possibly somewhat higher or lower. In fact, since $2 \times 1.75\% = 3.5\%$, we'd expect about 95% of polls to find results within 3.5% of 45%, and conclude that between 41.5% and 48.5% of American adults believe in ghosts. This is what we mean by **sampling error.** It's not really an *error* at all, but just *variability* you'd expect to see from one sample to another. A better term would be **sampling variability.**

Assumptions and Conditions

Before using a Normal model for the sampling distribution for sample proportions, we need to think about two assumptions and check some conditions.

The Independence Assumption: The sampled values must be independent of each other.

To think about the Independence Assumption, we wonder whether there is any reason to think that the data values might affect each other. Usually we're drawing samples, and can consider data values to be reasonably independent if two conditions are satisfied:

Randomization Condition: Your sample should be a simple random sample of the population. If some other sampling design was used, be sure the sampling method was not biased and that the data are representative of the population.

10% Condition: The sample size, n, must be no larger than 10% of the population. For national polls, the total population is usually very large, so the sample is a small fraction of the population.

The terms "success" and "failure" for the outcomes that have probability p and q are common in Statistics. But they are completely arbitrary labels. When we say that a disease occurs with probability p, we certainly don't mean that getting sick is a "success" in the ordinary sense of the word.

Our other assumption is:

> **The Sample Size Assumption:** The sample size, n, must be large enough.

To verify that, we check the

> **Success/Failure Condition:** The sample size has to be big enough so that we expect at least 10 successes and at least 10 failures. When np and nq are at least 10, we have enough data for sound conclusions. For the CBS survey, a "success" might be believing in ghosts. With $p = 0.45$, we expect $808 \times 0.45 = 364$ successes and $808 \times 0.55 = 444$ failures. Both are at least 10, so we certainly expect enough successes and enough failures for the condition to be satisfied.[1]

A Sampling Distribution Model for a Proportion

We have now answered the question raised at the start of the chapter. To know how variable a sample proportion is, we need to know the proportion and the size of the sample. That's all.

Creating this model represents an important change in our point of view. No longer is a proportion something we just compute for a set of data. We now see it as a random variable taking on a different value for each sample. We have a model describing that variable, called the **sampling distribution model** for the proportion $\hat{p}$.

THE SAMPLING DISTRIBUTION MODEL FOR A PROPORTION

Provided that the sampled values are independent and the sample size is large enough, the sampling distribution of $\hat{p}$ is modeled by a Normal model with mean $\mu(\hat{p}) = p$ and standard deviation $SD(\hat{p}) = \sqrt{\dfrac{pq}{n}}$.

This is a Big Deal. Sampling models are what make Statistics work. They act as a bridge from the sample data we know to the population truths we wish we knew. By informing us about the amount of error we should expect when we sample, they enable us to say something about the real world when all we have is data.

It's time to take the huge leap of Statistics. The sampling model for a proportion enables us to know how precise our sample's estimate of the true population proportion might be. That's the path to the *margin of error* you hear about in polls and surveys. Here we go.

---

[1] Two conditions may seem to conflict with each other. The **Success/ Failure Condition** wants sufficient data. How much? That depends on p. If p is near 0.5, we need a sample of only 20 or so. If p is only 0.01, however, we'd need 1000. But the **10% Condition** says that a sample should be no larger than 10% of the population. If you're thinking, "Wouldn't a larger sample be better?" you're right of course. It's just that if the sample were more than 10% of the population, we'd need to use different methods to analyze the data. Fortunately, this isn't usually a problem in practice. Often, as in polls that sample from all U.S. adults or industrial samples from a day's production, the populations are much larger than 10 times the sample size.

For Example Creating and Interpreting a Sampling Model

You plan to toss a coin 100 times to see whether it's fair or not. While you expect half the tosses to land heads, you wouldn't be surprised by a small deviation from perfection.

Question: Based on the sampling model for the proportion of heads, how far from a perfect 50-50 would the results have to be to surprise you?

I want a sampling model for $\hat{p}$, the proportion of heads I might get in $n = 100$ tosses of a coin, assuming that it's fair: $p = 0.5$.

Checking assumptions and conditions:

✔ **Independence Assumption:** No toss influences any other toss.
✔ **Success/Failure Condition:** I expect $np = 50$ heads and $nq = 50$ tails, both at least 10.

It's ok to use a Normal model.

The mean (expected value of $\hat{p}$) is $p = 0.5$.

The standard deviation of $\hat{p}$ is $\sqrt{\dfrac{pq}{n}} = \sqrt{\dfrac{(0.5)(0.5)}{100}} = 0.05$

The sampling model for $\hat{p}$ is $N(0.50, 0.05)$.

Based on the 68–95–99.7 Rule, the model indicates this distribution of the proportions of heads in 100 tosses:

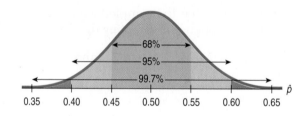

In 100 tosses of a fair coin, I'd expect usually to get between 40% and 60% heads. I'd be quite suspicious about the fairness of the coin if it landed heads under 40% or over 60% of the time, and would be pretty sure the coin wasn't fair if I saw under 35% or over 65% heads.

Just Checking

1. About 6% of Americans have Type O− blood, making them universal donors. The Red Cross expects about 250 people to donate at an upcoming blood drive. For planning purposes, the Red Cross wonders what percentage of these volunteers might be universal donors. Let's construct a sampling model.

   a) What is p?
   b) In this context, what does $\hat{p}$ represent?
   c) Is it okay to use a Normal model? Check the assumptions and conditions.
   d) Find the mean and standard deviation of the sampling model.
   e) Sketch the sampling model for $\hat{p}$ using the 68–95–99.7 Rule.
   f) Write a few sentences to tell the Red Cross what might happen at the blood drive.

(Check your answers on page 395.)

A Confidence Interval

WHO	Sea fans
WHAT	Percent infected
WHEN	June 2000
WHERE	Las Redes Reef, Akumal, Mexico, 40 feet deep
WHY	Research

Coral reef communities are home to one-quarter of all marine plants and animals worldwide. These reefs support large fisheries by providing breeding grounds and safe havens for young fish of many species. Coral reefs are seawalls that protect shorelines against tides, storm surges, and hurricanes, and are sand "factories" that produce the limestone and sand of which beaches are made. Beyond the beach, these reefs are major tourist attractions for snorkelers and divers, driving a tourist industry worth tens of billions of dollars.

But marine scientists say that 10% of the world's reef systems have been destroyed in recent times. At current rates of loss, 70% of the reefs could be gone in 40 years. Pollution, global warming, outright destruction of reefs, and increasing acidification of the oceans are all likely factors in this loss.

Dr. Drew Harvell's lab studies corals and the diseases that affect them. They sampled sea fans[2] at 19 randomly selected reefs along the Yucatan peninsula and diagnosed whether the animals were affected by the disease *aspergillosis*.[3] In specimens collected at a depth of 40 feet at the Las Redes Reef in Akumal, Mexico, these scientists found that 54 of 104 sea fans sampled were infected with that disease.

Of course, we care about much more than these particular 104 sea fans. We care about the health of coral reef communities throughout the Caribbean. What can this study tell us about the prevalence of the disease among sea fans?

We have a sample proportion, which we write as $\hat{p}$, of 54/104, or 51.9%. Our first guess might be that this observed proportion is close to the population proportion, p. But how close? The sampling model can suggest a **margin of error**, telling us how far off our estimate might be.

First we need to check the assumptions and conditions. Are the sampled fans independent? We'll need to assume that the ocean reef is pretty big and that the researchers selected fans at random from across a broad region. These 104 specimens are almost certainly far fewer than 10% of the sea fans living there. The disease infected 54 of them, and the other 50 were healthy—more than 10 successes and 10 failures.

We can proceed with a Normal model for the sampling distribution. But what do we know about it? We don't know its mean—that's the proportion of *all* infected sea fans on the Las Redes Reef. Is the infected proportion of *all* sea fans 51.9%? No, that's just $\hat{p}$, our estimate. We don't know the proportion, p, of all the infected sea fans; that's what we're trying to find out. We do know, though, that the sampling distribution model of $\hat{p}$ is centered at p, whatever it is.

We know that the standard deviation of the sampling distribution is $\sqrt{\dfrac{pq}{n}}$, but we have a problem: Since we don't know p, we can't find the true standard deviation of the sampling distribution model. We do know the observed proportion, $\hat{p}$, so, we'll use that to make an estimate. That may not seem like a big deal, but it gets a special name. Whenever we estimate the standard deviation of a sampling distribution, we call it a **standard error.**[4] For a sample proportion, $\hat{p}$, the standard error is

$$SE(\hat{p}) = \sqrt{\frac{\hat{p}\hat{q}}{n}}.$$

Remember that $\hat{p}$ is our sample-based estimate of the true proportion p. Recall also that q is just shorthand for $1 - p$, and $\hat{q} = 1 - \hat{p}$.

When we use $\hat{p}$ to estimate the standard deviation of the sampling distribution model, we call that the **standard error** and write $SE(\hat{p}) = \sqrt{\dfrac{\hat{p}\hat{q}}{n}}$.

---

[2]That's a sea fan in the picture. Although they look like trees, they are actually colonies of genetically identical animals.

[3]K. M. Mullen, C. D. Harvell, A. P. Alker, D. Dube, E. Jordán-Dahlgren, J. R. Ward, and L. E. Petes, "Host range and resistance to aspergillosis in three sea fan species from the Yucatan," *Marine Biology* (2006), Springer-Verlag.

[4]This isn't such a great name because it isn't standard and nobody made an error. But it's much shorter and more convenient than saying, "the estimated standard deviation of the sampling distribution of the sample statistic."

For the sea fans, then:

$$SE(\hat{p}) = \sqrt{\frac{\hat{p}\hat{q}}{n}} = \sqrt{\frac{(0.519)(0.481)}{104}} = 0.049 = 4.9\%.$$

Great. What does that tell us? Well, because the sampling model is Normal, in 95% of random samples, $\hat{p}$ will be no more than 2 SEs away from p. We don't know p, but we do know $\hat{p}$. So we know there's a 95% chance that p is no more than 2 SEs away from our $\hat{p}$. That's our margin of error: 2 SEs = 2(4.9%) = 9.8%. We're 95% sure that p is between $51.9 - 9.8 = 42.1\%$ and $51.9 + 9.8 = 61.7\%$. Of course, even if our interval does catch p, we still don't know the exact value. The best we can do is an interval, and even then we can't be positive it contains p.

FIGURE 17.3 The margin of error. Reaching out 2 SEs on either side of $\hat{p}$ makes us 95% confident that we'll trap the true proportion, p.

So what can we really say about p? Here's a list of things we'd like to be able to say, in order of strongest to weakest and the reasons we can't say most of them:

1. **"51.9% of *all* sea fans on the Las Redes Reef are infected."** It would be nice to be able to make absolute statements about population values with certainty, but we just don't have enough information to do that. There's no way to be sure that the population proportion is the same as the sample proportion; in fact, it almost certainly isn't. Observations vary. Another sample would yield a different sample proportion.
2. **"It is *probably* true that 51.9% of all sea fans on the Las Redes Reef are infected."** No. In fact, we can be pretty sure that whatever the true proportion is, it's not exactly 51.900%. So the statement is not true.
3. **"We don't know exactly what proportion of sea fans on the Las Redes Reef is infected, but we *know* that it's within the interval 51.9% ± 2 × 4.9%. That is, it's between 42.1% and 61.7%."** This is getting closer, but we still can't be certain. We can't know *for sure* that the true proportion is in this interval—or in any particular interval.
4. **"We don't know exactly what proportion of sea fans on the Las Redes Reef is infected, but the interval from 42.1% to 61.7% *probably* contains the true proportion."** We've now fudged twice—first by giving an interval and second by admitting that we only think the interval "probably" contains the true value. And this statement is true.

That last statement may be true, but it's a bit wishy-washy. We can tighten it up a bit by quantifying what we mean by "probably." We saw that 95% of the

"Far better an approximate answer to the right question, . . . than an exact answer to the wrong question."

—John W. Tukey

time when we reach out 2 *SEs* from $\hat{p}$ we capture *p, so we can be 95% confident that this is one of those times*. Now we can say:

>   5. **"We are 95% confident that between 42.1% and 61.7% of Las Redes sea fans are infected."** Statements like these are called **confidence intervals.** They're the best we can do.

The interval calculated and interpreted here is sometimes called a **one-proportion z-interval** or a **confidence interval for a proportion.**

95% CONFIDENCE INTERVAL FOR A PROPORTION

When the sampled values are independent and the sample size is large enough, we can be 95% confident that the interval $\hat{p} \pm 2SE(\hat{p})$ captures the true population proportion p $\left(\text{where the standard deviation of sample proportions is estimated by } SE(\hat{p}) = \sqrt{\dfrac{\hat{p}\hat{q}}{n}}\right).$

For Example   Creating and Interpreting a Confidence Interval

Many polling organizations track the public's approval of the president, which varies from month to month. For several months after his inauguration, Barack Obama's approval rating stood at well over 60%, but began to slide as the economy worsened.

In September 2010, with the midterm elections approaching, a Gallup Poll interviewed a sample of 3490 adults nationwide by randomly dialing both standard land-line and cell phone numbers. Of that sample, 1571 people (45%) said they approved of the job President Obama was doing.

Question: Based on a 95% confidence interval, is it possible Obama still had majority support at that point in his presidency?

First I'll check the assumptions and conditions:

✔ **Randomization Condition:** This is a random sample of U.S. adults.
✔ **10% Condition:** 3490 is fewer than 10% of all U.S. adults.
✔ **Success/Failure Condition:** 1571 people approved and 1919 did not, both at least 10.

It's okay to use a Normal model to describe the sampling distribution for $\hat{p}$.

For this sample, $\hat{p} = \dfrac{1571}{3490} = 45\%$.

$SE(\hat{p}) = \sqrt{\dfrac{(0.45)(0.55)}{3490}} \approx 0.0084$, so this poll's margin of error is $2(0.0084) = 1.7\%$.

The confidence interval is from $45\% - 1.7\% = 43.3\%$ to $45\% + 1.7\% = 46.7\%$.

I'm 95% confident that in late August 2009, between 43.3% and 46.7% of all American adults approved of President Obama's performance in office. It's virtually certain that he no longer had the support of the majority.

Just Checking

2. A Pew Research study regarding cell phones asked questions about cell phone experience. One growing concern is unsolicited advertising in the form of text messages. Pew asked cell phone owners, "Have you ever received unsolicited text messages on your cell phone from advertisers?" and 17% reported that they had. Pew estimates a 95% confidence interval to be 0.17 ± 0.04, or between 13% and 21%.

Are the following statements about people who have cell phones correct? Explain.

a) In Pew's sample, somewhere between 13% and 21% of respondents reported that they had received unsolicited advertising text messages.

b) We can be 95% confident that 17% of U.S. cell phone owners have received unsolicited advertising text messages.

c) We are 95% confident that between 13% and 21% of all U.S. cell phone owners have received unsolicited advertising text messages.

d) We know that between 13% and 21% of all U.S. cell phone owners have received unsolicited advertising text messages.

e) 95% of all U.S. cell phone owners have received unsolicited advertising text messages.

f) 95% of the time the true proportion of all U.S. cell phone owners who have received unsolicited advertising text messages is between 13% and 21%.

(Check your answers on page 395.)

■ What Does "95% Confidence" Really Mean?

What do we mean when we say we have 95% confidence that our interval contains the true proportion? Formally, what we mean is that "95% of samples of this size will produce confidence intervals that capture the true proportion." This is correct, but a little long winded, so we sometimes say, "we are 95% confident that the true proportion lies in our interval." But what does that mean?

If other researchers select their own samples of sea fans, they'll also find some infected by the disease, but each person's sample proportion will almost certainly differ from ours. Each of us will end up with a different confidence interval. Our interval guessed the true proportion of infected sea fans to be between about 42% and 62%. Other researchers might guess between 46% and 66%, or between 23% and 43%, and so on. Every possible sample would produce yet another confidence interval.

The figure below shows confidence intervals produced by simulating 20 different random samples. The red dots are the proportions of infected fans in each sample, and the blue segments show the confidence intervals found for each. The green line represents the true rate of infection in the population, so you can see that most of the intervals caught it—but a few missed. (And notice again that it is the *intervals* that vary from sample to sample; the green line doesn't move.)

TI-*nspire*

Confidence intervals. Generate confidence intervals from many samples to see how often they capture the true proportion.

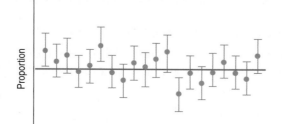

FIGURE 17.4 Confidence intervals work – usually. The horizontal green line shows the true percentage of all sea fans that are infected. Most of the 20 simulated samples produced confidence intervals that captured the true value, but a few missed.

These are just some of the huge number of possible samples that *could* be drawn. That's a large pile of possible confidence intervals, and ours is just one of those in the pile. Did *our* confidence interval "work"? We can never be sure, because we'll never know the true proportion of all the sea fans that are infected. All we know is that 95% of the intervals in the pile are winners, covering the true value, and only 5% are duds. *That's* why we're 95% confident that our interval is a winner!

For Example   Polls and Margin of Error

On January 30–31, 2007, Fox News/Opinion Dynamics polled 900 registered voters nationwide.[5] When asked, "Do you believe global warming exists?" 82% said "Yes." Fox reported their margin of error to be ±3%.

Question: It is standard among pollsters to use a 95% confidence level unless otherwise stated. Given that, what does Fox News mean by claiming a margin of error of ±3% in this context?

If this polling were done repeatedly, 95% of all random samples would yield estimates that come within ±3% of the true proportion of all registered voters who believe that global warming exists.

STEP–BY–STEP EXAMPLE   A Confidence Interval for a Proportion

In May 2006, the Gallup Poll[6] asked 510 randomly sampled adults the question "Generally speaking, do you believe the death penalty is applied fairly or unfairly in this country today?" Of these, 60% answered "Fairly," 35% said "Unfairly," and 4% said they didn't know.

WHO	Adults in the United States
WHAT	Response to a question about the death penalty
WHEN	May 2006
WHERE	United States
HOW	510 adults were randomly sampled and asked by the Gallup Poll
WHY	Public opinion research

Question: From this survey, what can we conclude about the opinions of *all* adults?

To answer this question, we'll build a confidence interval for the proportion of all U.S. adults who believe the death penalty is applied fairly. There are four steps to building a confidence interval for proportions: Plan, Model, Mechanics, and Conclusion.

(continued)

---
[5]www.foxnews.com, "Fox News Poll: Most Americans Believe in Global Warming," Feb 7, 2007.

[6]www.gallup.com

Plan State the problem and the W's.

Identify the *parameter* you wish to estimate.

Identify the *population* about which you wish to make statements.

Choose and state a confidence level.

Model Think about the assumptions and check the conditions.

I want to find an interval that is likely, with 95% confidence, to contain the true proportion, p, of U.S. adults who think the death penalty is applied fairly. I have a random sample of 510 U.S. adults.

✔ **Independence Assumption:** Gallup phoned a random sample of U.S. adults. It is very unlikely that any of their respondents influenced each other.

✔ **Randomization Condition:** Gallup drew a random sample from all U.S. adults. I don't have details of their randomization but assume that I can trust it.

✔ **10% Condition:** Although sampling was necessarily without replacement, there are many more U.S. adults than were sampled. The sample is certainly less than 10% of the population.

✔ **Success/Failure Condition:**
$n\hat{p} = 510(60\%) = 306 \geq 10$ and
$n\hat{q} = 510(40\%) = 204 \geq 10$,

so the sample appears to be large enough to use the Normal model.

State the sampling distribution model for the statistic.

Choose your method.

The conditions are satisfied, so I can use a Normal model to find a **one-proportion z-interval.**

Mechanics Construct the confidence interval.

First find the standard error. (Remember: It's called the "standard error" because we don't know p and have to use $\hat{p}$ instead.)

Next find the margin of error, 2 *SE*s.

Write the confidence interval (CI).

$n = 510, \hat{p} = 0.60$, so

$$SE(\hat{p}) = \sqrt{\frac{\hat{p}\hat{q}}{n}} = \sqrt{\frac{(0.60)(0.40)}{510}} = 0.022$$

The margin of error is

$$ME = 2 \times SE(\hat{p}) = 2(0.022) = 0.044$$

So the 95% confidence interval is

$$0.60 \pm 0.044 \text{ or } (0.556, 0.644)$$

Conclusion Interpret the confidence interval in the proper context. We're 95% confident that our interval captured the true proportion.

I am 95% confident that between 55.6% and 64.4% of all U.S. adults think that the death penalty is applied fairly.

TI Tips

Finding confidence intervals

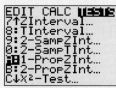

It will come as no surprise that your TI can calculate a confidence interval for a population proportion. Remember the sea fans? Of 104 sea fans, 54 were diseased. To find the resulting confidence interval, we first take a look at a whole new menu.

• Under **STAT** go to the **TESTS** menu. Quite a list! Commands are found here for the inference procedures you will learn through the coming chapters.

• We're using a Normal model to find a confidence interval for a proportion based on one sample. Scroll down the list and select **A:1-PropZInt**.

• Enter the number of successes observed and the sample size.

• Specify a confidence level 0.95 and then **Calculate**. (Yes, you may choose to be more—or less—confident.)

And there it is! Note that the TI calculates the sample proportion for you, but the important result is the interval itself, 42% to 62%. The calculator did the easy part—just Show. Tell is harder. It's your job to interpret that interval correctly.

Beware: You may run into a problem. When you enter the value of **x**, you need a *count*, not a percentage. Suppose the marine scientists had reported that 52% of the 104 sea fans were infected. You can enter **x:.52*104**, and the calculator will evaluate that as 54.08. Wrong. Unless you fix that result, you'll get an error message. Think about it—the number of infected sea fans must have been a whole number, evidently 54. When the scientists reported the results, they rounded off the actual percentage (54 ÷ 104 = 51.923%) to 52%. Simply change the value of **x** to 54 and you should be able to **Calculate** the correct interval.

Just Checking

3. In 2005 the Colorado Department of Public Health conducted a study of child safety.[7] A phone survey of randomly selected parents in Colorado estimated that almost three-quarters of children 5 to 14 years old engaged in bicycling or other wheeled sports. The researchers were specifically interested in the rate of helmet use, because helmets are highly effective in preventing serious head injuries, a common result of accidents.

   Among the parents contacted, 354 had a child who played with a skateboard, a scooter, or inline skates, of whom 162 said the child always wore a helmet.

   a) Can we use a Normal model for the sampling distribution of sample proportions? Check the assumptions and conditions.
   b) Find $\hat{p}$ and the standard error.
   c) Find the margin of error for a 95% confidence interval.
   d) Interpret your interval in this context.

(Check your answers on page 395.)

---

[7]www.cdphe.state.co.us/pp/injepi/HSbikefinal.pdf

Thinking About Margins of Error

Certainty vs. Precision

The margin of error for our 95% confidence interval was 2 *SE*. What if we wanted to be more confident? To be more confident, we'll need to capture *p* more often, and to do that we'll need to make the interval wider. For example, if we want to be 99.7% confident, the margin of error will have to be 3 *SE*.

FIGURE 17.5 Greater confidence. Reaching out 3 SEs on either side of $\hat{p}$ makes us 99.7% confident we'll trap the true proportion p. Compare with Figure 17.2.

The more confident we want to be, the larger the margin of error must be. We can be 100% confident that the proportion of infected sea fans is between 0% and 100%, but this isn't likely to be very useful. On the other hand, we could give a narrow confidence interval from 51.8% to 52.0%, but we can't be very confident about a guess that precise. Every confidence interval is a balance between certainty and precision.

The tension between certainty and precision is always there. Fortunately, in most cases we can be both sufficiently certain and sufficiently precise to make useful statements. You get to choose a confidence level yourself. The most commonly chosen confidence levels are 90%, 95%, and 99%.

Critical Values

In our sea fans example we used 2*SE* to give us a 95% confidence interval. To change the confidence level, we'd need to change the *number* of SEs so that the size of the margin of error corresponds to the new level. This number of SEs is called the **critical value.** Here it's based on the Normal model, so we denote it z^*. For any confidence level, we can find the corresponding critical value from a computer or a calculator.

For a 95% confidence interval, you'll find the precise critical value is $z^* = 1.96$. That is, 95% of a Normal model is found within ± 1.96 standard deviations of the mean. We've been using $z^* = 2$ from the 68–95–99.7 Rule because it's easy to remember.

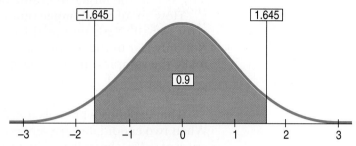

FIGURE 17.6 Smaller margin of error. For a 90% confidence interval, the critical value is 1.645, because, for a Normal model, 90% of the values are within 1.645 standard deviations from the mean.

Some Common Critical Values

Confidence Level	z^*
90%	1.645
95%	1.96
98%	2.33
99%	2.58

ONE-PROPORTION z-INTERVAL

When the conditions are met, we are ready to find the confidence interval for the population proportion, p. The confidence interval is $\hat{p} \pm z^* \times SE(\hat{p})$ where the standard deviation of sample proportions is estimated by $SE(\hat{p}) = \sqrt{\dfrac{\hat{p}\hat{q}}{n}}$.

For Example Finding the Margin of Error

Recap: In January 2007 a Fox News poll of 900 registered voters found that 82% of the respondents believed that global warming exists. Fox reported a 95% confidence interval with a margin of error of ±3%.

Questions: What would be the margin of error for a 90% confidence interval? What's good and bad about this change?

With $n = 900$ and

$\hat{p} = 0.82$, $SE(\hat{p}) = \sqrt{\dfrac{\hat{p}\hat{q}}{n}} = \sqrt{\dfrac{(0.82)(0.18)}{900}} = 0.0128$

For a 90% confidence level, $z^* = 1.645$, so $ME = 1.645(0.0128) = 0.021$

Now the margin of error is only about ±2%, producing a narrower interval. That makes for a more precise estimate of voter belief, but provides less certainty that the interval actually contains the true proportion of voters believing in global warming.

Sample Size

Ideally, of course, we'd like to have a small margin of error and high confidence, too. And that's possible—if when we plan our study we decide on a larger sample size. That makes sense, doesn't it? The larger a sample of a population we look at, the more precise we can be in describing characteristics of that population.

We can see how that works for confidence intervals by simply looking at the way we calculate the margin of error:

$$ME = z^*\sqrt{\frac{\hat{p}\hat{q}}{n}}.$$

See where the sample size shows up? It's that n in the denominator. That tells us that the larger the value of n, the smaller the margin of error.

How should we change our sample size if we want to cut the margin of error in half? Your first instinct might be to double n, but look more closely. Not only is n in the denominator, it's also inside the square root. We'd need to make the sample 4 times as large in order to halve the margin of error, because $\frac{1}{2} = \sqrt{\frac{1}{4}}$.

That explains why polls don't try to predict results in very close elections. When two candidates are separated by only 2% in a poll having a margin of error of ±4%, the pollsters just say the election is "too close to call." Why don't they just conduct a poll with a margin of error of only ±1%? Cutting their margin of error to only ¼ as big requires a sample $4^2 = 16$ times as large! It's simply not worth the extra time and expense to try to be that precise.

For Example Thinking About Sample Size

Recap: The 2007 Fox News poll about Americans' belief in global warming sampled opinions from 900 registered voters and had a margin of error of ±3%.

Question: If Fox wants to repeat the poll today, this time with a margin of error of ±1%, how many people must they contact?

Going from ±3% to ±1% would make the margin of error 1/3 as large. The sample would have to be $3^2 = 9$ times as large. That's 9 × 900 = 8100 people, probably too many for Fox to poll in a reasonable amount of time and at a reasonable cost.

Just Checking

4. Think some more about the 95% confidence interval Fox News created for the proportion of registered voters who believe that global warming exists.

   a) If Fox wanted to be 98% confident, would their confidence interval need to be wider or narrower?

   b) Fox's margin of error was about ±3%. If they reduced it to ±1%, would their level of confidence be higher or lower?

   c) If Fox News had polled more people, would the 95% confidence interval's margin of error have been larger or smaller?

   d) With their sample of 900 people, a 90% confidence interval gave fox a margin of error of ±2%. How large a sample would they have needed to cut that margin of error to only ±1%?

   (Check your answers on page 395.)

WHAT CAN GO WRONG?

WHAT *CAN* I SAY?

Confidence intervals are based on random samples, so the interval is random, too. We know that 95% of the random samples will yield intervals that capture the true value. That's what we mean by being 95% confident.

Technically, we should say, "I am 95% confident that the interval from 42.1% to 61.7% captures the true proportion of infected sea fans." But you may choose a more casual phrasing like "I am 95% confident that between 42.1% and 61.7% of the Las Redes sea fans are infected." You've made it clear that it's the interval that's random and is the focus of both our confidence and doubt.

Confidence intervals are powerful tools. Not only do they tell what we know about the parameter value, but—more important—they also tell what we *don't* know. In order to use confidence intervals effectively, you must be clear about what you say about them.

Don't Misstate What the Interval Means

- **Don't be certain.** Saying "Between 42.1% and 61.7% of sea fans are infected" asserts that the population proportion cannot be outside that interval. Of course, we can't be absolutely certain of that. (Just pretty sure.)

- **Don't suggest that the parameter varies.** A statement like "There is a 95% chance that the true proportion is between 42.7% and 51.3%" sounds as though you think the population proportion wanders around and sometimes happens to fall between 42.7% and 51.3%. When you interpret a confidence interval, make it clear that *you* know that the population parameter is fixed and that it is the interval that varies from sample to sample.

- **Don't claim that other samples will agree with yours.** Keep in mind that the confidence interval makes a statement about the true population proportion. An interpretation such as "In 95% of samples of U.S. adults, the proportion who think marijuana should be decriminalized will be between 42.7% and 51.3%" is just wrong. The interval is about the population proportion, not other samples.

- **Don't forget: It's about the parameter.** Don't say, "I'm 95% confident that $\hat{p}$ is between 42.1% and 61.7%." Of course you are—in fact, we calculated that $\hat{p} = 51.9\%$ of the fans in our sample were infected. So we already *know* the sample proportion. The confidence interval is about the (unknown) population parameter, p.

- **Don't claim to know too much.** Don't say, "I'm 95% confident that between 42.1% and 61.7% of all the sea fans in the world are infected." You didn't sample from all 500 species of sea fans found in coral reefs around the world. Just those of this type on the Las Redes Reef.

- **Do treat the whole interval equally.** Although a confidence interval is a set of plausible values for the parameter, don't think that the values in the middle of a confidence interval are somehow "more plausible" than the values near the edges. Your interval provides no information about where in your current interval (if at all) the parameter value is most likely to be hiding.

Margin of Error Too Large to Be Useful

We know we can't be exact, but how precise do we need to be? A confidence interval that says that the percentage of infected sea fans is between 10% and 90% wouldn't be of much use. What can you do?

One way to make the margin of error smaller is to reduce your level of confidence. But that may not be a useful solution. It's a rare study that reports confidence levels lower than 80%. Levels of 95% or 99% are more common.

To get a narrower interval without giving up confidence, plan your study to use a larger sample.

(continued)

Violations of Assumptions

Confidence intervals and margins of error are often reported along with poll results and other analyses. But it's easy to misuse them and wise to be aware of the ways things can go wrong.

- **Watch out for biased sampling.** Don't forget about the potential sources of bias in surveys that we discussed in Chapter 10. A questionnaire that finds that 85% of people enjoy filling out surveys still suffers from nonresponse bias. Don't put a confidence interval around this (biased) estimate.

- **Think about independence.** The assumption that the values in our sample are mutually independent is one that we usually cannot check. It always pays to think about it, though. For example, the disease affecting the sea fans might be contagious, so that fans growing near a diseased fan are more likely themselves to be diseased. Such contagion would violate the Independence Assumption and could severely affect our sample proportion. It could be that the proportion of infected sea fans on the entire reef is actually quite small, and the researchers just happened to find an infected area. To avoid this, the researchers should be careful to sample sites far enough apart to make contagion unlikely.

CONFIDENCE INTERVALS **IN YOUR WORLD**

Math Problematic for U.S. Teens

. . . according to the latest Gallup Youth Survey*, more teenagers name math than any other subject as the course they find most difficult in school. Twenty-nine percent name math generally, 6% specifically mention algebra, and 2% name geometry. About equal numbers of teens mention the sciences and English as the most difficult subject: 20% and 18%, respectively. Foreign languages, history, and social studies are each mentioned by less than 10% of the sample.

> *These results are based on telephone interviews with a randomly selected national sample of 1,028 teenagers in the Gallup Poll Panel of households, aged 13 to 17, conducted Jan. 17 to Feb. 6, 2005. For results based on this sample, one can say with 95% confidence that the maximum error attributable to sampling and other random effects is ±3 percentage points. In addition to sampling error, question

wording and practical difficulties in conducting surveys can introduce error or bias into the findings of public opinion polls.

In 2004, Gallup asked teens to name their favorite subject, and math ranked at the top—although by a much smaller margin (23%) than the percentage of teens who say math is their most difficult subject (37%)**. More students also identify English as an area they lag in rather than an area they enjoy (18% vs. 13%), and there is a gap for science (20% say it is the most difficult subject, while 12% say it is their favorite).

**The Gallup Youth Survey is conducted via an Internet methodology provided by Knowledge Networks, using an online research panel that is designed to be representative of the entire U.S. population. The current questionnaire was completed by 785 respondents, aged 13 to 17, between Jan. 22 and March 9, 2004. For results based on the total sample, one can say with 95% confidence that the maximum margin of sampling error is ±4 percentage points.*

http://www.gallup.com/poll/16360/math-problematic-us-teens.aspx

What should not get lost in the scramble to improve U.S. mathematical competency is that "right brain" skills—creative thinking and problem solving—as well as interpersonal skills, are rated most important to job success by American workers. More than four in five workers told Gallup in 2003 that these are highly important factors in doing their jobs well. Is it possible that in the new "flat" economy, Americans will have just enough mathematical competency to get by, but will continue to thrive based on a superior ability to invent, synthesize, organize, get along with people, and manage others?

WHAT HAVE WE LEARNED?

Way back in Chapter 1 we said that Statistics is about variation. We know that no sample fully and exactly describes the population; sample proportions and means will vary from sample to sample. That's sampling error (or, better, sampling variability). We know it will always be present—indeed, the world would be a boring place if variability didn't exist. You might think that sampling variability would prevent us from learning anything reliable about a population by looking at a sample, but that's just not so. The fortunate fact is that sampling variability is not just unavoidable—it's predictable!

We've learned how a Normal model describes the behavior of sample proportions—shape, center, and spread—as long as certain assumptions and conditions are met. The sample must be independent, random, and large enough that we expect at least 10 successes and failures. Then:

▶ The sampling distribution (the imagined histogram of the proportions from all possible samples) is shaped like a Normal model.

▶ The mean of the sampling model is p, the true proportion in the population.

▶ The standard deviation of the sample proportions is $\sqrt{\dfrac{pq}{n}}$.

We've learned (at last!) to use the sample we have at hand to say something about the *world at large*. This process, called statistical inference, is based on our understanding of sampling models and will be our focus for the rest of the book.

As our first step in statistical inference, we've learned to use our sample to make a *confidence interval* that estimates what proportion of a population has a certain characteristic.

We've learned that:

▶ Our best estimate of the true population proportion is the proportion we observed in the sample, so we center our confidence interval there.

▶ Samples don't represent the population perfectly, so we create our interval with a *margin of error*.

▶ This method successfully captures the true population proportion most of the time, providing us with a level of confidence in our interval.

▶ The higher the level of confidence we want, the *wider* our confidence interval becomes.

(continued)

▶ The larger the sample size we have, the *narrower* our confidence interval can be.

▶ There are important assumptions and conditions we must check before using this (or any) statistical inference procedure.

We've learned to interpret a confidence interval by *Telling* what we believe is true in the entire population from which we took our random sample. Of course, we can't be *certain*. We've learned not to overstate or misinterpret what the confidence interval says.

Terms

Sampling distribution model
Different random samples give different values for a statistic. The sampling distribution model shows the behavior of the statistic over all the possible samples for the same size *n*.

Sampling variability
Sampling error
The variability we expect to see from one random sample to another. It is sometimes called sampling error.

Sampling distribution model for a proportion
If assumptions of independence and random sampling are met, and we expect at least 10 successes and 10 failures, then the sampling distribution of a proportion is modeled by a Normal model with a mean equal to the true proportion value, *p*, and a standard deviation equal to $\sqrt{\dfrac{pq}{n}}$.

Standard error
When we estimate the standard deviation of a sampling distribution using statistics found from the data, the estimate is called a standard error.

$$SE(\hat{p}) = \sqrt{\dfrac{\hat{p}\hat{q}}{n}}$$

Confidence interval for a proportion
A 95% confidence interval for the true value of a proportion is

$$\hat{p} \pm 2*SE(\hat{p})$$

95% of all random samples will yield intervals that capture the true proportion, *p*.

Margin of error
In a confidence interval, the extent of the interval on either side of the observed statistic value. In a 95% confidence interval for a proportion

$$ME = 2\sqrt{\dfrac{\hat{p}\hat{q}}{n}}$$

Critical value
The number of standard errors to move away from the mean of the sampling distribution to correspond to a desired level of confidence. The critical value, denoted z^*, is usually found from a table or with technology, and the margin of error is

$$ME = z^*\sqrt{\dfrac{\hat{p}\hat{q}}{n}}$$

Skills

▶ Understand that the variability of a statistic depends on the size of the sample. Statistics based on larger samples are less variable.

▶ Understand confidence intervals as a balance between the precision and the certainty of a statement about a model parameter.

▶ Understand that the margin of error of a confidence interval for a proportion changes with the sample size and the level of confidence.

▶ Know how to examine your data for violations of conditions that would make inference about a population proportion unwise or invalid.

▶ Be able to use a sampling distribution model to make simple statements about the variability of proportions from sample to sample.

▶ Be able to construct a 95% confidence interval for a proportion.

▶ Be able to interpret a sampling model as describing the distribution of sample proportions from all possible samples.

▶ Be able to interpret a 95% confidence interval for a proportion in a simple sentence or two.

CONFIDENCE INTERVALS FOR PROPORTIONS ON THE COMPUTER

Confidence intervals for proportions are so easy and natural that many statistics packages don't offer special commands for them. Some statistics programs want the "raw data" for computations. For proportions, the raw data are the "success" and "failure" status for each case. Usually, these are given as 1 or 0, but they might be category names like "yes" and "no." Other software and graphing calculators allow you to create confidence intervals from summaries of the data—all you need to enter are the number of successes and the sample size.

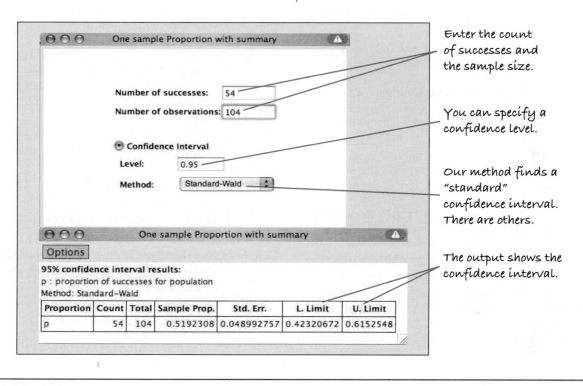

Enter the count of successes and the sample size.

You can specify a confidence level.

Our method finds a "standard" confidence interval. There are others.

The output shows the confidence interval.

EXERCISES

A

1. Dice. When rolling two fair dice, the probability of getting a total of 7 is $p = 1/6$. We plan to roll them 100 times.
a) What's the expected proportion of 7's we might observe?
b) What's the standard deviation of sample proportions?

2. Seeds. A packet of garden seeds contains 250 seeds. The information printed on the packet indicates that the germination rate for this variety of seed is 96%.

a) What proportion of these seeds should we expect to germinate?
b) What's the standard deviation of germination rates for these packets?

3. Dice again. Let's construct the sampling model for the proportion of totals of 7 observed in 100 rolls of a pair of fair dice.
a) Is a Normal model appropriate? Check the assumptions and conditions.
b) Use the 68–95–99.7 Rule to sketch the model.
c) Write a few sentences describing what the model suggests may happen if we roll fair dice 100 times.

4. **Seeds again.** Let's construct the sampling model for the proportion of seeds that might germinate in packets of 250 seeds described in Exercise 2.
   a) Is a Normal model appropriate? Check the assumptions and conditions.
   b) Use the 68–95–99.7 Rule to sketch the model.
   c) Write a few sentences describing what the model suggests about seed germination in these packets.

5. **Blood.** According to the Red Cross, about 42% of Americans have Type A blood. Suppose 80 people show up at a typical blood drive.
   a) What's the expected proportion of people who are Type A?
   b) What's the standard deviation of the Type A proportions found at such drives?

6. **Vision.** About 12% of children are nearsighted. A school district will be testing the vision of 170 incoming kindergarten children.
   a) What proportion of these kids should we expect to be nearsighted?
   b) What's the standard deviation of the proportion who are nearsighted?

7. **Blood again.** Let's construct the sampling model for the proportion of people with Type A blood among 80 people who come to a blood drive, knowing that 42% of Americans are Type A.
   a) Is a Normal model appropriate? Check the assumptions and conditions.
   b) Use the 68–95–99.7 Rule to sketch the model.
   c) Write a few sentences describing what the model suggests may happen at blood drives like these.

8. **Vision again.** Let's construct the sampling model for the proportion of nearsightedness among the kindergartners described in Exercise 6.
   a) Is a Normal model appropriate? Check the assumptions and conditions.
   b) Use the 68–95–99.7 Rule to sketch the model.
   c) Write a few sentences describing what the model suggests schools might find in kindergarten groups this size.

9. **Margin of error.** A TV newscaster reports the results of a poll of voters, and then says, "The margin of error is plus or minus 4%." Explain carefully what that means.

10. **Margin of error.** A medical researcher estimates the percentage of children exposed to lead-base paint, adding that he believes his estimate has a margin of error of about 3%. Explain what the margin of error means.

11. **Conclusions.** A catalog sales company promises to deliver orders placed on the Internet within 3 days. Follow-up calls to a few randomly selected customers show that a 95% confidence interval for the proportion of all orders that arrive on time is 88% ± 6%. What does this mean? Are these conclusions correct? Explain.
   a) Between 82% and 94% of all orders arrive on time.
   b) 95% of all random samples of customers will show that 88% of orders arrive on time.
   c) 95% of all random samples of customers will show that 82% to 94% of orders arrive on time.

   d) We are 95% sure that between 82% and 94% of the orders placed by the sampled customers arrived on time.
   e) On 95% of the days, between 82% and 94% of the orders will arrive on time.

12. **More conclusions.** In January 2002, two students made worldwide headlines by spinning a Belgian euro 250 times and getting 140 heads—that's 56%. That makes the 90% confidence interval (51%, 61%). What does this mean? Are these conclusions correct? Explain.
   a) Between 51% and 61% of all euros are unfair.
   b) We are 90% sure that in this experiment this euro landed heads on between 51% and 61% of the spins.
   c) We are 90% sure that spun euros will land heads between 51% and 61% of the time.
   d) If you spin a euro many times, you can be 90% sure of getting between 51% and 61% heads.
   e) 90% of all spun euros will land heads between 51% and 61% of the time.

13. **Cars.** What fraction of cars is made in Japan? The computer output below summarizes the results of a random sample of 125 autos. Explain carefully what it tells you.

   ```
   z-Interval for proportion
   With 95.00% confidence,
   0.29874 < p(japan) < 0.46926
   ```

14. **Parole.** A study of 902 decisions made by the Nebraska Board of Parole produced the following computer output. Assuming these cases are representative of all cases that may come before the Board, what can you conclude?

   ```
   z-Interval for proportion
   With 95.00% confidence,
   0.56100658 < p(parole) < 0.62524619
   ```

15. **Scrabble.** Using a computer to simulate games of Scrabble, researcher Charles Robinove found that the letter "A" appeared in 54% of the hands. He said his study had a margin of error of ±10%. What conclusion can you make about Scrabble?

16. **Diabetes.** Based on some screening tests a medical researcher reports that 22% of her adult subjects have high levels of blood glucose, a warning sign for diabetes. She says her study has a margin of error of ±7%. What can you conclude about all adults?

17. **Conditions.** For each situation described below, identify the population and the sample, explain what p and $\hat{p}$ represent, and tell whether the methods of this chapter can be used to create a confidence interval.
   a) Police set up an auto checkpoint at which drivers are stopped and their cars inspected for safety problems. They find that 14 of the 134 cars stopped have at least one safety violation. They want to estimate the percentage of all cars that may be unsafe.
   b) A TV talk show asks viewers to register their opinions on prayer in schools by logging on to a website. Of the 602 people who voted, 488 favored prayer in schools. We want to estimate the level of support among the general public.

c) A school is considering requiring students to wear uniforms. The PTA surveys parent opinion by sending a questionnaire home with all 1245 students; 380 surveys are returned, with 228 families in favor of the change.

d) A college admits 1632 freshmen one year, and four years later 1388 of them graduate on time. The college wants to estimate the percentage of all their freshman enrollees who graduate on time.

18. More conditions. Consider each situation described. Identify the population and the sample, explain what p and $\hat{p}$ represent, and tell whether the methods of this chapter can be used to create a confidence interval.

a) A consumer group hoping to assess customer experiences with auto dealers surveys 167 people who recently bought new cars; 3% of them expressed dissatisfaction with the salesperson.

b) What percent of college students have cell phones? 2883 students were asked as they entered a football stadium, and 2432 said they had phones with them.

c) 240 potato plants in a field in Maine are randomly checked, and only 7 show signs of blight. How severe is the blight problem for the U.S. potato industry?

d) 12 of the 309 employees of a small company suffered an injury on the job last year. What can the company expect in future years?

19. Smoking. Among 1815 randomly selected high school students surveyed by the Centers for Disease Control, 417 said they were current smokers.

a) Find the sample proportion $\hat{p}$.

b) Estimate the variability in such sample proportions by finding $SE(\hat{p})$.

20. Cheating. Rutgers University surveyed 4,500 randomly selected high school students nationwide; 3,329 of these students admitted they had cheated on a test at least once.

a) Find the sample proportion $\hat{p}$.

b) Estimate the variability in such sample proportions by finding $SE(\hat{p})$.

21. Smoking again. Consider the CDC study described in Exercise 19.

a) Is it appropriate to use a Normal model to describe the sampling distribution of $\hat{p}$? Check the appropriate assumptions and conditions.

b) Find the margin of error for a 95% confidence interval.

c) Interpret the 95% confidence interval in this context.

22. Cheating again. Consider the Rutgers study described in Exercise 20.

a) Is it appropriate to use a Normal model to describe the sampling distribution of $\hat{p}$? Check the appropriate assumptions and conditions.

b) Find the margin of error for a 95% confidence interval.

c) Interpret the 95% confidence interval in this context.

23. Speeding. A traffic safety study mounted a radar gun along a section of rural interstate highway where the traffic was moving smoothly, and found that 243 out of 355 cars were exceeding the posted speed limit by at least 5 miles per hour.

a) Find the sample proportion $\hat{p}$.

b) Estimate the variability in such sample proportions by finding $SE(\hat{p})$.

24. Vitamin D. A national nutrition study by the Centers for Disease Control reported that 649 of 1546 African-American women tested had vitamin D deficiency.

a) Find the sample proportion $\hat{p}$.

b) Estimate the variability in such sample proportions by finding $SE(\hat{p})$.

25. Speeding again. Consider the traffic safety study described in Exercise 23.

a) Is it appropriate to use a Normal model to describe the sampling distribution of $\hat{p}$? Check the appropriate assumptions and conditions.

b) Find the margin of error for a 95% confidence interval.

c) Interpret the 95% confidence interval in this context.

26. Vitamin D again. Consider the CDC study described in Exercise 24.

a) Is it appropriate to use a Normal model to describe the sampling distribution of $\hat{p}$? Check the appropriate assumptions and conditions.

b) Find the margin of error for a 95% confidence interval.

c) Interpret the 95% confidence interval in this context.

B

27. Coin tosses. In a large class of introductory Statistics students, the professor has each person toss a coin 16 times and calculate the proportion of his or her tosses that were heads. The students then report their results, and the professor plots a histogram of these several proportions.

a) What shape would you expect this histogram to be? Why?

b) Where do you expect the histogram to be centered?

c) How much variability would you expect among these proportions?

d) Explain why a Normal model should not be used here.

28. M&M's. The candy company claims that 10% of the M&M's it produces are green. Suppose that the candies are packaged at random in small bags containing about 50 M&M's. A class of elementary school students learning about percents opens several bags, counts the various colors of the candies, and calculates the proportion that are green.

a) If we plot a histogram showing the proportions of green candies in the various bags, what shape would you expect it to have?

b) Can that histogram be approximated by a Normal model? Explain.

c) Where should the center of the histogram be?

d) What should the standard deviation of the proportion be?

29. **More coins.** Suppose the class in Exercise 27 repeats the coin-tossing experiment.
    a) The students toss the coins 25 times each. Use the 68–95–99.7 Rule to describe the sampling distribution model.
    b) Confirm that you can use a Normal model here.
    c) They increase the number of tosses to 64 each. Draw and label the appropriate sampling distribution model. Check the appropriate conditions to justify your model.
    d) Explain how the sampling distribution model changes as the number of tosses increases.

30. **Bigger bag.** Suppose the class in Exercise 28 buys bigger bags of candy, with 200 M&M's each. Again the students calculate the proportion of green candies they find.
    a) Explain why it's appropriate to use a Normal model to describe the distribution of the proportion of green M&M's they might expect.
    b) Use the 68–95–99.7 Rule to describe how this proportion might vary from bag to bag.
    c) How would this model change if the bags contained even more candies?

31. **Safe food.** Some food retailers propose subjecting food to a low level of radiation in order to improve safety, but sale of such "irradiated" food is opposed by many people. Suppose a grocer wants to find out what his customers think. He has cashiers distribute surveys at checkout and ask customers to fill them out and drop them in a box near the front door. He gets responses from 122 customers, of whom 78 oppose the radiation treatments. What can the grocer conclude about the opinions of all his customers?

32. **Local news.** The mayor of a small city has suggested that the state locate a new prison there, arguing that the construction project and resulting jobs will be good for the local economy. A total of 183 residents show up for a public hearing on the proposal, and a show of hands finds only 31 in favor of the prison project. What can the city council conclude about public support for the mayor's initiative?

33. **Contaminated chicken.** In January 2007 *Consumer Reports* published their study of bacterial contamination of chicken sold in the United States. They purchased 525 broiler chickens from various kinds of food stores in 23 states and tested them for types of bacteria that cause food-borne illnesses. Laboratory results indicated that 83% of these chickens were infected with *Campylobacter*.
    a) Check the conditions for creating a confidence interval.
    b) Construct a 95% confidence interval.
    c) Explain what your confidence interval says about chicken sold in the United States.

34. **Contaminated chicken, second course.** The January 2007 *Consumer Reports* study described in Exercise 33 also found that 15% of the 525 broiler chickens tested were infected with *Salmonella*.
    a) Are the conditions for creating a confidence interval satisfied? Explain.

b) Construct a 95% confidence interval.
c) Explain what your confidence interval says about chicken sold in the United States.

35. **More speeders.** Suppose you wanted to estimate the proportion of cars exceeding the speed limit, and have a margin of error only one-third as large as the study described in Exercise 23. How many cars' speeds would you have to measure with your radar gun?

36. **Chicken again.** Suppose you wanted to estimate the proportion of contaminated chicken with only half the margin of error in the *Consumer Reports* study described in Exercise 33. How many chickens would you have to test?

37. **Contributions, please.** The Paralyzed Veterans of America is a philanthropic organization that relies on contributions. They send free mailing labels and greeting cards to potential donors on their list and ask for a voluntary contribution. To test a new campaign, they recently sent letters to a random sample of 100,000 potential donors and received 4781 donations.
    a) Give a 95% confidence interval for the true proportion of their entire mailing list who may donate.
    b) A staff member thinks that the true rate is 5%. Given the confidence interval you found, do you find that percentage plausible?

38. **Take the offer.** First USA, a major credit card company, is planning a new offer for their current cardholders. The offer will give double airline miles on purchases for the next 6 months if the cardholder goes online and registers for the offer. To test the effectiveness of the campaign, First USA recently sent out offers to a random sample of 50,000 cardholders. Of those, 1184 registered.
    a) Give a 95% confidence interval for the true proportion of those cardholders who will register for the offer.
    b) If the acceptance rate is only 2% or less, the campaign won't be worth the expense. Given the confidence interval you found, what would you say?

39. **Eggs.** Hens typically lay about 3 eggs every 4 days. An agricultural researcher hoping to increase egg production gives a food supplement to a test group of 280 hens, and finds they now produce 221 eggs each day.
    a) Discuss the assumptions and conditions required to create a confidence interval for the true proportion.
    b) Create a 95% confidence interval.
    c) Interpret your interval in this context.
    d) Encouraged by these results, the researcher would like to test this supplement further, this time estimating the production rate with only half the margin of error. How many hens must he use for this study?

40. **New school?** City voters will soon go to the polls to decide whether to support a tax increase to build a new high school. Approval of bond issues like this requires a 60% "super majority" of yes votes. A local radio station phones 148 randomly selected voters, and finds 96 in favor of building the school.
    a) Discuss the assumptions and conditions required to create a confidence interval for the true proportion.

b) Create a 95% confidence interval.

c) Explain why the station said the outcome was "too close to call."

d) The local newspaper wants to conduct a poll of its own, with a margin of error only one-third as large. How many voters must the paper contact?

41. Confidence intervals. Several factors are involved in the creation of a confidence interval. Among them are the sample size, the level of confidence, and the margin of error. Which statements are true?

a) For a given sample size, higher confidence means a smaller margin of error.

b) For a specified confidence level, larger samples provide smaller margins of error.

c) For a fixed margin of error, larger samples provide greater confidence.

d) For a given confidence level, halving the margin of error requires a sample twice as large.

42. Confidence intervals again. Several factors are involved in the creation of a confidence interval. Among them are the sample size, the level of confidence, and the margin of error. Which statements are true?

a) For a given sample size, reducing the margin of error will mean lower confidence.

b) For a certain confidence level, you can get a smaller margin of error by selecting a bigger sample.

c) For a fixed margin of error, smaller samples will mean lower confidence.

d) For a given confidence level, a sample 9 times as large will make a margin of error one-third as big.

C

43. Send money. When they send out their fund-raising letter, a philanthropic organization typically gets a return from about 5% of the people on their mailing list. To see what the response rate might be for future appeals, they did a simulation using samples of size 20, 50, 100, and 200. For each sample size, they simulated 1000 mailings with success rate $p = 0.05$ and constructed the histogram of the 1000 sample proportions, shown below. Explain what these histograms demonstrate about the sampling distribution model for sample proportions. Be sure to talk about shape, center, and spread.

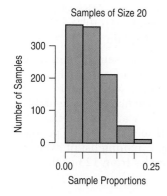

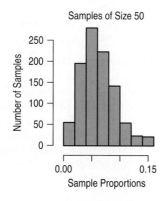

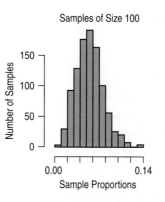

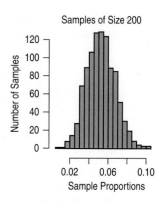

44. Character recognition. An automatic character recognition device can successfully read about 85% of handwritten credit card applications. To estimate what might happen when this device reads a stack of applications, the company did a simulation using samples of size 20, 50, 75, and 100. For each sample size, they simulated 1000 samples with success rate $p = 0.85$ and constructed the histogram of the 1000 sample proportions, shown here. Explain what these histograms demonstrate about the sampling distribution model for sample proportions. Be sure to talk about shape, center, and spread.

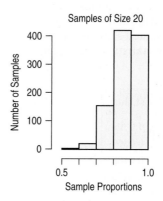

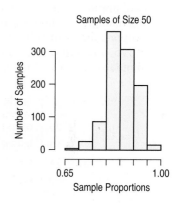

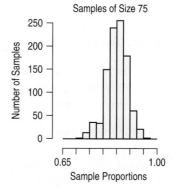

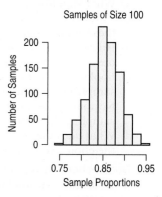

45. Send money again. The philanthropic organization in Exercise 43 expects about a 5% success rate when they send fund-raising letters to the people on their mailing list. In Exercise 43 you looked at the histograms showing distributions of sample proportions from 1000 simulated mailings for samples of size 20,

50, 100, and 200. The sample statistics from each simulation were as follows:

n	mean	st. dev.
20	0.0497	0.0479
50	0.0516	0.0309
100	0.0497	0.0215
200	0.0501	0.0152

a) What should the theoretical mean and standard deviation of the sampling distribution for $\hat{p}$ be for these sample sizes?
b) How close are those theoretical values to what was observed in these simulations?
c) Looking at the histograms in Exercise 43, at what sample size would you be comfortable using the Normal model as an approximation for the sampling distribution?
d) What does the Success/Failure Condition say about the choice you made in part c?

46. Character recognition, again. The automatic character recognition device discussed in Exercise 44 successfully reads about 85% of handwritten credit card applications. In Exercise 44 you looked at the histograms showing distributions of sample proportions from 1000 simulated samples of size 20, 50, 75, and 100. The sample statistics from each simulation were as follows:

n	mean	st. dev.
20	0.8481	0.0803
50	0.8507	0.0509
75	0.8481	0.0406
100	0.8488	0.0354

a) What should the theoretical mean and standard deviation of the sampling distribution for $\hat{p}$ be for these sample sizes?
b) How close are those theoretical values to what was observed in these simulations?
c) Looking at the histograms in Exercise 44, at what sample size would you be comfortable using the Normal model as an approximation for the sampling distribution?
d) What does the Success/Failure Condition say about the choice you made in part c?

47. Nonsmokers. While some nonsmokers do not mind being seated in a smoking section of a restaurant, about 60% of the customers demand a smoke-free area. A new restaurant with 120 seats is being planned. How many seats should be in the nonsmoking area in order to be very sure of having enough seating there? Comment on the assumptions and conditions that support your model, and explain what "very sure" means to you.

48. Meals. A restauranteur anticipates serving about 180 people on a Friday evening, and believes that about 20% of the patrons will order the chef's steak special. How many of those meals should he plan on serving in order to be pretty sure of having enough steaks on hand to meet customer demand? Justify your answer, including an explanation of what "pretty sure" means to you.

49. Baseball fans. In a poll taken in March of 2007, Gallup asked 1006 U.S. adults whether they were baseball fans. 36% said they were. A year previously, 37% of a similar-size sample had reported being baseball fans.
a) Find the margin of error for the 2007 poll if we want 90% confidence in our estimate of the percent of U.S. adults who are baseball fans.
b) Explain what that margin of error means.
c) If we wanted to be 99% confident, would the margin of error be larger or smaller? Explain.
d) Find that margin of error.
e) In general, if all other aspects of the situation remain the same, will smaller margins of error produce greater or less confidence in the interval?
f) Do you think there's been a change from 2006 to 2007 in the real proportion of U.S. adults who are baseball fans? Explain.

50. Cloning 2007. A May 2007 Gallup Poll found that only 11% of a random sample of 1003 adults approved of attempts to clone a human.
a) Find the margin of error for this poll if we want 99% confidence in our estimate of the percent of American adults who approve of cloning humans.
b) Explain what that margin of error means.
c) If we only need to be 90% confident, will the margin of error be larger or smaller? Explain.
d) Find that margin of error.
e) In general, if all other aspects of the situation remain the same, would smaller samples produce smaller or larger margins of error?

51. Teenage drivers. An insurance company checks police records on 582 accidents selected at random and notes that teenagers were at the wheel in 91 of them.
a) Create a 98% confidence interval for the percentage of all auto accidents that involve teenage drivers.
b) Explain what your interval means.
c) Explain what "98% confidence" means.
d) A politician urging tighter restrictions on drivers' licenses issued to teens says, "In one of every five auto accidents, a teenager is behind the wheel." Does your confidence interval support or contradict this statement? Explain.

52. Junk mail. Direct mail advertisers send solicitations (a.k.a. "junk mail") to thousands of potential customers in the hope that some will buy the company's product. The acceptance rate is usually quite low. Suppose a company wants to test the response to a new flyer, and sends it to 1000 people randomly selected from their mailing list of over 200,000 people. They get orders from 123 of the recipients.

a) Create a 90% confidence interval for the percentage of people the company contacts who may buy something.

b) Explain what this interval means.

c) Explain what "90% confidence" means.

d) The company must decide whether to now do a mass mailing. The mailing won't be cost-effective unless it produces at least a 5% return. What does your confidence interval suggest? Explain.

53. Deer ticks. Wildlife biologists inspect 153 deer taken by hunters and find 32 of them carrying ticks that test positive for Lyme disease.

a) Create a 90% confidence interval for the percentage of deer that may carry such ticks.

b) If the scientists want to cut the margin of error in half, how many deer must they inspect?

c) What concerns do you have about this sample?

54. Pregnancy. In 1998 a San Diego reproductive clinic reported 49 live births to 207 women under the age of 40 who had previously been unable to conceive.

a) Find a 90% confidence interval for the success rate at this clinic.

b) Interpret your interval in this context.

c) The clinic wants to cut the stated margin of error in half. How many patients' results must be used?

d) Do you have any concerns about this sample? Explain.

55. Payments. In a May 2007 Experian/Gallup Personal Credit Index poll of 1008 U.S. adults aged 18 and over, 8% of respondents said they were very uncomfortable with their ability to make their monthly payments on their current debt during the next three months. A more detailed poll surveyed 1288 adults, reporting similar overall results and also noting differences among four age groups: 18–29, 30–49, 50–64, and 65+.

a) Do you expect the 95% confidence interval for the true proportion of all 18- to 29-year-olds who are worried to be wider or narrower than the 95% confidence interval for the true proportion of all U.S. consumers? Explain.

b) Do you expect this second poll's overall margin of error to be larger or smaller than the Experian/Gallup poll's? Explain.

56. Back to campus again. In 2004 ACT, Inc., reported that 74% of 1644 randomly selected college freshmen returned to college the next year. The study was stratified by type of college—public or private. The retention rates were 71.9% among 505 students enrolled in public colleges and 74.9% among 1139 students enrolled in private colleges.

a) Will the 95% confidence interval for the true national retention rate in private colleges be wider or narrower than the 95% confidence interval for the retention rate in public colleges? Explain.

b) Do you expect the margin of error for the overall retention rate to be larger or smaller? Explain.

Answers

Just Checking

1. a) $p = 0.06$

b) $\hat{p}$ is the proportion of these 250 donors who have Type O-negative blood.

c) Yes. We can assume these donors are a representative sample of American adults. $250 < 10\%$ of all adults; $250(0.06) = 15$ and $250(0.94) = 235$ are both at least 10.

d) $\mu = 0.06$, $SD(\hat{p}) = \sqrt{\dfrac{(0.06)(0.94)}{250}} \approx 0.015$

e)

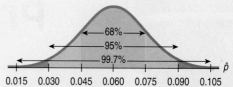

f) The Red Cross should expect to find between 4.5% and 7.5% Type O-negative donors at 68% of blood drives that attract 250 donors, between 3% and 9% at 95% of such drives, and between 1.5% and 10.5% at 99.7% of the drives.

2. a) No. We know that in the sample 17% said "yes"; there's no need for a margin of error.

b) No, we are 95% confident that the percentage falls in some interval, not exactly on a particular value.

c) Yes. That's what the confidence interval means.

d) No. We don't know for sure that's true; we are only 95% confident.

e) No. That's our level of confidence, not the proportion of people receiving unsolicited text messages. The sample suggests the proportion is much lower.

f) The true proportion isn't a variable. Either it's in our interval, or it isn't.

3. a) Yes; this is a random sample of less than 10% of Colorado parents and there are at least 10 successes (162) and failures (192).

b) $\hat{p} = \dfrac{162}{354} = 0.458$;

$SE(\hat{p}) = \sqrt{\dfrac{(0.458)(0.542)}{354}} \approx 0.0265$

c) $2SE = 2(0.0265) = 0.053$, or $\pm 5.3\%$

d) We can be 95% confident that in Colorado between 40.5% and 51.1% of 5–14 year-old children wear helmets when playing with a skateboard, scooter, or inline skates.

4. a) Wider.   b) Lower.   c) Smaller.

d) $4(900) = 3600$ registered voters.

Surprised? Testing Hypotheses About Proportions

Ingots are huge pieces of metal, often weighing more than 20,000 pounds, made in a giant mold. The metal, used for making parts for cars and planes, must be cast in one large piece. If it cracks while being made, the crack can ruin the part. Airplane manufacturers insist that metal for their planes be defect-free, so the ingot must be made over if any cracking is detected, a process costing thousands of dollars.

Metal manufacturers would like to avoid cracking if at all possible. But the casting process is complicated and not everything is completely under control. In one plant, only about 80% of the ingots have been free of cracks. Hoping to reduce cracking, the plant engineers and chemists recently tried out some changes in the casting process. Since then, 400 ingots have been cast and only 17% of them have cracked. Should management declare victory? Has the cracking rate really decreased, or was 17% just due to luck?

We can treat the 400 ingots cast with the new method as a random sample. We know that each random sample will have a somewhat different proportion of cracked ingots. Is the observed 17% merely a result of natural sampling variability, or is this lower cracking rate strong enough evidence to assure management that the true cracking rate now is really below 20%?

People want answers to questions like these all the time. Has the president's approval rating changed since last month? Has teenage smoking decreased in the past five years? Is the global temperature increasing? Did the Super Bowl ad we bought actually increase sales? To answer such questions, we test *hypotheses* about models.

Are the Results Convincing?

To test whether the changes made by the engineers have *reduced* the cracking rate, we assume that they have in fact made no difference and that any apparent improvement is just random fluctuation (sampling error). So, we start by assuming that the proportion of cracks is still 20%.

What would convince you that the cracking rate had actually gone down? If you observed a cracking rate of only 1% in your sample, you'd likely be convinced. But if the sample cracking rate is only slightly lower than 20%, you should be skeptical. After all, observed proportions do vary, so we wouldn't be surprised to see some difference. Is the reduction to 17% dramatic enough to be convincing? In other words, are these results **statistically significant?** To decide, we turn to the sampling model.

We have a large sample of 400 new ingots and there's no reason to think they're not independent, so we can use a Normal model. If the cracking rate is still $p = 0.20$, that's the proportion we'd expect to see in samples. Sample proportions could vary though, and the standard deviation of the sampling model is

WHY IS THIS A STANDARD DEVIATION AND NOT A STANDARD ERROR?

Because we haven't estimated anything. When we assume that the cracking rate is still 20%, it gives us a value for the model parameter p, so this is a standard deviation and not a standard error. (When we found a confidence interval we estimated the standard deviation using the sample value $\hat{p}$.)

$$SD(\hat{p}) = \sqrt{\frac{pq}{n}} = \sqrt{\frac{(0.20)(0.80)}{400}} = 0.02$$

Let's look at where the sample cracking rate of $\hat{p} = 0.17$ fits into the sampling model $N(0.20, 0.02)$:

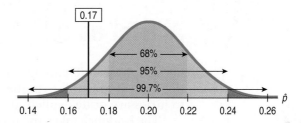

Sure, it's lower than what we expected, but not surprisingly low. If the cracking rate is really still 20%, it wouldn't be unusual to have a sample in which only 17% of the ingots crack. After all, that's less than 2 standard deviations below the mean—not statistically significant. The lower cracking rate may be just sampling error, not indicative of a genuine improvement. The company can't conclude that the new casting process helps.

The Reasoning of Hypothesis Testing

To decide whether the new casting process had helped, we asked ourselves whether it was reasonable to believe we could have seen results like our data if the cracking rate really had remained the same. Statisticians call this kind of reasoning **hypothesis testing,** and use it often. Testing a hypothesis requires a four-step procedure.

NOTATION ALERT

Capital H is the standard letter for hypotheses. H_0 always labels the null hypothesis, and H_A labels the alternative hypothesis.

1. **Hypothesis.** We begin with an assumption that nothing interesting has happened, called the **null hypothesis.** "Null" means "nothing"—the president's approval rating hasn't changed, teen smoking is the same as 5 years ago, global temperatures remain constant, the Super Bowl ads we bought didn't help sales.
2. **Model.** We check assumptions and conditions to confirm that our data will allow us to make meaningful decisions and to determine what model we should use.

3. **Mechanics.** We use our model to describe what might happen if the null hypothesis were true, and then ask ourselves whether our data look surprising.

4. **Conclusion.** If it's reasonable to think our data are an example of expected sample-to-sample variability, we can't claim that anything has happened. But if our data appear to be very unusual, we say the results are **statistically significant.** Based on this evidence we reject the null hypothesis concluding that something appears to have changed.

So, basically we start by assuming nothing has happened, then look for evidence to the contrary. Does that reasoning seem backward? Actually, you've seen hypothesis testing before in a different context. This is the logic of jury trials.

A Trial as a Hypothesis Test

Let's suppose a defendant has been accused of robbery. In the American justice system, the null hypothesis is that the defendant is innocent. Instructions to juries are quite explicit about this.

The evidence takes the form of facts (the trail's "data") that seem to contradict the presumption of innocence. To be admissible in court, evidence must satisfy a strict set of conditions. The judge is responsible for making everyone follow all the rules and procedures that model a fair trial for the defendant.

The mechanics of the trial take place in the courtroom. The prosecutor presents the evidence against the presumption of innocence. ("If the defendant were innocent, wouldn't it be remarkable that the police found him at the scene of the crime with a bag full of money in his hand, a mask on his face, and a getaway car parked outside?")

The jury judges whether the evidence against the defendant would be plausible *if the defendant were in fact innocent.* They must decide whether the evidence raises a reasonable doubt about the null hypothesis. If the evidence contradicts the hypothesis of innocence the jury rejects the null and declares the defendant to be guilty.

"If the People fail to satisfy their burden of proof, you must find the defendant not guilty."

— NY state jury instructions

If the evidence is not strong enough to reject the defendant's presumption of innocence, what verdict does the jury return? They say "not guilty." Notice that they do not say that the defendant is innocent. All they say is that they have not seen sufficient evidence to convict, to reject innocence. The defendant may, in fact, be innocent, but the jury has no way to be sure.

The Process of Hypothesis Testing

Hypothesis tests follow the carefully structured 4-step path that we outlined above. Now let's look at each step in detail.

1. Hypotheses

Some folks pronounce the hypothesis labels "Ho!" and "Ha!" (but it makes them seem overexcitable). We prefer to pronounce H_0 "H naught" (as in "all is for naught").

First we state the null hypothesis. That's usually the skeptical claim that nothing's different. Are we considering a (New! Improved!) possibly better method? The null hypothesis says, "Oh yeah? Convince me!" To convert a skeptic, enough evidence must pile up against the null hypothesis that we can reasonably reject it.

In statistical hypothesis testing, hypotheses are almost always about population parameters. To assess how unlikely our data may be, we need a null model. The **null hypothesis** specifies a particular parameter value to use in our model. In the usual shorthand, we write H_0: *parameter = hypothesized value*. The **alternative hypothesis**, H_A, contains the values of the population parameter we consider plausible if we reject the null.

That's not as complicated as it sounds. For the ingot manufacturer wondering if the new casting process has improved the 20% cracking rate, the hypotheses are:

$$H_0: p = 0.20$$
$$H_A: p < 0.20$$

For Example Writing Hypotheses

A large city's Department of Motor Vehicles claimed that 80% of candidates pass driving tests, but a newspaper reporter's survey of 90 randomly selected local teens who had taken the test found only 61 who passed.

Question: Does this finding suggest that the passing rate for teenagers is lower than the DMV reported? Write appropriate hypotheses.

I'll assume that the passing rate for teenagers is the same as the DMV's overall rate of 80%, unless there's strong evidence that it's lower.

$$H_O: p = 0.80$$
$$H_A: p < 0.80$$

Just Checking

1. A drug company is currently selling an allergy medication that gives relief to 75% of the people who use it. Now the company's scientists want to see if their new version of the drug works even better.

   a) What's their null hypothesis?
   b) What's the alternative hypothesis?

(Check your answers on page 418.)

2. Model

To plan a statistical hypothesis test, specify the *model* you will use to test the null hypothesis and the parameter of interest. Of course, all models require assumptions, so you will need to state them and check any corresponding conditions.

Watch out: Your Model step could end with

Because the conditions are not satisfied, I can't proceed with the test. (If that's the case, stop and reconsider.)

But usually your Model step will say:

Because the conditions are satisfied, I can use a Normal model for the sampling distribution of the proportion.

And then you should name the test procedure you'll use. The test about proportions is called a **one-proportion z-test.**

ONE-PROPORTION Z-TEST

The conditions for the one-proportion z-test are the same as for the one-proportion z-interval. We test the hypothesis $H_0: p = p_0$ using the statistic $z = \dfrac{(\hat{p} - p_0)}{SD(\hat{p})}$. We use the hypothesized proportion to find the standard deviation, $SD(\hat{p}) = \sqrt{\dfrac{p_0 q_0}{n}}$.

When the conditions are met we can use a Normal model.

WHEN THE CONDITIONS FAIL . . .

You might proceed with caution, explicitly stating your concerns. Or you may need to do the analysis with and without an outlier, or on different subgroups, or after re-expressing the response variable. Or you may not be able to proceed at all.

The ingot manufacturer cast a sample of $n = 400$ ingots using the new process.

✔ **Independence Assumption.** There's no reason to think these ingots aren't independent of one another.
✔ **Randomization Condition.** Although not a random sample, these ingots should be representative of all ingots.
✔ **10% Condition.** These 400 are fewer than 10% of all the ingots the new casting method might produce.
✔ **Success/Failure Condition.** If there's been no change in the cracking rate, the company would expect $np = 400(0.20) = 80$ "successes" and $nq = 320$ "failures".[1]

The assumptions and conditions are satisfied, so it's okay to use a Normal model and do a one-proportion z-test.

For Example   Checking the Conditions

Recap: A large city's DMV claimed that 80% of candidates pass driving tests. A reporter has results from a survey of 90 randomly selected local teens who had taken the test.

Question: Are the conditions for inference satisfied?

✔ **Randomization Condition:** The 90 teens surveyed were a random sample of local teenage driving candidates.
✔ **10% Condition:** 90 is fewer than 10% of the teenagers who take driving tests in a large city.
✔ **Success/Failure Condition:** We expect $np_0 = 90(0.80) = 72$ successes and $nq_0 = 90(0.20) = 18$ failures. Both are at least 10.

The conditions are satisfied, so it's okay to use a Normal model and perform a one-proportion z-test.

---

[1]Only a statistician would call *a cracked* ingot a success.

Just Checking

2. The drug company plans to test their new medication on a random sample of 80 patients with allergies. The old allergy drug offered relief to 75% of those who used it. Check the assumptions and conditions for inference.

(Check your answers on page 418.)

3. Mechanics

Now it's time to Do The Math. Could the observed results just be sampling error? Or are they so unusual that we can't continue to believe our null hypothesis that nothing has changed? To decide, we use the sampling model to calculate the z-score that shows how far our sample proportion $\hat{p}$ lies from the hypothesized value of p.

In samples of 400 ingots, the sampling model says we should expect the proportion with cracks to be $p = 0.20$, varying from sample to sample with a standard deviation of $SD(\hat{p}) = 0.02$. The test sample had a cracking rate of $\hat{p} = 0.17$, for which the z-score is:

$$z = \frac{\hat{p} - p}{SD(\hat{p})} = \frac{0.17 - 0.20}{0.02} = -1.5$$

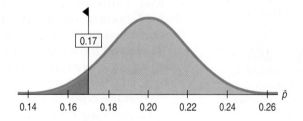

For Example Finding a z-Score

Recap: A large city's DMV claimed that 80% of candidates pass driving tests, but a survey of 90 randomly selected local teens who had taken the test found only 61 who passed.

Question: What's the z-score for the one-proportion z-test?

I have $n = 90$, $x = 61$, and a hypothesized $p = 0.80$.

$$\hat{p} = \frac{61}{90} \approx 0.678$$

$$SD(\hat{p}) = \sqrt{\frac{p_0 q_0}{n}} = \sqrt{\frac{(0.8)(0.2)}{90}} \approx 0.042$$

$$z = \frac{\hat{p} - p_0}{SD(\hat{p})} = \frac{0.678 - 0.800}{0.042} \approx -2.90$$

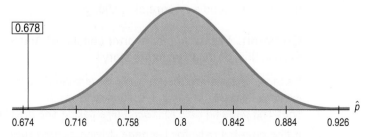

Just Checking

3. The drug company's current allergy medication offers relief to 75% of the people who take it. The new version of the drug relieved allergy symptoms for 69 of the 80 patients who tested it.

   a) What's $\hat{p}$ for these patients?
   b) What's the z-score for this sample?

(Check your answers on page 418.)

DON'T "ACCEPT" THE NULL HYPOTHESIS

Every child knows that he (or she) is at the "center of the universe," so it's natural to suppose that the sun revolves around the earth. The fact that the sun appears to rise in the east every morning and set in the west every evening is *consistent* with this hypothesis and *seems* to lend support to it, but it certainly doesn't prove it, as we all eventually come to understand.

4. Conclusion

Now we must make a decision. Can we still trust the null hypothesis? Is what happened statistically significant—too far from what we expected to be just sampling error? How far is too far? We base that decision on the **Rule of 2.**

> **The Rule of 2** A sample result that's more than 2 standard deviations from what should happen if the null hypothesis were true is strong evidence against the null hypothesis.

So, if $z > 2$ (or $z < -2$), we reject the null hypothesis. If z tells us our result is within 2 standard deviations of what the null suggests, we can't reject it. Notice, though, that we can't *accept* the null. Evidence that isn't strong enough to convince us that the null hypothesis is false doesn't prove it's true. Think about jury trials again. When there's reasonable doubt, juries find defendants "not guilty." The jury can't proclaim the person to be innocent just because the prosecution failed to prove guilt.

Once you have made that decision, it's time for your conclusion. Tell whether you have rejected or failed to reject the null hypothesis, and then, as always, state a conclusion in context.

When the company tested a new casting process for metal ingots, it found a cracking rate of 17%. Although that's less than the usual rate of 20%, the model finds $z = -1.5$—not unusual. Because the result is less than 2 standard deviations below the expected rate, the company fails to reject the null hypothesis. There is not enough evidence to say that the new casting process reduces the cracking rate.

For Example Stating the Conclusion

Recap: A large city's DMV claimed that 80% of candidates pass driving tests. Data from a reporter's survey of randomly selected local teens who had taken the test produced a z-score of -2.90.

Question: What can the reporter conclude? And how might the reporter explain this for the newspaper story?

Because the sample results were more than 2 standard deviations lower than expected, I reject the null hypothesis. These survey data provide strong evidence that the passing rate for teenagers taking the driving test is lower than 80%.

If the passing rate for teenage driving candidates were actually 80%, it would be extremely unusual to find a sample with a success rate this low, casting doubt that the DMV's stated success rate applies to teens.

Just Checking

4. When the drug company tested the new drug on 80 people, the success rate in providing relief from allergy symptoms was much higher than the 75% rate for the current version of this medicine. The sampling model for such sample proportions says this result has a z-score of $z = 2.32$. What should the company conclude?

5. Encouraged by this preliminary result, the company funds a more definitive large-scale study, testing the drug on over 1000 people. The z-score for the observed success rate among those patients turns out to be $z = 1.78$. Now what should the company conclude?

(Check your answers on page 418.)

STEP-BY-STEP EXAMPLE   Testing a Hypothesis

Anyone who plays or watches sports has heard of the "home field advantage." Teams tend to win more often when they play at home. Or do they?

If there were no home field advantage, the home teams would win about half of all games played. In the 2007 Major League Baseball season, there were 2431 regular-season games. (Tied at the end of the regular season, the Colorado Rockies and San Diego Padres played an extra game to determine who won the Wild Card playoff spot.) It turns out that the home team won 1319 of the 2431 games, or 54.26% of the time.

Question: Could this deviation from 50% be explained just from natural sampling variability, or is it evidence to suggest that there really is a home field advantage, at least in professional baseball?

Plan State what we want to know.

Define the variables and discuss the W's.

Hypotheses The null hypothesis makes the claim of no difference from the baseline. Here, that means no home field advantage.

We are interested in a home field *advantage,* so the alternative hypothesis is that the home team wins more often.

Model Think about the assumptions and check the appropriate conditions.

I want to know whether the home team in professional baseball is more likely to win. The data are all 2431 games from the 2007 Major League Baseball season. The variable is whether or not the home team won. The parameter of interest is the proportion of home team wins. If there's no advantage, I'd expect that proportion to be 0.50.

$$H_O: p = 0.50$$
$$H_A: p > 0.50$$

✔ **Independence Assumption:** Generally, the outcome of one game has no effect on the outcome of another game. But this may not be strictly true. For example, if a key player is injured, the probability that the team will win in the next couple of games may decrease slightly, but independence is still roughly true.

(continued)

✔ **Randomization Condition:** I have results for all 2431 games of the 2007 season. But I'm not just interested in 2007, and those games, while not randomly selected, should be a reasonable representative sample of all Major League Baseball games in the recent past and near future.

✔ **10% Condition:** We are interested in home field advantage for Major League Baseball for all seasons. While not a random sample, these 2431 games are fewer than 10% of all games played over the years.

✔ **Success/Failure Condition:** Both $np_O = 2431(0.50) = 1215.5$ and $nq_O = 2431(0.50) = 1215.5$ are at least 10.

Specify the sampling distribution model.

State what test you plan to use.

Because the conditions are satisfied, I'll use a Normal model for the sampling distribution of the proportion and do a **one-proportion z-test.**

Mechanics The null model gives us the mean, and (because we are working with proportions) the mean gives us the standard deviation.

The null model is a Normal distribution with a mean of 0.50 and a standard deviation of

$$SD(\hat{p}) = \sqrt{\frac{p_O q_O}{n}} = \sqrt{\frac{(0.5)(1 - 0.5)}{2431}}$$
$$= 0.01014$$

The observed proportion, $\hat{p}$, is 0.5426.

Now we find the z-score for the observed proportion, to find out how many standard deviations it is from the hypothesized proportion.

So the z-value is

$$z = \frac{0.5426 - 0.5}{0.01014} = 4.20$$

The sample proportion lies 4.20 standard deviations above the mean.

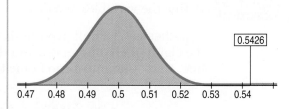

0.5426

| 0.47 | 0.48 | 0.49 | 0.5 | 0.51 | 0.52 | 0.53 | 0.54 |

Conclusion State your conclusion about the parameter—in context, of course!

If the true proportion of home team wins were 0.50, then an observed value of 0.5426 would be far more than 2 standard deviations above the mean. With a z-score so high, I reject H_O. I have evidence that the true proportion of home team wins is over 50%. It appears there is a home field advantage.

TI Tips | Testing a hypothesis

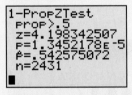

By now probably nothing surprises you about your calculator. Of course it can help you with the mechanics of a hypothesis test. But that's not much. It cannot write the correct hypotheses, check the appropriate conditions, interpret the results, or state a conclusion. You have to do the tough stuff!

Let's do the mechanics of the Step-By-Step example about home field advantage in baseball. We hypothesized that home teams would win 50% of all games, but during this 2431-game season they actually won 54.26% of the time.

- Go to the **STAT TESTS** menu. Scroll down the list and select **5:1-Prop ZTest**.
- Specify the hypothesized proportion **p0**.
- Enter x, the observed number of wins: **1319**.
- Specify the sample size: **2431**.
- Indicate that you want to see if the observed proportion is significantly greater than what was hypothesized.
- **Calculate** the result.

OK, the rest is up to you. The calculator reports a *z*-score of 4.20. Such a large value indicates that the high percentage of home team wins is highly unlikely to be sampling error. State your conclusion in the appropriate context.

And how big is the advantage for the home team? In the last chapter you learned to create a 95% confidence interval. Try it here.

Looks like we can be 95% confident that in major league baseball games the home team wins between 52.3% and 56.2% of the time. Over a full season, the low end of this interval, 52.3% of the 81 home games, is nearly 2 extra victories, on average. The upper end, 56.2%, is 5 extra wins.

Making Errors

Nobody's perfect. Even with lots of evidence, we can still make the wrong decision. In fact, when we perform a hypothesis test correctly, we can make mistakes in *two* ways:

  I. The null hypothesis is true, but we mistakenly reject it.
 II. The null hypothesis is false, but we fail to reject it.

These two types of errors are known as **Type I** and **Type II errors**. One way to keep the names straight is to remember that we start by assuming the null hypothesis is true, so a Type I error is the first kind of error we could make.

Here's an illustration of the situations:

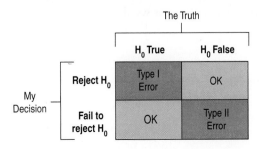

Which type of error is more serious depends on the situation. In the jury trial, a Type I error occurs if the jury convicts an innocent person. A Type II error occurs if the jury fails to convict a guilty person. Which seems more serious to you?

The company testing a new casting process in hopes of reducing the percentage of ingots that crack must be concerned about the possibility of making each type of error. The null hypothesis is that the new process doesn't help. If that's true, but just because of sampling error the test sample of ingots has very few cracks, then the company will commit a Type I error. It will reject the null hypothesis, falsely concluding that the problem is solved. The company may then waste a lot of money implementing a new process that doesn't work. On the other hand, maybe the new casting process really does help, but the test sample results don't show an improvement that's dramatic enough to be convincing. The company will fail to reject the null hypothesis, and miss an opportunity to solve the cracking problem.

For Example Describing the Errors

Recap: The city's DMV claims that 80% of teens pass drivers tests the first time they try. A reporter surveys a sample of local teens to see if the passing rate is really lower.

Question: What would it mean if the reporter's research leads to a Type I error? Or a Type II error?

It's a Type I error if the DMV claim is actually correct, but the reporter writes a story falsely asserting that fewer than 80% of teens pass the test on the first try.

It's a Type II error if the reporter fails to discover that the teen pass rate really is lower than the DMV's 80% claim.

Just Checking

6. Consider again the drug company's test to see if a new version of its allergy medication provides relief for more than the 75% of patients helped by the current drug.

   a) Explain what a Type I error would be in this context.
   b) Explain what a Type II error would be in this context.

(Check your answers on page 418.)

Alternative Alternatives

Tests on the ingot data can be viewed in three different ways. We know the old cracking rate is 20%, so the null hypothesis is

$$H_0: p = 0.20$$

Now we have a choice of alternative hypotheses. A metallurgist working for the company might be interested in *any* change in the cracking rate due to the new process. Even if the rate got worse, she might learn something useful from it. She's interested in possible changes on both sides of the null hypothesis. So she would write her alternative hypothesis as

$$H_A: p \neq 0.20$$

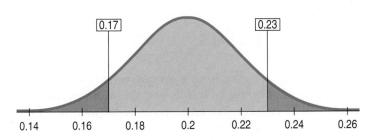

An alternative hypothesis such as this is known as a **two-tailed alternative**, because we are equally interested in deviations on either side of the null hypothesis value. When we are looking for changes in either direction, we say we're doing a **two-tailed test.**

But management is really interested only in *lowering* the cracking rate below 20%. The scientific value of knowing how to *increase* the cracking rate may not appeal to them. The only alternative of interest to them is that the cracking rate *decreases*. The null hypothesis is the same, but they would write their alternative hypothesis as

$$H_A: p < 0.20$$

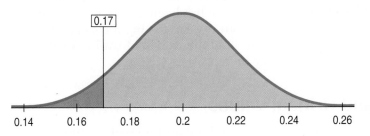

An alternative hypothesis that focuses on deviations from the null hypothesis value in only one direction is called a **one-tailed alternative.** When we're interested in knowing whether the parameter has decreased, we say we're doing a **lower-tail test.**

And there's one more possibility. Perhaps the company's scientists have developed a new casting process that's much cheaper. Management would like to switch to that process, but fears that doing so could increase the cracking rate. They plan to make the cost-saving change unless a test run of ingots reveals evidence of trouble. The null hypothesis is still the same, but now the alternative is:

$$H_A: p > 0.20$$

When we're interested in knowing whether the parameter has increased, we say we're doing an **upper-tail test.**

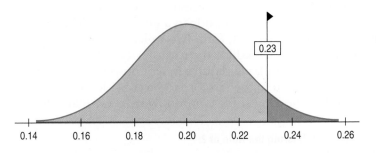

Just Checking

7. For each situation, decide whether you should do an upper-tail, lower-tail, or two-tailed test, then write the appropriate null and alternative hypotheses.

(continued)

a) In 2000 Florida changed its laws, no longer requiring motorcycle riders over the age of 21 to wear helmets. Younger riders are still legally required to wear helmets, but recent accident records raise concerns that helmet use among riders under 21 may have dropped below the 60% level seen in the past.

b) It's a medical fact that male babies are slightly more common than female babies; in the past, 51.7% of babies were male. Prenatal ultrasound exams now make it possible for couples to find out the sex of their child early in a pregnancy. There is a fear that some parents may use this technology to select the sex of their children.

c) People with Type 2 diabetes are known to have about a 20.2% chance of suffering a heart attack within a 7-year period. A 2007 article in the *New England Journal of Medicine* raised concerns that patients taking the diabetes drug *Avandia* might carry an increased risk.

(Check your answers on page 418.)

Alternative Decision Rules

So far we have used the Rule of 2 to decide whether an outcome is statistically significant. That's a pretty good rule of thumb, but as you know, it doesn't guarantee that our decision will be correct. We know that ±2 SDs captures the middle 95% of the sampling model, but that means even when the null hypothesis is true, about 5% of all samples will produce results beyond the $z = \pm 2$ threshold. If that happens we'll reject the null and we'll be wrong. Using the Rule of 2, then, means we run a 5% risk of Type I error.

Whether that level of risk is acceptable depends on the situation. If we wonder whether teen smoking has changed in the past 10 years, we might be comfortable with a 10% risk of thinking there has been a change when in fact teens are still smoking at the same rate. But if we're testing the steel bridge girders to see if they're strong enough to hold the weight of traffic, we'd want the risk of being wrong to be far lower than 5%.

Alpha Levels and Critical Values

Our willingness to risk a Type I error is called the **alpha level** (or the **significance level**) of the test. When we use the Rule of 2, the alpha level is approximately 5%; we write that $\alpha = 0.05$. Commonly used alpha levels are 0.10, 0.05, and 0.01. The z-score that corresponds to a given alpha level is called the **critical value,** written z^*. In fact, for a two-tailed test using $\alpha = 0.05$, $z^* = 1.96$—our Rule of 2. To reduce the risk of a Type I error (incorrectly rejecting a null that's true), we require more compelling evidence to reject it, and that means it takes a higher z-score to be considered statistically significant. Here's a table showing some commonly used critical values:

Adjusting the Rule of 2:

Acceptable Risk of Type I Error	Critical Values	
	One-tailed Test	Two-tailed Test
$\alpha = 0.10$	$z^* = 1.28$	$z^* = 1.645$
$\alpha = 0.05$	$z^* = 1.645$	$z^* = 1.96$
$\alpha = 0.01$	$z^* = 2.33$	$z^* = 2.58$

While we can't eliminate the chance of ever being wrong, we can pick alpha levels and critical values that help us take acceptable risks. We can also measure the size of that risk more precisely, using another guide to decision-making.

P-Values

We have many P's to keep straight. We use an upper-case *P* for probabilities, as in *P*(A), and for the special probability we care about in hypothesis testing, the P-value.

We use lowercase *p* to denote our model's under-lying proportion parameter and $\hat{p}$ to denote our observed proportion statistic.

The fundamental step in our reasoning is the question "Are the data surprising?" Once we know how unusual our results are (the *z*-score), we can take one more step. We find the probability of seeing data like these (or even more extreme) given that the null hypothesis is true, called the **P-value.**

- When the P-value is low enough, that tells us it's very unlikely we'd observe data like these if the null were true. That makes our results statistically significant, and we reject the null hypothesis.
- When the P-value is high, we haven't seen anything surprising. Events with a high probability of happening happen often. If our results could reasonably be explained as just sampling error, then they're not statistically significant, We won't reject the null hypothesis.

The ingot manufacturer's trial of a new casting process found a cracking rate of only 17%. Although lower than the usual rate of 20%, the *z*-score for this sample of 400 ingots was $z = -1.5$. The P-value is the probability of getting a sample with such a low (or even lower) failure rate just by chance:
$$\text{P-value} = P(z < -1.5) = 0.067.^2$$

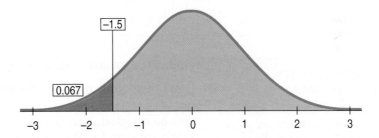

There's a 6.7% chance that this lower cracking rate could show up in a sample even if the new process isn't any better. This P-value is greater than an alpha level of 5%, so management won't be convinced that the new casting process helps.

For Example Finding a P-Value

Recap: The DMV said that 80% of candidates pass driving tests, but there was a lower success rate in a random sample of teenagers, yielding a *z*-score of $z = -2.90$.

Questions: What's the P-value of this test? What does it mean in this context? What conclusion can we make?

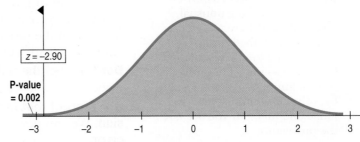

This is a lower-tail test, so the P-value = $P(z < -2.90) = 0.002$

If the passing rate for teenage drivers really were 80%, we'd expect to find a success rate this low in only 1 of 500 samples (0.2%). A P-value this small is strong evidence that the passing rate for teens is lower than 80%.

---

²If you need to refresh your memory about how your calculator or a computer can find Normal model probabilities, look back at Chapter 6 (TI Tips, p. 124).

Just Checking

8. The 2007 article in the *New England Journal of Medicine* investigating whether patients taking the diabetes drug *Avandia* might be at increased risk of having a heart attack found a P-value of 0.03. What should the researchers conclude?

9. Earlier research about *Avandia* had also noticed an increased risk in the patients studied, reporting a P-value of 0.27. Why didn't those researchers express alarm about the drug?

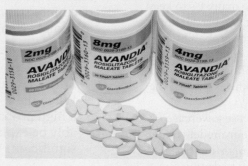

(Check your answers on page 418.)

🚫 WHAT CAN GO WRONG?

DON'T WE WANT TO REJECT THE NULL?

Often the folks who collect the data or perform the experiment hope to reject the null. (They hope the new drug is better than the placebo, or new ad campaign is better than the old one.) But when we practice Statistics, we can't allow that hope to affect our decision. The essential attitude for a hypothesis tester is skepticism. Until we become convinced otherwise, we cling to the null's assertion that there's nothing unusual, no effect, no difference, etc. As in a jury trial, the burden of proof rests with the alternative hypothesis—innocent until proven guilty. When you test a hypothesis, you must act as judge and jury, but you are not the prosecutor.

Hypothesis tests are widely used—and widely misused. Beware.

- **Don't base your null hypotheses on what you see in the data.** You are not allowed to look at the data first and create a null hypothesis that will be rejected. You should always *Think* about the situation you are investigating and make your null hypothesis describe the "nothing interesting" or "nothing has changed" scenario. No peeking at the data!

- **Don't base your alternative hypothesis on the data, either.** Again, you need to *Think* about the situation. Are you interested only in knowing whether something has *increased*? Then write a one-tailed (upper-tail) alternative. Or would you be equally interested in a change in either direction? Then you want a two-tailed alternative. You should decide whether to do a one- or two-tailed test based on what results would be of interest to you, not what you see in the data.

- **Don't forget to check the conditions.** The reasoning of inference depends on randomization. No amount of care in calculating a test result can recover from biased sampling. The probabilities we compute depend on the independence assumption. And our sample must be large enough to justify our use of a Normal model.

- **Don't accept the null hypothesis.** You may not have found enough evidence to reject it, but you surely have *not* proven it's true!

- **Don't forget that in spite of all your care, you might make a wrong decision.** We can never eliminate the possibility of a Type I error or a Type II error.

🚫

HYPOTHESIS TESTING **IN YOUR WORLD**

Retired NFL players have high rate of brain damage

A study commissioned by the National Football League reports that Alzheimer's disease or similar memory-related diseases appear to have been diagnosed in the league's former players vastly more often than in the national population.

. . . [T]he University of Michigan's Institute for Social Research, conducted a phone survey in late 2008 [asking] 1,063 retired players . . . if they had ever been diagnosed with "dementia, Alzheimer's disease, or other memory-related disease." The Michigan researchers found that 6.1 percent of players age 50 and above reported that they had received a dementia-related diagnosis, five times higher than the cited national average, 1.2 percent. Players ages 30 through 49 showed a rate of 1.9 percent, or 19 times that of the national average, 0.1 percent.

Dr. Daniel P. Perl, the director of neuropathology at the Mount Sinai School of Medicine in New York, . . . described the Michigan work as [statistically] significant.

The findings could ring loud at the youth and college levels, which often take cues from the N.F.L. on safety policies and whose players emulate the pros. Hundreds of on-field concussions are sustained at every level each week, with many going undiagnosed and untreated. Sean Morey, an Arizona Cardinals player who has been vocal in supporting research in this area, said: "This is about more than us—it's about the high school kid . . . who might not die on the field because he ignored the risks of concussions."

http://www.nytimes.com/2009/09/30/sports/football/30dementia.html?_r=2&hp

WHAT HAVE WE LEARNED?

We've learned to use what we see in a random sample to test a hypothesis about the world. This is our second step in statistical inference, complementing our use of confidence intervals.

We've learned that testing a hypothesis involves proposing a model, then seeing whether the data we observe are consistent with that model or are so unusual that we must reject it. We base our decision on the Rule of 2.

▶ We start with a null hypothesis specifying the parameter of a model we'll test using our data.

▶ Our alternative hypothesis can be one- or two-tailed, depending on what we want to learn.

▶ We must check the appropriate assumptions and conditions before proceeding with our test.

▶ If the data are far out of line with the null hypothesis model, we will reject the null hypothesis.

▶ If the data are consistent with the null hypothesis model, we will not reject the null hypothesis.

▶ We must always state our conclusion in the context of the original question.

(continued)

We've learned about the two kinds of errors we might make, and we've seen why in the end we're never sure we've made the right decision.

▶ If the null hypothesis is really true and we reject it, that's a Type I error; the alpha level of the test is the probability that this could happen.

▶ If the null hypothesis is really false but we fail to reject it, that's a Type II error.

Terms

Null hypothesis
The claim being assessed in a hypothesis test is called the null hypothesis. Usually, the null hypothesis is a statement of "no change from the traditional value," "no effect," "no difference," or "no relationship."

Alternative hypothesis
The alternative hypothesis proposes what we should conclude if we find the null hypothesis to be unlikely.

Two-tailed alternative
An alternative hypothesis is two-tailed ($H_A: p \neq p_0$) when we are interested in changes in *either* direction.

One-tailed alternative
An alternative hypothesis is one-tailed ($H_A: p > p_0$ or $H_A: p < p_0$) when we are interested in changes in *only one* direction.

One-proportion z-test
A test of the null hypothesis that a proportion equals a specified value ($H_0: p = p_0$) when a Normal model is appropriate, by calculating $z = \dfrac{\hat{p} - p_0}{SD(\hat{p})}$.

The Rule of 2
We consider a sample statistic more than 2 standard deviations from the hypothesized value ($|z| > 2$) to be strong evidence that the null hypothesis is false.

Statistically significant
Sample results are statistically significant if they are too unusual to be reasonably explained as sampling error, and thus cast doubt on the null hypothesis.

Type I error
The error of rejecting a null hypothesis that is actually true. The probability of Type I error is α.

Type II error
The error of failing to reject a null hypothesis that is actually false.

**Significance level
Alpha level (α)**
The alpha level of a test establishes the level of proof we'll require. That determines the critical value of z required to reject the null hypothesis. When we use the Rule of 2, $\alpha \approx 0.05$.

P-value
The probability that results at least as extreme as our data could have arisen if the null hypothesis were true. When the P-value is small (less than α), the results are statistically significant and we reject the null hypothesis. When the P-value is large, we lack evidence that the null hypothesis is false.

Skills

▶ Be able to state the null and alternative hypotheses for a one-proportion z-test.

▶ Know the conditions that must be true for a one-proportion z-test to be appropriate, and know how to examine your data for violations of those conditions.

▶ Be able to identify and use the alternative hypothesis when testing hypotheses. Understand how to choose between a one-tailed and two-tailed alternative hypothesis.

▶ Be able to perform a one-proportion z-test and make a decision based on the Rule of 2.

▶ Be able to write a sentence interpreting the results of a one-proportion z-test.

▶ Know that we do not "accept" a null hypothesis if we cannot reject it but, rather, that we can only "fail to reject" the hypothesis for lack of evidence against it.

HYPOTHESIS TESTS FOR A PROPORTION ON THE COMPUTER

You can conduct a hypothesis test for a proportion using a graphing calculator or a statistics software package on a computer. Often all you need to do is enter information about the hypotheses, the observed number of successes, and the sample size. Some programs want the original data, in which success and failure may be coded as 1 and 0 or "yes" and "no." The technology will report the z-score and the P-value.

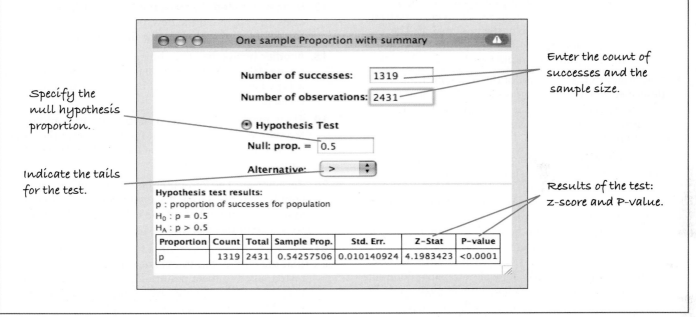

Specify the null hypothesis proportion.

Indicate the tails for the test.

Enter the count of successes and the sample size.

Results of the test: z-score and P-value.

EXERCISES

1. Hypotheses. Write the null and alternative hypotheses you would use to test each of the following situations:
a) A governor is concerned about his "negatives"—the percentage of state residents who express disapproval of his job performance. His political committee pays for a series of TV ads, hoping that they can lower the negatives to below 30%. They will use follow-up polling to assess the ads' effectiveness.
b) Is a coin fair?
c) Only about 20% of people who try to quit smoking succeed. Sellers of a motivational tape claim that listening to the recorded messages can help people quit.

2. More hypotheses. Write the null and alternative hypotheses you would use to test each situation.
a) In the 1950s only about 40% of high school graduates went on to college. Has the percentage changed?

b) 20% of cars of a certain model have needed costly transmission work after being driven between 50,000 and 100,000 miles. The manufacturer hopes that a redesign of a transmission component has solved this problem.
c) We field-test a new-flavor soft drink, planning to market it only if we are sure that over 60% of the people like the flavor.

3. Governor. Among the people polled in Exercise 1a, the governor's negative rating was below the hypothesized 30%, with a z-score of $z = -2.37$. What can the governor conclude about public approval of his job performance?

4. Graduates. Researchers investigating the question posed in Exercise 2a selected a random sample of recent high school graduates, and found the proportion attending college to be higher than the hypothesized 40%. If the z-score for this sample proportion was $z = 3.19$, what can we conclude?

5. **Fair coin?** To test the coin of Exercise 1b, someone tossed it many times. He got heads more than half of the time, but the z-score for his results was only $z = 0.73$. Can he conclude the coin is fair? Explain.

6. **Transmissions.** The auto manufacturer of Exercise 2b collected repair data on several of the redesigned transmissions. Although fewer of them needed repairs, the z-score for the sample proportion was only $z = -1.62$. What should the company conclude about the new design?

7. **Tape.** As evidence that the motivational tape described in Exercise 1c is effective, the company selling it advertised that among a sample of smokers who tried it, more than 20% were able to quit smoking. The sample proportion had a z-score of $z = 0.88$. What would you conclude about the tape?

8. **Soft drink.** In the taste test of the new soft drink described in Exercise 2c, lots of people liked it, but the sample proportion's z-score was only $z = 0.53$. Does this prove that the drink isn't popular enough to market? Explain.

9. **One tailed or two?** In each of the following situations, is the alternative hypothesis one-tailed or two-tailed? What are the hypotheses?
   a) PepsiCo recently reformulated Diet Pepsi in an attempt to appeal to teenagers. They run a taste test to see if the new formula appeals to more teenagers than the standard formula.
   b) A budget override in a small town requires a two-thirds majority to pass. A local newspaper conducts a poll to see if there's evidence it will pass.
   c) One financial theory states that the stock market will go up or down with equal probability. A student collects data over several years to test the theory.

10. **Which alternative?** In each of the following situations, is the alternative hypothesis one-tailed or two-tailed? What are the hypotheses?
   a) In recent years, 10% of college juniors have applied for study abroad. The dean's office conducts a survey to see if that's changed this year.
   b) A pharmaceutical company conducts a clinical trial to see if more patients who take a new drug experience headache relief than the 22% who claimed relief after taking the placebo.
   c) At a small computer peripherals company, only 60% of the hard drives produced passed all their performance tests the first time. Management recently invested a lot of resources into the production system and now conducts a test to see if it helped.

11. **Taste.** After the taste test described in Exercise 9a, the company said the results were not statistically significant. Explain what that means about the new formula.

12. **Abroad.** The results of the dean's survey described in Exercise 10a were not statistically significant. Explain what that means about college juniors.

13. **Another taste.** The taste test in Exercises 9a and 11 may have led the company to make a Type II error. Explain what that means in this context.

14. **Abroad again.** Perhaps the survey described in Exercises 10a and 12 led the dean to make a Type II error. Explain what that means in this context.

15. **Budget.** After conducting the poll described in Exercise 9b, the newspaper reported that the results were statistically significant. Explain what that means about the upcoming budget vote.

16. **Headaches.** After the clinical trial described in Exercise 10b, the pharmaceuticals company reported that the results were statistically significant. Explain what that means about the new headache drug.

17. **Budget revisited.** Perhaps the poll described in Exercises 9b and 15 led the newspaper to make a Type I error. Explain what that means in this context.

18. **Another headache.** The clinical trial described in Exercises 10b and 16 may have led the company to make a Type I error. Explain what that means in this context.

19. **Stock market.** The student in Exercise 9c could reach the wrong conclusion about the behavior of the stock market. Explain what a Type I and a Type II error would be in this context.

20. **Peripherals.** The computer company in Exercise 10c could reach the wrong decision about its production system. Explain what a Type I and a Type II error would be in this context.

21. **Pollution.** A company with a fleet of 150 cars found that the emissions systems of 7 out of the 22 they tested failed to meet pollution control guidelines. Is this strong evidence that more than 20% of the fleet might be out of compliance? Explain why the company can't test a hypothesis using these results.

22. **Scratch and dent.** An appliance manufacturer stockpiles washers and dryers in a large warehouse for shipment to retail stores. Sometimes in handling them the appliances get damaged. Even though the damage may be minor, the company must sell those machines at drastically reduced prices. The company goal is to keep the level of damaged machines below 2%. One day an inspector randomly checks 60 washers and finds that 5 of them have scratches or dents. Is this strong evidence that the warehouse is failing to meet the company goal? Explain why the company can't test a hypothesis using these results.

B

23. **Dropouts.** Some people are concerned that new tougher standards and high-stakes tests adopted in many states have driven up the high school dropout rate. The National Center for Education Statistics reported that the high school dropout rate for the year 2004 was 10.3%. One school district whose dropout rate has always been very close to the national average reports that 210 of their 1782 high school students dropped out last year. Is this evidence that their dropout rate may be increasing?
   a) Write the null hypothesis.
   b) Write the alternative hypothesis.

c) Check the assumptions and conditions for inference.
d) Find the *z*-score for this sample proportion.
e) State your conclusion (in context, of course).

24. Football 2006. During the 2006 season, the home team won 136 of the 240 regular-season National Football League games. Is this strong evidence of a home field advantage in professional football?
a) Write the null hypothesis.
b) Write the alternative hypothesis.
c) Check the assumptions and conditions for inference.
d) Find the *z*-score for this sample proportion.
e) State your conclusion (in context, of course).

25. Twins. In 2001 a national vital statistics report indicated that about 3% of all births produced twins. Is the rate of twin births the same among very young mothers? Data from a large city hospital found that only 7 sets of twins were born to 469 teenage girls.
a) Write the null hypothesis.
b) Write the alternative hypothesis.
c) Check the assumptions and conditions for inference.
d) Find the *z*-score for this sample proportion.
e) State your conclusion (in context, of course).

26. Seeds. A garden center wants to store leftover packets of vegetable seeds for sale the following spring, but the center is concerned that the seeds may not germinate at the same rate a year later. The manager finds a packet of last year's green bean seeds and plants them as a test. Although the packet claims a germination rate of 92%, only 171 of 200 test seeds sprout. Is this evidence that seeds lose viability during a year in storage?
a) Write the null hypothesis.
b) Write the alternative hypothesis.
c) Check the assumptions and conditions for inference.
d) Find the *z*-score for this sample proportion.
e) State your conclusion (in context, of course).

27. Women executives. A company is criticized because only 13 of 43 people in executive-level positions are women. The company explains that although this proportion is lower than it might wish, it's not surprising given that only 40% of all its employees are women. What do you think?
a) Write the null hypothesis.
b) Write the alternative hypothesis.
c) Check the assumptions and conditions for inference.
d) Find the *z*-score for this sample proportion.
e) Give the P-value for this test.
f) State your conclusion (in context, of course).

28. Jury. Census data for a certain county show that 19% of the adult residents are Hispanic. Suppose 72 people are called for jury duty and only 9 of them are Hispanic. Does this apparent underrepresentation of Hispanics call into question the fairness of the jury selection system?
a) Write the null hypothesis.
b) Write the alternative hypothesis.

c) Check the assumptions and conditions for inference.
d) Find the *z*-score for this sample proportion.
e) Give the P-value for this test.
f) State your conclusion (in context, of course).

29. Lost luggage. An airline's public relations department says that the airline rarely loses passengers' luggage. It further claims that on those occasions when luggage is lost, 90% is recovered and delivered to its owner within 24 hours. A consumer group that surveyed a large number of air travelers found that only 103 of 122 people who lost luggage on that airline were reunited with the missing items by the next day. Does this cast doubt on the airline's claim?
a) Write the null hypothesis.
b) Write the alternative hypothesis.
c) Check the assumptions and conditions for inference.
d) Find the *z*-score for this sample proportion.
e) Give the P-value for this test.
f) State your conclusion (in context, of course).

30. Acid rain. A study of the effects of acid rain on trees in the Hopkins Forest shows that 25 of 100 trees sampled exhibited some sort of damage from acid rain. This rate seemed to be higher than the 15% quoted in a recent *Environmetrics* article on the average proportion of damaged trees in the Northeast. Does the sample suggest that trees in the Hopkins Forest are more susceptible than trees from the rest of the region?
a) Write the null hypothesis.
b) Write the alternative hypothesis.
c) Check the assumptions and conditions for inference.
d) Find the *z*-score for this sample proportion.
e) Give the P-value for this test.
f) State your conclusion (in context, of course).

31. Homeowners 2005. In 2005 the U.S. Census Bureau reported that 68.9% of American families owned their homes. Census data reveal that the ownership rate in one small city is much lower. The city council is debating a plan to offer tax breaks to first-time home buyers in order to encourage people to become homeowners. They decide to adopt the plan on a 2-year trial basis and use the data they collect to make a decision about continuing the tax breaks. Since this plan costs the city tax revenues, they will continue to use it only if there is strong evidence that the rate of home ownership is increasing.
a) In words, what will their hypotheses be?
b) What would a Type I error be?
c) What would a Type II error be?
d) For each type of error, tell who would be harmed.

32. Alzheimer's. Testing for Alzheimer's disease can be a long and expensive process, consisting of lengthy tests and medical diagnosis. Recently, a group of researchers (Solomon *et al.*, 1998) devised a 7-minute test to serve as a quick screen for the disease for use in the general population of senior citizens. A patient who tested positive would then go through the more expensive battery of tests and medical diagnosis.

The authors reported a false positive rate of 4% and a false negative rate of 8%.
a) Put this in the context of a hypothesis test. What are the null and alternative hypotheses?
b) What would a Type I error mean?
c) What would a Type II error mean?
d) Which is worse here, a Type I or Type II error? Explain.

33. **Testing cars.** A clean air standard requires that vehicle exhaust emissions not exceed specified limits for various pollutants. Many states require that cars be tested annually to be sure they meet these standards. Suppose state regulators double-check a random sample of cars that a suspect repair shop has certified as okay. They will revoke the shop's license if they find significant evidence that the shop is certifying vehicles that do not meet standards.
a) In this context, what is a Type I error?
b) In this context, what is a Type II error?
c) Which type of error would the shop's owner consider more serious?
d) Which type of error might environmentalists consider more serious?

34. **Quality control.** Production managers on an assembly line must monitor the output to be sure that the level of defective products remains small. They periodically inspect a random sample of the items produced. If they find a significant increase in the proportion of items that must be rejected, they will halt the assembly process until the problem can be identified and repaired.
a) In this context, what is a Type I error?
b) In this context, what is a Type II error?
c) Which type of error would the factory owner consider more serious?
d) Which type of error might customers consider more serious?

35. **Equal opportunity?** A company is sued for job discrimination because only 19% of the newly hired candidates were minorities when 27% of all applicants were minorities. Is this strong evidence that the company's hiring practices are discriminatory?
a) Is this a one-tailed or a two-tailed test? Why?
b) In this context, what would a Type I error be?
c) In this context, what would a Type II error be?

36. **Stop signs.** Highway safety engineers test new road signs, hoping that increased reflectivity will make them more visible to drivers. Volunteers drive through a test course with several of the new- and old-style signs and rate which kind shows up the best.
a) Is this a one-tailed or a two-tailed test? Why?
b) In this context, what would a Type I error be?
c) In this context, what would a Type II error be?

37. **Discrimination.** The P-value for the hypothesis test in Exercise 35 was 0.022. What should the courts conclude about the company's hiring practices?

38. **Second stop.** When the engineers tested the hypothesis about the road signs described in Exercise 36, they got a P-value of 0.18. What should they conclude about the new signs?

39. **Relief.** A company's old antacid formula provided relief for 70% of the people who used it. The company tests a new formula to see if it is better and gets a P-value of 0.27. Is it reasonable to conclude that the new formula and the old one are equally effective? Explain.

40. **Cars.** A survey investigating whether the proportion of today's high school seniors who own their own cars is higher than it was a decade ago finds a P-value of 0.017. Is it reasonable to conclude that more high schoolers have cars? Explain.

41. **Negatives.** After the political ad campaign described in Exercise 1a, pollsters check the governor's negatives. They test the hypothesis that the ads produced no change against the alternative that the negatives are now below 30% and find a P-value of 0.22. Which conclusion is appropriate? Explain.
a) There's a 22% chance that the ads worked.
b) There's a 78% chance that the ads worked.
c) There's a 22% chance that their poll is correct.
d) There's a 22% chance that natural sampling variation could produce poll results like these if there's really no change in public opinion.

42. **Dice.** The seller of a loaded die claims that it will favor the outcome 6. We don't believe that claim, and roll the die 200 times to test an appropriate hypothesis. Our P-value turns out to be 0.03. Which conclusion is appropriate? Explain.
a) There's a 3% chance that the die is fair.
b) There's a 97% chance that the die is fair.
c) There's a 3% chance that a loaded die could randomly produce the results we observed, so it's reasonable to conclude that the die is fair.
d) There's a 3% chance that a fair die could randomly produce the results we observed, so it's reasonable to conclude that the die is loaded.

C

43. **P-value.** A medical researcher tested a new treatment for poison ivy against the traditional ointment. He concluded that the new treatment is more effective. Explain what the P-value of 0.047 means in this context.

44. **Another P-value.** Have harsher penalties and ad campaigns increased seat belt use among drivers and passengers? Observations of commuter traffic failed to find evidence of a significant change compared with three years ago. Explain what the study's P-value of 0.17 means in this context.

45. **Alpha.** A researcher developing scanners to search for hidden weapons at airports has concluded that a new device is significantly better than the current scanner. He made this decision based on a test using $\alpha = 0.05$. Would he have made the same decision at $\alpha = 0.10$? How about $\alpha = 0.01$? Explain.

46. **Alpha again.** Environmentalists concerned about the impact of high-frequency radio transmissions on birds found that there was no evidence of a higher mortality rate among hatchlings in nests near cell towers. They based this conclusion on a test using $\alpha = 0.05$. Would

they have made the same decision at $\alpha = 0.10$? How about $\alpha = 0.01$? Explain.

47. **WebZine.** A magazine is considering the launch of an online edition. The magazine plans to go ahead only if it's convinced that more than 25% of current readers would subscribe. The magazine contacted a simple random sample of 500 current subscribers, and 137 of those surveyed expressed interest. What should the company do? Test an appropriate hypothesis and state your conclusion. Be sure the appropriate assumptions and conditions are satisfied before you proceed.

48. **TV ads.** A start-up company is about to market a new computer printer. It decides to gamble by running commercials during the Super Bowl. The company hopes that name recognition will be worth the high cost of the ads. The goal of the company is that over 40% of the public recognize its brand name and associate it with computer equipment. The day after the game, a pollster contacts 420 randomly chosen adults and finds that 181 of them know that this company manufactures printers. Would you recommend that the company continue to advertise during Super Bowls? Explain.

49. **Law School.** According to the Law School Admission Council, in the fall of 2006, 63% of law school applicants were accepted to some law school.[3] The training program *LSATisfaction* claims that 163 of the 240 students trained in 2006 were admitted to law school. You can safely consider these trainees to be representative of the population of law school applicants. Has *LSATisfaction* demonstrated a real improvement over the national average?
a) What are the hypotheses?
b) Check the conditions and find the P-value.
c) Would you recommend this program based on what you see here? Explain.

50. **Med School.** According to the Association of American Medical Colleges, only 46% of medical school applicants were admitted to a medical school in the fall of 2006.[4] Upon hearing this, the trustees of Striving College expressed concern that only 77 of the 180 students in their class of 2006 who applied to medical school were admitted. The college president assured the trustees that this was just the kind of year-to-year fluctuation in fortunes that is to be expected and that, in fact, the school's success rate was consistent with the national average. Who is right?
a) What are the hypotheses?
b) Check the conditions and find the P-value.
c) Are the trustees right to be concerned, or is the president correct? Explain.

51. **John Wayne.** Like a lot of other Americans, John Wayne died of cancer. But is there more to this story? In 1955 Wayne was in Utah shooting the film *The Conqueror*. Across the state line, in Nevada, the United States military was testing atomic bombs. Radioactive fallout from those tests drifted across the filming location. A total of 46 of the 220 people working on the film eventually died of cancer. Cancer experts estimate that one would expect only about 30 cancer deaths in a group this size.
a) Is the death rate among the movie crew unusually high?
b) Does this prove that exposure to radiation increases the risk of cancer?

52. **Abnormalities.** In the 1980s it was generally believed that congenital abnormalities affected about 5% of the nation's children. Some people believe that the increase in the number of chemicals in the environment has led to an increase in the incidence of abnormalities. A recent study examined 384 children and found that 46 of them showed signs of an abnormality.
a) Is this strong evidence that the risk has increased?
b) Do environmental chemicals cause congenital abnormalities?

53. **He cheats!** A friend of yours claims that when he tosses a coin he can control the outcome. You are skeptical and want him to prove it. He tosses the coin, and you call heads; it's tails. You try again and lose again.
a) Do two losses in a row convince you that he really can control the toss? Explain.
b) You try a third time, and again you lose. What's the probability of losing three tosses in a row if the process is fair?
c) Would three losses in a row convince you that your friend cheats? Explain.
d) How many times in a row would you have to lose in order to be pretty sure that this friend really can control the toss? Justify your answer by calculating a probability and explaining what it means.

54. **Candy.** Someone hands you a box of a dozen chocolate-covered candies, telling you that half are vanilla creams and the other half peanut butter. You pick candies at random and discover the first three you eat are all vanilla.
a) If there really were 6 vanilla and 6 peanut butter candies in the box, what is the probability that you would have picked three vanillas in a row?
b) Do you think there really might have been 6 of each? Explain.
c) Would you continue to believe that half are vanilla if the fourth one you try is also vanilla? Explain.

55. **Loans.** Before lending someone money, banks must decide whether they believe the applicant will repay the loan. One strategy used is a point system. Loan officers assess information about the applicant, totaling points they award for the person's income level, credit history, current debt burden, and so on. The higher the point total, the more convinced the bank is that it's safe to make the loan. Any applicant with a lower point total than a certain cutoff score is denied a loan.

We can think of this decision as a hypothesis test. Since the bank makes its profit from the interest collected on repaid loans, their null hypothesis is that the

---
[3]As reported by the Cornell office of career services in their *Class of 2006 Postgraduate Report*.
[4]*Ibid*.

applicant will repay the loan and therefore should get the money. Only if the person's score falls below the minimum cutoff will the bank reject the null and deny the loan. This system is reasonably reliable, but, of course, sometimes there are mistakes.

a) When a person defaults on a loan, which type of error did the bank make?

b) Which kind of error is it when the bank misses an opportunity to make a loan to someone who would have repaid it?

c) Suppose the bank decides to lower the cutoff score from 250 points to 200. Is that analogous to choosing a higher or lower value of α for a hypothesis test? Explain.

d) What impact does this change in the cutoff value have on the chance of each type of error?

56. Spam. Spam filters try to sort your e-mails, deciding which are real messages and which are unwanted. One method used is a point system. The filter reads each incoming e-mail and assigns points to the sender, the subject, key words in the message, and so on. The higher the point total, the more likely it is that the message is unwanted. The filter has a cutoff value for the point total; any message rated lower than that cutoff passes through to your inbox, and the rest, suspected to be spam, are diverted to the junk mailbox.

We can think of the filter's decision as a hypothesis test. The null hypothesis is that the e-mail is a real message and should go to your inbox. A higher point total provides evidence that the message may be spam; when there's sufficient evidence, the filter rejects the null, classifying the message as junk. This usually works pretty well, but, of course, sometimes the filter makes a mistake.

a) When the filter allows spam to slip through into your inbox, which kind of error is that?

b) Which kind of error is it when a real message gets classified as junk?

c) Some filters allow the user (that's you) to adjust the cutoff. Suppose your filter has a default cutoff of 50 points, but you reset it to 60. Is that analogous to choosing a higher or lower value of α for a hypothesis test? Explain.

d) What impact does this change in the cutoff value have on the chance of each type of error?

Answers
Just Checking

1. a) $H_0: p = 0.75$, where p is the percentage of people who would get relief from the new version of the drug.
   b) $H_A: p > 0.75$

2. It's a random sample; 80 is less than 10% of all allergy sufferers; we expect $np = 80(0.75) = 60$ successes and $80(0.25) = 20$ failures, both at least 10. OK to use a Normal model.

3. a) $\hat{p} = 69/80 = 0.8625$
   b) $z = 2.32$

4. With $z > 2$, these results are unlikely to be sampling error; reject the null hypothesis. There's strong evidence that the new medication is more effective.

5. Since $z = 1.78$ is less than 2, the data can be reasonably explained as sampling error. The company lacks evidence to say the new medication is more effective.

6. a) The company concludes that the new drug is more effective, but it really isn't.
   b) The new drug really is more effective, but the company concludes it is not.

7. a) Lower-tail. $H_0: p = 0.60$, $H_A: p < 0.60$
   b) Two-tail. $H_0: p = 0.517$, $H_A: p \neq 0.517$
   c) Upper-tail. $H_0: p = 0.202$, $H_A: p > 0.202$

8. If *Avandia* users are not at greater risk of heart attack, only about 3 in 100 samples would show an increase at least this large. There is evidence the drug may present a danger.

9. We'd expect more than 1 in every 4 samples of patients to show an increase in heart attacks as large as that observed in this study, so it was reasonable to attribute tha apparent risk to sampling error.

CHAPTER 1

Inferences About a Mean

Motor vehicle crashes are the leading cause of death for people between 4 and 33 years old. In 2006, motor vehicle accidents claimed the lives of 43,300 people in the United States—on average 119 deaths each day, or 1 death every 12 minutes. Speeding is a factor in 31% of all fatal accidents, according to the National Highway Traffic Safety Administration.

Triphammer Road is a busy street that passes through a residential neighborhood. People who live there are concerned that vehicles traveling on Triphammer often exceed the speed limit of 30 miles per hour. The local police sometimes place a radar speed detector by the side of the road to display each vehicle's speed to its driver.

The local residents are not convinced that such a passive method is helping the problem. They want the village to add extra police patrols to enforce the speed limit. To help their case, one day a resident stood by the detector and recorded the speeds of vehicles passing by for 15 minutes. (When clusters of vehicles went by, he counted only the first one.) You can see his data and the histogram on the next page.

We're interested both in estimating the true mean speed of all vehicles on Triphammer Road and in testing whether it exceeds the posted speed limit.

You've learned how to create confidence intervals and test hypotheses about proportions. Now we went to do exactly the same thing for means. We'll get back to the speeds of these cars, but first we need to know how means vary from sample to sample.

WHO	Vehicles on Triphammer Road
WHAT	Speed
UNITS	Miles per hour
WHEN	April 11, 2000, 1 p.m.
WHERE	A small town in the northeastern United States
WHY	Concern over impact on residential neighborhood

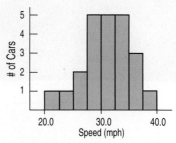

Speed		
29	29	24
34	34	34
34	32	36
28	31	31
30	27	34
29	37	36
38	29	21
31	26	

FIGURE 19.1 Speeds. The distribution of speeds of cars on Triphammer Road seems to be unimodal and symmetric.

Simulating the Sampling Distribution of a Mean

Here's a simple simulation. Let's start with one fair die. If we toss this die 10,000 times, what should the histogram of the numbers on the face of the die look like? Here are the results of a simulated 10,000 tosses:

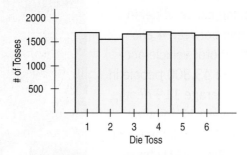

Now let's toss a *pair* of dice and record the average of the two. If we repeat this 10,000 times, what will the histogram of these 10,000 averages look like? Let's see:

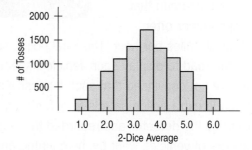

We're much more likely to get an average near 3.5 than we are to get one near 1 or 6. After all the *only* way to get an average of 1 is to get two 1's. To get a total of 7 (for an average of 3.5), though, there are many more possibilities.

What if we average 3 dice? Or 5? Or more?

It will get even harder to have averages near the ends. Getting an average of 1 or 6 with many dice requires all of them to come up 1 or 6, respectively. That's even less likely than for 2 dice. The distribution is pushed toward the middle.

As the sample size (number of dice) gets larger, each sample average is more likely to be closer to the population mean of 3.5. But the shape of the distribution is the surprising part. It's approaching a Normal model!

Let's try 20 dice. The histogram of averages for 10,000 throws of 20 dice looks like this:

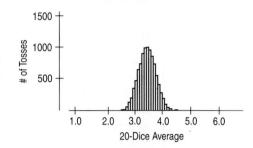

Now we see the Normal shape (and notice how much smaller the spread is). The good news is we can count on this happening for situations other than dice throws. It turns out that Normal models work well amazingly often.

The Fundamental Theorem of Statistics

We've just encountered a very important result: The sampling distribution of *any* mean becomes more nearly Normal as the sample size grows. All we need is for the observations to be independent and collected with randomization. We don't even care about the shape of the population distribution! This surprising fact is called the **Central Limit Theorem**[1] (CLT).

The CLT says that the sampling distribution of any mean is approximately Normal. But which Normal model? We know that any Normal is specified by its mean and standard deviation. For means, it's centered at the population mean. What else would we expect?

What about the standard deviations, though? We noticed in our dice simulation that the histograms got narrower as we averaged more and more dice together. This shouldn't be surprising. Means vary less than the individual observations. Think about it for a minute. Which would be more surprising, having *one* person in your Statistics class who is over 6'9" tall or having the *mean* height of all the students be over 6'9"? You may have seen somebody this tall in one of your classes sometime. But finding a whole class whose mean height is over 6'9" tall just won't happen. Why? Because *means have smaller standard deviations than individuals*.

How much smaller? The Normal model for the sampling distribution of the mean has a standard deviation equal to

$$SD(\bar{y}) = \frac{\sigma}{\sqrt{n}}$$

where σ is the standard deviation of the population. See n in the denominator? That confirms that means of larger samples vary less.

---

[1] The word "central" in the name of the theorem means "fundamental." It doesn't refer to the center of a distribution.

THE SAMPLING DISTRIBUTION MODEL FOR A MEAN (CLT)

When a random sample is drawn from any population with mean μ and standard deviation σ, its sample mean, $\bar{y}$, has a sampling distribution with the same *mean μ* but whose *standard deviation* is $\dfrac{\sigma}{\sqrt{n}}$ (and we write $\sigma(\bar{y}) = SD(\bar{y}) = \dfrac{\sigma}{\sqrt{n}}$). No matter what population the random sample comes from, the *shape* of the sampling distribution is approximately Normal as long as the sample size is large enough. The larger the sample used, the more closely the Normal approximates the sampling distribution for the mean. A Normal model won't be useful for small sample sizes (under 30) if the population distribution is strongly skewed.

Because this powerful insight lies at the heart of most of our methods of analysis, we consider it to be the Fundamental Theorem of Statistics.

For Example   Using the CLT

Based on weighing thousands of animals, the American Angus Association reports that mature Angus cows have a mean weight of 1309 pounds with a standard deviation of 157 pounds. This result was based on a very large sample of animals from many herds over a period of 15 years, so let's assume that these summaries are the population parameters.

Question: What does the CLT predict about the mean weight seen in random samples (herds) of 100 mature Angus cows?

It's given that weights of all mature Angus cows have $\mu = 1309$ and $\sigma = 157$ pounds. Because $n = 100$ animals is a fairly large sample, I can apply the Central Limit Theorem. I expect the resulting sample means $\bar{y}$ will average 1309 pounds and have a standard deviation of

$$SD(\bar{y}) = \frac{\sigma}{\sqrt{n}} = \frac{157}{\sqrt{100}} = 15.7 \text{ pounds.}$$

The CLT also says that the distribution of sample means follows a Normal model, so the 68–95–99.7 Rule applies. I'd expect that

- in 68% of random samples of 100 mature Angus cows, the mean weight will be between $1309 - 15.7 = 1293.3$ and $1309 + 15.7 = 1324.7$ pounds;
- in 95% of such samples, $1277.6 \le \bar{y} \le 1340.4$ pounds;
- in 99.7% of such samples, $1261.9 \le \bar{y} \le 1356.1$ pounds.

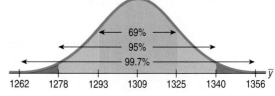

Sample distributions and sampling models Be careful. Now we have *two* distributions to deal with. The first is the real-world distribution of the data in the sample, which we might display with a histogram. The second is the math world *sampling distribution model* of the statistic, a Normal model based on the Central Limit Theorem. Don't confuse the two.

For example, don't mistakenly think the CLT says that the *data* are Normally distributed as long as the sample is large enough. In fact, as samples get larger, we expect the distribution of the data to look more and more like the population from which they are drawn—skewed, bimodal, whatever—but not necessarily Normal. You can collect a sample of CEO salaries for the next 1000 years, but the histogram will never look Normal. It will be strongly skewed to the right. The Central Limit Theorem doesn't talk about the distribution of the data from the sample. It talks about the sample *means* and sample *proportions* of many different random samples drawn from the same population. Then a Normal model is useful, a very surprising and powerful result.

Just Checking

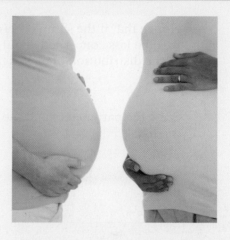

1. **a)** Human gestation times (pregnancies) have a mean of about 266 days, with a standard deviation of about 16 days. If we record the gestation times of a sample of 100 women, do we know that a histogram of the times will be well modeled by a Normal model?

    b) Suppose we look at the *average* gestation times for a sample of 100 women. If we imagined all the possible random samples of 100 women we could take and looked at the distribution of all the sample means, what shape would it have?

    c) Where would the center of that distribution be?

    d) What would be the standard deviation of that distribution?

    e) Use the 68–95–99.7 Rule to sketch the sampling model for the sample means.

    (Check your answers on page 456.)

STEP-BY-STEP EXAMPLE

Working with the Sampling Distribution Model for the Mean

The Centers for Disease Control and Prevention reports that the mean weight of adult men in the United States is 190 lb with a standard deviation of 59 lb.[2]

Question: An elevator in our building has a weight limit of 10 persons or 2500 lb. If 10 men get on the elevator, should we worry that they will overload its weight limit?

(continued)

[2]Cynthia L. Ogden, Cheryl D. Fryar, Margaret D. Carroll, and Katherine M. Flegal, *Mean Body Weight, Height, and Body Mass Index United States 1960–2002, Advance Data from Vital and Health Statistics Number 347*, Oct. 27, 2004. https://www.cdc.gov/nchs.

Plan State what we want to know.

Asking whether the total weight of a sample of 10 men might exceed 2500 pounds is equivalent to asking if their mean weight might be greater than 250 pounds.

Model Think about the assumptions and check the conditions to verify that it's okay to apply the Central Limit Theorem.

✔ **Independence Assumption:** It's reasonable to think that the weights of 10 randomly sampled men will be independent of each other. (But there could be exceptions—for example, if they were all from the same family or if the elevator were in a building with a diet clinic!)

✔ **Randomization Condition:** I'll assume that the 10 men getting on the elevator are a random sample from the population.

✔ **10% Condition:** 10 men is surely less than 10% of the population of possible elevator riders.

✔ **Large Enough Sample Condition:** I suspect the distribution of population weights is roughly unimodal and symmetric, so my sample of 10 men seems large enough.

Notice that if the sample were larger we'd be less concerned about the shape of the distribution of all weights.

State the parameters and the sampling model.

The mean for all weights is $\mu = 190$ and the standard deviation is $\sigma = 59$ pounds. Since the conditions are satisfied, the CLT says that the sampling distribution of $\bar{y}$ has a Normal model with mean 190 and standard deviation

$$SD(\bar{y}) = \frac{\sigma}{\sqrt{n}} = \frac{59}{\sqrt{10}} \approx 18.66$$

Plot Make a picture. Sketch the model and shade the area we're interested in. Here the mean weight of 250 pounds appears to be far out on the right tail of the curve.

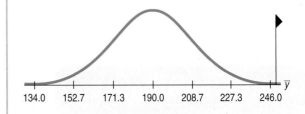

Mechanics Use the standard deviation as a ruler to find the z-score of the cutoff mean weight. We see that an average of 250 pounds is far more than 2 standard deviations above the mean.

$$z = \frac{\bar{y} - \mu}{SD(\bar{y})} = \frac{250 - 190}{18.66} = 3.21$$

Conclusion Interpret your result in the proper context, being careful to relate it to the original question.

An average of over 250 pounds would be more than 3 standard deviations above the mean. So, if they are a random sample, it is quite unlikely that 10 people will exceed the total weight allowed on the elevator.

Trouble! Gosset's *t* to the Rescue

The CLT says that all we need to model the sampling distribution of $\bar{y}$ is a random sample of quantitative data.

And the true population standard deviation, σ.

Uh oh. That's a problem. How are we supposed to know σ? We can't know the population standard deviation. So what should we do? We do what any sensible person would do: We estimate the population parameter σ with s, the sample standard deviation based on the data. The resulting standard error is $SE(\bar{y}) = \dfrac{s}{\sqrt{n}}$.

A century ago, people used this standard error with the Normal model, assuming it would work. And for large sample sizes it *did* work pretty well. But they began to notice problems with smaller samples. The sample standard deviation, s, like any other statistic, varies from sample to sample. And this extra variation in the standard error was messing up the margins of error.

William S. Gosset is the man who first investigated this fact. He realized that to allow for the extra variation we need a new sampling distribution model. Gosset's work transformed Statistics, but most people who use his work don't even know his name.

He published his discovery under the pseudonym "Student," and ever since, the model he found has been known as **Student's *t*.**

Gosset's model is always bell-shaped, but the smaller the sample the fatter the tails. So the Student's *t*-models form a whole *family* of related distributions that depend on a parameter known as **degrees of freedom.** We often denote degrees of freedom as *df* and the model as t_{df}, with the degrees of freedom as a subscript.

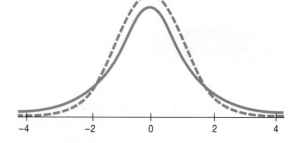

FIGURE 19.2 **Gosset's *t*.** The *t*-model (solid curve) on 2 degrees of freedom has fatter tails than the Normal model (dashed curve). So the 68–95–99.7 Rule doesn't work for *t*-models with only a few degrees of freedom.

Student's *t*-models are unimodal, symmetric, and bell-shaped, just like the Normal. But *t*-models with only a few degrees of freedom have much fatter tails than the Normal. (That's what makes the margin of error bigger.) As the degrees of freedom increase, the *t*-models look more and more like the Normal. For large sample sizes, then, our Rule of 2 still works well. For smaller samples we'll need to use slightly larger cutoffs, called **critical values.** Fortunately, technology makes that adjustment easy.

Assumptions and Conditions

These are the assumptions we need to use the Student's *t*-models.

Independence Assumption

Independence Assumption: The data values should be independent. There's really no way to check independence of the data by looking at the sample, but we should think about whether the assumption is reasonable.

Because we estimate the standard deviation of the sampling distribution model from the data, it's a *standard error*. So we use the $SE(\bar{y})$ notation. Remember, though, that it's just the estimated standard deviation of the sampling distribution model for means.

To find the sampling distribution of $\dfrac{\bar{y}}{s/\sqrt{n}}$, Gosset simulated it *by hand*. He drew paper slips of small samples from a hat *hundreds of times* and computed the means and standard deviations with a mechanically cranked calculator.

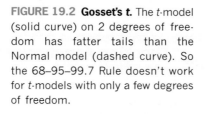

TI-*nspire*

The *t*-models. See how *t*-models change as you change the degrees of freedom.

Randomization Condition: The data arise from a random sample or suitably randomized experiment. Randomly sampled data—and especially data from a Simple Random Sample—are ideal.

10% Condition: The sample is no more than 10% of the population.

Normal Population Assumption

Student's *t*-models won't work for data that are badly skewed. How skewed is too skewed? Well, formally, we assume that the data are from a population that follows a Normal model. Practically speaking, there's no way to be certain this is true. The good news, however, is that even for small samples, it's sufficient to check the . . .

Nearly Normal Condition: The data come from a distribution that is unimodal and symmetric.

Check this condition by making a histogram. The importance of Normality for Student's *t* depends on the sample size. Just our luck: It matters most when it's hardest to check.

- For very small samples ($n < 15$ or so), the data should follow a Normal model pretty closely. Of course, with so little data, it's rather hard to tell. But if you do find outliers or strong skewness, don't use these methods.
- For moderate sample sizes (*n* between 15 and 40 or so), the *t* methods will work well as long as the data are unimodal and reasonably symmetric.
- When the sample size is larger than 40 or 50, the *t* methods are safe to use unless the data are extremely skewed.

WE DON'T *WANT* TO STOP

We check conditions hoping that we can make a meaningful analysis of our data. The conditions serve as *disqualifiers*—we keep going unless there's a serious problem. If we find minor issues, we note them and express caution about our results.

- If the sample is not an SRS, but we believe it's representative of some populations, we limit our conclusions accordingly.
- If there are outliers, rather than stop, we perform the analysis both with and without them.
- If the sample looks bimodal, we try to analyze subgroups separately.

Only when there's major trouble—like a strongly skewed small sample or an obviously nonrepresentative sample—are we unable to proceed at all.

For Example | Checking Assumptions and Conditions for Student's *t*

In 2004 a team of researchers published a study of contaminants in farmed salmon.[3] One of those was the insecticide mirex which has been shown to cause cancer and may be toxic to the liver, kidneys, and endocrine system. The histogram shows the concentrations of mirex in 150 farmed salmon.

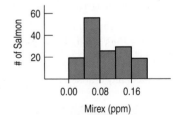

Question: Are the assumptions and conditions for inference satisfied?

✔ **Independence Assumption:** The fish were raised in many different places, and samples were purchased independently from several sources.

✔ **Randomization Condition:** The fish were selected randomly from those available for sale.

✔ **10% Conditions:** There are lots of fish in the sea (and at the fish farms); 150 is certainly far fewer than 10% of the population.

✔ **Nearly Normal Condition:** The histogram of the data is unimodal. Although it may be somewhat skewed to the right, this is not a concern with a sample size of 150.

It's okay to use these data for inference about farm-raised salmon.

---

[3]Ronald A. Hites, Jeffery A. Foran, David O. Carpenter, M. Coreen Hamilton, Barbara A. Knuth, and Steven J. Schwager, "Global Assessment of Organic Contaminants in Farmed Salmon," *Science* 9 January 2004: Vol. 303., no. 5655, pp. 226–229.

A Confidence Interval for Means

To make confidence intervals or test hypotheses for means, we need to use Gosset's model. Which one? Well, for means, it turns out the right value for degrees of freedom is $df = n - 1$.

> ### ONE-SAMPLE t-INTERVAL FOR THE MEAN
> When the assumptions and conditions are met, we are ready to find the confidence interval for the population mean, μ. The confidence interval is
> $$\bar{y} \pm t^*_{n-1} \times SE(\bar{y}),$$
> where the standard error of the mean is $SE(\bar{y}) = \dfrac{s}{\sqrt{n}}$.
>
> The critical value t^*_{n-1} depends on the particular confidence level, C, that you specify and on the number of degrees of freedom, $n - 1$, which we get from the sample size.
> For 95% confidence intervals based on large samples (over 30) our Rule of 2 still works well.

When Gosset corrected the model for the extra uncertainty, the margin of error got bigger, as you might have guessed. When you use Gosset's model instead of the Normal model's Rule of 2, your confidence intervals will be just a bit wider, especially for small samples. That's the correction you need. By using the t-model, you've compensated for the extra variability in precisely the right way.

For Example   A One-Sample t-interval for the Mean

Recap: Researchers purchased farm-raised salmon from 51 farms, testing for contaminants. One of those was the insecticide mirex. Summaries for the mirex concentrations (in parts per million) in the farmed salmon are:

$$n = 150 \quad \bar{y} = 0.0913 \text{ ppm} \quad s = 0.0495 \text{ ppm}.$$

Question: What does a 95% confidence interval say about mirex?

$$df = 150 - 1 = 149$$

$$SE(\bar{y}) = \frac{s}{\sqrt{n}} = \frac{0.0495}{\sqrt{150}} = 0.0040$$

$$t^*_{149} \approx 1.976 \text{ (from technology)}$$

So the confidence interval for μ is $\bar{y} \pm t^*_{149} \times SE(\bar{y}) = 0.0913 \pm 1.976(0.0040)$

$$= 0.0913 \pm 0.0079$$
$$= (0.0834, 0.0992)$$

I'm 95% confident that the mean level of mirex concentration in farm-raised salmon is between 0.0834 and 0.0992 parts per million.

TI Tips   Finding a confidence interval for a mean (Part 1)

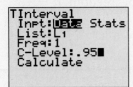

Yes, your calculator can create a confidence interval for a mean. Let's try it out on the speed of the cars on Triphammer Road (p. 420).

Find a confidence interval given a set of data

- Type the speeds of the 23 Triphammer cars into L1. Go ahead; we'll wait.

29	34	34	28	30	29	38	31	29	34	32	31
27	37	29	26	24	34	36	31	34	36	21	

- Set up a **STATPLOT** to create a histogram of the data so you can check the Nearly Normal condition. Looks okay—unimodal and roughly symmetric.
- Under **STAT TESTS** choose 8:TInterval.
- Choose Inpt:Data, then specify that your data is List:L1.
- For these data the frequency is 1. (If your data have a frequency distribution stored in another list, you would specify that.)
- Choose the confidence level you want.
- Calculate the interval.

There's the 95% confidence interval. That was easy—but remember, the calculator only does the *Show*. Now you have to *Tell* what it means.

STEP-BY-STEP EXAMPLE   A One-Sample *t*-Interval for the Mean

Let's build a 95% confidence interval for the mean speed of all vehicles traveling on Triphammer Road. The interval that we'll make is called the **one-sample *t*-interval.**

Question: What can we say about the mean speed of all cars on Triphammer Road?

Plan State what we want to know. Identify the parameter of interest.

Identify the variables and review the W's.

I want to find a 95% confidence interval for the mean speed, μ, of vehicles driving on Triphammer Road. I have data on the speeds of 23 cars there, sampled on April 11, 2000.

Make a picture. Check the distribution shape and look for skewness, multiple modes, and outliers.

Here's a histogram of the 23 observed speeds.

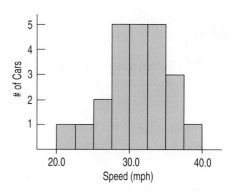

REALITY CHECK The histogram centers around 30 mph, and the data lie between 20 and 40 mph. We'd expect a confidence interval to place the population mean within a few mph of 30.

Model Think about the assumptions and check the conditions.

✔ **Independence Assumption:** This is a convenience sample, but care was taken to select cars that were not driving near each other, so their speeds are plausibly independent.

✔ **Randomization Condition:** Not really met. This is a convenience sample, but I have reason to believe that it is representative.

Note that with this small sample we probably didn't need to check the 10% Condition. On the other hand, doing so gives us a chance to think about what the population is.

✔ **10% Condition:** The cars I observed were fewer than 10% of all cars that travel Triphammer Road.

✔ **Nearly Normal Condition:** The histogram of the speeds is unimodal and symmetric.

State the sampling distribution model for the statistic.

The conditions are satisfied, so I will use a Student's t-model with

$$(n - 1) = 22 \text{ degrees of freedom}$$

Choose your method.

and find a **one-sample t-interval for the mean**.

Mechanics Construct the confidence interval.

Be sure to include the units along with the statistics.

Calculating from the data (see page 420):

$$n = 23 \text{ cars}$$
$$\bar{y} = 31.0 \text{ mph}$$
$$s = 4.25 \text{ mph}.$$

The standard error of $\bar{y}$ is

$$SE(\bar{y}) = \frac{s}{\sqrt{n}} = \frac{4.25}{\sqrt{23}} = 0.886 \text{ mph}.$$

The 95% critical value is $t^*_{22} = 2.074$, so the margin of error is

$$ME = t^*_{22} \times SE(\bar{y})$$
$$= 2.074(0.886)$$
$$= 1.84 \text{ mph}.$$

Because this is a small sample, the critical value we need to make a 95% interval will be slightly greater than 2. Using the Rule of 2 would underestimate the margin of error as 1.77.

The 95% confidence interval for the mean speed is 31.0 ± 1.84, or

$$(29.2 \text{ mph}, 32.8 \text{ mph})$$

(continued)

Conclusion Interpret the confidence interval in the proper context.

When we construct confidence intervals in this way, we expect 95% of them to cover the true mean and 5% to miss the true value. That's what "95% confident" means.

I am 95% confident that the interval from 29.2 mph to 32.8 mph contains the true mean speed of all vehicles on Triphammer Road.

Caveat: This was not a random sample of vehicles. It was a convenience sample taken at one time on one day. And the participants were not blinded. Drivers could see the police device, and some may have slowed down. I'm reluctant to extend this inference to other situations.

More Cautions About Interpreting Confidence Intervals

SO WHAT *SHOULD* WE SAY?

Since 95% of random samples yield an interval that captures the true mean, we *should* say, "I am 95% confident that the interval from 29.2 to 32.8 mph contains the mean speed of all the vehicles on Triphammer Road." It's also okay to say something less formal: "I am 95% confident that the average speed of all vehicles on Triphammer Road is between 29.2 and 32.8 mph." Remember: *our uncertainty is about the interval, not the true mean.* The interval varies randomly. The true mean speed is neither variable nor random—just unknown.

Confidence intervals for means offer new tempting wrong interpretations. Here are some things you *shouldn't* say:

- **Don't say,** "*95% of all the vehicles* on Triphammer Road drive at a speed between 29.2 and 32.8 mph." The confidence interval is about the *mean* speed, not about the speeds of *individual* vehicles.
- **Don't say,** "We are 95% confident that *a randomly selected vehicle* will have a speed between 29.2 and 32.8 mph." This false interpretation is also about individual vehicles rather than about the *mean* of the speeds. We are 95% confident that the *mean* speed of all vehicles on Triphammer Road is between 29.2 and 32.8 mph.
- **Don't say,** "The mean speed of the vehicles is 31.0 mph *95% of the time*." That's about means, but still wrong. It implies that the true mean varies, when in fact it is the confidence interval that would have been different had we gotten a different sample.
- Finally, **don't say,** "*95% of all samples* will have mean speeds between 29.2 and 32.8 mph." That statement suggests that *this* interval somehow sets a standard for every other interval. In fact, this interval is no more (or less) likely to be correct than any other. You could say that 95% of all possible samples will produce intervals that actually do contain the true mean speed. (The problem is that, because we'll never know where the true mean speed really is, we can't know if our sample was one of those 95%.)
- **Do say,** "95% of intervals that could be found in this way would cover the true value." Or make it more personal and say, "I am 95% confident that the true mean speed is between 29.2 and 32.8 mph."

TI Tips Finding a confidence interval for a mean (Part 2)

No data? Find a confidence interval given the sample's mean and standard deviation

Sometimes instead of the original data you just have the summary statistics. For instance, suppose a random sample of 53 lengths of fishing line had a mean strength of 83 pounds and standard deviation of 4 pounds. Let's make a 95% confidence interval for the mean strength of this kind of fishing line.

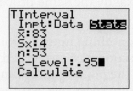

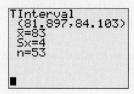

- Without the data you can't check the Nearly Normal Condition. But 53 is a moderately large sample, so assuming there were no outliers, it's okay to proceed. You need to say that.
- Go back to STAT TESTS and choose 8:TInterval again. This time indicate that you wish to enter the summary statistics. To do that, select Stats, then hit ENTER.
- Specify the sample mean, standard deviation, and sample size.
- Choose a confidence level and Calculate the interval.
- If (repeat, IF . . .) strengths of fishing lines follow a Normal model, we are 95% confident that this kind of line has a mean strength between 81.9 and 84.1 pounds.

Just Checking

2. A consumer watchdog organization received customer complaints about the performance of one brand of batteries. To investigate, the researchers purchased 24 of these batteries and ran them until they died. This sample lasted a mean of 47 hours with a standard deviation of 3 hours.

  a) What assumption must we make about battery life in order to use a *t* model?
  b) Find a 95% confidence interval for the mean lifespan of this brand of batteries.
  c) Write a conclusion based on your confidence interval.

(Check your answers on page 456.)

A Hypothesis Test for the Mean

The residents along Triphammer Road are concerned. It appears that the mean speed along the road is higher than it ought to be. To get the police to patrol more frequently, though, they'll need to show that the true mean speed is *in fact greater* than the 30 mph speed limit. This calls for a hypothesis test called the **one-sample *t*-test for the mean.**

ONE-SAMPLE *t*-TEST FOR THE MEAN

The assumptions and conditions for the one-sample *t*-test for the mean are the same as for the one-sample *t*-interval. We test the hypothesis $H_0: \mu = \mu_0$ using the statistic

$$t_{n-1} = \frac{\bar{y} - \mu_0}{SE(\bar{y})}.$$

The standard error of $\bar{y}$ is $SE(\bar{y}) = \dfrac{s}{\sqrt{n}}.$

When the conditions are met and the null hypothesis is true, this statistic follows a Student's *t*-model with $n - 1$ degrees of freedom.

You already know enough to construct this test; it's exactly the same process you used for proportions. Let's review the steps.

1. **Hypotheses.** The null hypothesis assumes there's nothing unusual going on, and the alternative hypothesis identifies the issue we're concerned about. The only difference: now the parameter is the population mean μ. On Triphammer Road, until the evidence proves otherwise, we assume that on average cars observe the speed limit, making the null hypothesis $H_0: \mu = 30$. Local residents are worried about speeding, so there's an upper-tail alternative hypothesis $H_A: \mu > 30$.

2. **Test.** Check the assumptions and conditions to be sure it's okay to use a t-model and perform a one-sample t-test.

3. **Mechanics.** Find the value of t for the observed data. The additional variability arising from using the sample standard deviation s (because we can't know σ) makes the Rule of 2 a bit shaky, especially for small samples. Therefore, with the help of technology, we'll also find the **P-value.** A P-value tells us the probability that sample results at least as extreme as ours could have been observed if the null hypothesis really is true. The lower the P-value, the stronger the evidence against the null hypothesis.

4. **Conclusion.** Make a decision about the null hypothesis, and then write a conclusion in context. For large samples (over 30, say) it's okay to base the decision on the Rule of 2, but even then we get more information from the P-value.
   - If the P-value is very low (less than 0.05, say) our data are unlikely to be just chance variation. They provide strong evidence that H_0 is false, so we reject the null hypothesis.
   - If the P-value is high, then our data are not unusual. Lacking sufficient evidence to say that H_0 is false, we fail to reject the null hypothesis.

For Example A One-Sample *t*-Test for the Mean

Recap: Researchers tested 150 farm-raised salmon for organic contaminants. They found the mean concentration of the carcinogenic insecticide mirex to be 0.0913 parts per million, with standard deviation 0.0495 ppm. As a safety recommendation to recreational fishers, the Environmental Protection Agency's (EPA) recommended "screening value" for mirex is 0.08 ppm.

Question: Are farmed salmon contaminated beyond the level permitted by the EPA? (We've already checked the assumptions and conditions; see page 426.)

$$H_O: \mu = 0.08$$
$$H_A: \mu > 0.08$$

These data satisfy the conditions for inference; I'll do a one-sample t-test for the mean:

$n = 150, df = 149$

$\bar{y} = 0.0913, s = 0.0495$

$$SE(\bar{y}) = \frac{0.0495}{\sqrt{150}} = 0.0040$$

$$t_{149} = \frac{0.0913 - 0.08}{0.0040} = 2.796$$

$P(t_{149} > 2.825) = 0.0029$ (from technology).

With a P-value that low, I reject the null hypothesis and conclude that, in farm-raised salmon, the mirex contamination level does exceed the EPA screening value.

STEP-BY-STEP EXAMPLE

A One-Sample *t*-Test for the Mean

Let's apply the one-sample *t*-test to the Triphammer Road car speeds. The speed limit is 30 mph, so we'll use that as the null hypothesis value.

Question: Does the mean speed of all cars exceed the posted speed limit?

THINK

Plan State what we want to know. Make clear what the population and parameter are.

Identify the variables and review the W's.

Hypotheses The null hypothesis is that the true mean speed is equal to the limit. Because we're interested in whether the vehicles are speeding, the alternative is one-sided.

Make a picture. Check the distribution for skewness, multiple modes, and outliers.

REALITY CHECK The histogram of the observed speeds is clustered around 30, so we'd be surprised to find that the mean was much higher than that. (The fact that 30 is within the confidence interval that we've just found confirms this suspicion.)

Model Think about the assumptions and check the conditions.

(We won't worry about the 10% Condition—it's a small sample.)

I want to know whether the mean speed of vehicles on Triphammer Road exceeds the posted speed limit of 30 mph. I have a sample of 23 car speeds on April 11, 2000.

H_O: Mean speed, $\mu = 30$ mph
H_A: Mean speed, $\mu > 30$ mph

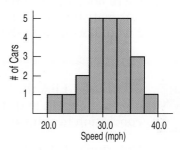

✔ **Independence Assumption:** These cars are a convenience sample, but they were selected so no two cars were driving near each other, so I am justified in believing that their speeds are independent.

✔ **Randomization Condition:** Although I have a convenience sample, I have reason to believe that it is a representative sample.

✔ **Nearly Normal Condition:** The histogram of the speeds is unimodal and reasonably symmetric.

State the sampling distribution model; be sure to include the degrees of freedom.

Choose your method.

The conditions are satisfied, so I'll use a Student's *t*-model with $(n - 1) = 22$ degrees of freedom to do a **one-sample *t*-test for the mean.**

(continued)

Mechanics Be sure to include the units when you write down what you know from the data.

From the data,

$$n = 23 \text{ cars}$$
$$\bar{y} = 31.0 \text{ mph}$$
$$s = 4.25 \text{ mph}$$
$$SE(\bar{y}) = \frac{s}{\sqrt{n}} = \frac{4.25}{\sqrt{23}} = 0.886 \text{ mph.}$$

We use the null model to find the P-value. Make a picture of the t-model centered at $\mu = 30$. Since this is an upper-tail test, shade the region to the right of the observed mean speed.

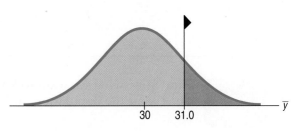

The t-statistic calculation is just a standardized value, like z. We subtract the hypothesized mean and divide by the standard error.

$$t = \frac{\bar{y} - \mu_0}{SE(\bar{y})} = \frac{31.0 - 30.0}{0.886} = 1.13$$

The P-value is the probability of observing a sample mean as large as 31.0 (or larger) *if* the true mean were 30.0, as the null hypothesis states.

(The observed mean is only 1.13 standard errors above the hypothesized value for less than 2.)

 REALITY CHECK We're not surprised that the difference isn't statistically significant. The Rule of 2 would bring us to the same conclusion.

$$P\text{-value} = P(t_{22} > 1.13) = 0.136$$

Conclusion Link the P-value to your decision about H_0, and state your conclusion in context.

Unfortunately for the residents, there is no course of action associated with failing to reject this particular null hypothesis.

The P-value of 0.136 says that if the true mean speed of vehicles on Triphammer Road were 30 mph, samples of 23 vehicles can be expected to have an observed mean of at least 31.0 mph 13.6% of the time. That P-value is not small enough for me to reject the hypothesis that the true mean is 30 mph. I conclude that there is not enough evidence to say the average speed is too high.

TI Tips · Testing a hypothesis about a mean

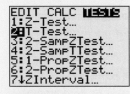

Testing a hypothesis given a set of data

Still have the Triphammer Road auto speeds in L1? Good. Let's use the TI calculator to see if the mean is significantly higher than 30 mph (you've already checked the histogram to verify the nearly Normal condition, of course).

- Go to the **STAT TESTS** menu, and choose **2:T-Test**.
- Tell it you want to use the stored **Data**.
- Enter the mean of the null model, and indicate where the data are.
- Since this is an upper-tail test, choose the $>\mu_0$, option.
- **Calculate**.

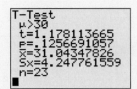

```
T-Test
 µ>30
 t=1.178113665
 p=.1256691057
 x̄=31.04347826
 Sx=4.247761559
 n=23
■
```

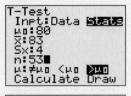

```
T-Test
 Inpt:Data Stats
 µo:80
 x̄:83
 Sx:4
 n:53■
 µ:≠µo <µo >µo
Calculate Draw
```

```
T-Test
 µ>80
 t=5.460082417
 p=6.7566262E-7
 x̄=83
 Sx=4
 n=53
■
```

There's everything you need to know: the summary statistics, the calculated value of *t*, and the P-value of 0.126. (*t* and P differ slightly from the values in our worked example because when we did it by hand we rounded off the mean and standard deviation. No harm done.)

As always, the *Tell* is up to you.

Testing a Hypothesis Given the Sample's Mean and Standard Deviation

Don't have the actual data? Just summary statistics? No problem, assuming you can verify the necessary conditions. In the last TI Tips we created a confidence interval for the strength of fishing line. We had test results for a random sample of 53 lengths of line showing a mean strength of 83 pounds and a standard deviation of 4 pounds. Is there evidence that this kind of fishing line exceeds the "80-lb test" as labeled on the package?

We bet you know what to do even without our help. Try it before you read on.

- Go back to 2:T-Test.
- You're entering Stats this time.
- Specify the hypothesized mean and the sample statistics.
- Choose the alternative being tested (upper tail here).
- Calculate.

The results of the calculator's mechanics show a large *t* and a really small P-value (0.0000007). We have very strong evidence that the mean breaking strength of this kind of fishing line is over the 80 pounds claimed by the manufacturer.

Just Checking

3. The sample of 24 batteries a consumer watchdog organization tested ran for a mean of 47 hours with a standard deviation of 3 hours. Is this strong evidence that this brand fails to meet the manufacturer's advertised "50-Hour Life!"?

a) What are the null and alternative hypotheses?
b) Assuming the assumptions and conditions are met, find *t* and the P-value.
c) What's your conclusion about the batteries?

(Check your answers on page 456.)

Paired Data

Speed-skating races are run in pairs. Two skaters start at the same time, one on the inner lane and one on the outer lane. Halfway through the race, they cross over, switching lanes so that each will skate the same distance in each lane. Even though this seems fair, at the 2006 Olympics some fans thought there might have been an advantage to starting on the outside. After all, the winner, Cindy Klassen, started on the outside and skated a remarkable 1.47 seconds faster than the silver medalist.

Here are the data for the women's 1500-m race:

Inner Lane		Outer Lane	
Name	**Time**	**Name**	**Time**
OLTEAN Daniela	129.24	(no competitor)	
ZHANG Xiaolei	125.75	NEMOTO Nami	122.34
ABRAMOVA Yekaterina	121.63	LAMB Maria	122.12
REMPEL Shannon	122.24	NOH Seon Yeong	123.35
LEE Ju-Youn	120.85	TIMMER Marianne	120.45
ROKITA Anna Natalia	122.19	MARRA Adelia	123.07
YAKSHINA Valentina	122.15	OPITZ Lucille	122.75
BJELKEVIK Hedvig	122.16	HAUGLI Maren	121.22
ISHINO Eriko	121.85	WOJCICKA Katarzyna	119.96
RANEY Catherine	121.17	BJELKEVIK Annette	121.03
OTSU Hiromi	124.77	LOBYSHEVA Yekaterina	118.87
SIMIONATO Chiara	118.76	JI Jia	121.85
ANSCHUETZ THOMS Daniela	119.74	WANG Fei	120.13
BARYSHEVA Varvara	121.60	van DEUTEKOM Paulien	120.15
GROENEWOLD Renate	119.33	GROVES Kristina	116.74
RODRIGUEZ Jennifer	119.30	NESBITT Christine	119.15
FRIESINGER Anni	117.31	KLASSEN Cindy	115.27
WUST Ireen	116.90	TABATA Maki	120.77

WHO	Olympic speed-skaters
WHAT	Time for women's 1500 m
UNITS	Seconds
WHEN	2006
WHERE	Torino, Italy
WHY	To see whether one lane is faster than the other

Data such as these are called **paired**. We have the times for skaters in each lane for each race. We can focus on the *differences* in times for each racing pair.

Paired data arise in a number of ways. Perhaps the most common way is to compare subjects with themselves before and after a treatment. Or we might test two brands of tires on the same cars. Or collect data from husbands and wives. Regardless of how the data are paired, they're not independent, so we can't use 2-sample *t*-methods.

For Example Identifying Paired Data

Do flexible schedules reduce the demand for resources? The Lake County, Illinois, Health Department experimented with a flexible four-day workweek. For a year, the department recorded the mileage driven by 11 field workers on an ordinary five-day workweek. Then it changed to a flexible four-day workweek and recorded mileage for another year.[4] The data are shown.

Question: Why are these data paired?

The mileage data are paired because each driver's mileage is measured before and after the change in schedule. I'll examine the differences in mileages for each driver to see if the new workweek results in a change.

Name	5-Day mileage	4-Day mileage	Difference
Jeff	2798	2914	−116
Betty	7724	6112	1612
Roger	7505	6177	1328
Tom	838	1102	−264
Aimee	4592	3281	1311
Greg	8107	4997	3110
Larry G.	1228	1695	−467
Tad	8718	6606	2112
Larry M.	1097	1063	34
Leslie	8089	6392	1697
Lee	3807	3362	445

---

[4]Charles S. Catlin, "Four-day Work Week Improves Environment," *Journal of Environmental Health*, Denver, 59:7.

Pairing isn't a problem; it's an opportunity. Once we recognize that the speed-skating data are matched pairs, it makes sense to consider the difference in times for each two-skater race. So we look at the *pairwise* differences:

Skating Pair	Inner Time	Outer Time	Inner − Outer
1	129.24		•
2	125.75	122.34	3.41
3	121.63	122.12	−0.49
4	122.24	123.35	−1.11
5	120.85	120.45	0.40
6	122.19	123.07	−0.88
7	122.15	122.75	−0.60
8	122.16	121.22	0.94
9	121.85	119.96	1.89
10	121.17	121.03	0.14
11	124.77	118.87	5.90
12	118.76	121.85	−3.09
13	119.74	120.13	−0.39
14	121.60	120.15	1.45
15	119.33	116.74	2.59
16	119.30	119.15	0.15
17	117.31	115.27	2.04
18	116.90	120.77	−3.87

The first skater raced alone, so we'll omit that race. Because it is the *differences* we care about, we'll treat them as if *they* were the data, ignoring the original two columns. Now that we have only one column of values to consider, we can use simple one-sample *t*-methods. The sample size is the number of pairs.

- **A paired *t*-test** is just a one-sample *t*-test for the mean of these pairwise differences. That will tell us whether there is a significant difference in times for racers in the two skating lanes.
- **A paired *t*-interval** is just a one-sample *t*-interval for the mean of these pairwise differences. That will tell us how large an advantage may exist for skaters in one of the lanes.

THE PAIRED *t*-TEST

When the conditions are met, we are ready to test whether the mean of paired differences is significantly different from zero. We test the hypothesis

$$H_0: \mu_d = 0,$$

where the *d*'s are the pairwise differences.

We use the statistic

$$t_{n-1} = \frac{\overline{d} - 0}{SE(\overline{d})}$$

where $\overline{d}$ is the mean of the pairwise differences, *n* is the number of *pairs*, and

$$SE(\overline{d}) = \frac{s_d}{\sqrt{n}}.$$

(continued)

$SE(\bar{d})$ is the ordinary standard error for the mean, applied to the differences.

When the conditions are met and the null hypothesis is true, we can model the sampling distribution of this statistic with a Student's t-model with $n-1$ degrees of freedom, and use that model to obtain a P-value.

PAIRED t-INTERVAL

When the conditions are met, we are ready to find the confidence interval for the mean of the paired differences. The confidence interval is

$$\bar{d} \pm t^*_{n-1} \times SE(\bar{d}),$$

where the standard error of the mean difference is $SE(\bar{d}) = \dfrac{s_d}{\sqrt{n}}$.

The critical value t^* from the Student's t-model depends on the particular confidence level, C, that you specify and on the degrees of freedom, $n-1$, which is based on the number of pairs, n.

We check the usual assumptions and conditions: independence, randomization, and normality (of the differences), plus one more.

Paired Data Condition: The data are paired. You must think about how the data were collected.

STEP-BY-STEP EXAMPLE   A Paired t-Test and Interval

Questions: Was there a difference in speeds between the inner and outer speed-skating lanes at the 2006 Winter Olympics? How big may any advantage have been?

Plan State what we want to know.

Identify the *parameter* we care about. Here our parameter is the mean difference in race times.

Identify the variables and check the W's.

Hypotheses State the null and alternative hypotheses.

Although fans suspected one lane was faster, we can't look at the data we have and specify the direction of a test. We (and Olympic officials) would be interested in a difference in either direction, so we'd better test a two-sided alternative.

I want to know whether there really was a difference in the speeds of the two lanes for speed skating at the 2006 Olympics. I have data for the women's 1500-m race.

H_0: Neither lane offered an advantage:

$$\mu_d = 0.$$

H_A: The mean difference is different from zero:

$$\mu_d \neq 0.$$

Model Think about the assumptions and check the conditions.

State why you think the data are paired. Simply having the same number of individuals in each group and displaying them in side-by-side columns doesn't make them paired.

Think about what we hope to learn and where the randomization comes from. Here, the randomization comes from the racer pairings and lane assignments.

Make a picture—just one. Don't plot separate distributions of the two groups—that entirely misses the pairing. For paired data, it's the Normality of the *differences* that we care about. Treat those paired differences as you would a single variable, and check the Nearly Normal Condition with a histogram.

✔ **Independence Assumption:** Each race is independent of the others, so the differences are mutually independent.

✔ **Paired Data Assumption:** The data are paired because racers compete in pairs.

✔ **Randomization Condition:** Skaters are assigned to pairings and lanes at random.

✔ **Nearly Normal Condition:** The histogram of the differences is unimodal and symmetric:

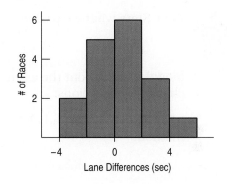

The conditions are met, so I'll use a Student's t-model with $(n - 1) = 16$ degrees of freedom, and perform a **paired t-test**.

Specify the sampling distribution model.

Choose the method.

Mechanics
n is the number of *pairs*—in this case, the number of races.

$\bar{d}$ is the mean difference.

s_d is the standard deviation of the differences.

Find the standard error and the *t*-score of the observed mean difference. There is nothing new in the mechanics of the paired-*t* methods. These are the mechanics of the *t*-test for a mean applied to the differences.

Make a picture. Sketch a *t*-model centered at the hypothesized mean of 0. Because this is a two-tail test, shade both the region to the right of the observed mean difference of 0.499 seconds and the corresponding region in the lower tail.

Find the P-value, using technology.

The data give

$$n = 17 \text{ pairs}$$
$$\bar{d} = 0.499 \text{ seconds}$$
$$s_d = 2.333 \text{ seconds.}$$

I estimate the standard deviation of $\bar{d}$ using

$$SE(\bar{d}) = \frac{s_d}{\sqrt{n}} = \frac{2.333}{\sqrt{17}} = 0.5658$$

$$\text{So } t_{16} = \frac{\bar{d} - 0}{SE(\bar{d})} = \frac{0.499}{0.5658} = 0.882$$

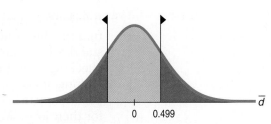

$$P\text{-value} = 2P(t_{16} > 0.882) = 0.39$$

(continued)

Conclusion Link the P-value to your decision about H_0, and state your conclusion in context.

The P-value is large. Events that happen more than a third of the time are not remarkable. So, even though there is an observed difference between the lanes, I can't conclude that it isn't due simply to random chance. It appears the fans may have interpreted a random fluctuation in the data as favoring one lane. I'll construct a confidence interval to see how much difference there may be.

THINK AGAIN

Plan State what we wish to estimate.

I want to estimate the mean difference in skaters' times between racers in Lane 1 and Lane 2.

Model Think about the assumptions and check the conditions.

The conditions for inference are satisfied.

Identify the model and choose your method.

I'll use a t-model with 16 df to construct a paired t-interval.

SHOW MORE

Mechanics Use the known values of n, $\bar{d}$, and $SE(\bar{d})$ to construct a 95% confidence interval. For a larger sample we could use the Rule of 2 for the critical value, but here we'll let technology provide t^*.

The 95% confidence interval is

$$\bar{d} \pm t^*_{16} SE(\bar{d})$$
$$0.499 \pm t^*_{16}(0.5658)$$
$$0.499 \pm 1.199$$
$$(-0.70, 1.70)$$

TELL ALL

Conclusion Interpret the confidence interval in context.

In fact, except for the Gold-Silver gap, the time differences between each skater and the next-faster one were all less than this.

The hypothesis test provided insufficient evidence to declare any lack of fairness, but the confidence interval says it's possible that the true mean difference between lanes could have been as large as 1.70 seconds. In tight competitions like the Olympics, that could make a difference.

Working with Paired Data

Inference using paired data is straightforward mechanically; after all, once we find the differences between the paired measurements we just use the standard one-sample t-methods. However, preparing the data for inference, checking conditions, and knowing what to say are also important, and worth looking at again. Let's examine the question about whether putting health department workers on a 4-day workweek changes how many miles they drive for their jobs. We'll see how to work with those data on your calculator, write the appropriate hypothesis test, and create and interpret a confidence interval.

TI Tips   Working with paired data

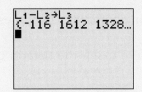

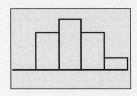

Enter paired data

Since the inference procedures for matched data are essentially just the one-sample *t* procedures, you already know what to do . . . once you have the list of paired differences, that is. That list is not hard to create.

- Enter the driving data from page 436 into two lists, say *5-Day mileage* in L1, *4-Day mileage* in L2.
- Create a list of the differences. We want to take each value in L1, subtract the corresponding value in L2, and store the paired difference in L3. The command is L1–L2 → L3. (The arrow is the STO button.) Now take a look at L3. See—it worked!

Check for normality

- Make a histogram of the differences, L3, to check the nearly Normal condition. Notice that we do not look at the histograms of the *5-day mileage* or the *4-day mileage*. Those are not the data that we care about when using a paired procedure. Note also that the calculator's first histogram is not close to Normal. More work to do . . .
- As you have seen before, small samples often produce ragged histograms, and these may look very different after a change in bar width. Reset the WINDOW to Xmin=-3000, Xmax=4500, and Xscl=1500. The new histogram looks okay.

For Example   Checking Assumptions and Conditions

Recap: Field workers for a health department compared driving mileage on a 5-day work schedule with mileage on a new 4-day schedule. To see if the new schedule changed the amount of driving they did, we'll look at paired differences in mileages before and after.

Question: Is it okay to use these data to test whether the new schedule changed the amount of driving?

✔ **Paired Data Assumption:** The data are paired because each value is the mileage driven by the same person before and after a change in work schedule.

✔ **Independence Assumption:** The driving behavior of any individual worker is independent of the others.

✔ **Randomization Condition:** The mileages are the sums of many individual trips, each of which experienced random events that arose while driving.

✔ **Nearly Normal Condition:** The histogram of the mileage differences is unimodal and symmetric:

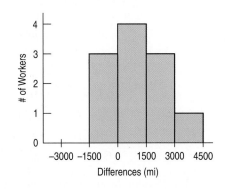

Since the assumptions and conditions are satisfied, it's okay to use paired-*t* methods for inference.

TI Tips

Testing a hypothesis with paired data

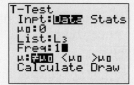

- Under **STAT TESTS** simply use **2:T-Test**, as you've done before for hypothesis tests about a mean.
- Specify that the hypothesized difference is 0, you're using the **Data** in **L3**, and it's a two-tailed test.
- **Calculate**.

The small P-value shows strong evidence that on average the change in the workweek reduces the number of miles workers drive.

For Example Doing a Paired *t*-Test

Recap: We want to test whether a change from a 5-day workweek to a 4-day workweek could change the amount driven by field workers of a health department. We've already confirmed that the assumptions and conditions for a paired *t*-test are met.

Question: Is there evidence that a 4-day workweek would change how many miles workers drive?

H_0: The change in the health department workers' schedules didn't change the mean mileage driven; the mean difference is zero:

$$\mu_d = 0.$$

H_A: The mean difference is different from zero:

$$\mu_d \neq 0.$$

The conditions are met, so I'll use a Student's *t*-model with $(n - 1) = 10$ degrees of freedom and perform a **paired *t*-test.**

The data give

$$n = 11 \text{ pairs}$$
$$\bar{d} = 982 \text{ miles}$$
$$s_d = 1139.6 \text{ miles.}$$
$$SE(\bar{d}) = \frac{s_d}{\sqrt{n}} = \frac{1139.6}{\sqrt{11}} = 343.6$$
$$\text{So } t_{10} = \frac{\bar{d} - 0}{SE(\bar{d})} = \frac{982.0}{343.6} = 2.86$$
$$P\text{-value} = 0.017$$

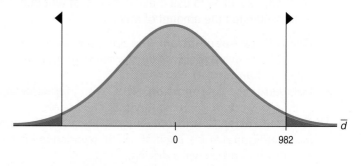

The P-value is small, so I reject the null hypothesis and conclude that the change in workweek did lead to a change in average driving mileage. It appears that changing the work schedule may reduce the mileage driven by workers.

TI Tips

Creating a confidence interval

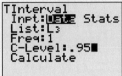

So, it appears that workers on a 4-day workweek would drive fewer miles. Let's create a 95% confidence interval to estimate the size of this change. We'll base our calculation on the paired differences, still stored in L3.

- Just as we did for data from one sample, use the **STATS TEST** one-sample procedure 8:TInterval.
- Specify Inpt:Data, List:L3, and 95% confidence.
- Calculate.

Done. Finding the interval was the easy part. Now it's time for you to *Tell* what it means—in context, of course!

For Example   Interpreting a Paired-*t* Confidence Interval

Recap: We know that, on average, the switch from a five-day workweek to a four-day workweek reduced the amount driven by field workers in that Illinois health department. However, finding that there is a significant difference doesn't necessarily mean that difference is meaningful or worthwhile. To assess the size of the effect, we need a confidence interval. We already know the assumptions and conditions are met.

Question: By how much, on average, might a change in workweek schedule reduce the amount driven by workers?

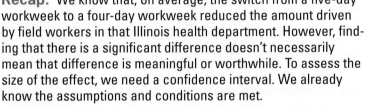

$$\bar{d} = 982 \text{ mi} \qquad SE(\bar{d}) = 343.6$$
$$\bar{d} \pm t^*_{10} \times SE(\bar{d}) = 982 \pm t^*_{10}(343.6)$$

So the 95% confidence interval for μ_d is (216.46, 1747.54) fewer miles.

With 95% confidence, I estimate that by switching to a four-day workweek employees would drive an average of between 216 and 1748 fewer miles per year. With high gas prices, this could save a lot of money.

⊘ WHAT CAN GO WRONG?

- **Don't confuse the sampling distribution with the distribution of the sample.** When you take a sample, you always look at the distribution of the values, usually with a histogram, and you may calculate summary statistics. Examining the distribution of the sample data is wise. But that's not the *sampling distribution*. The sampling distribution is an imaginary collection of all the values that a statistic *might* have taken for all possible random samples—the one you got and the ones that you didn't get. We use the sampling distribution model to make statements about how the *statistic* varies.

(continued)

- **Don't get confused about what's Normal (and what isn't).** The population from which we draw the sample may or may not be Normal; we'll never know. And the distribution of data in random samples tends to look like the population, more so the larger the sample. But the Central Limit Theorem tells us that regardless of the shape of the population (or samples drawn from it), the sampling model for *means* of those samples is essentially Normal (for sufficiently large samples).

- **Beware of observations that are not independent.** The CLT and Student's *t* methods require that sampled values be mutually independent. We think about sampling methods to be sure they don't introduce bias. We check for random sampling and the 10% Condition. If the sample isn't random (the elevator riders are all related, say) then statements we try to make about the mean are likely to be wrong.

- **Beware of skewed data and outliers.** Make a histogram of the data. Unless your sample size is large, you can't use the methods of this chapter to make inferences based on skewed data. If you find outliers, consider doing the analysis twice, once including the outliers in the data and then excluding them, to get a sense of how much they affect the results.

- **Interpret your confidence interval correctly.** Many statements that sound tempting are, in fact, misinterpretations of a confidence interval for means. Keep in mind that a confidence interval is a statement about the unknown mean of a population; it's not about means of samples, individuals in samples, or individuals in the population. Have another look at the list of common mistakes, explained on page 430.

- **Don't use paired *t*-methods when the data aren't paired.** Just because two groups have the same number of observations doesn't make them paired (even if you see them side-by-side in a table). Data based on 25 men and 25 women might be paired if they are husband and wife, say, but may be completely independent of each other.

- **Don't confuse proportions and means.** When you treat your data as categorical, counting successes and summarizing with a proportion, make inferences using the Normal model methods you learned in Chapter 18. When you treat your data as quantitative, summarizing measurements with a sample mean, make inferences using Student's *t* methods.

SIGNIFICANT DIFFERENCES **IN YOUR WORLD**

Marked Improvement in Performance for Students in Smaller Classes

In 1983 . . . [T]he [Tennessee] legislature created Project STAR to test whether class sizes that averaged 15 students in kindergarten through grade 3 would result in improved student performance,

when compared to class sizes that averaged 24 students.

The experiment required that children be randomly assigned to either a small size class (13 to 17 students), to a regular size class (22 to 25 students), or to a regular size class with a full-time aide. Teachers also were randomly assigned to one of the three class types.

Schools volunteered to participate in the project and agreed to abide by the rules of the demonstration. Seventy-nine schools were randomly selected from among the volunteers. These schools were divided into four geographic groups: inner-city, suburban, urban (small cities over 2,500), and rural. Two tests—the nationally normed Stanford Achievement Test and the curriculum-based Tennessee Basic Skills First Test—were given at the end of each year to students who participated in the demonstration.

http://www.library.ca.gov/CRB/clssz/index.html#RTFToC4

The Project STAR research team reported several [statistically] significant findings, including:

- Children in small classes consistently outperformed children in large classes [and] classes with aides.
- At the end of third grade, students in small classes in inner city schools, on average, scored 18 points higher on the SAT Reading Test than did their counterparts in regular-sized classes. This compared to differences in suburban schools of +6 points, rural schools of +7 points, and urban schools of +4 points. Comparable differences also existed for the SAT Math Test.
- At the end of third grade, inner-city children (about 97 percent of whom are minorities) in small classes . . . closed some of the performance gap between themselves and children in large classes elsewhere.

WHAT HAVE WE LEARNED?

Again we see that Statistics is about variation. We know that no sample fully and exactly describes the population; like proportions, means will vary from sample to sample. That's sampling error (or, better, sampling variability). We know that sampling variability is not just unavoidable—it's predictable!

We've learned to describe the behavior of sample means based on the amazing result known as the Center Limit Theorem—the Fundamental Theorem of Statistics. Again the sample must be independent and random—no surprise there—and needs to be larger if our data come from a population that's not roughly unimodal and symmetric. Then:

▶ Regardless of the shape of the original population, the shape of the distribution of the means of all possible samples can be described by a Normal model, provided the samples are large enough.

▶ The center of the sampling model will be the true mean of the population from which we took the sample.

▶ The standard deviation of the sample means is the population's standard deviation divided by the square root of the sample size, $\frac{\sigma}{\sqrt{n}}$.

We first learned to create confidence intervals and test hypotheses about proportions. Now we've turned our attention to means, and learned that statistical inference for means relies on the same concepts; only the mechanics and our model have changed.

▶ We've learned that what we can say about a population mean is inferred from data, using the mean of a representative random sample.

▶ We've learned to describe the sampling distribution of sample means using a new model we select from the Student's *t* family based on our degrees of freedom.

(continued)

▶ We've learned that our ruler for measuring the variability in sample means is the standard error $SE(\bar{y}) = \dfrac{s}{\sqrt{n}}$.

▶ We've learned to find the margin of error for a confidence interval using that ruler and critical values based on a Student's t-model.

▶ And we've also learned to use that ruler to test hypotheses about the population mean.

▶ We've learned that paired t-methods look at pairwise differences. Based on these differences, we test hypotheses and generate confidence intervals. These procedures are mechanically identical to the one-sample t-methods.

▶ We've also learned to be careful to recognize pairing when it is present but not assume it when it is not.

Above all, we've learned that the reasoning of inference, the need to verify that the appropriate assumptions are met, and the proper interpretation of confidence intervals and P-values all remain the same regardless of whether we are investigating means or proportions.

Terms

Sampling distribution model
Different random samples give different values for a statistic. The sampling distribution model shows the behavior of the statistic over all the possible samples for the same size n.

Sampling variability
Sampling error
The variability we expect to see from one random sample to another. It is sometimes called sampling error, but sampling variability is the better term.

Central Limit Theorem
The Central Limit Theorem (CLT) states that the sampling distribution model of the sample mean (and proportion) from a random sample is approximately Normal for large n, *regardless of the distribution of the population, as long as the observations are independent.*

Sampling distribution model for a mean
If assumptions of independence and random sampling are met, and the sample size is large enough, the sampling distribution of the sample mean is modeled by a Normal model with a mean equal to the population mean, μ, and a standard deviation equal to $\dfrac{\sigma}{\sqrt{n}}$.

Student's t
Degrees of freedom (df)
A family of distributions, the t-models are unimodal symmetric, and bell shaped, but generally have fatter tails and a narrower center than the Normal model. As the degrees of freedom increase, t-distributions approach the Normal.

Critical value
The number of SEs to move away from the mean of a sampling distribution to attain a desired level of confidence.

One-sample t-interval for the mean
A one-sample t-interval for the population mean is

$$\bar{y} \pm t^*_{n-1} \times SE(\bar{y}), \text{ where } SE(\bar{y}) = \frac{s}{\sqrt{n}}$$

The critical value t^*_{n-1} depends on the particular confidence level, C, that you specify and on the number of degrees of freedom, $n - 1$.

One-sample t-test for the mean
The one-sample t-test for the mean tests the hypothesis $H_0: \mu = \mu_0$ using the statistic

$$t_{n-1} = \frac{\bar{y} - \mu_0}{SE(\bar{y})}.$$

The standard error of $\bar{y}$ is

$$SE(\bar{y}) = \frac{s}{\sqrt{n}}.$$

Paired data
Data are paired when the observations are collected in pairs or the observations in one group are naturally related to observations in the other. The simplest form of pairing is to measure each subject twice—often before and after a treatment is applied. Pairing in observational and survey data is a form of matching.

Paired *t*-test A hypothesis test for the mean of the pairwise differences of two groups. It tests the null hypothesis $H_0: \mu_d = 0$, using the statistic

$$t = \frac{\bar{d} - 0}{SE(\bar{d})}$$

with $n - 1$ degrees of freedom, where $SE(\bar{d}) = \frac{s_d}{\sqrt{n}}$, and n is the number of pairs.

Paired-*t* confidence interval A confidence interval for the mean of the pairwise differences between paired groups found as

$$\bar{d} \pm t^*_{n-1} \times SE(\bar{d}), \text{ where } SE(\bar{d}) = \frac{s_d}{\sqrt{n}} \text{ and } n \text{ is the number of pairs.}$$

Skills

▶ Understand that the variability of a statistic (as measured by the standard deviation of its sampling distribution) depends on the size of the sample. Statistics based on larger samples are less variable.

▶ Understand that the Central Limit Theorem gives the sampling distribution model of the mean for sufficiently large samples regardless of the underlying population.

▶ Know the assumptions required for *t*-tests and *t*-based confidence intervals.

▶ Know how to examine your data for violations of conditions that would make inference about the population mean unwise or invalid.

▶ Be able to recognize whether a design that compares two groups is paired.

▶ Be able to use a sampling distribution model to make simple statements about the distribution of a mean under repeated sampling.

▶ Be able to compute and interpret a *t*-test for the population mean.

▶ Be able to compute and interpret a *t*-based confidence interval for the population mean.

▶ Be able to find a paired confidence interval, recognizing that it is mechanically equivalent to doing a one-sample *t*-interval applied to the differences.

▶ Be able to perform a paired *t*-test, recognizing that it is mechanically equivalent to a one-sample *t*-test applied to the differences.

▶ Be able to interpret a sampling distribution model as describing the possible values taken by a statistic in repeated samples or experiments.

▶ Be able to explain the meaning of a confidence interval for a population mean. Make clear that the randomness associated with the confidence level is a statement about the interval bounds and not about the population parameter value.

▶ Understand that a 95% confidence interval does not trap 95% of the sample values.

▶ Be able to interpret the result of a test of a hypothesis about a population mean.

▶ Know that we do not "accept" a null hypothesis if we cannot reject it. We say that we fail to reject it.

▶ Understand that the P-value of a test represents the probability that results like ours could be explained as merely sampling error. It does not give the probability that the null hypothesis is correct.

▶ Be able to interpret a paired *t*-test or interval, recognizing that these inferences are about the mean of the differences between paired values.

INFERENCE FOR A MEAN ON THE COMPUTER

Statistics packages can compute hypothesis tests and confidence intervals for a mean. You'll need to enter your data and then specify your hypotheses or desired confidence level. The calculated results appear in an output table.

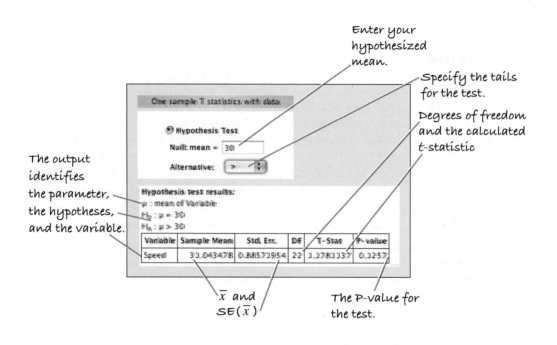

Enter your hypothesized mean.

Specify the tails for the test.

Degrees of freedom and the calculated *t*-statistic

The output identifies the parameter, the hypotheses, and the variable.

$\bar{x}$ and $SE(\bar{x})$

The P-value for the test.

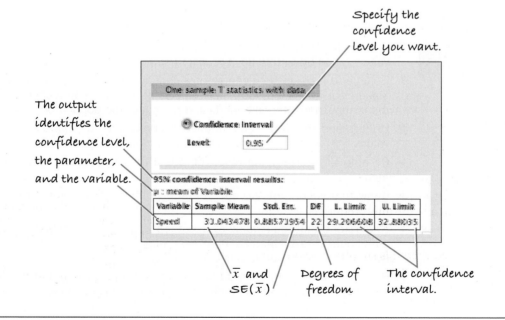

Specify the confidence level you want.

The output identifies the confidence level, the parameter, and the variable.

$\bar{x}$ and $SE(\bar{x})$

Degrees of freedom

The confidence interval.

EXERCISES

A

1. **Potato chips.** A packaging machine fills bags of potato chips by weight. The amount of chips in each small bag is believed to follow a Normal model with a mean of 10 ounces and a standard deviation of 0.20 ounces. Stores sell these chips in a "Value Pack" containing 3 bags. We can treat any Value Pack as a random sample of all the bags of chips.
   a) What's the expected mean weight of the 3 bags?
   b) What's the expected variability in that mean weight? Find $SD(\bar{x})$.

2. **Tires.** A delivery company has just installed new tires on 6 of their vans. Based on past experience, they believe sets of these tires can be driven an average of 38,000 miles, with a standard deviation of 4000 miles.
   a) What's the expected mean lifetime for the tires on these 6 vans?
   b) What's the expected variability in that mean mileage? Find $SD(\bar{x})$.

3. **More chips.** The company that manufactures the potato chips described in Exercise 1 also ships them to grocery stores in cases containing 24 bags.
   a) What's the expected mean weight of the 24 bags in a case?
   b) What's the expected variability in that mean weight? Find $SD(\bar{x})$.

4. **More tires.** Based on good experience with the tires described in Exercise 2, the company decides to install them on all 40 of their delivery vans.
   a) What's the expected mean lifetime for the tires on their fleet?
   b) What's the expected variability in that mean mileage? Find $SD(\bar{x})$.

5. **Cattle.** Livestock are given a special feed supplement to see if it will promote weight gain. Researchers report that the 77 cows studied gained an average of 56 pounds, and that a 95% confidence interval for the mean weight gain this supplement produces has a margin of error of ±11 pounds. Some students wrote the following conclusions. Did anyone interpret the interval correctly? Explain any misinterpretations.
   a) 95% of the cows studied gained between 45 and 67 pounds.
   b) We're 95% sure that a cow fed this supplement will gain between 45 and 67 pounds.
   c) We're 95% sure that the average weight gain among the cows in this study was between 45 and 67 pounds.
   d) The average weight gain of cows fed this supplement will be between 45 and 67 pounds 95% of the time.
   e) If this supplement is tested on another sample of cows, there is a 95% chance that their average weight gain will be between 45 and 67 pounds.

6. **Teachers.** Software analysis of the salaries of a random sample of 288 Nevada teachers produced the confidence interval shown below. Which conclusion is correct? What's wrong with the others?

   t-Interval for μ: with 90.00% Confidence,
   $38944 < \mu(\text{TchPay}) < 42893$

   a) If we took many random samples of 288 Nevada teachers, about 9 out of 10 of them would produce this confidence interval.
   b) If we took many random samples of Nevada teachers, about 9 out of 10 of them would produce a confidence interval that contained the mean salary of all Nevada teachers.
   c) About 9 out of 10 Nevada teachers earn between $38,944 and $42,893.
   d) About 9 out of 10 of the teachers surveyed earn between $38,944 and $42,893.
   e) We are 90% confident that the average teacher salary in the United States is between $38,944 and $42,893.

7. **Meal plan.** After surveying students at Dartmouth College, a campus organization calculated that a 95% confidence interval for the mean cost of food for one term (of three in the Dartmouth trimester calendar) is ($1102, $1290). Now the organization is trying to write its report and is considering the following interpretations. Comment on each.
   a) 95% of all students pay between $1102 and $1290 for food.
   b) 95% of the sampled students paid between $1102 and $1290.
   c) We're 95% sure that students in this sample averaged between $1102 and $1290 for food.
   d) 95% of all samples of students will have average food costs between $1102 and $1290.
   e) We're 95% sure that the average amount all students pay is between $1102 and $1290.

8. **Snow.** Based on meteorological data for the past century, a local TV weather forecaster estimates that the region's average winter snowfall is 23", with a margin of error of ±2 inches. Assuming he used a 95% confidence interval, how should viewers interpret this news? Comment on each of these statements:
   a) During 95 of the last 100 winters, the region got between 21" and 25" of snow.
   b) There's a 95% chance the region will get between 21" and 25" of snow this winter.
   c) There will be between 21" and 25" of snow on the ground for 95% of the winter days.
   d) Residents can be 95% sure that the area's average snowfall is between 21" and 25".
   e) Residents can be 95% confident that the average snowfall during the last century was between 21" and 25" per winter.

9. Pulse rates. A medical researcher measured the pulse rates (beats per minute) of a sample of randomly selected adults and found the following Student's *t*-based confidence interval:

$$\text{With 95.00\% Confidence,}$$
$$70.887604 < \mu(\text{Pulse}) < 74.497011$$

a) Explain carefully what the software output means in context.
b) What's the margin of error for this interval?

10. Crawling. Data collected by child development scientists produced this confidence interval for the average age (in weeks) at which babies begin to crawl:

$$t\text{-Interval for } \mu$$
$$(95.00\% \text{ Confidence}): \quad 29.202 < \mu(\text{age}) < 31.844$$

a) Explain carefully what the software output means in context.
b) What is the margin of error for this interval?

11. More eggs? A farmer wonders whether a food additive will increase egg production above the average of 340 eggs per day that his hens have been producing. He plans to try feeding them the food additive for a month to see if they lay more eggs. Write his null and alternative hypotheses.

12. Be friendly. A waiter wonders whether he'll get bigger tips if he takes more time for friendly chatting with the restaurant patrons. His records show that his average tip has been $3.20 per person at a table. He plans to try the friendly approach for a month to see what happens. Write his null and alternative hypothesis.

13. Rain. Long-term weather records indicate that during the 20th century Denver averaged 5.2 inches of rain during summers (June, July, and August). We want to look at rainfall for summers since 2000 to see if there has been any change in the mean summer rainfall, perhaps attributable to global warming. Write the null and alternative hypothesis.

14. Purchasing power. Teen Research Unlimited reported that in 2003 people aged 12–19 spent an average of $103 per week. To see if that has changed, we'll survey a random sample of teenagers. Write the null and alternative hypotheses.

15. Snow tires. One of the authors bought a new 2010 Prius. During the first few months he owned it, the car averaged 48.2 miles per gallon. As winter arrived, though, he had snow tires installed. While the increased traction is important, snow tires can reduce fuel efficiency, so he plans to check the car's performance for several tankfuls. Write the null and alternative hypotheses.

16. Pollution. U.S. National Air Quality Standards set a level of 9 parts per million for carbon monoxide in outdoor air, though higher levels (up to 35 ppm) can be considered safe for short periods of time (less than one hour). We plan to take several measurements at various times and places around our city to see if

there is evidence that our average CO level meets the guidelines. Write the null and alternative hypotheses.

17. Pizza. A researcher tests whether the mean cholesterol level among those who eat frozen pizza exceeds the value considered to indicate a health risk. She gets a P-value of 0.07. Explain in this context what the "7%" represents.

18. Golf balls. The United States Golf Association (USGA) sets performance standards for golf balls. For example, the initial velocity of the ball may not exceed 250 feet per second when measured by an apparatus approved by the USGA. Suppose a manufacturer introduces a new kind of ball and provides a sample for testing. Based on the mean speed in the test, the USGA comes up with a P-value of 0.34. Explain in this context what the "34%" represents.

19. Popcorn. A company wonders whether a newly designed microwave popcorn bag can reduce the number of un-popped kernels. They test several bags and get a P-value of 0.19. Explain what that means in this context.

20. Wind power. Before installing a wind turbine to produce electricity, a landowner puts out instruments to record windspeeds. She won't buy the turbine unless there's evidence that the winds are strong enough to make enough electricity to justify the expense. Her data produced a P-value of 0.02. Explain what that means in this context.

21. Paired data? In the following studies, would you consider the data to be paired? Why or why not?
a) Can a study guide help students raise their ACT scores? A researcher has some students who took the exam recently use the guide and then retake the ACT to see whether there is a significant increase in their scores.
b) To see if there is a gender difference in running speed at age 4, a researcher times 25 boys and 25 girls in a 40-yard dash.
c) To estimate how braking is affected by wet rain, a researcher measures the stopping distances for 10 cars traveling 40 miles per hour on a dry test track, then wets the pavement and measures these cars' stopping distances again.

22. More pairs? In the following studies, would you consider the data to be paired? Why or why not?
a) To see whether cars can get better gas mileage by using a different grade of fuel, a researcher gets a taxi company to run 50 cars on regular gasoline and 50 identical cars on premium.
b) Many people think that students tend to gain weight during their first semester at college. To find out, a researcher weighed a random sample of 100 students when they arrived on campus, then weighed them again at the end of the semester.
c) To compare health care costs in the United States. and Canada, a consumer group identified 20 commonly used drugs and looked at prescription prices in each country.

B _____

23. Sampling, part I. We've chosen a fairly small random sample from a population that is somewhat bimodal. After looking at the preliminary results, we decide to repeat our investigation, this time using a very large random sample.
a) What effect would you expect the increase in sample size to have on the distribution (shape, center, and spread) of the sample data?
b) What effect will the increase in sample size have on the sampling distribution model (shape, center, and spread) for means of such samples?

24. Sampling, part II. We plan to choose a random sample from a population that is strongly skewed to the left. Explain why it would be better to use a fairly large sample than a smaller (and less expensive) one.
a) What effect would you expect the larger sample size to have on the distribution (shape, center, and spread) of the sample data?
b) What effect would a larger sample size have on the sampling distribution model (shape, center, and spread) for means of such samples?

25. GPAs. A college's data about the incoming freshmen indicates that the mean of their high school GPAs was 3.4, with a standard deviation of 0.35; the distribution was roughly mound-shaped and only slightly skewed. The students are randomly assigned to freshman writing seminars in groups of 25. What might the mean GPA of one of these seminar groups be? Describe the appropriate sampling distribution model—shape, center, and spread—with a sketch using the 68–95–99.7 Rule.

26. Home values. Assessment records indicate that the value of homes in a small city is skewed right, with a mean of $140,000 and standard deviation of $60,000. To check the accuracy of the assessment data, officials plan to conduct a detailed appraisal of 100 homes selected at random. Using the 68–95–99.7 Rule, draw and label an appropriate sampling model for the mean value of the homes selected.

T 27. Departures. What are the chances your flight will leave on time? The U.S. Bureau of Transportation Statistics of the Department of Transportation publishes information about airline performance. Here are a histogram and summary statistics for the percentage of flights departing on time each month from 1995 thru 2006.

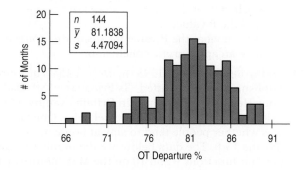

a) Check the assumptions and conditions for inference.
b) Find a 95% confidence interval for the true percentage of flights that depart on time.
c) Interpret this interval for a traveler planning to fly.

T 28. Late arrivals. Will your flight get you to your destination on time? The U.S. Bureau of Transportation Statistics reported the percentage of flights that were late each month from 1995 through 2006. Here's a histogram, along with some summary statistics:

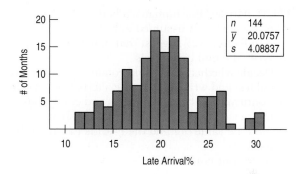

We can consider these data to be a representative sample of all months.
a) Check the assumptions and conditions for inference about the mean.
b) Find a 95% confidence interval for the true percentage of flights that arrive late.
c) Interpret this interval for a traveler planning to fly.

T 29. Normal temperature. The researcher described in Exercise 9 also measured the body temperatures of that randomly selected group of adults. Here are summaries of the data he collected. We wish to estimate the average (or "normal") temperature among the adult population.

Summary	Temperature
Count	52
Mean	98.285
StdDev	0.6824

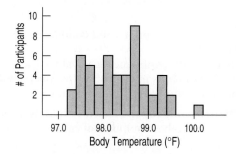

a) Check the conditions for creating a *t*-interval.
b) Find a 95% confidence interval for mean body temperature.
c) Explain the meaning of that interval.

30. **Parking.** Hoping to lure more shoppers downtown, a city builds a new public parking garage in the central business district. The city plans to pay for the structure through parking fees. During a two-month period (44 weekdays), daily fees collected averaged $126, with a standard deviation of $15.
    a) What assumptions must you make in order to use these statistics for inference?
    b) Write a 95% confidence interval for the mean daily income this parking garage will generate.
    c) Interpret this confidence interval in context.

31. **What's normal?** For humans, a body temperature of 98.6° is commonly assumed to be "normal." Do the data in Exercise 29 suggest that may not be correct?
    a) Write the null hypothesis.
    b) Decide whether you should conduct a one- or two-tail test and write appropriate alternative hypothesis.
    c) Confirm that the assumptions and conditions for inference based on t-models are satisfied.
    d) Find the value of t and the P-value based on these data.
    e) State your conclusion.

32. **Parking, part II.** The consultant who advised the city on building the parking garage described in Exercise 30 predicted that the parking revenues would average $130 per day. Do the data from these two months suggest the consultant's estimate may have been incorrect?
    a) Write the null hypothesis.
    b) Decide whether you should conduct a one- or two-tail test and write an appropriate alternative hypothesis.
    c) Confirm that the assumptions and conditions for inference based on t-models are satisfied.
    d) Find the value of t and the P-value based on these data.
    e) State your conclusion.

33. **Chips Ahoy!** In 1998, as an advertising campaign, the Nabisco Company announced a "1000 Chips Challenge," claiming that every 18-ounce bag of their Chips Ahoy! cookies contained at least 1000 chocolate chips. Dedicated Statistics students at the Air Force Academy (no kidding) purchased some randomly selected bags of cookies, and counted the chocolate chips. Some of their data are given below. (*Chance*, 12, no. 1[1999])

1219	1214	1087	1200	1419	1121	1325	1345
1244	1258	1356	1132	1191	1270	1295	1135

    a) Check the assumptions and conditions for inference.
    b) Create a 95% confidence interval for the average number of chips in bags of Chips Ahoy! cookies.
    c) Interpret your interval in this context.

34. **Yogurt.** *Consumer Reports* tested 14 brands of vanilla yogurt and found these numbers of calories per serving:

160	200	220	230	120	180	140
130	170	190	80	120	100	170

    a) Check the assumptions and conditions for inference.
    b) Create a 95% confidence interval for the average calorie content of vanilla yogurt.
    c) Interpret your interval in this context.

35. **Another cookie.** Do the data collected by the intrepid Air Force Academy investigators in Exercise 33 provide evidence to support Nabisco's claim of at least 1000 chips in each bag of cookies?
    a) Write the null hypothesis.
    b) Decide whether you should conduct a one- or two-tail test and write an appropriate alternative hypothesis.
    c) Confirm that the assumptions and conditions for inference based on t-models are satisfied.
    d) Find the value of t and the P-value based on these data.
    e) State your conclusion.

36. **More yogurt.** A diet guide indicates that you will get 120 calories from a serving of vanilla yogurt. Do the *Consumer Reports* data in Exercise 34 provide evidence that the diet guide's estimate may be too low?
    a) Write the null hypothesis.
    b) Decide whether you should conduct a one- or two-tail test and write an appropriate alternative hypothesis.
    c) Confirm that the assumptions and conditions for inference based on t-models are satisfied.
    d) Find the value of t and the P-value based on these data.
    e) State your conclusion.

37. **Marriage.** In 1960, census results indicated that the age at which American men first married had a mean of 23.3 years. It is widely suspected that young people today are waiting longer to get married. We want to find out if the mean age of first marriage has increased during the past 40 years.
    a) Write appropriate hypotheses.
    b) We plan to test our hypothesis by selecting a random sample of 40 men who married for the first time last year. Do you think the necessary assumptions for inference are satisfied? Explain.
    c) The men in our sample married at an average age of 24.2 years, with a standard deviation of 5.3 years. What's the P-value for this result?
    d) Explain (in context) what this P-value means.
    e) What's your conclusion?

38. **Fuel economy.** A company with a large fleet of cars hopes to keep gasoline costs down and sets a goal of attaining a fleet average of at least 26 miles per gallon. To see if the goal is being met, they check the gasoline usage for 50 company trips chosen at random, finding a mean of 25.02 mpg and a standard deviation of 4.83 mpg. Is this strong evidence that they have failed to attain their fuel economy goal?
    a) Write appropriate hypotheses.
    b) Are the necessary assumptions to make inferences satisfied?
    c) Find the P-value.
    d) Explain what the P-value means in this context.
    e) State an appropriate conclusion.

39. **Friday the 13th, I.** In 1993 the *British Medical Journal* published an article titled, "Is Friday the 13th Bad for Your Health?" Researchers in Britain examined how Friday the 13th affects human behavior. One question was whether people tend to stay at home more on Friday the 13th. The data below are the number of cars passing Junctions 9 and 10 on the M25 motorway for

consecutive Fridays (the 6th and 13th) for five different periods.

Year	Month	6th	13th
1990	July	134,012	132,908
1991	September	133,732	131,843
1991	December	121,139	118,723
1992	March	124,631	120,249
1992	November	117,584	117,263

State a conclusion based on this analyses:

```
Paired t-Test of mu(1 − 2) = 0 vs. mu(1 − 2)>0
Mean of Paired Differences: 2022.4
t-Statistic = 2.9377 w/4 df
P = 0.0212
```

40. Friday the 13th, II: The researchers in Exercise 39 also examined the number of people admitted to emergency rooms for vehicular accidents on 12 Friday evenings (6 each on the 6th and 13th).

Year	Month	6th	13th
1989	October	9	13
1990	July	6	12
1991	September	11	14
1991	December	11	10
1992	March	3	4
1992	November	5	12

Based on these data, is there evidence that more people are admitted, on average, on Friday the 13th? Use this analysis to state your conclusion:

```
Paired t-Test of mu(1 − 2) = 0 vs. mu(1 − 2) < 0
Mean of Paired Differences = 3.333
t-Statistic = 2.7116 w/5 df
P = 0.0211
```

41. Online insurance. After seeing countless commercials claiming one can get cheaper car insurance from an online company, a local insurance agent was concerned that he might lose some customers. To investigate, he randomly selected profiles (type of car, coverage, driving record, etc.) for 10 of his clients and checked online price quotes for their policies. The comparisons are shown in the table below.

Local	Online	PriceDiff
568	391	177
872	602	270
451	488	−37
1229	903	326
605	677	−72
1021	1270	−249
783	703	80
844	789	55
907	1008	−101
712	702	10

a) Write appropriate hypotheses.
b) Are the assumptions and conditions satisfied? Here's a histogram for the difference (*Local – Online*):

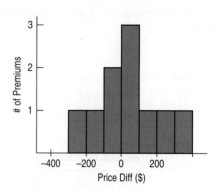

c) Find *t* and the P-value.
d) Is there evidence that drivers might save money by switching to the online company?

42. Windy, part I. To select the site for an electricity-generating wind turbine, wind speeds were recorded at several potential sites every 6 hours for a year. Two sites not far from each other looked good. Each had a mean wind speed high enough to qualify, but we should choose the site with a higher average daily wind speed. Because the sites are near each other and the wind speeds were recorded at the same times, we should view the speeds as paired. Here are the summaries of the speeds (in miles per hour):

Variable	Count	Mean	StdDev
site2	1114	7.452	3.586
site4	1114	7.248	3.421
site2 – site4	1114	0.204	2.551

a) Write appropriate hypotheses.
b) Are the assumptions and conditions satisfied? Here's a histogram for the difference in wind speeds at the two sites:

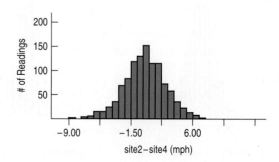

c) Find *t* and the P-value.
d) Is there evidence that either of these sites has a higher average wind speed?

43. Waist size. A study measured the *Waist Size* of 250 men, finding a mean of 36.33 inches and a standard deviation of 4.02 inches. Here is a histogram of these measurements

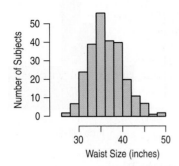

a) Describe the histogram of *Waist Size*.
b) To explore how the mean might vary from sample to sample, we simulated by drawing many samples of size 2, 5, 10, and 20, with replacement, from the 250 measurements. Here are histograms of the sample means for each simulation. Explain how these histograms demonstrate what the Central Limit Theorem says about the sampling distribution model for sample means.

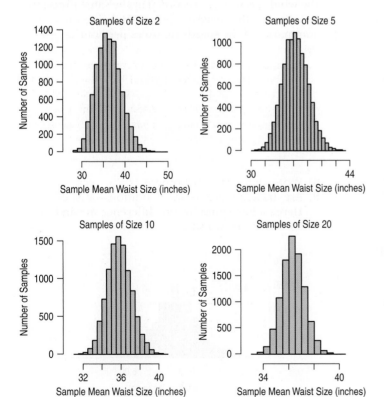

44. CEO compensation. In Chapter 6 we saw the distribution of the total compensation of the chief executive officers (CEOs) of the 800 largest U.S. companies (the Fortune 800). The average compensation (in thousands of dollars) is 10,307.31 and the standard deviation is 17,964.62. Here is a histogram of their annual compensations (in $1000):

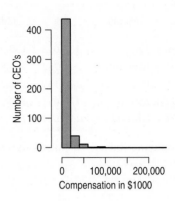

a) Describe the histogram of *Total Compensation*. A research organization simulated sample means by drawing samples of 30, 50, 100, and 200, with replacement, from the 800 CEOs. The histograms show the distributions of means for many samples of each size.

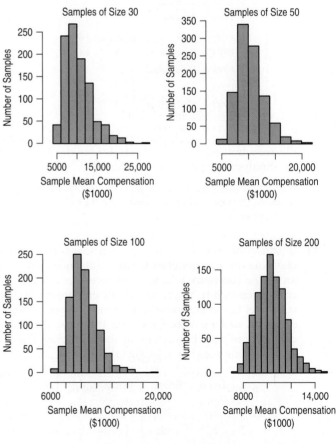

b) Explain how these histograms demonstrate what the Central Limit Theorem says about the sampling distribution model for sample means. Be sure to talk about shape, center, and spread.
c) Comment on the "rule of thumb" that "With a sample size of at least 30, the sampling distribution of the mean is Normal."

45. t-models, part I. Describe how the shape, center, and spread of t-models change as the number of degrees of freedom increases.

46. t-models, part II. Describe how the critical value of t for a 95% confidence interval changes as the number of degrees of freedom increases.

47. Ruffles. Students investigating the packaging of potato chips purchased 6 bags of Lay's Ruffles marked with a net weight of 28.3 grams. They carefully weighed the contents of each bag, recording the following weights (in grams): 29.3, 28.2, 29.1, 28.7, 28.9, 28.5.
a) Do these data satisfy the assumptions for inference? Explain.
b) Find the mean and standard deviation of the weights.
c) Create a 95% confidence interval for the mean weight of such bags of chips.
d) Explain in context what your interval means.
e) Comment on the company's stated net weight of 28.3 grams.

48. Doritos. Some students checked 6 bags of Doritos marked with a net weight of 28.3 grams. They carefully weighed the contents of each bag, recording the following weights (in grams): 29.2, 28.5, 28.7, 28.9, 29.1, 29.5.
a) Do these data satisfy the assumptions for inference? Explain.
b) Find the mean and standard deviation of the weights.
c) Create a 95% confidence interval for the mean weight of such bags of chips.
d) Explain in context what your interval means.
e) Comment on the company's stated net weight of 28.3 grams.

49. Normal temperatures, part II. Consider again the statistics about human body temperature in Exercise 29.
a) Would a 90% confidence interval be wider or narrower than the 95% confidence interval you calculated before? Explain. (Don't compute the new interval.)
b) What are the advantages and disadvantages of the 95% confidence interval?
c) If we conduct further research, this time using a sample of 500 adults, how would you expect the 95% confidence interval to change? Explain.

50. Parking II. Suppose that, for budget planning purposes, the city in Exercise 30 needs a better estimate of the mean daily income from parking fees.
a) Someone suggests that the city use its data to create a 98% confidence interval instead of the 95% interval first created. How would this interval be better for the city? (You need not actually create the new interval.)
b) How would the 98% interval be worse for the planners?

51. TV safety. The manufacturer of a metal stand for home TV sets must be sure that its product will not fail under the weight of the TV. Since some larger sets weigh nearly 300 pounds, the company's safety inspectors have set a standard of ensuring that the stands can support an average of over 500 pounds. Their inspectors regularly subject a random sample of the stands to increasing weight until they fail. They test the hypothesis $H_0: \mu = 500$ against $H_A: \mu > 500$. If the sample of stands fail to pass this safety test, the inspectors will not certify the product for sale to the general public.
a) Is this an upper-tail or lower-tail test? In the context of the problem, why do you think this is important?
b) Explain what will happen if the inspectors commit a Type I error.
c) Explain what will happen if the inspectors commit a Type II error.

52. Catheters. During an angiogram, heart problems can be examined via a small tube (a catheter) threaded into the heart from a vein in the patient's leg. It's important that the company that manufactures the catheter maintain a diameter of 2.00 mm. (The standard deviation is quite small.) Each day, quality control personnel make several measurements to test $H_0: \mu = 2.00$ against $H_A: \mu \neq 2.00$. If they discover a problem, they will stop the manufacturing process until it is corrected.
a) Is this a one-sided or two-sided test? In the context of the problem, why do you think this is important?
b) Explain in this context what happens if the quality control people commit a Type I error.
c) Explain in this context what happens if the quality control people commit a Type II error.

53. Job satisfaction. A company institutes an exercise break for its workers to see if it will improve job satisfaction, as measured by a questionnaire that assesses workers' satisfaction. Scores for 10 randomly selected workers before and after the implementation of the exercise program are shown in the following table.
a) Identify the procedure you would use to assess the effectiveness of the exercise program, and check to see if the conditions allow the use of that procedure.
b) Test an appropriate hypothesis and state your conclusion.
c) If your conclusion turns out to be incorrect, what kind of error did you commit?

Worker Number	Job Satisfaction Index	
	Before	After
1	34	33
2	28	36
3	29	50
4	45	41
5	26	37
6	27	41
7	24	39
8	15	21
9	15	20
10	27	37

54. Summer school. Having done poorly on their Math final exams in June, six students repeat the course in summer school and take another exam in August.

June	54	49	68	66	62	62
Aug.	50	65	74	64	68	72

a) If we consider these students to be representative of all students who might attend this summer school in other years, do these results provide evidence that the program is worthwhile? Test an appropriate hypothesis and state your conclusion.
b) This conclusion, of course, may be incorrect. If so, which type of error was made?

55. Temperatures. The following table gives the average high temperatures in January and July for several European cities. Write a 90% confidence interval for the mean temperature difference between summer and winter in Europe. Be sure to check conditions for inference, and clearly explain what your interval means.

	Mean High Temperatures (°F)	
City	**Jan.**	**July**
Vienna	34	75
Copenhagen	36	72
Paris	42	76
Berlin	35	74
Athens	54	90
Rome	54	88
Amsterdam	40	69
Madrid	47	87
London	44	73
Edinburgh	43	65
Moscow	21	76
Belgrade	37	84

56. Marathons 2006. The following table shows the winning times (in minutes) for men and women in the New York City Marathon between 1978 and 2006. Assuming that performances in the Big Apple resemble performances elsewhere, we can think of these data as a sample of performance in marathon competitions. Create a 90% confidence interval for the mean difference in winning times for male and female marathon competitors. (www.nycmarathon.org)

Year	Men	Women	Year	Men	Women
1978	132.2	152.5	1993	130.1	146.4
1979	131.7	147.6	1994	131.4	147.6
1980	129.7	145.7	1995	131.0	148.1
1981	128.2	145.5	1996	129.9	148.3
1982	129.5	147.2	1997	128.2	148.7
1983	129.0	147.0	1998	128.8	145.3
1984	134.9	149.5	1999	129.2	145.1
1985	131.6	148.6	2000	130.2	145.8
1986	131.1	148.1	2001	127.7	144.4
1987	131.0	150.3	2002	128.1	145.9
1988	128.3	148.1	2003	130.5	142.5
1989	128.0	145.5	2004	129.5	143.2
1990	132.7	150.8	2005	129.5	144.7
1991	129.5	147.5	2006	130.0	145.1
1992	129.5	144.7			

Answers

Just Checking

1. a) No, this is a histogram of the individuals. We don't know if it's Normal.
b) A Normal model (approximately).
c) 266 days
d) 1.6 days
e)

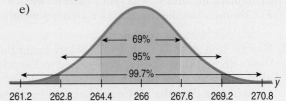

2. a) We assume battery lifetimes are (at least approximately) Normally distributed.
b) (45.7, 48.3)
c) We can be 95% confident that this brand of batteries averages between 45.7 and 48.3 hours of useful life.

3. a) $H_0: \mu = 50$; $H_A: \mu < 50$
b) $t = -4.9$; P-value = 0.00003
c) Based on such a tiny P-value, I reject the null hypothesis. These data provide strong evidence that the average life of these batteries is less than the 50 hours the manufacturer advertises.

CHAPTER 20

Comparing Proportions or Means

Do men take more risks than women? Psychologists have documented that in many situations, men choose riskier behavior than women do. But what if a woman is by their side? A recent seatbelt observation study in Massachusetts[1] found that, not surprisingly, male drivers wear seatbelts less often than women do. The study also noted that men's belt-wearing jumped more than 16 percentage points when they had a female passenger. Seatbelt use was recorded at 161 locations in Massachusetts, using random-sampling methods developed by the National Highway Traffic Safety Administration (NHTSA). Female drivers wore belts more than 70% of the time, regardless of the sex of their passengers. Of 4208 male drivers with female passengers, 2777 (66.0%) were belted. But among 2763 male drivers with male passengers only, 1363 (49.3%) wore seatbelts. This was only a random sample, but it suggests there may be a shift in men's risk-taking behavior when women are present. What would we estimate the true size of that gap to be?

Comparisons between two groups are much more common than questions about just one. And they are more interesting. We often want to know how percentages or means differ, whether a treatment is better than a placebo control, or whether this year's results are better than last year's.

---

[1]Massachusetts Traffic Safety Research Program [June 2007].

Another Ruler

The difference between the proportions of men wearing seat belts seen in the *sample* is 16.7%. But what's the *true* difference for all men? We know that our estimate probably isn't exactly right; each proportion will vary from sample to sample. To say more about the difference, we need a new ruler—the standard deviation of the sampling distribution model for the difference in the proportions.

That standard deviation comes to us from the Pythagorean Theorem of Statistics:

The variance of the sum or difference of two independent random variables is the sum of their variances.

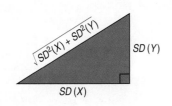

Combining independent random quantities always *increases* the overall variation, so even for *differences* of independent random variables, **variances add.**

This is such an important (and powerful) idea in Statistics that it's worth pausing a moment to look at the reasoning. Here's some intuition about why variation increases even when we subtract two random quantities.

Grab a full box of cereal. The box claims to contain 16 ounces of cereal. We know that's not exact: There's some small variation from box to box. Now pour a bowl of cereal. Of course, your 2-ounce serving will not be exactly 2 ounces. There'll be some variation there, too. How much cereal would you guess was left in the box? Do you think your guess will be as close as your guess for the full box? *After* you pour your bowl, the amount of cereal in the box is still a random quantity (with a smaller mean than before), but it is even *more variable* because of the additional variation in the amount you poured.

According to our rule, the variance of the amount of cereal left in the box would now be the *sum* of the two *variances*.

We want a standard deviation, not a variance, but that's just a square root away. We can write symbolically what we've just said:

$$Var(X - Y) = Var(X) + Var(Y), \text{ so}$$
$$SD(X - Y) = \sqrt{SD^2(X) + SD^2(Y)} = \sqrt{Var(X) + Var(Y)}.$$

Be careful, though—this simple formula applies only when X and Y are independent. Just as the Pythagorean Theorem[2] works only for right triangles, our formula works only for independent random variables. Always check for independence before using the methods of this chapter.

The Standard Deviation of the Difference Between Two Proportions

Fortunately, proportions observed in independent random samples *are* independent, so we can put the two proportions in for X and Y and add their variances. We just need to use careful notation to keep things straight.

When we have two samples, each can have a different size and proportion value, so we keep them straight with subscripts. Often we choose subscripts that remind us of the groups. For our example, we might use "$_M$" and "$_F$", but generically we'll just use "$_1$" and "$_2$". We will represent the two sample proportions as $\hat{p}_1$ and $\hat{p}_2$, and the two sample sizes as n_1 and n_2.

[2]If you don't remember the formula, don't rely on the Scarecrow's version from *The Wizard of Oz*. He may have a brain and have been awarded his Th.D. (Doctor of Thinkology), but he gets the formula wrong.

The standard deviations of the sample proportions are $SD(\hat{p}_1) = \sqrt{\dfrac{p_1 q_1}{n_1}}$ and $SD(\hat{p}_2) = \sqrt{\dfrac{p_2 q_2}{n_2}}$, so the variance of the difference in the proportions is

$$Var(\hat{p}_1 - \hat{p}_2) = \left(\sqrt{\dfrac{p_1 q_1}{n_1}}\right)^2 + \left(\sqrt{\dfrac{p_2 q_2}{n_2}}\right)^2 = \dfrac{p_1 q_1}{n_1} + \dfrac{p_2 q_2}{n_2}.$$

The standard deviation is the square root of that variance:

$$SD(\hat{p}_1 - \hat{p}_2) = \sqrt{\dfrac{p_1 q_1}{n_1} + \dfrac{p_2 q_2}{n_2}}.$$

We usually don't know the true values of p_1 and p_2. When we have the sample proportions in hand from the data, we use them to estimate the variances. So the standard error is

$$SE(\hat{p}_1 - \hat{p}_2) = \sqrt{\dfrac{\hat{p}_1 \hat{q}_1}{n_1} + \dfrac{\hat{p}_2 \hat{q}_2}{n_2}}.$$

For Example   Finding the Standard Error of a Difference in Proportions

A recent survey of 886 randomly selected teenagers (aged 12–17) found that more than half of them had online profiles.[3] Some researchers and privacy advocates are concerned about the possible access to personal information about teens in public places on the Internet. There appear to be differences between boys and girls in their online behavior. Among teens aged 15–17, 57% of the 248 boys had posted profiles, compared to 70% of the 256 girls. Let's start the process of estimating how large the true gender gap might be.

Question: What's the standard error of the difference in sample proportions?

Because the boys and girls were selected at random, it's reasonable to assume their behaviors are independent, so it's okay to use the Pythagorean Theorem of Statistics and add the variances:

$$SE(\hat{p}_{boys}) = \sqrt{\dfrac{0.57 \times 0.43}{248}} = 0.0314 \qquad SE(\hat{p}_{girls}) = \sqrt{\dfrac{0.70 \times 0.30}{256}} = 0.0286$$

$$SE(\hat{p}_{girls} - \hat{p}_{boys}) = \sqrt{0.0314^2 + 0.0286^2} = 0.0425$$

Assumptions and Conditions

As always we need to check assumptions and conditions. One of them is new—and very important:

Independent Groups Assumption: The two groups we're comparing must be independent *of each other*. Usually, the independence of the groups from each other is evident from the way the data were collected.

Why is the Independent Groups Assumption so important? If we compare husbands with their wives, or a group of subjects before and after

---

[3]Princeton Survey Research Associates International for the Pew Internet & American Life Project.

some treatment, we can't just add the variances. Paired data are not independent, so the Pythagorean-style variance formula does not work to find the standard deviation we need.

You'll recognize the rest of the assumptions and conditions:

- **Independence Assumption:** Within each group, the data should be based on results for independent individuals.
- **Randomization Condition:** The data in each group should come from a random sample or a randomized experiment.
- **The 10% Condition:** If the data are from a sample, the sample should not exceed 10% of the population.
- **Success/Failure Condition:** Both groups are big enough that at least 10 successes and at least 10 failures have been observed in each.

For Example Checking Assumptions and Conditions

Recap: Among randomly sampled teens aged 15–17, 57% of the 248 boys had posted online profiles, compared to 70% of the 256 girls.

Question: Can we use these results to make inferences about all 15–17-year-olds?

✔ **Randomization Condition:** The sample of boys and the sample of girls were both chosen randomly.

✔ **10% Condition:** 248 boys and 256 girls are each less than 10% of all teenage boys and girls.

✔ **Independent Groups Assumption:** Because the samples were selected at random, it's reasonable to believe the boys' online behaviors are independent of the girls' online behaviors.

✔ **Success/Failure Condition:** Among the boys, 248(0.57) = 141 had online profiles and the other 248(0.43) = 107 did not. For the girls, 256(0.70) = 179 successes and 256(0.30) = 77 failures. All counts are at least 10.

Because all the assumptions and conditions are satisfied, it's okay to proceed with inference for the difference in proportions.

(Note that when we find the *observed* counts of successes and failures, we round off to whole numbers. We're using the reported percentages to figure out what the actual counts must have been.)

A Confidence Interval

We're almost there. We just need one more fact about proportions. We already know that for large enough samples, each of our proportions has an approximately Normal sampling distribution. The same is true of their difference.

That gives us all we need to find a margin of error for the difference in proportions—much like a one-proportion z-interval.

A TWO-PROPORTION z-INTERVAL

When the conditions are met, we are ready to find the confidence interval for the difference of two proportions, $p_1 - p_2$. The confidence interval is

$$(\hat{p}_1 - \hat{p}_2) \pm z^* \times SE(\hat{p}_1 - \hat{p}_2)$$

where we find the standard error of the difference,

$$SE(\hat{p}_1 - \hat{p}_2) = \sqrt{\frac{\hat{p}_1\hat{q}_1}{n_1} + \frac{\hat{p}_2\hat{q}_2}{n_2}}$$

from the observed proportions.

The critical value z^* depends on the particular confidence level, C, that we specify. For a 95% confidence intervals, we can use the Rule of 2.

For Example   Finding a Two-Proportion z-Interval

Recap: Among randomly sampled teens aged 15–17, 57% of the 248 boys had posted online profiles, compared to 70% of the 256 girls. We calculated the standard error for the difference in sample proportions to be $SE(\hat{p}_{girls} - \hat{p}_{boys}) = 0.0425$ and found that the assumptions and conditions required for inference checked out okay.

Question: What does a confidence interval say about the difference in online behavior?

A 95% confidence interval for $p_{girls} - p_{boys}$ is $(\hat{p}_{girls} - \hat{p}_{boys}) \pm 2SE(\hat{p}_{girls} - \hat{p}_{boys})$

$$(0.70 - 0.57) \pm 2(0.0425)$$
$$0.13 \pm 0.085$$
$$(4.5\%, 21.5\%)$$

We can be 95% confident that among teens aged 15–17, the proportion of girls who post online profiles is between 4.5 and 21.5 percentage points higher than the proportion of boys who do. It seems clear that teen girls are more likely to post profiles than are boys the same age.

STEP-BY-STEP EXAMPLE   A Two-Proportion z-Interval

Now we are ready to be more precise about the passenger-based gap in male drivers' seat belt use. We'll estimate the difference with a **two-proportion z-interval** by following the four confidence interval steps.

Question: How much difference is there in the proportion of male drivers who wear seat belts when sitting next to a male passenger and the proportion who wear seat belts when sitting next to a female passenger?

(continued)

Plan State what you want to know. Discuss the variables and the W's.

Identify the parameter you wish to estimate. (It usually doesn't matter in which direction we subtract, so, for convenience, we usually choose the direction with a positive difference.)

Choose and state a confidence level.

Model Think about the assumptions and check the conditions.

The Success/Failure Condition must hold for each group.

State the sampling distribution model for the statistic.

Choose your method.

I want to know the true difference in the population proportion, p_M, of male drivers who wear seat belts when sitting next to a man and p_F, the proportion who wear seat belts when sitting next to a woman. The data are from a random sample of drivers in Massachusetts in 2007, observed according to procedures developed by the NHTSA. The parameter of interest is the difference $p_F - p_M$.

I will find a 95% confidence interval for this difference.

✔ **Independence Assumption:** Driver behavior was independent from car to car.

✔ **Randomization Condition:** The NHTSA methods result in a suitable random sample.

✔ **10% Condition:** The samples include far fewer than 10% of all male drivers accompanied by male or by female passengers.

✔ **Independent Groups Assumption:** There's no reason to believe that seat belt use among drivers with male passengers and those with female passengers are not independent.

✔ **Success Failure Condition:** Among male drivers with female passengers, 2777 wore seat belts and 1431 did not; of those driving with male passengers, 1363 wore seat belts and 1400 did not. Each group contained far more than 10 successes and 10 failures.

Under these conditions, the sampling distribution of the difference between the sample proportions is approximately Normal, so I'll find a **two-proportion z-interval.**

Mechanics Construct the confidence interval.

As often happens, the key step in finding the confidence interval is estimating the standard deviation of the sampling distribution model of the statistic. Here the statistic is the difference in the proportions of men who wear seat belts when they have a female passenger and the proportion who do so with a male passenger. Substitute the data values into the SE formula.

I know
$n_F = 4208, n_M = 2763.$

The observed sample proportions are
$\hat{p}_F = \frac{2777}{4208} = 0.660, \hat{p}_M = \frac{1363}{2763} = 0.493$

I'll estimate the SD of the difference with

$$SE(\hat{p}_F - \hat{p}_M) = \sqrt{\frac{\hat{p}_F \hat{q}_F}{n_F} + \frac{\hat{p}_M \hat{q}_M}{n_M}}$$

$$= \sqrt{\frac{(0.660)(0.340)}{4208} + \frac{(0.493)(0.507)}{2763}}$$

$$= 0.012$$

The sampling distribution is Normal, so we use the Rule of 2 to find the margin of error. (The critical value for a 95% confidence interval, z^*, is actually 1.96.)	$ME = 2 \times SE(\hat{p}_F - \hat{p}_M)$ $= 2(0.012) = 0.024$
The confidence interval is the statistic $\pm ME$.	The observed difference in proportions is $\hat{p}_F - \hat{p}_M = 0.660 - 0.493 = 0.167$, so the 95% confidence interval is 0.167 ± 0.024 or 14.3% to 19.1%

 Conclusion Interpret your confidence interval in the proper context. (Remember: We're 95% confident that our interval captured the true difference.)

I am 95% confident that the proportion of male drivers who wear seat belts when driving next to a female passenger is between 14.3 and 19.1 percentage points higher than the proportion who wear seat belts when driving next to a male passenger.

This is an interesting result—but be careful not to try to say too much! In Massachusetts, overall seat belt use is lower than the national average, so we can't be certain that these results generalize to other states. And these were two different groups of men, so we can't say that, individually, men are more likely to buckle up when they have a woman passenger. You can probably think of several alternative explanations; we'll suggest just a couple. Perhaps age is a lurking variable: Maybe older men are more likely to wear seat belts and also more likely to be driving with their wives. Or maybe men who don't wear seat belts have trouble attracting women!

TI Tips — Finding a confidence interval

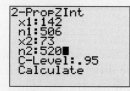
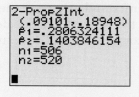

You can use a routine in the **STAT TESTS** menu to create confidence intervals for the difference of two proportions. Remember, the calculator can do only the mechanics—checking conditions and writing conclusions are still up to you.

A Gallup Poll asked whether the attribute "intelligent" described men in general. The poll revealed that 28% of 506 men thought it did, but only 14% of 520 women agreed. We want to estimate the true size of the gender gap by creating a 95% confidence interval.

- Go to the **STAT TESTS** menu. Scroll down the list and select **B:2-PropZInt**.
- Enter the observed number of males: **.28*506**. Remember that the actual number of males must be a whole number, so be sure to round off.
- Enter the sample size: **506** males.
- Repeat those entries for women: **.14*520** agreed, and the sample size was **520**.
- Specify the desired confidence level.
- **Calculate** the result.

And now explain what you see: We are 95% confident that the proportion of all men who think the attribute "intelligent" describe males in general is between 9 and 19 percentage points higher than the proportion of women who think so.

Just Checking

1. A public broadcasting station plans to launch a special appeal for additional contributions from current members. Unsure of the most effective way to contact people, they run an experiment. They randomly select two groups of current members. They send the same request for donations to everyone, but it goes to one group by e-mail and to the other group by regular mail. The station was successful in getting contributions from 26% of the members they e-mailed but only from 15% of those who received the request by regular mail. A 90% confidence interval estimated the difference in donation rates to be 11% ± 7%. Interpret the confidence interval in this context.

(Check your answers on page 492.)

Will I Snore When I'm 64?

WHO	Randomly selected U.S. adults over age 18
WHAT	Proportion who snore, categorized by age (less than 30, 30 or older)
WHEN	2001
WHERE	United States
WHY	To study sleep behaviors of U.S. adults

The National Sleep Foundation asked a random sample of 1010 U.S. adults questions about their sleep habits. The sample was selected in the fall of 2001 from random telephone numbers, stratified by region and sex, guaranteeing that an equal number of men and women were interviewed (2002 Sleep in America Poll, National Sleep Foundation, Washington, DC).

One of the questions asked about snoring. Of the 995 respondents, 37% of adults reported that they snored at least a few nights a week during the past year. Would you expect that percentage to be the same for all age groups? Split into two age categories, 26% of the 184 people under 30 snored, compared with 39% of the 811 in the older group. Is this difference of 13% real, or due only to natural fluctuations in the sample we've chosen?

The question calls for a hypothesis test. Now the parameter of interest is the true *difference* between the (reported) snoring rates of the two age groups.

What's the appropriate null hypothesis? That's easy here. We hypothesize that there is no difference in the proportions. This is such a natural null hypothesis that we rarely consider any other. But instead of writing $H_0: p_1 = p_2$, we usually express it in a slightly different way. To make it relate directly to the *difference*, we hypothesize that the difference in proportions is zero:

$$H_0: p_1 - p_2 = 0.$$

Naturally, we'll reject our null hypothesis if we see a large enough difference in the two proportions. How can we decide whether the difference we see, $\hat{p}_1 - \hat{p}_2$, is large? The answer is the same as always: we'll use the standard error to get a z-score. The z-score will tell us how many standard errors the observed difference is away from 0. We can then use the Rule of 2 to decide whether this is large (or some technology to get an exact P-value). The result is a **two-proportion z-test.**

TWO-PROPORTION *z*-TEST
The conditions for the two-proportion *z*-test are the same as for the two-proportion *z*-interval. We are testing the hypothesis

$$H_0: p_1 - p_2 = 0.$$

We find the test statistic,

$$z = \frac{(\hat{p}_1 - \hat{p}_2) - 0}{SE(\hat{p}_1 - \hat{p}_2)}$$

where

$$SE(\hat{p}_1 - \hat{p}_2) = \sqrt{\frac{\hat{p}_1\hat{q}_1}{n} + \frac{\hat{p}_2\hat{q}_2}{n_2}}$$

STEP-BY-STEP EXAMPLE   A Two-Proportion *z*-Test

Question: Are the snoring rates of the two age groups really different?

THINK

Plan State what you want to know. Discuss the variables and the W's.

I want to know whether snoring rates differ for those under and over 30 years old. The data are from a random sample of 1010 U.S. adults surveyed in the 2002 Sleep in America Poll. Of these, 995 responded to the question about snoring, indicating whether or not they had snored at least a few nights a week in the past year.

Hypotheses The study simply broke down the responses by age, so there is no sense that either alternative was preferred. A two-sided alternative hypothesis is appropriate.

H_0: There is no difference in snoring rates in the two age groups:

$$p_{old} - p_{young} = 0.$$

H_A: The rates are different: $p_{old} - p_{young} \neq 0.$

Model Think about the assumptions and check the conditions.

✔ **Independence Assumption:** The National Sleep Foundation selected respondents at random, so they should be independent.

✔ **Randomization Condition:** The respondents were randomly selected by telephone number and stratified by sex and region.

✔ **10% Condition:** The number of adults surveyed in each age group is certainly far less than 10% of that population.

✔ **Independent Groups Assumption:** The two groups are independent of each other because the sample was selected at random.

(continued)

✔ **Success/Failure Condition:** In the younger age group, 48 snored and 136 didn't. In the older group, 318 snored and 493 didn't. The observed numbers of both successes and failures are much more than 10 for both groups.[4]

State the null model.

Choose your method.

Because the conditions are satisfied, I'll use a Normal model and perform a **two-proportion z-test.**

Mechanics

Use the SE to estimate $SD(p_{old} - p_{young})$.

$$n_{young} = 184, y_{young} = 48, \hat{p}_{young} = 0.261$$
$$n_{old} = 811, \quad y_{old} = 318, \quad \hat{p}_{old} = 0.392$$

$$SE(\hat{p}_{old} - \hat{p}_{young})$$
$$= \sqrt{\frac{\hat{p}_{old}\hat{q}_{old}}{n_{old}} + \frac{\hat{p}_{young}\hat{q}_{young}}{n_{young}}}$$
$$= \sqrt{\frac{(0.392)(0.608)}{811} + \frac{(0.261)(0.739)}{184}}$$
$$\approx 0.0366$$

The observed difference in sample proportions is
$\hat{p}_{old} - \hat{p}_{young} = 0.392 - 0.261 = 0.131$

Make a picture. Sketch a Normal model centered at the hypothesized difference of 0. Shade the region to the right of the observed difference, and because this is a two-tailed test, also shade the corresponding region in the other tail.

Find the z-score for the observed difference in proportions, 0.131.

The Rule of 2 tells us this difference is statistically significant. If you want the P-value, use technology.

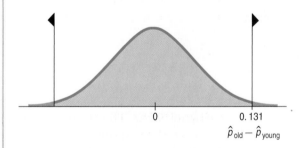

$$z = \frac{(\hat{p}_{old} - \hat{p}_{young}) - 0}{SE(\hat{p}_{old} - \hat{p}_{young})} = \frac{0.131 - 0}{0.0366} = 3.58$$
$$P\text{-value} = 2P(z \geq 3.58) = 0.0003$$

Conclusion Explain your decision about the null hypothesis, and state your conclusion in context.

If there really were no difference in (reported) snoring rates between the two age groups, then the difference observed in this study would be over 3 standard errors above 0. This is so unusual that I reject the null hypothesis of no difference and conclude that there is a difference in the rate of snoring between older adults and younger adults. It appears that older adults are more likely to snore.

---

[4]This is one of those situations in which the traditional term "success" seems a bit weird. A success here is that a person snores. "Success" and "failure" are terms left over from studies of gambling games.

TI Tips | Testing the hypothesis

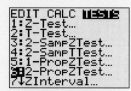

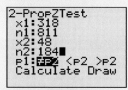

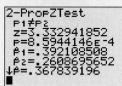

Yes, of course, there's a **STAT TESTS** routine to test a hypothesis about the difference of two proportions. Let's do the mechanics for the test about snoring. Of 811 people over 30 years old, 318 snored, while only 48 of the 184 people under 30 did.

- In the **STAT TESTS** menu select **6:2-PropZTest**.
- Enter the observed numbers of snorers and the sample sizes for both groups.
- Since this is a two-tailed test, indicate that you want to see if the proportions are unequal. When you choose this option, the calculator will automatically include both tails as it determines the P-value.
- **Calculate** the result.

Now it is up to you to interpret the result and state a conclusion. We see a z-score of 3.33 and the P-value is 0.0008. (Don't worry about the minor differences from what we got in the Step-by-Step Example. The calculator uses a slightly more sophisticated formula.) Such a small P-value indicates that the observed difference is unlikely to be sampling error. What does that mean about snoring and age? Here's a great opportunity to follow up with a confidence interval so you can Tell even more!

For Example Another 2-Proportion *z*-Test

Recap: One concern of the study on teens' online profiles was safety and privacy. In the random sample, girls were less likely than boys to say that they are easy to find online from their profiles. Only 19% (62 girls) of 325 teen girls with profiles say that they are easy to find, while 28% (75 boys) of the 268 boys with profiles say the same.

Question: Are these results evidence of a real difference between boys and girls? Perform a two-proportion *z*-test and discuss what you find.

$$H_O: p_{boys} - p_{girls} = 0$$
$$H_A: p_{boys} - p_{girls} \neq 0$$

✔ **Randomization Condition:** The sample of boys and the sample of girls were both chosen randomly.

✔ **10% Condition:** 268 boys and 325 girls are each less than 10% of all teenage boys and girls with online profiles.

✔ **Independent Groups Assumption:** Because the samples were selected at random, it's reasonable to believe the boys' perceptions are independent of the girls'.

✔ **Success/Failure Condition:** Among the girls, there were 62 "successes" and 263 failures, and among boys, 75 successes and 193 failures. These counts are at least 10 for each group.

(continued)

Because all the assumptions and conditions are satisfied, it's okay to do a **two-proportion z-test:**

$$SE(\hat{p}_{boys} - \hat{p}_{girls}) = \sqrt{\frac{0.28 \times 0.72}{268} + \frac{0.19 \times 0.81}{325}} = 0.035$$

$$z = \frac{(0.28 - 0.19) - 0}{0.035} = 2.57$$

$$P\text{-value} = 2P(z > 2.57) = 0.01$$

Because this $z > 2$, I reject the null hypothesis. This study provides strong evidence that there really is a difference in the proportions of teen girls and boys who say they are easy to find online.

Just Checking

2. It's estimated that 50,000 pregnant women worldwide die each year of eclampsia, a condition involving high blood pressure and seizures. In 2002 the medical journal *The Lancet* published results of a randomized experiment that involved nearly 10,000 at-risk women from 175 hospitals in 33 countries. Researchers found that treating women with magnesium sulfate significantly reduced the occurrence of eclampsia, an important advance in women's health.

   However, 11 of the 40 women (27.5%) who developed eclampsia despite receiving magnesium sulfate died. In the placebo group, 20 of the 96 women who developed eclampsia died—only 20.8%. Is there cause for concern that even though this treatment dramatically reduces the occurrence of eclampsia, women who do develop the condition may face a greater risk of death?

   a) Write appropriate hypotheses.
   b) Check the conditions for inference.
   c) Find the *z*-score for these data.
   d) State your conclusion.

(Check your answers on page 492.)

Comparing Two Means

WHO	AA alkaline batteries
WHAT	Length of battery life while playing a CD continuously
UNITS	Minutes
WHY	Class project
WHEN	1998

Should you buy generic rather than brand-name batteries? A Statistics student designed a study to test battery life. He wanted to know whether there was any real difference between brand-name batteries and a generic brand. To estimate the difference in mean lifetimes, he kept a battery-powered CD player[5] continuously playing the same CD, with the volume control fixed at 5, and measured the time until no more music was heard through the headphones. For his trials he used six sets of AA alkaline batteries from two major battery manufacturers: a well-known brand name and a generic brand. He measured the time in minutes until the sound stopped. The table shows his data (times in minutes):

Brand Name	Generic
194.0	190.7
205.5	203.5
199.2	203.5
172.4	206.5
184.0	222.5
169.5	209.4

---

[5]Once upon a time, not so very long ago, there were no iPods. At the turn of the century, people actually carried CDs around—and devices to play them. We bet you can find one in your parents' closet.

Studies that compare two means are common throughout both science and industry. We might want to compare the effects of a new drug with the traditional therapy, the fuel efficiency of two car engine designs, or the sales of new products in two different test cities. In fact, battery manufacturers do research like this on their products and competitors' products themselves.

Comparing two means is not very different from comparing two proportions. Now the population model parameter of interest is the difference between the *mean* battery lifetimes of the two brands, $\mu_1 - \mu_2$.

The rest is the same as before. The statistic of interest is the difference in the two observed means, $\bar{y}_1 - \bar{y}_2$. We'll start with this statistic to build our confidence interval, but we'll need to know its standard deviation. We turn once more to the Pythagoream Theorem of statistics:

For independent random variables, the variance of their *difference* is the *sum* of their individual variances, $Var(Y - X) = Var(Y) + Var(X)$.

To find the standard deviation of the difference between the two independent sample means, we add their variances and then take a square root:

$$SD(\bar{y}_1 - \bar{y}_2) = \sqrt{Var(\bar{y}_1) + Var(\bar{y}_2)}$$

$$= \sqrt{\left(\frac{\sigma_1}{\sqrt{n_1}}\right)^2 + \left(\frac{\sigma_2}{\sqrt{n_2}}\right)^2}$$

$$= \sqrt{\frac{\sigma_1^2}{n_1} + \frac{\sigma_2^2}{n_2}}.$$

Of course, we still don't know the true standard deviations of the two groups, σ_1 and σ_2. As usual, we'll use the estimates, s_1 and s_2 to find the *standard error*:

$$SE(\bar{y}_1 - \bar{y}_2) = \sqrt{\frac{s_1^2}{n_1} + \frac{s_2^2}{n_2}}.$$

We'll use the standard error to see how big the difference really is. Because we are working with means and estimating the standard error of their difference using the data, we shouldn't be surprised that the sampling model is a Student's t.

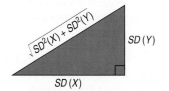

$SD(Y)$

$SD(X)$

The Pythagorean Theorem of Statistics

z OR t?

If you know σ, use z. (That's rare!) Whenever you use s to estimate σ, use t.

For Example — Finding the Standard Error of the Difference in Independent Sample Means

Can you tell how much you are eating from how full you are? Or do you need visual cues? Researchers[6] constructed a table with two ordinary 18 oz soup bowls and two identical-looking bowls that had been modified to slowly, imperceptibly, refill as they were emptied. They assigned experiment participants to the bowls randomly and served them tomato soup. Those eating from the ordinary bowls had their bowls refilled by ladle whenever they were one-quarter full. If people judge their portions by internal cues, they should eat about the same amount. How big a difference was there in the amount of soup consumed? The table summarizes their results.

(continued)

[6]Brian Wansink, James E. Painter, and Jill North, "Bottomless Bowls: Why Visual Cues of Portion Size May Influence Intake," *Obesity Research*, Vol. 13, No. 1, January 2005.

Question: How much variability do we expect in the difference between the two means? Find the standard error.

Participants were randomly assigned to bowls, so the two groups should be independent. It's okay to add variances.

	Ordinary bowl	Refilling bowl
n	27	27
$\bar{y}$	8.5 oz	14.7 oz
s	6.1 oz	8.4 oz

$$SE(\bar{y}_{refill} - \bar{y}_{ordinary}) = \sqrt{\frac{s_r^2}{n_r} + \frac{s_o^2}{n_o}} = \sqrt{\frac{8.4^2}{27} + \frac{6.1^2}{27}} = 2.0 \text{ oz.}$$

Assumptions and Conditions

Now we've got everything we need. Before we can make a two-sample *t*-interval or perform a two-sample *t*-test, though, we have to check the assumptions and conditions. You knew that. And you'll recognize them.

- **Independent Groups Assumption:** To use the two-sample *t* methods, the two groups we are comparing must be independent of each other. In fact, this test is sometimes called the two *independent samples t*-test. Think about how the data were collected. The assumption would be violated, if the data are paired—husbands and wives, before and after a treatment for example. In such cases, use the matched pairs methods of Chapter 19.
- **Independence Assumption:** The data within each group must be drawn independently.
- **Randomization Condition:** Were the data collected with suitable randomization? For surveys, are they a representative random sample? For experiments, was the experiment randomized?
- **10% Condition:** We usually don't check this condition for differences of means. We'll check it only if we have a very small population or an extremely large sample. We needn't worry about it at all for randomized experiments.
- **Nearly Normal Condition:** Student's *t*-models assume that the underlying populations are *each* Normally distributed. We must check this for *both* groups; a violation by either one violates the condition. As we saw for single sample means, the Normality Assumption matters most when sample sizes are small. For samples of $n < 15$ in either group, you should not use these methods if the histogram shows severe skewness. For *n*'s closer to 40, a mildly skewed histogram is OK, but you should remark on any outliers you find and not work with severely skewed data. When both groups are bigger than 40, the Central Limit Theorem starts to kick in no matter how the data are distributed, so the Nearly Normal Condition for the data matters less. Even in large samples, however, you should still be on the lookout for outliers, extreme skewness, and multiple modes.

For Example   Checking Assumptions and Conditions

Recap: Researchers randomly assigned people to eat soup from one of two bowls: 27 got ordinary bowls that were refilled by ladle, and 27 others got bowls that secretly refilled slowly as the people ate.

Question: Can the researchers use their data to make inferences about the role of visual cues in determining how much people eat?

✔ **Independence Assumption:** The amount consumed by one person should be independent of the amount consumed by others.

✔ **Randomization Condition:** Subjects were randomly assigned to the treatments.

✔ **Nearly Normal Condition:** The histograms for both groups look unimodal but somewhat skewed to the right. I believe both groups are large enough (27) to allow use of t-methods.

✔ **Independent Groups Assumption:** Randomization to treatment groups guarantees this.

It's okay to construct a two-sample t-interval for the difference in means.

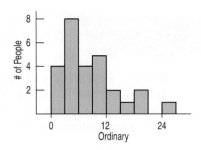

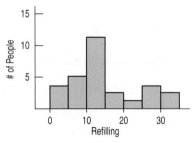

A Confidence Interval for the Difference of Two Means

When the assumptions and conditions are met, we can proceed with inference.

The confidence interval we build is called a **two-sample *t*-interval** (for the difference in means). The interval looks just like all the others we've seen—the statistic plus or minus an estimated margin of error:

$$(\bar{y}_1 - \bar{y}_2) \pm ME$$
$$\text{where } ME = t^* \times SE(\bar{y}_1 - \bar{y}_2).$$

What are we missing? Only the degrees of freedom for the Student's *t*-model. Unfortunately, *that* is strange. We'll rely on technology to tell us how many degrees of freedom there are.

TWO-SAMPLE *t*-INTERVAL FOR THE DIFFERENCE BETWEEN MEANS

When the conditions are met, we are ready to find the confidence interval for the difference between means of two independent groups, $\mu_1 - \mu_2$. The confidence interval is

$$(\bar{y}_1 - \bar{y}_2) \pm t^*_{df} \times SE(\bar{y}_1 - \bar{y}_2),$$

where the standard error of the difference of the means

$$SE(\bar{y}_1 - \bar{y}_2) = \sqrt{\frac{s_1^2}{n_1} + \frac{s_2^2}{n_2}}.$$

The critical value t^*_{df} depends on the particular confidence level, C, that you specify and on the number of degrees of freedom, which we get from technology.

For Example — Finding a Confidence Interval for the Difference in Sample Means

Recap: Researchers studying the role of internal and visual cues in determining how much people eat conducted an experiment in which some people ate soup from bowls that secretly refilled. The results are summarized in the table.

	Ordinary bowl	Refilling bowl
n	27	27
$\bar{y}$	8.5 oz	14.7 oz
s	6.1 oz	8.4 oz

We've already checked the assumptions and conditions, and have found the standard error for the difference in means to be $SE(\bar{y}_{refill} - \bar{y}_{ordinary}) = 2.0$ oz.

Question: What does a 95% confidence interval say about the difference in mean amounts eaten?

The observed difference in means is $\bar{y}_{refill} - \bar{y}_{ordinary} = (14.7 - 8.5) = 6.2$ oz

(from technology, $df = 47.46$ and $t^*_{47.46} = 2.011$; we'll use the Rule of 2.)

$$ME = t^* \times SE(\bar{y}_{refill} - \bar{y}_{ordinary}) = 2(2.0) = 4 \text{ oz}$$

The 95% confidence interval for $\mu_{refill} - \mu_{ordinary}$ is 6.2 ± 4, or $(2.2, 10.2)$ oz.

I am 95% confident that people eating from a subtly refilling bowl will eat an average of between 2.2 and 10.2 more ounces of soup than those eating from an ordinary bowl.

STEP-BY-STEP EXAMPLE — A Two-Sample t-Interval

The generic batteries seem to have lasted longer than the brand-name batteries. Before we change our buying habits, what should we expect to happen with the next batteries we buy?

Question: How much longer might the generic batteries last?

THINK

Plan State what we want to know.

Identify the *parameter* you wish to estimate. Here our parameter is the difference in the means, not the individual group means.

Identify the *population(s)* about which you wish to make statements. We hope to make decisions about purchasing batteries, so we're interested in all the AA batteries of these two brands.

I have measurements of the lifetimes (in minutes) of 6 sets of generic and 6 sets of brand-name AA batteries from a randomized experiment. I want to find an interval that is likely, with 95% confidence, to contain the true difference $\mu_G - \mu_B$ between the mean lifetime of the generic AA batteries and the mean lifetime of the brand-name batteries.

Model Think about the appropriate assumptions and check the conditions to be sure that a Student's *t*-model for the sampling distribution is appropriate.

For very small samples like these, we often don't worry about the 10% Condition.

✔ **Randomization Condition:** The batteries were selected at random from those available for sale. Not exactly an SRS, but a reasonably representative random sample.

? **Independence Assumption:** The batteries were packaged together, so they may not be independent. For example, a storage problem might affect all the batteries in the same pack. Repeating the study for several different packs of batteries would make the conclusions stronger.

✔ **Independent Groups Assumption:** Batteries manufactured by two different companies and purchased in separate packages should be independent.

✔ **Nearly Normal Condition:** The samples are small, but the histograms look unimodal and symmetric:

Make histograms to check the *shape* of distribution of each group.

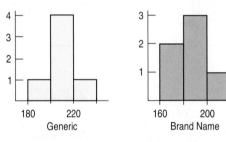

State the sampling distribution model for the statistic. Here the degrees of freedom will come from technology.

Under these conditions, it's okay to use a Student's t-model.

Specify your method.

I'll use a **two-sample *t*-interval.**

SHOW

Mechanics Construct the confidence interval.

Be sure to include the units along with the statistics. Use meaningful subscripts to identify the groups.

Use the sample standard deviations to find the standard error of the sampling distribution for the difference in means.

I know $n_G = 6$ $n_B = 6$

$\bar{y}_G = 206.0$ min $\bar{y}_B = 187.4$ min

$s_G = 10.3$ min $s_B = 14.6$ min

The groups are independent, so

$$SE(\bar{y}_G - \bar{y}_B) = \sqrt{SE^2(\bar{y}_G) + SE^2(\bar{y}_B)}$$

$$= \sqrt{\frac{s_G^2}{n_G} + \frac{s_B^2}{n_B}}$$

$$= \sqrt{\frac{10.3^2}{6} + \frac{14.6^2}{6}}$$

$$= \sqrt{\frac{106.09}{6} + \frac{213.16}{6}}$$

$$= \sqrt{53.208}$$

$$= 7.29 \text{ min.}$$

(continued)

Let technology deal with degrees of freedom and the critical value t^*.

df (from technology[7]) = 8.98

The corresponding critical value for a 95% confidence level is $t^* = 2.263$.

So the margin of error is

$$ME = t^* \times SE(\bar{y}_G - \bar{y}_B)$$
$$= 2.263(7.29)$$
$$= 16.50 \text{ min.}$$

The 95% confidence interval is

$$(206.0 - 187.4) \pm 16.5 \text{ min.}$$
$$\text{or } 18.6 \pm 16.5 \text{ min.}$$
$$= (2.1, 35.1) \text{ min.}$$

Conclusion Interpret the confidence interval in the proper context.

I'm 95% confident that generic batteries last an average of 2.1 to 35.1 minutes longer than brand-name batteries.

Another One Just Like the Other Ones?

Yes. That's been our point all along. Once again we see a statistic plus or minus the margin of error. And the ME is just a critical value times the standard error. Just look out for those crazy degrees of freedom!

TI Tips — Creating the confidence interval

If you have been successful using your TI to make confidence intervals for proportions and 1-sample means, then you can probably already use the 2-sample function just fine. But humor us while we do one. Please?

Find a confidence interval for the difference in means, given data from two independent samples.

- Let's do the batteries. Always think about whether the samples are independent. If not, stop right here. These procedures are appropriate only for independent groups.
- Enter the data into two lists.

NameBrand in L1:	194.0	205.5	199.2	172.4	184.0	169.5
Generic in L2:	190.7	203.5	203.5	206.5	222.5	209.4

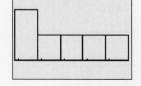

- Make histograms of the data to check the Nearly Normal Condition. We see that L1's histogram doesn't look so good. But remember—this is a very small

---

[7]Yes, it looks strange to have fractional degrees of freedom. That comes from a *very* messy formula. Trust us—you don't want to know.

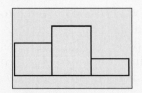

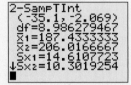

data set. The bars represent only one or two values each. It's not unusual for the histogram to look a little ragged. Try resetting the **WINDOW** to a range of 160 to 220 with **XScl=20**, and **Ymax=4**. Redraw the **GRAPH**. Looks better.

- It's your turn to try this. Check **L2**. Go on, do it.
- Under **STAT TESTS** choose **0:2-SampTint**.
- Specify that you are using the **Data** in **L1** and **L2**, specify 1 for both frequencies, and choose the confidence level you want.
- **Pooled?** Just Say No.
- To **Calculate** the interval, you need to scroll down one more line.

Now you have the 95% confidence interval. See **df**? The calculator did a messy degrees of freedom calculation for you. You have to love that!

Notice that the interval bounds are negative. That's because the TI is doing $\mu_1 - \mu_2$, and the generic batteries (**L2**) lasted longer. No harm done—you just need to be careful to interpret that result correctly when you *Tell* what the confidence interval means.

No data? Find a confidence interval using the sample statistics.

In many situations we don't have the original data, but must work with the summary statistics from the two groups. As we saw in the last chapter, you can still have your TI create the confidence interval with **0:2-SampTint** by choosing the **Inpt:Stats** option. Enter both means, standard deviations, and sample sizes, then **Calculate**. We show you the details in the next TI Tips.

Just Checking

3. Carpal tunnel syndrome (CTS) causes pain and tingling in the hand, sometimes bad enough to keep sufferers awake at night and restrict their daily activities. Researchers studied the effectiveness of two alternative surgical treatments for CTS (Mackenzie, Hainer, and Wheatley, *Annals of Plastic Surgery*, 2000). Patients were randomly assigned to have endoscopic or open-incision surgery. Four weeks later the endoscopic surgery patients demonstrated a mean pinch strength of 9.1 kg compared to 7.6 kg for the open-incision patients.

   a) Why is the randomization of the patients into the two treatments important?

   b) A 95% confidence interval for the difference in mean strength is about (0.04 kg, 2.96 kg). Explain what this interval means.
   c) Why might you want to see the data before trusting the confidence interval?

(Check your answers on page 492.)

Testing the Difference Between Two Means

If you bought a used camera in good condition from a friend, would you pay the same as you would if you bought the same item from a stranger? A researcher at Cornell University (J. J. Halpern, "The Transaction Index: A Method for Standardizing Comparisons of Transaction Characteristics Across

Different Contexts," *Group Decision and Negotiation,* 6: 557–572) wanted to know how friendship might affect simple sales such as this. She randomly divided subjects into two groups and gave each group descriptions of items they might want to buy. One group was told to imagine buying from a friend whom they expected to see again. The other group was told to imagine buying from a stranger.

Here are the prices they offered for a used camera in good condition:

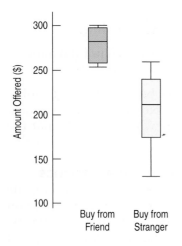

WHO	University students
WHAT	Prices offered for a used camera
UNITS	$
WHY	Study of the effects of friendship on transactions
WHEN	1990s
WHERE	U.C. Berkeley

Price Offered for a Used Camera ($)	
Buying from a Friend	**Buying from a Stranger**
275	260
300	250
260	175
300	130
255	200
275	225
290	240
300	

The researcher who designed this study had a specific concern. Previous theories had doubted that friendship had a measurable effect on pricing. She hoped to find an effect of friendship. This calls for a hypothesis test—in this case a **two-sample *t*-test for the difference between means.**[8]

You already know enough to construct this test. The test statistic looks just like the others we've seen. It finds the observed difference between the group means and compares this with the hypothesis of no difference. We then use the standard error of that difference to find *t* and the P-value.

TWO-SAMPLE *t*-TEST FOR THE DIFFERENCE BETWEEN MEANS

The conditions for the two-sample *t*-test for the difference between the means of two independent groups are the same as for the two-sample *t*-interval. We test the hypothesis

$$H_0: \mu_1 - \mu_2 = 0$$

using the statistic

$$t = \frac{(\bar{y}_1 - \bar{y}_2) - 0}{SE(\bar{y}_1 - \bar{y}_2)}.$$

The standard error of $\bar{y}_1 - \bar{y}_2$ is

$$SE(\bar{y}_1 - \bar{y}_2) = \sqrt{\frac{s_1^2}{n_1} + \frac{s_2^2}{n_2}}.$$

We use technology to find the degrees of freedom and the P-value.

---

[8]Because it is performed so often, this test is usually just called a "two-sample *t*-test."

STEP-BY-STEP EXAMPLE

A Two-Sample *t*-Test for the Difference Between Two Means

The null hypothesis is that there's no difference in means for the camera purchase prices.

Question: Is there a difference in the price people would offer a friend rather than a stranger?

THINK

Plan State what we want to know.

Identify the *parameter* you wish to estimate. Here our parameter is the difference in the means, not the individual group means.

Identify the variables and check the W's.

Hypotheses State the null and alternative hypotheses. The natural null hypothesis is that friendship makes no difference.

We didn't start with any knowledge of whether friendship might increase or decrease the price, so we choose a two-sided alternative.

Model Think about the assumptions and check the conditions. (Note that, because this is a randomized experiment, we haven't sampled at all, so the 10% Condition does not apply.)

Make histograms to check the shapes of both distributions.

I want to know whether people are likely to offer a different amount for a used camera when buying from a friend than when buying from a stranger. I wonder whether the difference between mean amounts is zero. I have bid prices from 8 subjects buying from a friend and 7 buying from a stranger, found in a randomized experiment.

H_O: The difference in mean price offered to friends and the mean price offered to strangers is zero:

$$\mu_F - \mu_S = O.$$

H_A: The difference in mean prices is not zero:

$$\mu_F - \mu_S \neq O.$$

✔ **Randomization Condition:** The experiment was randomized. Subjects were assigned to treatment groups at random.

✔ **Independence Assumption:** This is an experiment, so there is no need for the subjects to be randomly selected from any particular population. All we need to check is whether they were assigned randomly to treatment groups.

✔ **Independent Groups Assumption:** Randomizing the experiment gives independent groups.

✔ **Nearly Normal Condition:** Histograms of the two sets of prices are roughly unimodal and symmetric:

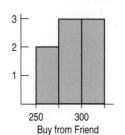
Buy from Friend

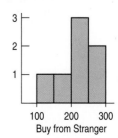
Buy from Stranger

(continued)

State the sampling distribution model. Specify your method.	The assumptions are reasonable and the conditions are okay, so I'll use a Student's t-model to perform a **two-sample t-test.**

Mechanics List the summary statistics. Be sure to use proper notation.	From the data: $n_F = 8$ $\qquad$ $n_S = 7$ $\bar{y}_F = \$281.88$ $\quad$ $\bar{y}_S = \$211.43$ $s_F = \$18.31$ $\quad$ $s_S = \$46.43$
Use the null model to find the P-value. First determine the standard error of the difference between sample means.	For independent groups, $$SE(\bar{y}_F - \bar{y}_S) = \sqrt{SE^2(\bar{y}_F) + SE^2(\bar{y}_S)}$$ $$= \sqrt{\frac{s_F^2}{n_F} + \frac{s_S^2}{n_S}}$$ $$= \sqrt{\frac{18.31^2}{8} + \frac{46.43^2}{7}}$$ $$= 18.70$$ The observed difference is $$(\bar{y}_F - \bar{y}_S) = 281.88 - 211.43 = \$70.45$$
Make a picture. Sketch the t-model centered at the hypothesized difference of zero. Because this is a two-tailed test, shade the region to the right of the observed difference and the corresponding region in the other tail.	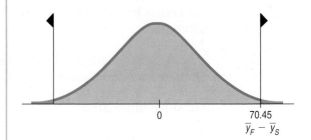
Find the t-value. That' a very large t, so we'll reject H$_0$. For more information, technology can tell us df and the P-value.	$$t = \frac{(\bar{y}_F - \bar{y}_S) - (0)}{SE(\bar{y}_F - \bar{y}_S)} = \frac{70.45}{18.70} = 3.77$$ $df = 7.62$ (from technology) $$\text{P-value} = 2P(t_{7.62} > 3.77) = 0.006$$

Conclusion Link t or P-value to your decision about the null hypothesis, and state the conclusion in context. Be cautious about generalizing to items whose prices are outside the range of those in this study.	If there were no difference in the mean prices, a difference this large would occur only 6 times in 1000. That's too rare to believe, so I reject the null hypothesis and conclude that prices offered to friends and strangers are not the same. It appears that people are likely to offer a friend more than they'd offer a stranger for a used camera (and possibly for other, similar items).

TI Tips   Testing a hypothesis about a difference in means

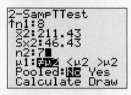

Now let's use the TI to do a hypothesis test for the difference of two means—independent, of course! (Have we said that enough times yet?)

Test a hypothesis when you know the sample statistics.

We'll demonstrate by using the statistics from the camera-pricing example. A sample of 8 people suggested they'd sell the camera to a friend for an average price of $281.88 with standard deviation $18.31. An independent sample of 7 other people would charge a stranger an average of $211.43 with standard deviation $46.43. Does this represent a significant difference in prices?

- From the STAT TESTS menu select 4:2-SampTTest.
- Specify Inpt:Stats, and enter the appropriate sample statistics.

- You have to scroll down to complete the specifications. This is a two-tailed test, so choose alternative ≠μ2.
- Pooled? Just say No.
- Ready . . . set . . . Calculate!

The TI reports a calculated value of $t = 3.77$ and a P-value of 0.006. It's hard to tell who your real friends are.

By now we probably don't have to tell you how to do a 2-SampTTest starting with data in lists.

So we won't.

Just Checking

Recall the experiment comparing patients 4 weeks after surgery for carpal tunnel syndrome. The patients who had endoscopic surgery demonstrated a mean pinch strength of 9.1 kg compared to 7.6 kg for the open-incision patients.

4. What hypotheses would you test?

5. The P-value of the test was less than 0.05. State a brief conclusion.

(Check your answers on page 492.)

For Example A Two-Sample *t*-Test

Many office "coffee stations" collect voluntary payments for the food consumed. Researchers at the University of Newcastle upon Tyne performed an experiment to see whether the image of eyes watching would change employee behavior.[9] They alternated pictures (seen here) of eyes looking at the viewer with pictures of flowers each week on the cupboard behind the "honesty box." They measured the consumption of milk to approximate the amount of food consumed and recorded the contributions (in £) each week per liter of milk. The table summarizes their results.

Question: Do these results provide evidence that there really is a difference in honesty even when it's only photographs of eyes that are "watching"?

	Eyes	Flowers
n (# weeks)	5	5
$\bar{y}$	0.417 £/l	0.151 £/l
s	0.1811 £/l	0.067 £/l

$$H_O: \mu_{eyes} - \mu_{flowers} = 0$$
$$H_A: \mu_{eyes} - \mu_{flowers} \neq 0$$

✔ **Independence Assumption:** The amount paid by one person should be independent of the amount paid by others.

✔ **Randomization Condition:** This study was observational. Treatments alternated a week at a time and were applied to the same group of office workers.

✔ **Nearly Normal Condition:** I don't have the data to check, but it seems unlikely there would be outliers in either group. I could be more certain if I could see histograms for both groups.

✔ **Independent Groups Assumption:** The same workers were recorded each week, but week-to-week independence is plausible.

It's okay to do a two-sample *t*-test for the difference in means:

$$SE(\bar{y}_{eyes} - \bar{y}_{flowers}) = \sqrt{\frac{s^2_{eyes}}{n_{eyes}} + \frac{s^2_{flowers}}{n_{flowers}}} = \sqrt{\frac{0.1811^2}{5} + \frac{0.067^2}{5}} = 0.0864$$

$$df = 5.07$$

$$t_5 = \frac{(\bar{y}_{eyes} - \bar{y}_{flowers}) - 0}{SE(\bar{y}_{eyes} - \bar{y}_{flowers})} = \frac{0.417 - 0.151}{0.0864} = 3.08$$

$$P\text{-value} = 2P(t_5 > 3.08) = 0.027$$

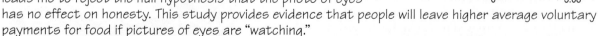

Assuming the data were free of outliers, the very low P-value leads me to reject the null hypothesis that the photo of eyes has no effect on honesty. This study provides evidence that people will leave higher average voluntary payments for food if pictures of eyes are "watching."

[9]Melissa Bateson, Daniel Nettle, and Gilbert Roberts, "Cues of Being Watched Enhance Cooperation in a Real-World Setting," *Biol. Lett. doi*:10.1098/rsbl.2006.0509.

WHAT CAN GO WRONG?

- **Don't use two-sample methods when the samples aren't independent.** These methods give wrong answers when this assumption of independence is violated. Good random sampling or random assignment of subjects to experimental treatments are the best insurance of independent groups.

- **Watch out for paired data.** The Independent Groups Assumption deserves special attention. If the samples are not independent, you can't use these two-sample methods. This is probably the main thing that can go wrong when using these two-sample methods. The methods of this chapter can be used *only* if the observations in the two groups are *independent*. Matched-pairs designs in which the observations are deliberately related arise often and are important, as we saw in Chapter 19.

- **Look at the plots.** The usual (by now) cautions about checking for outliers and non-Normal distributions apply, of course. The simple defense is to make and examine boxplots. You may be surprised how often this simple step saves you from the wrong or even absurd conclusions that can be generated by a single undetected outlier. You don't want to conclude that two methods have very different means just because one observation is atypical.

- **Don't apply inference methods where there was no randomization.** If the data do not come from representative random samples or from a properly randomized experiment, then the inference about the differences in proportions will be wrong.

- **Don't interpret a significant difference as cause-and-effect.** It turns out that people with higher incomes are more likely to snore. Does that mean money affects sleep patterns? Probably not. We have seen that older people are more likely to snore, and they are also likely to earn more. In a prospective or retrospective study, there is always the danger that other lurking variables not accounted for are the real reason for an observed difference. Be careful not to jump to conclusions about causality.

COMPARING MEANS **IN YOUR WORLD**

Do Pets Reduce Stress?

Several, but not all, studies suggest that those of us who own pets tend to be somewhat happier than those of us who do not. In addition, research . . . at the Maryland School of Nursing shows that pet ownership [. . . improves . . .] survival rates among victims of heart attacks.

Though interesting and potentially important, studies such as these are difficult to interpret because pet owners may differ in unmeasured ways from people who do not own pets. For example, pet

(continued)

owners may be better adjusted psychologically and have fewer cardiac risk factors (they may eat healthier diets and experience lower levels of hostility) than non–pet owners.

To unravel the potential influences of pets on well-being, researchers must conduct experiments that randomly assign some people, but not others, to receive a pet, either in the laboratory or in their home. Studies by psychologists Karen Allen of the University at Buffalo and James Blascovich of the University of California, Santa Barbara, and their colleagues demonstrate that the presence of a favorite pet during a stressful task—such as performing difficult mental arithmetic—largely prevents spikes in participants' blood pressure. In contrast, the presence of a friend does not. In addition, Allen's work shows that stressed-out, hypertensive stockbrokers who were randomly assigned to adopt either a pet dog or cat ended up with lower blood pressure than those who were not. These studies suggest that the presence of pets may lower our blood pressure and stress levels, although they do not tell us the reasons for this effect.

This *Scientific American* story doesn't give a lot of information about the research suggesting pets may reduce stress in their owners, so we looked up the details for you:

- The subjects were 48 people who lived alone and had high-stress jobs (stockbrokers).
- They were randomly divided into two groups. Both groups received the same blood pressure medication. People in one group also got a dog or a cat.
- Researchers assessed stress in the subjects by measuring how much their blood pressure increased when they tried to do mental arithmetic. Initially there was no difference between the two groups.
- During another mental arithmetic test six months later, people with pets had a significantly lower mean increase in blood pressure than people without pets (P-value < 0.001).

http://www.scientificamerican.com/article.cfm?id=is-animal-assisted-therapy&offset=2

WHAT HAVE WE LEARNED?

In the last few chapters we began our exploration of statistical inference; we learned how to create confidence intervals and test hypotheses. Now we've looked at inference for the difference in two proportions or two means.

Are proportions the same in two groups? If not, how different are they? We've learned to use statistical inference to compare proportions in two independent groups.

▶ We've learned that confidence intervals and hypothesis tests about the difference between two proportions, like those for an individual proportion, are based on Normal models.

▶ We've seen again that we must check assumptions to be sure our method will work.

▶ We've learned that the Pythagorean Theorem of Statistics allows us to find the standard error of the difference between two proportions, provided the two groups are independent.

Are the means of two groups the same? If not, how different are they? We've learned to use statistical inference to compare the means of two independent groups.

▶ We've seen that confidence intervals and hypothesis tests about the difference between two means, like those for an individual mean, use *t*-models.

▶ Once again we've seen the importance of checking assumptions that tell us whether our method will work.

▶ We've seen that, as when comparing proportions, finding the standard error for the difference in sample means depends on believing that our data come from independent groups.

▶ And we've seen once again that we can add variances of independent random variables to find the standard deviation of the difference in two independent means.

Finally, we've learned that the reasoning of statistical inference remains the same; only the mechanics change.

Terms

Pythagorean Theorem of Statistics
The variance of a sum or difference of independent random variables is the sum of the variances of those variables.

$$Var(X \pm Y) = Var(X) + Var(Y)$$

Sampling distribution of the difference between two proportions
The sampling distribution of $\hat{p}_1 - \hat{p}_2$ is, under appropriate assumptions, modeled by a Normal model with mean $\mu = p_1 - p_2$ and standard error

$$SE(\hat{p}_1 - \hat{p}_2) = \sqrt{\frac{\hat{p}_1\hat{q}_1}{n_1} + \frac{\hat{p}_2\hat{q}_2}{n_2}}.$$

Two-proportion z-interval
A two-proportion z-interval gives a confidence interval for the true difference in proportions, $p_1 - p_2$, in two independent groups.

The confidence interval is $(\hat{p}_1 - \hat{p}_2) \pm z^* \times SE(\hat{p}_1 - \hat{p}_2)$, where z^* is a critical value. For 95% confidence we'll use $z^* = 2$.

Two-proportion z-test
Test the null hypothesis $H_0: p_1 - p_2 = 0$ by finding the statistic

$$z = \frac{\hat{p}_1 - \hat{p}_2}{SE(\hat{p}_1 - \hat{p}_2)}.$$

Two-sample t-interval for the difference between means
A confidence interval for the difference between the means of two independent groups found as

$$(\bar{y}_1 - \bar{y}_2) \pm t^*_{df} \times SE(\bar{y}_1 - \bar{y}_2)$$

where

$$SE(\bar{y}_1 - \bar{y}_2) = \sqrt{\frac{s_1^2}{n_1} + \frac{s_2^2}{n_2}}$$

and the number of degrees of freedom is given by technology.

Two-sample t-test for the difference between means
A hypothesis test for the difference between the means of two independent groups. It tests the null hypothesis

$$H_0: \mu_1 - \mu_2 = 0,$$

using the statistic

$$t_{df} = \frac{(\bar{y}_1 - \bar{y}_2) - 0}{SE(\bar{y}_1 - \bar{y}_2)},$$

with the number of degrees of freedom given by technology.

Skills

▶ Be able to state the null and alternative hypotheses for testing the difference between two population proportions or two population means.

▶ Know how to examine your data for violations of conditions that would make inference unwise or invalid.

▶ Understand that the formula for the standard error of the difference between two independent sample proportions or means is based on the principle that when finding the sum or difference of two independent random variables, their variances add.

▶ Know how to find a confidence interval for the difference between two proportions or means.

▶ Be able to perform a significance test of the natural null hypothesis that two population proportions or means are equal.

(continued)

▶ Know how to write a sentence describing what is said about the difference between two population proportions or means by a confidence interval.

▶ Know how to write a sentence interpreting the results of a significance test of the null hypothesis that two population proportions or means are equal.

▶ Be able to interpret the meaning of a P-value in context.

▶ Know that we do not "accept" a null hypothesis if we fail to reject it.

INFERENCE FOR DIFFERENCES ON THE COMPUTER

Here's some typical computer software output for confidence intervals or hypothesis tests for the difference in proportions or means based on two independent groups.

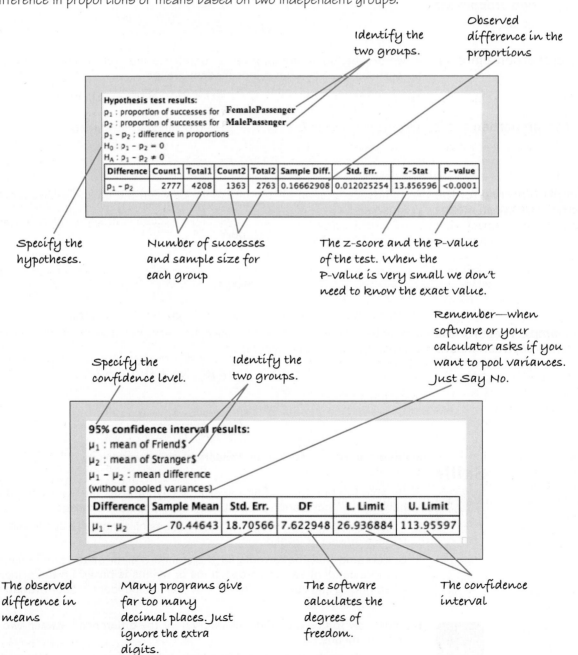

Identify the two groups.

Observed difference in the proportions

Specify the hypotheses.

Number of successes and sample size for each group

The z-score and the P-value of the test. When the P-value is very small we don't need to know the exact value.

Specify the confidence level.

Identify the two groups.

Remember—when software or your calculator asks if you want to pool variances. Just Say No.

The observed difference in means

Many programs give far too many decimal places. Just ignore the extra digits.

The software calculates the degrees of freedom.

The confidence interval

EXERCISES

A

1. **Online social networking.** The Parents & Teens 2006 Survey of 935 12- to 17-year-olds found that, among teens aged 15–17, girls were significantly more likely to have used social networking sites and online profiles. 70% of the girls surveyed had used an online social network, compared to 54% of the boys. What does it mean to say that the difference in proportions is "significant"?

2. **Science news.** In 2007 a Pew survey asked 1447 Internet users about their sources of news and information about science. Among those who had broadband access at home, 34% said they would turn to the Internet for most of their science news. The report on this survey claims that this is not significantly different from the percentage (33%) who said they ordinarily get their science news from television. What does it mean to say that the difference is not significant?

3. **Name recognition.** A political candidate runs a week-long series of TV ads designed to attract public attention to his campaign. Polls taken before and after the ad campaign show some increase in the proportion of voters who now recognize this candidate's name, with a P-value of 0.033. Is it reasonable to believe the ads may be effective?

4. **Origins.** In a 1993 Gallup Poll, 47% of the respondents agreed with the statement "*God created human beings pretty much in their present form at one time within the last 10,000 years or so.*" When Gallup asked the same question in 2001, only 45% of those respondents agreed. Is it reasonable to conclude that there was a change in public opinion given that the P-value is 0.37? Explain.

5. **Pets.** Researchers at the National Cancer Institute released the results of a study that investigated the effect of weed-killing herbicides on house pets. They examined 827 dogs from homes where an herbicide was used on a regular basis, diagnosing malignant lymphoma in 473 of them. Of the 130 dogs from homes where no herbicides were used, only 19 were found to have lymphoma. What's the standard error of the difference in the two proportions?

6. **Carpal tunnel.** The painful wrist condition called carpal tunnel syndrome can be treated with surgery or less invasive wrist splints. In September 2002, *Time* magazine reported on a study of 176 patients. Among the half that had surgery, 80% showed improvement after three months, but only 54% of those who used the wrist splints improved. What's the standard error of the difference in the two proportions?

7. **Another pet.** Data from the cancer study described in Exercise 5 were analyzed using a computer, producing this output:

   `95% CI for P(Herb) - P(noHerb) = (0.36,0.49)`

   Explain what this interval means about dogs, herbicides, and lymphoma.

8. **Carpal tunnel revisited.** Data from the experiment described in Exercise 6 were analyzed using a computer, producing this output:

   `95% CI for P(Surgery) - P(Splints) = (0.12,0.38)`

   Explain what this interval means about these two treatments for carpal tunnel.

9. **Dogs and calories.** In July 2007, *Consumer Reports* examined the calorie content of two kinds of hot dogs: meat (usually a mixture of pork, turkey, and chicken) and all beef. The researchers purchased samples of several different brands. The meat hot dogs averaged 111.7 calories, compared to 135.4 for the beef hot dogs. A test of the null hypothesis that there's no difference in mean calorie content yields a P-value of 0.124. Explain what that means.

10. **Dogs and sodium.** The *Consumer Reports* article described in Exercise 9 also listed the sodium content (in mg) for the various hot dogs tested. A test of the null hypothesis that beef hot dogs and meat hot dogs don't differ in the mean amounts of sodium yields a P-value of 0.11. Explain what that means.

11. **Dogs and fat.** The *Consumer Reports* article described in Exercise 9 also listed the fat content (in grams) for samples of beef and meat hot dogs. The resulting 90% confidence interval for $\mu_{Meat} - \mu_{Beef}$ is $(-6.5, -1.4)$. Explain what that means about the two types of hot dogs.

12. **Washers.** In June 2007, *Consumer Reports* examined top-loading and front-loading washing machines, testing samples of several different brands of each type. One of the variables the article reported was "cycle time," the number of minutes it took each machine to wash a load of clothes. Among the machines rated good to excellent, the 98% confidence interval for the difference in mean cycle time ($\mu_{Top} - \mu_{Front}$) is $(-40, -22)$. Explain what that means about the two types of washers.

13. **Dogs and fat, second helping.** In Exercise 11, we saw a 90% confidence interval of $(-6.5, -1.4)$ grams for $\mu_{Meat} - \mu_{Beef}$, the difference in mean fat content for meat vs. all-beef hot dogs. Explain why you think each of the following statements is true or false:
    a) If I eat a meat hot dog instead of a beef dog, there's a 90% chance I'll consume less fat.
    b) 90% of meat hot dogs have between 1.4 and 6.5 grams less fat than a beef hot dog.
    c) I'm 90% confident that meat hot dogs average 1.4–6.5 grams less fat than the beef hot dogs.
    d) If I were to get more samples of both kinds of hot dogs, 90% of the time the meat hot dogs would average 1.4–6.5 grams less fat than the beef hot dogs.
    e) If I tested many samples, I'd expect about 90% of the resulting confidence intervals to include the true difference in mean fat content between the two kinds of hot dogs.

14. Second load of wash. In Exercise 12, we saw a 98% confidence interval of $(-40, -22)$ minutes for $\mu_{Top} - \mu_{Front}$, the difference in time it takes top-loading and front-loading washers to do a load of clothes. Explain why you think each of the following statements is true or false:
   a) 98% of top loaders are 22 to 40 minutes faster than front loaders.
   b) If I choose the laundromat's top loader, there's a 98% chance that my clothes will be done faster than if I had chosen the front loader.
   c) If I tried more samples of both kinds of washing machines, in about 98% of these samples I'd expect the top loaders to be an average of 22 to 40 minutes faster.
   d) If I tried more samples, I'd expect about 98% of the resulting confidence intervals to include the true difference in mean cycle time for the two types of washing machines.
   e) I'm 98% confident that top loaders wash clothes an average of 22 to 40 minutes faster than front-loaders.

15. Learning math. The Core Plus Mathematics Project (CPMP) is an innovative approach to teaching Mathematics that engages students in group investigations and mathematical modeling. After field tests in 36 high schools over a three-year period, researchers compared the performances of CPMP students with those taught using a traditional curriculum. In one test, students had to solve applied Algebra problems using calculators. Scores for 320 CPMP students were compared to those of a control group of 273 students in a traditional Math program. Computer software was used to create a confidence interval for the difference in mean scores. (*Journal for Research in Mathematics Education*, 31, no. 3[2000])

```
Conf level: 95% Variable: Mu(CPMP) — Mu(Ctrl)
Interval: (5.573, 11.427)
```

Explain what the calculated interval means about students who learn Mathematics with CPMP compared to those in traditional programs.

16. Ad campaign. In June 2002, the *Journal of Applied Psychology* reported on a study that examined whether the content of TV shows influenced how well viewers could recall brand names of items in the commercials. The researchers randomly assigned volunteers to watch one of three programs, each with the same nine commercials. One of the programs had violent content, another sexual content, and the third neutral content. After the shows ended, the subjects were asked to recall the brands of products that were advertised. Here are summaries of the results:

	Program Type		
	Violent	**Sexual**	**Neutral**
No. of subjects	108	108	108
Brands Recalled			
Mean	2.08	1.71	3.17
SD	1.87	1.76	1.77

You are a consultant to the marketing department of a business preparing to launch an ad campaign for a new product. The company can afford to run ads during one TV show, and has decided not to sponsor a show with sexual content. You read the study, then use a computer to create a confidence interval for the difference in mean number of brand names remembered between the groups watching violent shows and those watching neutral shows.

```
TWO-SAMPLE T
95% CI FOR MUviol-MUneut: (1.578,-0.602)
```

   a) At the meeting of the marketing staff, you have to explain what this output means. What will you say?
   b) What advice would you give the company about the upcoming ad campaign?

17. CPMP, again. During the study described in Exercise 15, students in both CPMP and traditional classes took another Algebra test that did not allow them to use calculators. The table below shows the results. Are the mean scores of the two groups significantly different?

Math Program	n	Mean	SD
CPMP	312	29.0	18.8
Traditional	265	38.4	16.2

Performance on Algebraic Symbolic Manipulation Without Use of Calculators

Here is computer output for this hypothesis test.

```
2-Sample t-Test of μC - μT ≠ 0
t-Statistic = -6.451 w/574.8761 df
P < 0.0001
```

State a conclusion about the CPMP program.

18. Streams. Researchers collected samples of water from streams in the Adirondack Mountains to investigate the effects of acid rain. They measured the pH (acidity) of the water and classified the streams with respect to the kind of substrate (type of rock over which they flow). A lower pH means the water is more acidic. Here is a plot of the pH of the streams by substrate (limestone, mixed, or shale):

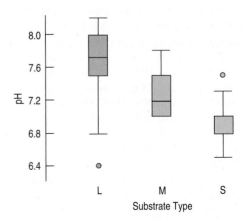

Here are selected parts of a software analysis comparing the pH of streams with limestone and shale substrates:

```
2-Sample t-Test of μ1 - μ2
Difference Between Means = 0.735
t-Statistic = 16.30 w/133 df
p ≤ 0.0001
```

What conclusion would you draw?

19. Another ad. Based on the experiment described in Exercise 16, what's the standard error of the difference in the mean number of brand names viewers might remember from commercials in shows with neutral content and shows with sexual content?

20. Pulse rates. A researcher wanted to see whether there is a significant difference in resting pulse rates for men and women. The data she collected are summarized below.

	Sex	
	Male	**Female**
Count	28	24
Mean	72.75	72.625
StdDev	5.37225	7.69987

What's the standard error of the difference in mean pulse rates?

 B

21. Arthritis. The Centers for Disease Control and Prevention reported a survey of randomly selected Americans age 65 and older, which found that 411 of 1012 men and 535 of 1062 women suffered from some form of arthritis.
a) Are the assumptions and conditions necessary for inference satisfied? Explain.
b) Create a 95% confidence interval for the difference in the proportions of senior men and women who have this disease.
c) Interpret your interval in this context.

22. Graduation. In October 2000 the U.S. Department of Commerce reported the results of a large-scale survey on high school graduation. Researchers contacted more than 25,000 Americans aged 24 years to see if they had finished high school; 84.9% of the 12,460 males and 88.1% of the 12,678 females indicated that they had high school diplomas.
a) Are the assumptions and conditions necessary for inference satisfied? Explain.
b) Create a 95% confidence interval for the difference in graduation rates between males and females.
c) Interpret your confidence interval.

23. Ear infections. A new vaccine was recently tested to see if it could prevent the painful and recurrent ear infections that many infants suffer from. *The Lancet*, a medical journal, reported a study in which babies about a year old were randomly divided into two groups. One group received vaccinations; the other did not. During the following year, only 333 of 2455 vaccinated children had ear infections, compared to 499 of 2452 unvaccinated children in the control group.
a) Are the conditions for inference satisfied?
b) Find a 95% confidence interval for the difference in rates of ear infection.
c) Use your confidence interval to explain whether you think the vaccine is effective.

24. Anorexia. The *Journal of the American Medical Association* reported on an experiment intended to see if the drug Prozac® could be used as a treatment for the eating disorder anorexia nervosa. The subjects, women being treated for anorexia, were randomly divided into two groups. Of the 49 who received Prozac, 35 were deemed healthy a year later, compared to 32 of the 44 who got the placebo.
a) Are the conditions for inference satisfied?
b) Find a 95% confidence interval for the difference in outcomes.
c) Use your confidence interval to explain whether you think Prozac is effective.

25. Teen smoking, part I. A Vermont study published in December 2001 by the American Academy of Pediatrics examined parental influence on teenagers' decisions to smoke. A group of students who had never smoked were questioned about their parents' attitudes toward smoking. These students were questioned again two years later to see if they had started smoking. The researchers found that, among the 284 students who indicated that their parents disapproved of kids smoking, 54 had become established smokers. Among the 41 students who initially said their parents were lenient about smoking, 11 became smokers. Do these data provide strong evidence that parental attitude influences teenagers' decisions about smoking?
a) Create a 95% confidence interval for the difference in the proportion of children who may smoke and have approving parents and those who may smoke and have disapproving parents.
b) Interpret your interval in this context.

26. Depression. A study published in the *Archives of General Psychiatry* in March 2001 examined the impact of depression on a patient's ability to survive cardiac disease. Researchers identified 450 people with cardiac disease, evaluated them for depression, and followed the group for 4 years. Of the 361 patients with no depression, 67 died. Of the 89 patients with minor or major depression, 26 died. Among people who suffer from cardiac disease, are depressed patients more likely to die than non-depressed ones?
a) Create a 95% confidence interval for the difference in survival rates.
b) Interpret your interval in this context.

27. Twins. In 2001, one county reported that, among 3132 white women who had babies, 94 were multiple births. There were also 20 multiple births to 606 black women. Does this indicate any racial difference in the likelihood of multiple births?
a) Write appropriate hypotheses.
b) Check the assumptions and conditions.
c) Find the z-score and P-value for these results.
d) State your conclusion in context.

28. Shopping. A survey of 430 randomly chosen adults found that 21% of the 222 men and 18% of the 208 women had purchased books online. Is there evidence that men are more likely than women to make online purchases of books?
a) Write appropriate hypotheses.
b) Check the assumptions and conditions.
c) Find the z-score and P-value for these results.
d) State your conclusion in context.

29. Gender gap. Candidates for political office realize that different levels of support among men and women may be a crucial factor in determining the outcome of an election. One candidate finds that 52% of 473 men polled say they will vote for him, but only 45% of the 522 women in the poll express support. Is there really a gender gap?
a) Write appropriate hypotheses.
b) Check the assumptions and conditions.
c) Find the z-score and P-value for these results.
d) State your conclusion in context.

30. Pain. Researchers comparing the effectiveness of two pain medications randomly selected a group of patients who had been complaining of a certain kind of joint pain. They randomly divided these people into two groups, then administered the pain killers. Of the 112 people in the group who received medication A, 84 said this pain reliever was effective. Of the 108 people in the other group, 66 reported that pain reliever B was effective. Is there a significant difference?
a) Write appropriate hypotheses.
b) Check the assumptions and conditions.
c) Find the z-score and P-value for these results.
d) State your conclusion in context.

31. Commuting. A man who moves to a new city sees that there are two routes he could take to work. A neighbor who has lived there a long time tells him Route A will average 5 minutes faster than Route B. The man decides to experiment. Each day he flips a coin to determine which way to go, driving each route 20 days. He finds that Route A takes an average of 40 minutes, with standard deviation 3 minutes, and Route B takes an average of 43 minutes, with standard deviation 2 minutes. Histograms of travel times for the routes are roughly symmetric and show no outliers.
a) Find a 95% confidence interval for the difference in average commuting time for the two routes.
b) Should the man believe the old-timer's claim that he can save an average of 5 minutes a day by always driving Route A? Explain.

32. Handy. A factory hiring people to work on an assembly line gives job applicants a test of manual agility. This test counts how many strangely shaped pegs the applicant can fit into matching holes in a one-minute period. The following table summarizes the data by sex of the job applicant. Assume that all conditions necessary for inference are met.

	Male	Female
Number of subjects	50	50
Pegs placed:		
Mean	19.39	17.91
SD	2.52	3.39

a) Find a 95% confidence interval for the difference in the mean number of pegs that could be placed by men and women.
b) What does this interval suggest about manual agility in men and women?

33. Cereal. The following data show the sugar content (as a percentage of weight) of several national brands of children's and adults' cereals. Create and interpret a 95% confidence interval for the difference in mean sugar content. Be sure to check the necessary assumptions and conditions.

Children's cereals: 40.3, 55, 45.7, 43.3, 50.3, 45.9, 53.5, 43, 44.2, 44, 47.4, 44, 33.6, 55.1, 48.8, 50.4, 37.8, 60.3, 46.6

Adults' cereals: 20, 30.2, 2.2, 7.5, 4.4, 22.2, 16.6, 14.5, 21.4, 3.3, 6.6, 7.8, 10.6, 16.2, 14.5, 4.1, 15.8, 4.1, 2.4, 3.5, 8.5, 10, 1, 4.4, 1.3, 8.1, 4.7, 18.4

34. Baseball 2006. American League baseball teams play their games with the designated hitter rule, meaning that pitchers do not bat. The league believes that replacing the pitcher, traditionally a weak hitter, with another player in the batting order produces more runs and generates more interest among fans. Below are the average numbers of runs scored in American League and National League stadiums for the 2006 season.

American		National	
11.4	9.9	10.5	9.5
10.5	9.7	10.3	9.4
10.4	9.1	10.0	9.1
10.3	9.0	10.0	9.0
10.2	9.0	9.7	9.0
10.0	8.9	9.7	8.9
9.9	8.8	9.6	8.9
		9.5	7.9

a) Create an appropriate display of these data. What do you see?
b) With a 95% confidence interval, estimate the difference in the mean number of runs scored in the two leagues.
c) Does your interval indicate that the AL's designated hitter rule may lead to more scoring?

35. Tees. Does it matter what kind of tee a golfer places the ball on? The company that manufactures

"Stinger" tees claims that the thinner shaft and smaller head will lessen drag, reducing spin and allowing the ball to travel farther. In August 2003, Golf Laboratories, Inc., compared the distance traveled by golf balls hit off regular wooden tees to those hit off Stinger tees. All the balls were struck by the same golf club using a robotic device set to swing the club head at approximately 95 miles per hour. Summary statistics from the test are shown in the table. Assume that 6 balls were hit off each tee and that the data were suitable for inference.

		Total Distance (yards)	Ball Velocity (mph)	Club Velocity (mph)
Regular tee	Mean	227.17	127.00	96.17
	SD	2.14	0.89	0.41
Stinger tee	Mean	241.00	128.83	96.17
	SD	2.76	0.41	0.52

Is there evidence that balls hit off the Stinger tees would travel farther?
a) Write appropriate hypotheses.
b) Find t and the P-value for these results.
c) State your conclusion in context.

T 36. **Hard water.** In an investigation of environmental causes of disease, data were collected on the annual mortality rate (deaths per 100,000) for males in 61 large towns in England and Wales. In addition, the water hardness was recorded as the calcium concentration (parts per million, ppm) in the drinking water. The data set also notes, for each town, whether it was south or north of Derby. Is there a significant difference in mortality rates in the two regions? Here are the summary statistics.

```
Summary of:          mortality
For categories in:   Derby
Group   Count   Mean      Median   StdDev
North   34      1631.59   1631     138.470
South   27      1388.85   1369     151.114
```

You may assume the data were suitable for inference.
a) Write appropriate hypotheses.
b) Find t and the P-value for these results.
c) State your conclusion in context.

37. **Hungry?** Researchers investigated how the size of a bowl affects how much ice cream people tend to scoop when serving themselves.[10] At an "ice cream social," people were randomly given either a 17 oz or a 34 oz bowl (both large enough that they would not be filled to capacity). They were then invited to scoop as much ice cream as they liked. Did the bowl size change the selected portion size? Here are the summaries:

Small Bowl		Large Bowl	
n	26	n	22
$\bar{y}$	5.07 oz	$\bar{y}$	6.58 oz
s	1.84 oz	s	2.91 oz

Assume any assumptions and conditions are satisfied.
a) Write appropriate hypotheses.
b) Find t and the P-value for these results.
c) State your conclusion in context.

38. **Thirsty?** Researchers randomly assigned participants either a tall, thin "highball" glass or a short, wide "tumbler," each of which held 355 ml. Participants were asked to pour a shot (1.5 oz = 44.3 ml) into their glass. Did the shape of the glass make a difference in how much liquid they poured?[11] Here are the summaries:

highball		tumbler	
n	99	n	99
$\bar{y}$	42.2 ml	$\bar{y}$	60.9 ml
s	16.2 ml	s	17.9 ml

Test an appropriate hypothesis and state your conclusions. Assume any assumptions and conditions are satisfied.
a) Write appropriate hypotheses.
b) Find t and the P-value for these results.
c) State your conclusion in context.

T 39. **Running heats.** In Olympic running events, preliminary heats are determined by random draw, so we should expect that the abilities of runners in the various heats to be about the same, on average. Here are the times (in seconds) for the 400-m women's run in the 2004 Olympics in Athens for preliminary heats 2 and 5. Is there any evidence that the mean time to finish is different for randomized heats?

Country	Name	Heat	Time
USA	HENNAGAN Monique	2	51.02
BUL	DIMITROVA Mariyana	2	51.29
CHA	NADJINA Kaltouma	2	51.50
JAM	DAVY Nadia	2	52.04
BRA	ALMIRAO Maria Laura	2	52.10
FIN	MYKKANEN Kirsi	2	52.53
BAH	WILLIAMS-DARLING Tonique	5	51.20
BLR	USOVICH Svetlana	5	51.37
UKR	YEFREMOVA Antonina	5	51.53
CMR	NGUIMGO Mireille	5	51.90
JAM	BECKFORD Allison	5	52.85
TOG	THIEBAUD-KANGNI Sandrine	5	52.87
SRI	DHARSHA K V Damayanthi	5	54.58

---

[10]Brian Wansink, Koert van Ittersum, and James E. Painter, "Ice Cream Illusions: Bowls, Spoons, and Self-Served Portion Sizes," *Am J Prev Med* 2006.

[11]Brian Wansink and Koert van Ittersum, "Shape of Glass and Amount of Alcohol Poured: Comparative Study of Effect of Practice and Concentration," *BMJ* 2005;331;1512–1514.

a) Write appropriate hypotheses.
b) Check the assumptions and conditions.
c) Find t and the P-value for these results.
d) State your conclusion in context.

T 40. Swimming heats. In Exercise 39 we looked at the times in two different heats for the 400-m women's run from the 2004 Olympics. Unlike track events, swimming heats are *not* determined at random. Instead, swimmers are seeded so that better swimmers are placed in later heats. Here are the times (in seconds) for the women's 400-m freestyle from heats 2 and 5. Do these results suggest that the mean times of seeded heats are not equal?

Country	Name	Heat	Time
ARG	BIAGIOLI Cecilia Elizabeth	2	256.42
SLO	CARMAN Anja	2	257.79
CHI	KOBRICH Kristel	2	258.68
MKD	STOJANOVSKA Vesna	2	259.39
JAM	ATKINSON Janelle	2	260.00
NZL	LINTON Rebecca	2	261.58
KOR	HA Eun-Ju	2	261.65
UKR	BERESNYEVA Olga	2	266.30
FRA	MANAUDOU Laure	5	246.76
JPN	YAMADA Sachiko	5	249.10
ROM	PADURARU Simona	5	250.39
GER	STOCKBAUER Hannah	5	250.46
AUS	GRAHAM Elka	5	251.67
CHN	PANG Jiaying	5	251.81
CAN	REIMER Brittany	5	252.33
BRA	FERREIRA Monique	5	253.75

a) Write appropriate hypotheses.
b) Check the assumptions and conditions.
c) Find t and the P-value for these results.
d) State your conclusion in context.

C

41. Gender gap. A presidential candidate fears he has a problem with women voters. His campaign staff plans to run a poll to assess the situation. They'll randomly sample 300 men and 300 women, asking if they have a favorable impression of the candidate. Obviously, the staff can't know this, but suppose the candidate has a positive image with 59% of males but with only 53% of females.
a) What sampling design is his staff planning to use?
b) What difference would you expect the poll to show?
c) Of course, sampling error means the poll won't reflect the difference perfectly. What's the standard deviation for the difference in the proportions?
d) Sketch a sampling model for the size difference in proportions of men and women with favorable impressions of this candidate that might appear in a poll like this.

e) Could the campaign be misled by the poll, concluding that there really is no gender gap? Explain.

42. Buy it again? A consumer magazine plans to poll car owners to see if they are happy enough with their vehicles that they would purchase the same model again. They'll randomly select 450 owners of American-made cars and 450 owners of Japanese models. Obviously, the actual opinions of the entire population couldn't be known, but suppose 76% of owners of American cars and 78% of owners of Japanese cars would purchase another.
a) What sampling design is the magazine planning to use?
b) What difference would you expect their poll to show?
c) Of course, sampling error means the poll won't reflect the difference perfectly. What's the standard deviation for the difference in the proportions?
d) Sketch a sampling model for the difference in proportions that might appear in a poll like this.
e) Could the magazine be misled by the poll, concluding that owners of American cars are much happier with their vehicles than owners of Japanese cars? Explain.

43. Pregnancy. In 1998, a San Diego reproductive clinic reported 42 births to 157 women under the age of 38, but only 7 births for 89 clients aged 38 and older. Is this strong evidence of a difference in the effectiveness of the clinic's methods for older women?
a) Was this an experiment? Explain.
b) Test an appropriate hypothesis and state your conclusion in context.
c) If you concluded there was a difference, estimate that difference with a confidence interval and interpret your interval in context.

44. Birth weight. In 2003 the *Journal of the American Medical Association* reported a study examining the possible impact of air pollution caused by the 9/11 attack on New York's World Trade Center on the weight of babies. Researchers found that 8% of 182 babies born to mothers who were exposed to heavy doses of soot and ash on September 11 were classified as having low birth weight. Only 4% of 2300 babies born in another New York City hospital whose mothers had not been near the site of the disaster were similarly classified. Does this indicate a possibility that air pollution might be linked to a significantly higher proportion of low-weight babies?
a) Was this an experiment? Explain.
b) Test an appropriate hypothesis and state your conclusion in context.
c) If you concluded there is a difference, estimate that difference with a 90% confidence interval and interpret that interval in context.

45. Teen smoking, part II. Consider again the study described in Exercise 25 investigating whether parental attitudes about smoking influence teenage behavior.
a) What kind of design did the researchers use?
b) Write appropriate hypotheses.
c) Are the assumptions and conditions necessary for inference satisfied?

d) Test the hypothesis and state your conclusion.

e) Explain in this context what your P-value means.

f) If that conclusion is actually wrong, which type of error did you commit?

46. Depression revisited. Consider again the study described in Exercise 26 investigating whether depression is a factor in deaths from heart disease.

a) What kind of design was used to collect these data?

b) Write appropriate hypotheses.

c) Are the assumptions and conditions necessary for inference satisfied?

d) Test the hypothesis and state your conclusion.

e) Explain in this context what your P-value means.

f) If your conclusion is actually incorrect, which type of error did you commit?

47. Music and memory. Is it a good idea to listen to music when studying for a big test? In a study conducted by some Statistics students, 62 people were randomly assigned to listen to rap music, music by Mozart, or no music while attempting to memorize objects pictured on a page. They were then asked to list all the objects they could remember. Here are summary statistics:

	Rap	Mozart	No Music
Count	29	20	13
Mean	10.72	10.00	12.77
SD	3.99	3.19	4.73

a) Does it appear that it is better to study while listening to Mozart than to rap music? Test an appropriate hypothesis and state your conclusion.

b) Create a 90% confidence interval for the mean difference in memory score between students who study to Mozart and those who listen to no music at all. Interpret your interval.

48. Rap. Using the results of the experiment described in Exercise 47, does it matter whether one listens to rap music while studying, or is it better to study without music at all?

a) Test an appropriate hypothesis and state your conclusion.

b) If you concluded there is a difference, estimate the size of that difference with a confidence interval and explain what your interval means.

49. Another commercial. Consider again the study described in Exercise 16 investigating whether the content (violent, sexual, or neutral) of a TV program is a factor in how well viewers will recall the brand names of products that were advertised. Here are the results:

Program Type		
	Violent	Sexual	Neutral
No. of subjects	108	108	108
Brands Recalled		
Mean	2.08	1.71	3.17
SD	1.87	1.76	1.77

Is there evidence that viewer memory for ads may differ between programs with sexual content and those with neutral content? If there is a difference, how big is it?

50. Commercials, part III. In the study described in Exercise 16, the researchers also contacted the subjects again, 24 hours later, and asked them to recall the brands advertised. Results are summarized below.

Program Type		
	Violent	Sexual	Neutral
No. of subjects	101	106	103
Brands Recalled		
Mean	3.02	2.72	4.65
SD	1.61	1.85	1.62

a) Is there a significant difference in viewers' abilities to remember brands advertised in shows with violent vs. neutral content?

b) Find a 95% confidence interval for the difference in mean number of brand names remembered between the groups watching shows with sexual content and those watching neutral shows. Interpret your interval in this context.

T 51. Cuckoos, part I. Cuckoos lay their eggs in the nests of other (host) birds. The eggs are then adopted and hatched by the host birds. But the potential host birds lay eggs of different sizes. Does the cuckoo change the size of her eggs for different foster species? The numbers in the table are lengths (in mm) of cuckoo eggs found in nests of three different species of other birds. The data are drawn from the work of O. M. Latter in 1902 and were used in a fundamental textbook on statistical quality control by L.H.C. Tippett (1902–1985), one of the pioneers in that field.

Cuckoo Egg Length (MM)		
Foster Parent Species		
Sparrow	Robin	Wagtail
---------	-------	---------
20.85	21.05	21.05
21.65	21.85	21.85
22.05	22.05	21.85
22.85	22.05	21.85
23.05	22.05	22.05
23.05	22.25	22.45
23.05	22.45	22.65
23.05	22.45	23.05
23.45	22.65	23.05
23.85	23.05	23.25
23.85	23.05	23.45
23.85	23.05	24.05
24.05	23.05	24.05
25.05	23.05	24.05
	23.25	24.85
	23.85	

Investigate the question of whether the mean length of cuckoo eggs is the same for sparrows and robins, and state your conclusion.

52. **Cuckoos, part II.** Use the data from the study in Exercise 51 to investigate whether the mean length of cuckoo eggs is the same for wagtails as for the other two species and state your conclusion.

Answers

Just Checking

1. We're 90% confident that if members are contacted by e-mail, the donation rate will be between 4 and 18 percentage points higher than if they received regular mail.

2. a) Compare the proportions of women in each group who died after contracting eclampsia. $H_0: p_{MS} - p_p = 0$ $H_A: p_{MS} - p_p > 0$
   b) Women were randomly assigned to treatments; 11 and 20 "successes," and 29 and 76 failures are all at least 10.
   c) $z = 0.81$
   d) Fail to reject H_0 These results are less than one standard deviation from what could be expected if the risk of death is the same among women treated with magnesium sulfate, so there's no evidence of a greater risk.

3. a) Randomization should balance unknown sources of variability in the two groups of patients and helps us believe the two groups are independent.
   b) We can be 95% confident that after 4 weeks endoscopic surgery patients will have a mean pinch strength between 0.04 kg and 2.96 kg higher than open-incision patients.
   c) Without data, we can't check the Nearly Normal Condition.

4. H_0: Mean pinch strength is the same after both surgeries. ($\mu_E - \mu_O = 0$)
   H_A: Mean pinch strength is different after the two surgeries. ($\mu_E - \mu_O \neq 0$)

5. With a P-value this low, we reject the null hypothesis. We can conclude that mean pinch strength differs after 4 weeks in patients who undergo endoscopic surgery vs. patients who have open-incision surgery. Results suggest that the endoscopic surgery patients may be stronger, on average.

Comparing Counts

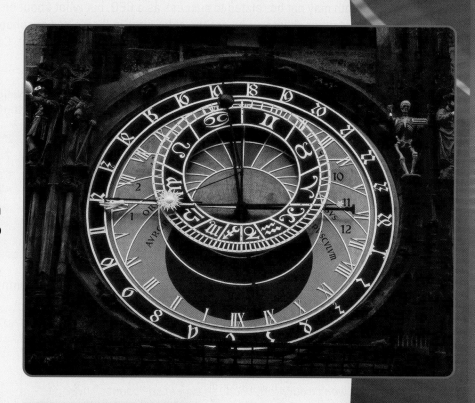

Does your zodiac sign predict how successful you will be later in life? *Fortune* magazine collected the zodiac signs of 256 heads of the largest 400 companies. The table shows the number of births for each sign.

We can see some variation in the number of births per sign, and there *are* more Pisces, but is that enough to claim that successful people are more likely to be born under some signs than others?

Births	Sign
23	Aries
20	Taurus
18	Gemini
23	Cancer
20	Leo
19	Virgo
18	Libra
21	Scorpio
19	Sagittarius
22	Capricorn
24	Aquarius
29	Pisces

Birth totals by sign for 256 Fortune 400 executives.

WHO	Executives of Fortune 400 companies
WHAT	Zodiac birth sign
WHY	Maybe the researcher was a Gemini and naturally curious?

Goodness-of-Fit

If births were distributed uniformly across the year, we would expect about 1/12 of them to occur under each sign of the zodiac. That suggests 256/12, or about 21.3 executives born under each sign. How closely do the observed numbers of births per sign fit this simple "null" model?

A hypothesis test to address this question is called a test of **"goodness-of-fit."** The name suggests a certain badness-of-grammar, but it is quite standard. After all, we are asking whether the model that births are uniformly distributed over the signs fits the data good, . . . er, well. Goodness-of-fit involves testing a hypothesis. We have specified a model for the distribution and want to know whether it fits. There is no single parameter to estimate, so a confidence interval wouldn't make much sense. And a one-proportion z-test won't work, because we have 12 hypothesized proportions, one for each sign. We need a test that considers all of them together.

For Example Finding Expected Counts

Birth month may not be related to success as a CEO, but what about on the ball field? Some researchers wonder whether children who are the older ones in their class at school naturally perform better in sports and that these children then get more coaching and encouragement. Could that make a difference in who makes it to the professional level in sports?

Baseball is a remarkable sport, in part because so much data are available. We have the birth dates of every one of the 16,804 players who ever played in a major league game. Since the effect we're suspecting may be due to relatively recent school policies, we'll consider the birth months of the 1478 major league players born since 1975 and who have played through 2006. We can also look up the national statistics to find what percentage of people were born in each

Month	Ballplayer Count	National Birth %	Month	Ballplayer Count	National Birth %
1	137	8%	7	102	9%
2	121	7%	8	165	9%
3	116	8%	9	134	9%
4	121	8%	10	115	9%
5	126	8%	11	105	8%
6	114	8%	12	122	9%
			Total	**1478**	**100%**

month. Let's test whether the observed distribution of ballplayers' birth months shows just random fluctuations or whether it's really different in some way.

Question: How can we find the expected counts?

There are 1478 players in this set of data. I'd expect 8% of them to have been born in January, and 1478(0.08) = 118.24. I won't round off, because expected "counts" needn't be integers. Multiplying 1478 by each of the birth percentages gives the expected counts shown in the table.

Month	Expected	Month	Expected
1	118.24	7	133.02
2	103.46	8	133.02
3	118.24	9	133.02
4	118.24	10	133.02
5	118.24	11	118.24
6	118.24	12	133.02

Assumptions and Conditions

These data are organized in tables showing summary counts in categories. (In our example, we don't see the birth signs of each of the 256 executives, only the totals for each sign.) We can't proceed with inference until we check several assumptions and conditions:

- **Counted Data Condition:** The values in each **cell** must be *counts* for the categories of a categorical variable. This might seem a simplistic, even silly condition. But we can't apply these methods to proportions, percentages, or measurements just because they happen to be organized in a table.
- **Independence Assumption:** The counts in the cells should be independent of each other. The easiest case is when the individuals who are counted in the cells are sampled independently from some population. That's what we'd like to have if we want to draw conclusions about that population. Randomness can arise in other ways, though. For example, the Fortune 400 executives are not a random sample, but it's reasonable to think that their birth dates should be randomly distributed throughout the year.

If we want to generalize to a large population, we should check two more conditions:

- **Randomization Condition:** The individuals who have been counted should be a random sample from the population of interest.
- **10% Condition:** Our sample is less than 10% of the population.

Finally, we must have enough data for the methods to work. We check the:

Expected Cell Frequency Condition: Our group should be large enough that we expect to see at least 5 individuals in each cell.

This is quite similar to the condition that np and nq be at least 10 when we tested proportions. In our astrology example, assuming equal births in each month leads us to expect 21.3 births per month, so the condition is easily met here.

For Example Checking Assumptions and Conditions

Recap: Are professional baseball players more likely to be born in some months than in others? We have observed and expected counts for the 1478 players born since 1975.

Question: Are the assumptions and conditions met for performing a goodness-of-fit test?

✔ **Counted Data Condition:** I have month-by-month counts of ballplayer births.
✔ **Independence Assumption:** These births were independent.
✔ **Randomization Condition:** Although they are not a random sample, we can take these players to be representative of players past and future.
✔ **10% Condition:** These 1478 players are less than 10% of the population of 16,804 players who have ever played (or will play) major league baseball.
✔ **Expected Cell Frequency Condition:** The expected counts range from 103.46 to 133.02, all much greater than 5.

It's okay to use these data for a goodness-of-fit test.

Just Checking

A Biology class of 124 students collected data on themselves to check the genetic theory about the frequency of tongue-rolling and free-hanging earlobes. Their results are summarized in the table.

Tongue	Earlobes	Observed Count	Expected Count
Non-curling	Attached	12	7.75
Non-curling	Free	22	23.25
Curling	Attached	31	23.25
Curling	Free	59	69.75

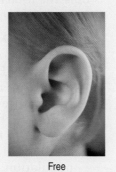

Free   Attached

2. Is it okay to proceed with inference? Check the assumptions and conditions.

(Check your answers on page 517.)

▢ Calculations

NOTATION ALERT

We compare the counts *observed* in each cell with the counts we *expect* to find. The usual notation uses O's and E's or abbreviations such as those we've used here. The method for finding the expected counts depends on the model.

Are the differences between what we saw and what we expected just natural sampling variability, or are they so large that they indicate something important? It's natural to look at the *differences* between these observed and expected counts, denoted ($Obs - Exp$). Just adding these differences won't work because some are positive, others negative. We've been in this predicament before, and we handle it the same way now: We square them. That gives us positive values and focuses attention on any cells with large differences from what we expected. Because the differences between observed and expected counts generally get larger the more data we have, we also need to get an idea of the *relative* sizes of the differences. To do that, we divide each squared difference by the expected count for that cell.

The test statistic, called the **chi-square** (or chi-squared) **statistic,** is found by adding up the sum of the squares of the deviations between the observed and expected counts divided by the expected counts:

$$\chi^2 = \sum_{all\ cells} \frac{(Obs - Exp)^2}{Exp}.$$

The chi-square statistic is denoted χ^2, where χ is the Greek letter chi (pronounced "ky" as in "sky"). It refers to a family of sampling distribution models we have not seen before called (remarkably enough) the **chi-square models.**

This family of models, like the Student's t-models, differ only in the number of degrees of freedom. The number of degrees of freedom for a goodness-of-fit test is $n - 1$. Here, however, n is *not* the sample size, but instead is the number of categories. For the zodiac example, we have 12 signs, so our χ^2 statistic has 11 degrees of freedom.

If the observed counts don't match the expected, the statistic will be large. It can't be "too small." That would just mean that our model *really* fit the data well. So the chi-square test is always one-sided. If the calculated statistic value is large enough, we'll reject the null hypothesis.

But what's large enough? The Rule of 2 doesn't work here; these aren't standard deviations units. It would be great if there were some other simple rule, but alas, how large the chi-square value must be in order to be statistically significant depends on the number of degrees of freedom. The table shows these critical values for small degrees of freedom.

Isn't there an easier way? Sure—use technology to find a P-value. Remember, that's the probability we'd see outcomes at least as strange as ours if the null hypothesis were true. The smaller the P-value, the greater our suspicion that the null may be false. If our data have a P-value that's less than 0.05, we'll reject the null hypothesis.

The Rule of, well, *not* 2	
Degrees of Freedom	**Reject H₀ if . . .**
1	$\chi^2 > 3.84$
2	$\chi^2 > 5.99$
3	$\chi^2 > 7.81$
4	$\chi^2 > 9.49$
5	$\chi^2 > 11.07$
6	$\chi^2 > 12.59$
7	$\chi^2 > 14.07$
8	$\chi^2 > 15.51$
9	$\chi^2 > 16.92$
10	$\chi^2 > 18.31$
11	$\chi^2 > 19.68$
12	$\chi^2 > 21.03$

TI-*nspire*

The χ^2 Models. See what a χ^2 model looks like, and watch it change as you change the degrees as freedom.

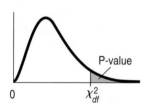

0   χ^2_{df} P-value

For Example   Doing a Goodness-of-fit Test

Recap: We're looking at data on the birth months of major league baseball players. We've checked the assumptions and conditions for performing a χ^2 test.

Questions: What are the hypotheses, and what does the test show?

H₀: The distribution of birth months for major league ballplayers is the same as that for the general population.

H_A: The distribution of birth months for major league ballplayers differs from that of the rest of the population.

(continued)

$$df = 12 - 1 = 11$$

$$\chi^2 = \sum \frac{(Obs - Exp)^2}{Exp}$$

$$= \frac{(137 - 118.24)^2}{118.24} + \frac{(121 - 103.46)^2}{103.46} + \cdots$$

$$= 26.48 \text{ (by technology)}$$

$$P\text{-value} = P(\chi^2_{11} \geq 26.48) = 0.0055 \text{ (by technology)}$$

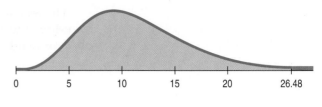

Because of the small P-value, I reject H_0; there's evidence that birth months of major league ballplayers have a different distribution from the rest of us.

Note: We found $\chi^2 = 26.48$ for our data, greater than 19.68, the critical value for 11 df in the table. That's another way you could decide to reject H_0.

Just Checking

Here's the table summarizing the frequency of two traits in that Biology class. Students are checking the genetic theory that the ratio of people with none, one, or both traits is 1:3:3:9.

3. Write the null and alternative hypothesis.

4. How many degrees of freedom are there?

5. Calculate the component of χ^2 for the bottom cell.

6. For these data $\chi^2 = 6.64$. Should the Bio class reject H_0? (Use the table of critical values.)

7. What should the students conclude?

Tongue	Earlobes	Observed Count	Expected Count
Non-curling	Attached	12	7.75
Non-curling	Free	22	23.25
Curling	Attached	31	23.25
Curling	Free	59	69.75

(Check your answers on page 517.)

STEP–BY–STEP EXAMPLE   **A Chi-Square Test for Goodness-of-Fit**

We have counts of 256 executives in 12 zodiac sign categories. The natural null hypothesis is that birth dates of executives are divided equally among all the zodiac signs.

Question: Are CEOs more likely to be born under some zodiac signs than others?

THINK

Plan State what you want to know.

Identify the variables and check the W's.

I want to know whether births of successful people are uniformly distributed across the signs of the zodiac. I have counts of 256 Fortune 400 executives, categorized by their birth sign.

Hypotheses State the null and alternative hypotheses. For χ^2 tests, it's usually easier to do that in words than in symbols.

H_0: Births are uniformly distributed over zodiac signs.

H_A: Births are not uniformly distributed over zodiac signs.

Model Make a picture. The null hypothesis is that the frequencies are equal, so a bar chart (with a line at the hypothesized "equal" value) is a good display.

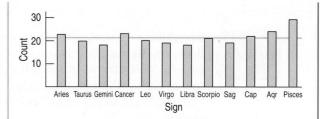

The bar chart shows some variation from sign to sign, and Pisces is the most frequent. But it is hard to tell whether the variation is more than I'd expect from random variation.

Think about the assumptions and check the conditions.

✔ **Counted Data Condition:** I have counts of the number of executives in 12 categories.

✔ **Independence Assumption:** The birth dates of executives should be independent of each other.

✔ **Randomization Condition:** This is a convenience sample of executives, but there's no reason to suspect bias.

✔ **Expected Cell Frequency Condition:** The null hypothesis expects that 1/12 of the 256 births, or 21.333, should occur in each sign. These expected values are all at least 5, so the condition is satisfied.

Specify the sampling distribution model.

Name the test you will use.

The conditions are satisfied, so I'll use a χ^2 model with $12 - 1 = 11$ degrees of freedom and do a **chi-square goodness-of-fit test.**

Mechanics Each cell contributes an $\dfrac{(Obs - Exp)^2}{Exp}$ value to the chi-square sum. We add up these components for each zodiac sign. If you do it by hand, it can be helpful to arrange the calculation in a table. We show that after this Step-By-Step.

The expected value for each zodiac sign is 21.333.

$$\chi^2 = \sum \frac{(Obs - Exp)^2}{Exp} = \frac{(23 - 21.333)^2}{21.333}$$
$$+ \frac{(20 - 21.333)^2}{21.333} + \cdots$$
$$= 5.094 \text{ for all 12 signs.}$$

We can decide what to do based on the table of critical values, or the P-value.

The P-value is the area in the upper tail of the χ^2 model above the computed χ^2 value.

For 11 df, we'd reject H_0 if $\chi^2 > 19.68$

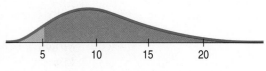

P-value $= P(\chi^2 > 5.094) = 0.926$

(continued)

Conclusion Link the P-value to your decision. Remember to state your conclusion in terms of what the data mean, rather than just making a statement about the distribution of counts.

Our calculated χ^2 is far less than the 19.68 that's statistically significant. The P-value of 0.926 says that if the zodiac signs of executives were in fact distributed uniformly, data like these would occur in about 93% of samples. This certainly isn't unusual, so I fail to reject the null hypothesis. There's virtually no evidence that executives are more likely to have certain zodiac signs.

The Chi-Square Calculation

Let's make the chi-square procedure very clear. Here are the steps:

1. **Find the expected values.** These come from the null hypothesis model. Every model gives a hypothesized proportion for each cell. The expected value is the product of the total number of observations times this proportion.

   For our example, the null model hypothesizes *equal* proportions. With 12 signs, 1/12 of the 256 executives should be in each category. The expected number for each sign is 21.333.
2. **Compute the deviations.** Once you have expected values for each cell, find the deviations, *Observed − Expected*.
3. **Square the deviations.**
4. **Compute the components.** Now find the component, $\dfrac{(Observed\ -\ Expected)^2}{Expected}$, for each cell.
5. **Find the sum of the components.** That's the chi-square statistic.
6. **Find the degrees of freedom.** It's equal to the number of cells minus one. For the zodiac signs, that's $12 - 1 = 11$ degrees of freedom.
7. **Test the hypothesis.** Large chi-square values mean lots of deviation from the hypothesized model, so they give small P-values. Look up the critical value from a table of chi-square values, or use technology to find the P-value directly.

The steps of the chi-square calculations are often laid out in tables like this:

Sign	Observed	Expected	Deviation = (Obs − Exp)	(Obs − Exp)²	Component = $\dfrac{(Obs\ -\ Exp)^2}{Exp}$
Aries	23	21.333	1.667	2.778889	0.130262
Taurus	20	21.333	−1.333	1.776889	0.083293
Gemini	18	21.333	−3.333	11.108889	0.520737
Cancer	23	21.333	1.667	2.778889	0.130262
Leo	20	21.333	−1.333	1.776889	0.083293
Virgo	19	21.333	−2.333	5.442889	0.255139
Libra	18	21.333	−3.333	11.108889	0.520737
Scorpio	21	21.333	−0.333	0.110889	0.005198
Sagittarius	19	21.333	−2.333	5.442889	0.255139
Capricorn	22	21.333	0.667	0.444889	0.020854
Aquarius	24	21.333	2.667	7.112889	0.333422
Pisces	29	21.333	7.667	58.782889	2.755491
					$\Sigma = 5.094$

Just Checking

8. Once more, here's the summary of the distribution of two traits among Biology students. Show that $\chi^2 = 6.64$ by completing a table of calculations like the one on the previous page.

Tongue	Earlobes	Observed Count	Expected Count
Non-curling	Attached	12	7.75
Non-curling	Free	22	23.25
Curling	Attached	31	23.25
Curling	Free	59	69.75

(Check your answers on page 517.)

TI Tips   Testing goodness of fit

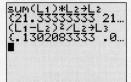

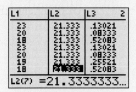

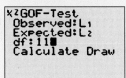

As always, the TI makes doing the mechanics of a goodness-of-fit test pretty easy, but it does take a little work to set it up. Let's use the zodiac data to run through the steps for a χ^2 GOF-Test.

- Enter the counts of executives born under each star sign in **L1**.
  Those counts were: 23 20 18 23 20 19 18 21 19 22 24 29
- Enter the expected percentages (or fractions, here 1/12) in **L2**. In this example they are all the same value, but that's not always the case.
- Convert the expected percentages to expected counts by multiplying each of them by the total number of observations. We use the calculator's summation command in the **LIST MATH** menu to find the total count for the data summarized in **L1** and then multiply that sum by the percentages stored in **L2** to produce the expected counts. The command is **sum(L1)*L2 → L2**. (We don't ever need the percentages again, so we can replace them by storing the expected counts in **L2** instead.)

- Choose **D: χ^2 GOF-Test** from the **STATS TESTS** menu.
- Specify the lists where you stored the observed and expected counts, and enter the number of degrees of freedom, here 11.
- Ready, set, **Calculate** . . .

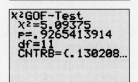

- . . . and there are the calculated value of χ^2 and your P-value, a whopping 0.93! There's nothing at all unusual about these data. (So much for the zodiac's predictive power.)

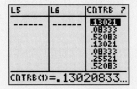

Notice, too, there's a list of values called **CNTRB**. You can scroll across them, or use **LIST NAMES** to display them as a datalist. Those are the cell-by-cell components of the χ^2 calculation. We aren't very interested in them this time, because our data failed to provide evidence that the zodiac sign mattered. However, in a situation where we rejected the null hypothesis, we'd want to look at the components to see where the biggest effects occurred. You'll read more about doing that later in this chapter.

Looking for Significance in Two-Way Tables

Many states and localities now collect data on traffic stops regarding the race of the driver. The initial concern was that Black drivers were being stopped more often (the "crime" ironically called "Driving While Black"). With more data in hand, attention has turned to other issues. For example, data from 2533 traffic stops in Cincinnati report the race of the driver (Black, White, or Other) and whether the traffic stop resulted in a search of the vehicle. Here's a **two-way table** summarizing those data with each **cell** showing the count of drivers in that category.

WHO	2533 drivers stopped by police
WHAT	Race and searches
WHERE	Cincinnati
WHY	To look for race-based differences in the treatment of drivers

		Race			
		Black	**White**	**Other**	**Total**
Search	**No**	787	594	27	**1408**
	Yes	813	293	19	**1125**
	Total	**1600**	**887**	**46**	**2533**

Is there any evidence of race-based differences in whether or not police searched the vehicles they stopped? Sometimes when we look at a table like this we wonder if two distributions are the same, and other times we wonder if two variables are independent. To find out, we'll do a **chi-square two-way table test.** The mechanics of this test are *identical* to the goodness-of-fit test we just looked at, as are the assumptions and conditions. We just count the degrees of freedom slightly differently. Let's work through the test.

Calculations in Two-Way Tables

The null hypothesis says that the likelihood a vehicle was searched should be the same regardless of the race of the driver. We can estimate the probability that a stopped vehicle gets searched by looking at the totals:

$$\frac{1125}{2533} = 44.4\% \text{ were searched, and } \frac{1408}{2533} = 55.6\% \text{ were not.}$$

Within each racial group, the expected proportion of searches should be the same. If race doesn't matter, we'd expect 44.4% of the 1600 Black drivers' cars to be searched ($0.444 \times 1600 = 710.40$) and the rest ($0.556 \times 1600 = 889.60$) not to be. Because these are theoretical values, they don't have to be whole numbers. We (or more likely, technology) repeat this for the other two racial groups, filling in expected values for each cell:

TABLE 21.1 **Expected counts.**

		Race		
		Black	**White**	**Other**
Search	**No**	889.60	493.17	25.58
	Yes	710.40	393.83	20.42

Now we can check the **Expected Cell Frequency Condition.** Indeed, there are at least 5 individuals expected in each cell.

Following the pattern of the goodness-of-fit test, compute the component for each cell of the table. For the highlighted cell, Black drivers whose vehicles were not searched, that's

$$\frac{(Obs - Exp)^2}{Exp} = \frac{(787 - 889.60)^2}{889.60} = 11.83$$

Summing these components across all cells gives

$$\chi^2 = \sum_{all\ cells} \frac{(Obs - Exp)^2}{Exp} = 73.25$$

How about the degrees of freedom? We don't really need to calculate all the expected values in the table. We know there is a total of 1408 cars that weren't searched. Once we find the expected values for two of the racial groups, we can determine the expected number for the other group by just subtracting. Similarly, after filling in the top row, we can find the expected values for the remaining row by subtracting. To fill out the table, we need to know the counts in only $R - 1$ rows and $C - 1$ columns. So the table has $(R - 1)(C - 1)$ degrees of freedom. In our example, with 2 rows and 3 columns we have a total of $1 \times 2 = 2$ degrees of freedom.

For a contingency table, R represents the number of rows and C the number of columns.

Just Checking

Tiny black potato flea beetles can damage potato plants in a vegetable garden. These pests chew holes in the leaves, causing the plants to wither or die. They can be killed with an insecticide, but a canola oil spray has been suggested as a non-chemical "natural" method of controlling the beetles. To conduct an experiment to test the effectiveness of the natural spray, we gather 500 beetles and place them in three Plexiglas® containers. Two hundred beetles go in the first container, where we spray them with the canola oil mixture. Another 200 beetles go in the second container; we spray them with the insecticide. The remaining 100 beetles in the last container serve as a control group; we simply spray them with water. Then we wait 6 hours and count the number of surviving beetles in each container.

9. Why do we need the control group?

10. What would our null hypothesis be?

11. After the experiment is over, we could summarize the results in a table as shown. How many degrees of freedom does our χ^2 test have?

12. Suppose that, all together, 125 beetles survived. (That's the first-row total.) What's the expected count in the first cell—survivors among those sprayed with the natural spray?

13. If it turns out that only 40 of the beetles in the first container survived, what's the calculated component of χ^2 for that cell?

	Natural spray	Insecticide	Water	Total
Survived				
Died				
Total	200	200	100	500

(Check your answers on page 517.)

STEP–BY–STEP EXAMPLE A Chi-Square Test for a Two-Way Table

We have reports on 2533 traffic stops in Cincinnati.

Question: Are there race-based differences in whether or not police search vehicles?

THINK

Plan State what you want to know.

Identify the variables and check the W's.

Hypotheses State the null and alternative hypotheses.

Model Think about the assumptions and check the conditions.

State the sampling distribution model and name the test you will use.

I want to know whether or not police search a stopped vehicle is related to the race of the driver. I have data from 2533 traffic stops in Cincinnati.

H_O: The chances that a stopped vehicle is searched are the same regardless of the race of the driver.

H_A: Police are more likely to search cars driven by some races than others.

✔ **Counted Data Condition:** I have counts of traffic stops categorized by race and searches.

✔ **Independence Assumption:** It's reasonable to assume each traffic stop is independent of the others.

✗ **Randomization Condition:** These data come only from Cincinnati, so I won't be able to draw conclusions about other parts of the country.

✔ **Expected Cell Frequency Condition:** The expected counts (shown below) are all at least 5.

The conditions seem to be met, so I can use a χ^2 model with $(2 - 1) \times (3 - 1) = 2$ degrees of freedom and do a **chi-square two-way table test.**

SHOW

Mechanics Show the expected counts for each cell of the data table. You could make separate tables for the observed and expected counts, or put both counts in each cell as shown here. While observed counts must be whole numbers, expected counts rarely are—don't be tempted to round those off.

	Black	White	Other
No Search	787 / 889.60	594 / 493.17	27 / 25.58
Search	813 / 710.40	293 / 393.83	19 / 20.42

Calculate χ^2.

$$\chi^2 = \sum_{\text{All cells}} \frac{(Obs - Exp)^2}{Exp}$$
$$= \frac{(787 - 889.60)^2}{889.60}$$
$$+ \frac{(594 - 493.17)^2}{493.17} + \cdots$$
$$= 73.25$$

This χ^2 value is for greater than the critical value of 5.99, so the P-value is quite small.

P-value $= P(\chi^2 > 73.25) < 0.0001$

TELL

Conclusion State your conclusion in the context of the data.

The P-value is quite small, so I reject the null hypothesis. There's strong evidence that the chances police search a car they stop are not the same for drivers of different races.

If you find that simply rejecting the null hypothesis is a bit unsatisfying, you're in good company. Ok, so the races are treated differently. What we'd really like to know is what the differences are, where they're the greatest, and where they're smallest. The chi-square test doesn't answer these interesting questions, but it does provide some evidence that can help us Read on.

Examining the Components

Whenever we reject the null hypothesis, it's a good idea to examine the components. (We don't need to do that when we fail to reject because when the χ^2 value is small, all of its components must have been small.) The components give us a chance to think about the underlying patterns and to consider the ways in which the decision to search a vehicle may differ from race to race.

Here are the components for the Cincinnati traffic stop data:

	Black	White	Other
No search	11.76	20.70	0.08
Search	14.75	25.91	0.10

TABLE 21.2 A closer look. Components can help show where the table differs from the null hypothesis pattern.

We see right away that two of the components are quite small. The chances that police search stopped cars driven by drivers of "other" races (not Black or White) don't seem to differ much from what we'd expect. But the components for Black drivers and for White drivers are huge. To see what that means, we need to look back at the table that compares the observed counts to what we expected.

The largest component, 25.91, is in the cell for searches of cars with White drivers. The table in the Step-by-Step shows that there were only 293 such searches where the null hypothesis predicted over 100 more. Among Black drivers, there were 813 traffic stop searches, but the null hypothesis predicted only about 710. Our data suggest that, in Cincinnati, when police stop a car with a Black driver they are more likely to search the vehicle than they would be if the driver were White.

For Example Looking at χ^2 Components

Recap: Some people suggest that school children who are the older ones in their class naturally perform better in sports and therefore get more coaching and encouragement. To see if there's any evidence for this, we looked at major league baseball players born since 1975. A goodness-of-fit test found their birth months to have a distribution that's significantly different from the rest of us. The table shows the components.

Question: What's different about the distribution of birth months among major league ballplayers?

It appears that, compared to the general population, fewer ballplayers than expected were born in July and more than expected in August. Either month would make them the younger kids in their grades in school, so these data don't offer support for the conjecture that being older is an advantage in terms of a career as a pro athlete.

Month	Observed	Expected	Component
1	137	118.24	2.99
2	121	103.46	2.96
3	116	118.24	0.04
4	121	118.24	0.63
5	126	118.24	0.50
6	114	118.24	0.15
7	102	133.02	7.24
8	165	133.02	7.67
9	134	133.02	0.01
10	115	133.02	2.43
11	105	118.24	1.49
12	122	133.02	0.92

Tattoos and Hepatitis

A study from the University of Texas Southwestern Medical Center examined whether the risk of hepatitis C was related to whether people had tattoos and to where they got their tattoos. Hepatitis C causes about 10,000 deaths each year in the United States, but often lies undetected for years after infection.

The data from this study can be summarized in a two-way table, as follows:

	Hepatitis C	No Hepatitis C	Total
Tattoo, parlor	17	35	**52**
Tattoo, elsewhere	8	53	**61**
None	22	491	**513**
Total	**47**	**579**	**626**

TABLE 21.3 **Counts of patients** classified by their hepatitis C test status according to whether they had a tattoo from a tattoo parlor or from another source, or had no tattoo.

WHO	Patients being treated for non-blood-related disorders
WHAT	Tattoo status and hepatitis C status
WHEN	1991, 1992
WHERE	Texas

Are the chances of developing hepatitis related to one's tattoo status? Let's use technology to see.

TI Tips

Testing a two-way table

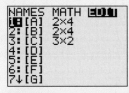

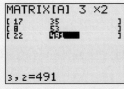

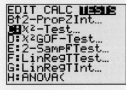

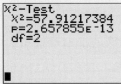

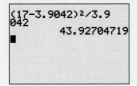

Yes, the TI will do chi-square two-way table tests. We'll show you how using the tattoo data. Here goes.

Stage 1: You need to enter the data as a matrix. A "matrix" is just a formal mathematical term for a table of numbers.

- Push the **MATRIX** button, and choose to **EDIT** matrix **[A]**.
- First specify the dimensions of the table, rows × columns.
- Enter the appropriate counts, one cell at a time. The calculator automatically asks for them row by row.

Stage 2: Do the test.
- In the **STAT TESTS** menu choose **C:χ^2-Test**.
- The TI now confirms that you have placed the observed frequencies in **[A]**. It also tells you that when it finds the expected frequencies it will store those in **[B]** for you. Now **Calculate** the mechanics of the test.

The TI reports a calculated value of $\chi^2 = 57.91$ and an exceptionally small P-value. That's strong evidence that getting hepatitis might be related to getting a tattoo. Maybe. Read on.

Stage 3: Check the expected counts.

- Go back to **MATRIX EDIT** and choose **[B]**.

Notice that two of the cells fail to meet the condition that expected counts be at least 5. This problem enters into our analysis and conclusions.

Stage 4: And now some bad news. There's no easy way to calculate the components. Look at the two matrices, **[A]** and **[B]**. Large components will happen when the corresponding entries differ greatly, especially when the expected count in **[B]** is small (because you will divide by that). The first cell is a good candidate, so we show you the calculation of its component.

A component of over 40 is pretty large—possibly an indication that you're more likely to get hepatitis in a tattoo parlor, but the expected count is smaller than 5. We're pretty sure that hepatitis status is not independent of having a tattoo, but we should be wary of saying anything more.

⃠ WHAT CAN GO WRONG?

- **Don't use chi-square methods unless you have counts.** All three of the chi-square tests apply only to counts. Other kinds of data can be arrayed in two-way tables. Just because numbers are in a two-way table doesn't make them suitable for chi-square analysis. Data reported as proportions or percentages can be suitable for chi-square procedures, *but only after they are converted to counts.* If you try to do the calculations without first finding the counts, your results will be wrong.

- **Don't interpret a small P-value as proof of causation.** Just as correlation between quantitative variables does not demonstrate causation, a failure of independence evidence of an association between two categorical

(continued)

variables does not show a cause-and-effect relationship between them, nor should we say that one variable *depends* on the other.

In our example, it's easy to imagine that lurking variables are responsible for the association between tattoos and hepatitis. Perhaps the lifestyles of some people include both tattoos and behaviors that put them at increased risk of hepatitis C, such as body piercings or even drug use. Even a small subpopulation of people with such a lifestyle among those with tattoos might be enough to create the observed result. After all, there were only 25 patients with both tattoos and hepatitis.

GOODNESS OF FIT **IN YOUR WORLD**

13 Really is the Unluckiest Lotto Number, so far!

Although the reverse applies in China where it is revered, the number 13 has been considered universally unlucky for anyone living in the [United Kingdom]. Although that may be a superstition dating back to an ancient association with the devil, it certainly appears to be relevant when it comes to the UK national lottery. That's because since the very first draw on November 19, 1994 the number 13 has been drawn the least number of times in total.

Indeed, as [of] July 29, 2009 the number 13 had been selected from the revolving lotto mixing ball a miserly total of 170 times; the lowest of all the 49 numbers entered into the draw lagging two selections behind the next least drawn number of 41. Also, for an amazing 49 consecutive draws between July and December 1997 it was a particularly unlucky number to choose on the play slip as it was not drawn once!

Conversely, the most frequently drawn number is 38, drawn a staggering 251 times during the same period. However, the national lottery is a game of pure chance, there is no skill involved. So, whether you pick 13 on your play slip or not the frequency with which numbers are drawn has no bearing on the probability of winning the top prize. Whatever selection of six numbers chosen the chances of winning the jackpot will still be 13,983,816 to one.

U.K. LOTTO HISTORY			
Number	Times Drawn	Number	Times Drawn
38	256	29	206
25	235	4	205
31	235	27	205
23	234	2	204
9	228	22	203
11	228	26	203
43	227	37	203
6	223	46	203
33	221	49	203
12	220	1	201
44	220	8	201
40	214	14	200
45	214	7	199
47	214	21	199
48	213	18	198
28	212	5	196
35	212	19	195
10	211	34	195
30	211	15	194
32	210	36	193
42	209	20	185
24	208	16	179
39	208	41	176
3	207	13	171
17	206		

Of course, except for silly superstitions there's nothing "unlucky" about the number 13. But are these results a bit strange? Could there be something unfair about the way the United Kingdom's lottery system works, making some numbers more likely to occur than others? The table shows how frequently each of the numbers 1–49 was drawn. Sure enough, 13 is a winner least often. (Our numbers don't quite match the counts in the previous story because we checked later that year after some more drawings had taken place.)

By now you know that randomness means we should expect some variation from perfection, but does the table show differences among the numbers that are so large we should be suspicious? Let's check, using a goodness-of-fit test.

- Our null hypothesis is that all 49 numbers are equally likely to be drawn.
- Our data are random and plentiful, so we can proceed with a chi-square analysis.
- Technology reports $\chi^2 = 56.37$ and a P-value of 0.19.

At first glance, the chi-square value may look large, but remember: there are 48 degrees of freedom. The P-value tells us that even if numbers are drawn in a way that makes them all equally likely, there's almost 1 chance in 5 of seeing discrepancies at least this large. That's no reason to suggest there's anything unfair about the UK lotto, or anything unlucky about the number 13.

WHAT HAVE WE LEARNED?

We've learned how to test hypotheses about categorical variables. We use two related methods. Both look at counts of data in categories and rely on chi-square models, a new family based in degrees of freedom.

▶ Goodness-of-fit tests compare the observed distribution of a single categorical variable to an expected distribution based on a theory or model.

▶ Two-way table tests compare the distributions of several groups for the same categorical variable, or examine counts from a single group for evidence of an association between two categorical variables.

We've seen that, mechanically, these tests are almost identical. If a table of counts deviates significantly from what we hypothesized, we reject the null hypothesis. When that happens we've learned to examine components in order to better understand patterns as in the table.

Terms

Cell	One element of a table corresponding to a specific row and a specific column.
Chi-square model	Chi-square models are skewed to the right, with critical values that depend on their degrees of freedom.
Critical value	We reject the null hypothesis when our calculated χ^2 is greater than a critical value. The larger the number of degrees of freedom, the greater that critical value.
Chi-square statistic	The chi-square statistic can be used to test whether the observed counts in a frequency distribution or two-way table match the counts we would expect according to some model. It is calculated as

$$\chi^2 = \sum_{\text{all cells}} \frac{(Obs - Exp)^2}{Exp}.$$

Chi-square test of goodness-of-fit	A test of whether the distribution of counts in one categorical variable matches the distribution predicted by a model. In a chi-square goodness-of-fit test, the expected counts come from the predicting model. The test finds a P-value from a chi-square model with $n - 1$ degrees of freedom, where n is the number of categories in the categorical variable.

(continued)

Two-way table	Each *cell* of a two-way table shows counts of individuals classified according to categorical variables.
Chi-square two-way table test	A test that checks for patterns in a two-way table. We find expected counts based on the margin totals. We find a P-value from a chi-square distribution with $(\#Rows - 1) \times (\#Cols - 1)$ degrees of freedom.
Chi-square component	The components of a chi-square calculation are

$$\frac{(Observed - Expected)^2}{Expected}$$

found for each cell of the table. When we reject a chi-square test, an examination of the components can sometimes reveal more about how the data deviate from the null model.

Skills

▶ Be able to recognize when a goodness-of-fit test or a two-way table test would be appropriate for a table of counts.

▶ Understand that the degrees of freedom for a chi-square test depend on the dimensions of the table and not on the sample size.

▶ Be able to display and interpret counts in a frequency table or two-way table.

▶ Know how to perform chi-square tests.

▶ Know how to compute a chi-square test using your statistics software or calculator.

▶ Be able to examine the components to explain the nature of the deviations from the null hypothesis.

TELL

▶ Know how to interpret the results of a chi-square test in a few sentences.

CHI-SQUARE ON THE COMPUTER

Software packages perform two-way table tests, and many also do goodness-of-fit tests. Some packages require the actual data, while others can also base the test on a table of summary counts.

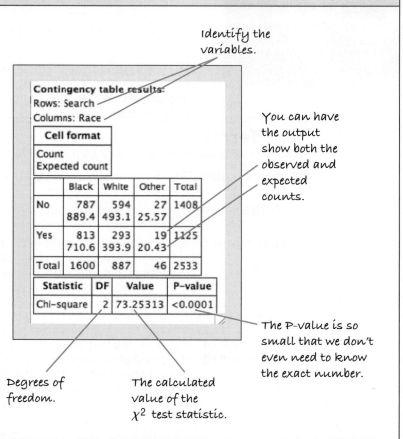

EXERCISES

A

1. **Dice.** After getting trounced by your little brother in a children's game, you suspect the die he gave you to roll may be unfair. To check, you plan to roll it 60 times, recording the number of times each face appears.
   a) What's your null hypothesis?
   b) Write the alternative hypothesis.

2. **Pi.** Many people know the mathematical constant π is approximately 3.14. But that's not exact. To be more precise, here are 20 decimal places: 3.14159265358979323846. Still not exact, though. In fact, the actual value is irrational, a decimal that goes on forever without any repeating pattern. But notice that there are no 0's and only one 7 in the 20 decimal places above. Does that pattern persist, or do all the digits show up with equal frequency? We'll look at the first million digits in the decimal representation of π.
   a) What's your null hypothesis?
   b) Write the alternative hypothesis.

3. **Dice, part II.** You're going to collect data from 60 rolls of the die in Exercise 1, counting the number of times each face appears. What are the expected values?

4. **Pi, part II.** Using a computer-generated list of the first million decimal digits of π, you'll count the number of times each of the digits 0, 1, 2, . . . , 9 appears. What are the expected values?

5. **Dice, part III.** To test the die in Exercise 1, you collect data from 60 rolls. Check the assumptions and conditions for analyzing the data with a chi-square goodness-of-fit test.

6. **Pi, part III.** To see if all digits are equally to appear in π, you look at the first million decimal places. Check the assumptions and conditions for analyzing the data with a chi-square goodness-of-fit test.

7. **Dice, part IV.** The table shows how many times the die in Exercise 1 actually showed each face in 60 rolls.

Face	Count
1	11
2	7
3	9
4	15
5	12
6	6

   a) How many degrees of freedom are there?
   b) Using the expected values you found in Exercise 3, what's the component of χ^2 for face 4?
   c) Show that $\chi^2 = 5.6$ for this table.

8. **Pi, part IV.** The table shows counts for the first million digits of π.

The first million digits of π	
Digit	**Count**
0	99,959
1	99,758
2	100,026
3	100,229
4	100,230
5	100,359
6	99,548
7	99,800
8	99,985
9	100,106

   a) How many degrees of freedom are there?
   b) Using the expected values you found in Exercise 4, what's the component of χ^2 for the digit 5?
   c) Show that $\chi^2 = 5.51$ for this table.

9. **Dice: the end!** In Exercise 1 we started examining your little brother's die for fairness. In Exercise 7 we saw the results from 60 rolls, and calculated $\chi^2 = 5.6$. What can we say about the die?
   a) The P-value = 0.35 for this test. What does that mean?
   b) State your conclusion about the die.

10. **Pi, at last!** In Exercise 2 we started wondering if some digits are more likely than others to show up in the decimal representation of π. In Exercise 8 we saw what happens in the first million digits, and calculated $\chi^2 = 5.51$.
    a) The P-value = 0.79 for this test. What does that mean?
    b) State your conclusion about π.

11. **Childbirth, part 1.** There is some concern that if a woman has an epidural to reduce pain during childbirth, the drug can get into the baby's bloodstream, making the baby sleepier and less willing to breastfeed. In December 2006, the *International Breastfeeding Journal* published results of a study conducted at Sydney University. Researchers followed up on 1178 births, noting whether the mother had an epidural and whether the baby was still nursing after 6 months. Here are their results:

Breastfeeding at 6 months?		Epidural? Yes	No	Total
	Yes	206	498	**704**
	No	190	284	**474**
	Total	**396**	**782**	**1178**

State the null and alternative hypotheses for a two-way table test.

12. Does your doctor know? A survey[1] of articles from the *New England Journal of Medicine* (*NEJM*) classified them based on whether statistics methods were used.

	Publication Year 1978–79	1989	2004–05	Total
No stats	90	14	40	**144**
Stats	242	101	271	**614**
Total	**332**	**115**	**311**	**758**

Has there been a change in the use of Statistics? State the null and alternative hypotheses for a two-way table test.

13. Childbirth, part 2. In Exercise 11, the table shows results of a study investigating whether aftereffects of epidurals administered during childbirth might interfere with successful breastfeeding. We're planning to do a chi-square test.
a) How many degrees of freedom are there?
b) The smallest expected count will be in the epidural/ no breastfeeding cell. What is it?
c) Check the assumptions and conditions for inference.

14. Does your doctor know? (part 2). The table in Exercise 12 shows whether *NEJM* medical articles during various time periods included statistics or not. We're planning to do a chi-square test.
a) How many degrees of freedom are there?
b) The smallest expected count will be in the 1989/No cell. What is it?
c) Check the assumptions and conditions for inference.

15. Childbirth, part 3. In Exercises 11 and 13, we've begun to examine the possible impact of epidurals on successful breastfeeding.
a) Calculate the component of chi-square for the epidural/no breastfeeding cell.
b) For this test, $\chi^2 = 14.87$ with P-value < 0.005. State your conclusion.

16. Does your doctor know? (part 3). In Exercises 12 and 14, we've begun to examine whether the use of statistics in *NEJM* medical articles has changed over time.
a) Calculate the component of chi-square for the 1989/No cell.
b) For this test, $\chi^2 = 25.28$ with P-value < 0.001. State your conclusion.

17. Childbirth, part 4. In Exercises 11, 13, and 15, we've tested a hypothesis about the impact of epidurals on successful breastfeeding. The tables show the expected counts and the test's components.

Expected		Epidural? Yes	No
Breastfeeding at 6 months?	Yes	236.66	467.34
	No	159.34	314.66

Components		Epidural? Yes	No
Breastfeeding at 6 months?	Yes	3.97	2.01
	No	5.90	2.99

What can you conclude from the components?

18. Does your doctor know? (part 4). In Exercises 12, 14, and 16, we've tested a hypothesis about whether the use of statistics in *NEJM* medical articles has changed over time. The tables show expected counts and the test's components.

Expected	1978–79	1989	2004–05
No stats	63.07	21.85	59.08
Stats	268.93	93.15	251.92

Components	1978–79	1989	2004–05
No stats	11.50	2.82	6.16
Stats	2.70	0.66	1.45

What can you conclude from the patterns in the components?

[1]Suzanne S. Switzer and Nicholas J. Horton, "What Your Doctor Should Know about Statistics (but Perhaps Doesn't)" *Chance*, 20:1, 2007.

B

T 19. Maryland lottery. In the Maryland Pick-3 Lottery, three random digits are drawn each day. A fair game depends on every value (0 to 9) being equally likely to show up in all three positions. If not, someone who detects a pattern could take advantage of that. Let's investigate. The table shows how many times each of the digits was drawn during a recent 32-week period, and some of them—4 and 7, for instance—seem to come up a lot. Could this just be a result of randomness, or is there evidence the digits aren't equally likely to occur?

Digit	Count
0	62
1	55
2	66
3	64
4	75
5	57
6	71
7	74
8	69
9	61

a) State the null and alternative hypotheses.
b) In all, 654 digits were drawn. Find the expected count for each.
c) Check the assumptions and conditions.
d) How many degrees of freedom are there?
e) Find the value of χ^2 for this test.
f) Are the results statistically significant? (Use the table on page 497, or the P-value.)
g) State your conclusion.

T 20. Stock market. Some investors believe that stock prices show weekly patterns, claiming for example that Fridays are more likely to be "up" days. From the trading sessions since October 1, 1928 we selected a random sample of 1000 days on which the Dow Jones Industrial Average (DJIA) showed a gain in stock prices. The table shows how many of these fell on each day of the week. Sure enough, more of them are Fridays—and Tuesday looks like a bad day to own stocks. Can this be explained as just randomness, or is there evidence here to help an investor?

Day of the week	Number of "up" days
Mon	192
Tues	189
Wed	202
Thu	199
Fri	218

a) State the null and alternative hypotheses.
b) Find the expected count for each day of the week.
c) Check the assumptions and conditions.
d) How many degrees of freedom are there?
e) Find the value of χ^2 for this test.
f) Are the results statistically significant? (Use the table on page 497, or the P-value.)
g) State your conclusion.

21. Nuts. A company says its premium mixture of nuts contains 10% Brazil nuts, 20% cashews, 20% almonds, and 10% hazelnuts, and the rest are peanuts. You buy a large can and separate the various kinds of nuts. Upon counting them, you find there are 112 Brazil nuts, 183 cashews, 207 almonds, 71 hazelnuts, and 446 peanuts. You wonder whether your mix is significantly different from what the company advertises.
a) State the null and alternative hypotheses.

b) Based on the company's stated percentages, how many nuts of each kind did you expect to get?
c) Check the assumptions and conditions.
d) How many degrees of freedom are there?
e) Find the value of χ^2 for this test.
f) Are the results statistically significant? (Use the table on page 497, or the P-value.)
g) State your conclusion.

22. M&M's. The Master-foods Company says that yellow candies make up 20% of its milk chocolate M&M's, red another 20%, and orange, blue, and green 10% each. The rest are brown. On his way home from work the day he was writing these exercises, one of the authors bought a bag of plain M&M's. He got 29 yellow ones, 23 red, 12 orange, 14 blue, 8 green, and 20 brown. Is this sample consistent with the company's stated proportions? Test an appropriate hypothesis and state your conclusion.
a) State the null and alternative hypotheses.
b) Based on the company's stated percentages, how many M&M's of each color did you expect to get?
c) Check the assumptions and conditions.
d) How many degrees of freedom are there?
e) Find the value of χ^2 for this test.
f) Are the results statistically significant? (Use the table on page 497, or the P-value.)
g) State your conclusion.

T 23. Titanic. Here is a table we first saw in Chapter 3 showing who survived the sinking of the *Titanic* based on whether they were crew members, or passengers booked in first-, second-, or third-class staterooms:

	Crew	First	Second	Third	Total
Alive	212	202	118	178	710
Dead	673	123	167	528	1491
Total	885	325	285	706	2201

a) If we draw an individual at random, what's the probability that we will draw a member of the crew?
b) What's the probability of randomly selecting a third-class passenger who survived?
c) What's the probability of a randomly selected passenger surviving, given that the passenger was a first-class passenger?
d) If someone's chances of surviving were the same regardless of their status on the ship, how many members of the crew would you expect to have lived?
e) State the null and alternative hypotheses.
f) Give the degrees of freedom for the test.
g) The chi-square value for the table is 187.8, and the corresponding P-value is barely greater than 0. State your conclusions about the hypotheses.

T 24. NYPD and sex discrimination. The following table shows the rank attained by male and female officers in the New York City Police Department (NYPD).

Do these data indicate that men and women are equitably represented at all levels of the department?

Rank	Male	Female
Officer	21,900	4,281
Detective	4,058	806
Sergeant	3,898	415
Lieutenant	1,333	89
Captain	359	12
Higher ranks	218	10

a) What's the probability that a person selected at random from the NYPD is a female?
b) What's the probability that a person selected at random from the NYPD is a detective?
c) Assuming no bias in promotions, how many female detectives would you expect the NYPD to have?
d) State the hypotheses.
e) Check the conditions.
f) How many degrees of freedom are there?
g) The chi-square value for the table is 290.1 and the P-value is less than 0.0001. State your conclusion about the hypotheses.

25. Titanic again. Examine and comment on these tables of the expected counts and components for the chi-square test you looked at in Exercise 23.

Expected	Crew	First	Second	Third
Alive	285.48	104.84	91.94	227.74
Dead	599.50	220.16	193.06	478.26

Components	Crew	First	Second	Third
Alive	18.91	90.05	7.39	10.86
Dead	9.01	42.88	3.52	5.17

26. NYPD again. Examine and comment on these tables of the expected counts and components for the chi-square test you looked at in Exercise 24.

Expected	Male	Female
Officer	22,250	3,931.5
Detective	4,133.6	730.4
Sergeant	3,655.3	647.7
Lieutenant	1,208.5	213.5
Captain	315.3	55.7
Higher ranks	193.8	34.2

Components	Male	Female
Officer	5.49	31.08
Detective	1.38	7.82
Sergeant	14.77	83.58
Lieutenant	12.83	72.63
Captain	6.06	34.30
Higher ranks	3.03	17.16

27. Cranberry juice. It's common folk wisdom that drinking cranberry juice can help prevent urinary tract infections in women. In 2001 the *British Medical Journal* reported the results of a Finnish study in which three groups of 50 women were monitored for these infections over 6 months. One group drank cranberry juice daily, another group drank a lactobacillus drink, and the third drank neither of those beverages, serving as a control group. In the control group, 18 women developed at least one infection, compared to 20 of those who consumed the lactobacillus drink and only 8 of those who drank cranberry juice. Does this study provide supporting evidence for the value of cranberry juice in warding off urinary tract infections?
a) Is this a survey, a retrospective study, a prospective study, or an experiment? Explain.
b) State the hypotheses.
c) Check the conditions.
d) How many degrees of freedom are there?
e) Find χ^2 and the P-value.
f) State your conclusion.

28. Cars. A random survey of autos parked in the student lot and the staff lot at a large university classified the brands by country of origin, as seen in the table. Are there differences in the national origins of cars driven by students and staff?

Origin	Driver	
	Student	Staff
American	107	105
European	33	12
Asian	55	47

a) Write appropriate hypotheses.
b) Check the necessary assumptions and conditions.
c) How many degrees of freedom are there?
d) Find χ^2 and the P-value of your test.
e) State your conclusion.

29. Montana. A poll conducted by the University of Montana classified respondents by whether they were male or female and political party, as shown in the table. We wonder if there is evidence of an association between being male or female and party affiliation.

	Democrat	Republican	Independent
Male	36	45	24
Female	48	33	16

a) Write an appropriate hypothesis.
b) Are the conditions for inference satisfied?
c) How many degrees of freedom are there?
d) Find χ^2 and the P-value for your test.
e) State your conclusion.

30. Fish diet. Medical researchers followed 6272 Swedish men for 30 years to see if there was any association between the amount of fish in their diet and prostate cancer. ("Fatty Fish Consumption and Risk of Prostate Cancer," *Lancet*, June 2001)

	Prostate Cancer	
Fish Consumption	**Yes**	**No**
Never/seldom	14	110
Small part of diet	201	2420
Moderate part	209	2769
Large part	42	507

a) Is this a survey, a retrospective study, a prospective study, or an experiment? Explain.
b) Write appropriate hypotheses.
c) Check the assumptions and conditions.
d) Find the value of χ^2 and the P-value for this test.
e) State your conclusion about diet and prostate cancer.
f) Does this study prove that eating fish does not prevent prostate cancer? Explain.

31. Montana revisited. The poll described in Exercise 29 also investigated the respondents' party affiliations based on what area of the state they lived in. Does this indicate significant geographical differences in political affiliation?

	Democrat	Republican	Independent
West	39	17	12
Northeast	15	30	12
Southeast	30	31	16

a) Write appropriate hypotheses.
b) Check the assumptions and conditions.
c) Find the value of χ^2 and the P-value for this test.
d) State your conclusion about Montana politics.

32. Working parents. In July 1991 and again in April 2001, the Gallup Poll asked random samples of 1015 adults about their opinions on working parents. The table summarizes responses to the question "Considering the needs of both parents and children, which of the following do you see as the ideal family in today's

society?" Based on these results, do you think there was a change in people's attitudes during the 10 years between these polls?

	1991	**2001**
Both work full time	142	131
One works full time, other part time	274	244
One works, other works at home	152	173
One works, other stays home for kids	396	416
No opinion	51	51

a) Is this a survey, a retrospective study, a prospective study, or an experiment? Explain.
b) Write appropriate hypotheses.
c) Check the assumptions and conditions.
d) Find the value of χ^2 and the P-value for this test.
e) State your conclusion about any change in attitudes.

33. Fruit flies. Offspring of certain fruit flies may have yellow or ebony bodies and normal wings or short wings. Genetic theory predicts that these traits will appear in the ratio 9:3:3:1 (9 yellow, normal: 3 yellow, short: 3 ebony, normal: 1 ebony, short). A researcher checks 100 such flies and finds the distribution of the traits to be 59, 20, 11, and 10, respectively. Are the results this researcher observed consistent with the theoretical distribution predicted by the genetic model?

34. Violence against women 2005. In its study *When Men Murder Women*, the Violence Policy Center (www.vpc.org) reported that 1857 women were murdered by men in 2005. Of these victims, a weapon could be identified for 1752 of them. Of those for whom a weapon could be identified, 966 were killed by guns, 390 by knives or other cutting instruments, 136 by other weapons, and 260 by personal attack (battery, strangulation, etc.). The FBI's Uniform Crime Report says that, among all murders nationwide, the weapon use rates were as follows: guns 63.4%, knives 13.1%, other weapons 16.8%, personal attack 6.7%. Is there evidence that violence against women involves different weapons than other violent attacks in the United States?

35. Racial steering. A subtle form of racial discrimination in housing is "racial steering." Racial steering occurs when real estate agents show prospective buyers only homes in neighborhoods already dominated by that family's race. This violates the Fair Housing Act of 1968. According to an article in *Chance* magazine (Vol. 14, no. 2 [2001]), tenants at a large apartment complex recently filed a lawsuit alleging racial steering. The complex is divided into two parts: Section A and Section B. The plaintiffs claimed that white potential renters were steered to Section A, while African-Americans were steered to Section B. The table displays the data that were presented in court to show the locations of recently rented apartments. Do you think there is evidence of racial steering?

New Renters

	White	Black	Total
Section A	87	8	95
Section B	83	34	117
Total	170	42	212

36. *Titanic*, **redux.** Newspaper headlines at the time, and traditional wisdom in the succeeding decades, have held that women and children escaped the *Titanic* in greater proportions than men. Here's a table with the relevant data. Do you think that survival was independent of whether the person was male or female? Explain.

	Female	Male	Total
Alive	343	367	710
Dead	127	1364	1491
Total	470	1731	2201

37. Pregnancies. Most pregnancies result in live births, but some end in miscarriages or stillbirths. A June 2001 National Vital Statistics Report examined those outcomes in the United States during 1997, broken down by the age of the mother. The table shows counts consistent with that report. Is there evidence that the distribution of outcomes is not the same for these age groups?

Age of Mother	Live Births	Fetal Losses
Under 20	49	13
20–29	201	41
30–34	88	21
35 or over	49	21

38. Education by age. Use the survey results in the table to investigate differences in education level attained among different age groups in the United States.

Education level	Age Group 25–34	35–44	45–54	55–64	≥65
Not HS grad	27	50	52	71	101
HS	82	19	88	83	59
1–3 years college	43	56	26	20	20
≥4 years college	48	75	34	26	20

39. Grades. Two different professors teach an introductory Statistics course. The table shows the distribution of final grades they reported. We wonder whether one of these professors is an "easier" grader.

	Prof. Alpha	Prof. Beta
A	3	9
B	11	12
C	14	8
D	9	2
F	3	1

a) Write appropriate hypotheses.
b) Find the expected counts for each cell, and explain why the chi-square procedures are not appropriate.
c) Here we create a new table displaying the same data, but combining D's and F's as "Below C":

	Prof. Alpha	Prof. Beta
A	3	9
B	11	12
C	14	8
Below C	12	3

Find the expected counts for each cell in this new table, and explain why a chi-square procedure is now appropriate.
d) With this change in the table, what has happened to the number of degrees of freedom?
e) Test your hypothesis about the two professors, and state an appropriate conclusion.

40. Full moon. Some people believe that a full moon elicits unusual behavior in people. The table shows the number of arrests made in a small town during weeks of six full moons and six other randomly selected weeks in the same year. We wonder if there is evidence of a difference in the types of illegal activity that take place.

	Full Moon	Not Full
Violent (murder, assault, rape, etc.)	2	3
Property (burglary, vandalism, etc.)	17	21
Drugs/Alcohol	27	19
Domestic abuse	11	14
Other offenses	9	6

a) Write appropriate hypotheses.
b) Find the expected counts for each cell, and explain why the chi-square procedures are not appropriate.
c) Find a sensible way to combine some categories that will make the expected counts acceptable.
d) Test a hypothesis about the full moon and state your conclusion.

Answers

Just Checking

1. 7.75, 23.25, 23.25, 69.75

2. These are counts; student traits are independent of each other; although not a random sample, these Biology students should be representative of all students; 124 is fewer than 10% of all students; all expected counts are at least 5.

3. H_0: The traits occur in the predicted proportions 1:3:3:9.
   H_A: The proportions of some of the traits are not as predicted.

4. 3 df

5. 1.66

6. No, don't reject H_0. (6.64 < 7.82)

7. There's no evidence to suggest the genetic theory is incorrect.

8. $\chi^2 = 2.33 + 0.07 + 2.58 + 1.66 = 6.64$

Obs	Exp	(Obs − Exp)	(Obs − Exp^2)	$\dfrac{(Obs − Exp)^2}{Exp}$
12	7.75	4.25	18.0625	2.33
22	23.25	0.75	1.5625	0.07
31	23.25	7.75	60.0625	2.58
59	69.75	−10.75	115.5625	1.66

9. We need to know how well beetles can survive 6 hours in a Plexiglas® box so that we have a baseline to compare the treatments.

10. There's no difference in survival rate in the three groups.

11. $(2 − 1)(3 − 1) = 2\ df$

12. 50

13. 2

Selected Formulas

$Range = Max - Min$

$IQR = Q3 - Q1$

Outlier Rule-of-Thumb: $y < Q1 - 1.5 \times IQR \quad$ or $\quad y > Q3 + 1.5 \times IQR$

$$\bar{y} = \frac{\sum y}{n}$$

$$s = \sqrt{\frac{\sum (y - \bar{y})^2}{n - 1}}$$

$$z = \frac{y - \mu}{\sigma} \text{ (model based)}$$

$$z = \frac{y - \bar{y}}{s} \text{ (data based)}$$

$$r = \frac{\sum z_x z_y}{n - 1}$$

$$\hat{y} = a + bx \qquad \text{where } b = \frac{rs_y}{s_x} \text{ and } a = \bar{y} - b\bar{x}$$

$$P(\mathbf{A}) = 1 - P(\mathbf{A^C})$$

$$P(\mathbf{A} \cup \mathbf{B}) = P(\mathbf{A}) + P(\mathbf{B}) - P(\mathbf{A} \cap \mathbf{B})$$

$$P(\mathbf{A} \cap \mathbf{B}) = P(\mathbf{A}) \times P(\mathbf{B}|\mathbf{A})$$

$$P(\mathbf{B}|\mathbf{A}) = \frac{P(\mathbf{A} \cap \mathbf{B})}{P(\mathbf{A})}$$

If $\mathbf{A}$ and $\mathbf{B}$ are independent, $P(\mathbf{B}|\mathbf{A}) = P(\mathbf{B})$

$E(X) = \mu = \sum x \cdot P(x)$

$Var(X) = \sigma^2 = \sum (x - \mu)^2 P(x)$

Binomial: $P(x) = {}_nC_x \, p^x q^{n-x} \qquad \mu = np \qquad \sigma = \sqrt{npq}$

Sampling distribution of sample proportion $\hat{p} = \dfrac{x}{n}$:

As n grows, the sampling distribution approaches the Normal model with

$$\mu(\hat{p}) = p \qquad SD(\hat{p}) = \sqrt{\dfrac{pq}{n}}$$

Sampling distribution of sample mean $\bar{y}$:

(CLT) As n grows, the sampling distribution approaches the Normal model with

$$\mu(\bar{y}) = \mu_y \qquad SD(\bar{y}) = \dfrac{\sigma}{\sqrt{n}}$$

Inference:

Confidence interval for parameter = ***statistic ± critical value × SD(statistic)***

$$\text{Test statistic} = \dfrac{Statistic - Parameter}{SD(statistic)}$$

Parameter	Statistic	SD(statistic)	SE(statistic)
p	$\hat{p}$	$\sqrt{\dfrac{pq}{n}}$	$\sqrt{\dfrac{\hat{p}\hat{q}}{n}}$
μ	$\bar{y}$	$\dfrac{\sigma}{\sqrt{n}}$	$\dfrac{s}{\sqrt{n}}$
μ_d	$\bar{d}$	$\dfrac{\sigma_d}{\sqrt{n}}$	$\dfrac{s_d}{\sqrt{n}}$
$p_1 - p_2$	$\hat{p}_1 - \hat{p}_2$	$\sqrt{\dfrac{p_1 q_1}{n_1} + \dfrac{p_2 q_2}{n_2}}$	$\sqrt{\dfrac{\hat{p}_1\hat{q}_1}{n_1} + \dfrac{\hat{p}_2\hat{q}_2}{n_2}}$
$\mu_1 - \mu_2$	$\bar{y}_1 - \bar{y}_2$	$\sqrt{\dfrac{\sigma_1^2}{n_1} + \dfrac{\sigma_2^2}{n_2}}$	$\sqrt{\dfrac{s_1^2}{n_1} + \dfrac{s_2^2}{n_2}}$

Chi-square: $\chi^2 = \sum \dfrac{(obs - exp)^2}{exp}$

Guide to Statistical Software

Chapter 3. Stories Categorical Data Tell

STATCRUNCH

To make a bar chart or pie chart

1. Click on **Graphics**.

2. Choose the type of plot **»** **with data** or **»** **with summary**.

3. Choose the variable name from the list of **Columns**; if using summaries, also choose the counts.

4. Click on **Next**.

5. Choose **Frequency/Counts** or (usually) **Relative frequency/Percents**. Note that you may elect to group categories under a specified percentage as "Other."

6. Click on **Create Graph**.

TI-NSPIRE

The TI-Nspire Handheld does not display plots for categorical variables.

TI-89

The TI-89 won't do displays for categorical variables.

EXCEL

First make a pivot table (Excel's name for a frequency table). From the **Data** menu, choose **Pivot Table** and **Pivot Chart Report**. When you reach the Layout window, drag your variable to the row area and drag your variable again to the data area. This tells Excel to count the occurrences of each category.

Once you have an Excel pivot table, you can construct bar charts and pie charts.

Click inside the Pivot Table.

Click the Pivot Table Chart Wizard button. Excel creates a bar chart.

A longer path leads to a pie chart; see your Excel documentation.

COMMENTS

Excel uses the pivot table to specify the category names and find counts within each category. If you already have that information, you can proceed directly to the Chart Wizard.

EXCEL 2007

To make a bar chart

▶ Select the variable in Excel you want to work with.

▶ Choose the **Column** command from the Insert tab in the Ribbon.

▶ Select the appropriate chart from the drop-down dialog.

To change the bar chart into a pie chart

▶ Right-click the chart and select **Change Chart Type...** from the menu. The Chart type dialog opens.

▶ Select a pie chart type.

▶ Click the **OK** button. Excel changes your bar chart into a pie chart.

Chapter 4. Exploring Quantitative Data

STATCRUNCH

To make a histogram, dotplot, or stem-and-leaf plot

1. Click on **Graphics**.
2. Choose the type of plot.
3. Choose the variable name from the list of **Columns**.
4. Click on **Next**.
5. (For a histogram) Choose **Frequency** or (usually) **Relative frequency**, and (if desired) set the axis scale by entering the **Start** value and **Bin width**.
6. Click on **Create Graph**.

To calculate summaries

1. Click on **Stat**.
2. Choose **Summary Stats » Columns**.
3. Choose the variable name from the list of **Columns**.
4. Click on **Calculate**.

COMMENTS

▶ You may need to hold down the control or command key to choose more than one variable to summarize.

▶ Before calculating, click on **Next** to choose additional summary statistics.

TI-NSPIRE

To plot a histogram using a named list, press ▲ several times so that the entire list is highlighted. Press (menu), ③ for Data, and ④ for Quick Graph. Then press (menu), ① for Plot Type, and ③ for Histogram.

To create the plot on a full page, press (⌂), and then ⑤ for Data & Statistics. Move the cursor to "Click to add variable," and then press (🔅) and select the list name. Then press (menu), ① for Plot Type, and ③ for Histogram.

To compute summary statistics using a named list, press (⌂), ① for Calculator, (menu), ⑥ for Statistics, ① for Stat Calculations, and ① for One-Variable Statistics. Complete the dialog boxes.

TI-89

To make a histogram

▶ Select [F2] **(Plots)**, then 1: **Plot Setup**. Select a plot and press [F1] to define it.

▶ Select plot type 4: **Histogram**. Use VAR-LINK to select the data list.

▶ Enter a number for the histogram bucket (bar) width.

▶ Press [ENTER] to complete the plot definition. Pres [F5] to display the histogram.

▶ Press ◆[F2] to adjust the window appropriately, then press ◆[F3] **(Graph)**.

To calculate summary statistics

▶ To compute summary statistics, press [F4] **(Calc)**. Input the name of the list using VAR-LINK. Press [ENTER].

▶ Use the down arrow to scroll through the output.

▶ To create a boxplot, press [F2] **(Plots)** then [ENTER]. Select a plot to define and press [F1]. Select either 3: **Box Plot** or

4: **Mod Box Plot** (to identify outliers). Select the mark type of your choice (for outliers). Press [ENTER] to finish.

▶ Press [F5] to display the graph.

COMMENTS

If the data are stored as a frequency table (say, with data values in list1 and frequencies in list2), change Use Freq and Categories to YES and use VAR-LINK to select list2 as the frequency variable on the plot definition screen.

If the data are stored as a frequency table (say, with data values in list1 and frequencies in list2), use VAR-LINK to select list2 as the frequency variable in 1-Var Stats.

For the plot, change Use Freq and Categories to YES and use VAR-LINK to select list2 as the frequency variable on the plot definition screen.

EXCEL

Excel cannot make histograms or dotplots without a third-party add-in.

To calculate summaries

Click on an empty cell. Type an equal sign and choose "**Average**" from the popup list of functions that appears to the left of the text-editing box. Enter the data range in the box that says "**Number 1.**" Click the **OK** button.

COMMENTS

Excel's Data Analysis add-in does offer something called a histogram, but it just makes a crude frequency table, and the Chart Wizard cannot then create a statistically appropriate histogram. The DDXL add-in provided on our DVD adds these and other capabilities to Excel.

To compute the standard deviation of a column of data directly, use the **STDEV** from the popup list of functions in the same way.

Excel's STDEV function should not be used for data values larger in magnitude than 100,000 or for lists of more than a few thousand values. It is programmed with an unstable formula that can generate rounding errors when these limits are exceeded.

EXCEL 2007

In Excel 2007 there is another way to find some of the standard summary statistics. For example, to compute the mean
▶ Click on an empty cell.
▶ Go to the Formulas tab in the Ribbon. Click on the drop-down arrow next to "AutoSum" and choose "**Average**".
▶ Enter the data range in the formula displayed in the empty box you selected earlier.
▶ Press **Enter.** This computes the mean for the values in that range.

To compute the standard deviation:
▶ Click on an empty cell.
▶ Go to the Formulas tab in the Ribbon and click the drop-down arrow next to "AutoSum" and select "**More functions. . ."**
▶ In the dialog window that opens, select "**STDEV**" from the list of functions and click **OK**. A new dialog window opens. Enter a range of fields into the text fields and click **OK**.

Excel 2007 computes the standard deviation for the values in that range and places it in the specified cell of the spreadsheet.

Chapter 5. Stories Quantitative Data Tell

STATCRUNCH

To make a boxplot
1. Click on **Graphics**.
2. Choose **Boxplot**.
3. Choose the variable name from the list of **Columns**.
4. Click on **Next**.
5. Indicate that you want to **identify outliers**.
6. Click on **Create Graph**.

To make side-by-side boxplots
1. Click on **Graphics**.
2. Choose **Boxplot**.
3. Choose the variable name from the list of **Columns**.
4. Choose the column that holds the categories to **Group by**.
5. Indicate that you want to **Plot groups for each column**.
6. Click on **Next**.
7. Indicate that you want to **identify outliers**.
8. Click on **Create Graph**.

TI-NSPIRE

To create a box plot using a named list, press ▲ several times so that the entire list is highlighted. Press (menu), ③ for Data, and ④ for Quick Graph. Then press (menu), ① for Plot Type, and ② for Box Plot.

To create the plot on a full page, press (⌂), and then press ⑤ for Data & Statistics. Move the cursor to "Click to add variable," and then press (✕) and select the list name. Then press (menu), ① for Plot Type, and ② for Box Plot.

TI-89

For the plot, change Use Freq and Categories to YES and use VAR-LINK to select list2 as the frequency variable on the plot definition screen.

To create a boxplot, press [F2] **(Plots)**, then [ENTER]. Select a plot to define and press [F1]. Select either 3: **Box Plot** or 4: **Mod Box Plot** (to identify outliers). Select the mark type of your choice (for outliers). Press [ENTER] to finish.
Press [F5] to display the graph.

EXCEL

Excel cannot make boxplots.

COMMENT

The DDXL add-on provided on the DVD adds the ability to make boxplots to Excel.

Chapter 6. What's Normal?

STATCRUNCH

To work with Normal percentiles

1. Click on **Stat**.
2. Choose **Calculators » Normal**.
3. Choose a lower tail (<=) or upper tail (>=) region.

4. Enter the z-score cutoff, then click on **Compute** to find the probability.

OR

Enter the desired probability, then click on **Compute** to find the z-score cutoff.

TI-NSPIRE

To compute the area under a normal curve, press (menu), ① for Calculator, (menu), ⑤ for Probability, ⑤ for Distributions, and ② for Normal Cdf. Complete the dialog box.

To compute the value for a given percentile, press (menu), ① for Calculator, (menu), ⑤ for Probability, ⑤ for Distributions, ③ for Inverse Normal. Complete the dialog box.

TI-89

▶ To find what percent of a Normal model lies between two z-scores, press F5 **(Distr)**. Then select 4: **Normal Cdf**. Enter the lower and upper z-scores, specify mean 0 and standard deviation 1, and press ENTER.

▶ To find the z-score for a given percentile, press F5 **(Distr)**. Then arrow down to 2: **Inverse** press the right arrow to see the sub menu and select 1: **Inverse Normal**. Enter the area to the left of the desired point, mean 0 and standard deviation 1, and press ENTER.

COMMENTS

Normal models strictly go to infinity on either end, which is 1EE99 on the calculator. In practice, any "large" number will work. For example, the percentage of the Normal model over two standard deviations above the mean can use Lower Value 2 and Upper Value 99. To find area more than 2 standard deviations below the mean, use Lower Value −99, and Upper value −2.

EXCEL

Excel does not calculate Normal probabilities.

Chapter 7. A Tale of Two Variables

STATCRUNCH

To make a scatterplot

1. Click on **Graphics.**
2. Choose **Scatter Plot**.
3. Choose X and Y variable names from the list of **Columns**.
4. Click on **Create Graph.**

To find a correlation

1. Click on **Stat**.
2. Choose **Summary Stats » Correlation**.
3. Choose two variable names from the list of **Columns**. (You may need to hold down the control or command key to choose the second one.)
4. Click on **Calculate**.

TI-NSPIRE

To create a scatterplot using named lists, press ▲ several times so that the first list is highlighted. Then press ⊙ ▸ so that the second list is highlighted. Press (menu), ③ for Data, and ④ for Quick Graph.

To create the plot on a full page, press (ctrl), then ⑤ for Data & Statistics. Move the cursor to "Click to add variable," and then press (click) and select the list name. Repeat for the other axis.

To find the correlation, press (ctrl), ① for Calculator, (menu), ⑥ for Statistics, ① for Stat Calculations, and ④ for Linear Regression. Complete the dialog boxes.

TI-89

To create a scatterplot, press F2 **(Plots)**. Select choice 1: **Plot Setup.** Select a plot to define and press F1. Select **Plot Type 1: Scatter.** Select a mark type. Specify the lists where the data are stored as Xlist and Ylist, using VAR-LINK. Press ENTER to finish. Press F5 to display the plot.

To find the correlation, press F4 **(CALC)**, then arrow to **3: Regressions,** press the right arrow, and select **1:LinReg(a+bx).** Then specify the lists where the data are stored. You can also select a *y*-function to store the equation of the line.

COMMENTS

Notice that if you **TRACE** (press F3) the scatterplot, the calculator will tell you the *x*- and *y*-value at each point.

EXCEL

To make a Scatterplot with the Excel Chart Wizard
► Click on the **Chart Wizard** Button in the menu bar. Excel opens the Chart Wizard's Chart Type Dialog window.

► Make sure the **Standard Types** tab is selected, and select **XY (Scatter)** from the choices offered.

► Specify the **scatterplot without lines** from the choices offered in the Chart subtype selections. The **Next** button takes you to the Chart Source Data dialog.

► If it is not already frontmost, click on the **Data Range** tab, and enter the data range in the space provided.

► By convention, we always represent variables in columns. The Chart Wizard refers to variables as Series. Be sure the **Column** option is selected.

► Excel places the leftmost column of those you select on the *x*-axis of the scatterplot. If the column you wish to see on the *x*-axis is not the leftmost column in your spreadsheet, click on the **Series** tab and edit the specification of the individual axis series.

► Click the **Next** button. The Chart Options dialog appears.

► Select the **Titles** tab. Here you specify the title of the chart and names of the variables displayed on each axis.

► Type the chart title in the **Chart title:** edit box.

► Type the *x*-axis variable name in the **Value (X) Axis:** edit box. Note that you must name the columns correctly here. Naming another variable will not alter the plot, only mislabel it.

► Type the *y*-axis variable name in the **Value (Y) Axis:** edit box.

► Click the **Next** button to open the chart location dialog.

► Select the **As new sheet:** option button.

► Click the **Finish** button.

Often, the resulting scatterplot will not be useful. By default, Excel includes the origin in the plot even when the data are far from zero. You can adjust the axis scales.
To change the scale of a plot axis in Excel:
► Double-click on the axis. The **Format Axis Dialog** appears.

► If the **scale tab** is not the frontmost, select it.

► Enter new minimum or new maximum values in the spaces provided. You can drag the dialog box over the scatterplot as a straightedge to help you read the maximum and minimum values on the axes.

► Click the **OK** button to view the rescaled scatterplot.

► Follow the same steps for the *x*-axis scale.

Compute a correlation in Excel with the **CORREL** function from the drop-down menu of functions. If CORREL is not on the menu, choose **More Functions** and find it among the statistical functions in the browser.
In the dialog that pops up, enter the range of cells holding one of the variables in the space provided.
Enter the range of cells for the other variable in the space provided.

EXCEL 2007

To make a scatterplot in Excel 2007
► Select the columns of data to use in the scatterplot. You can select more than one column by holding down the control key while clicking.

► In the Insert tab, click on the **Scatter** button and select the **Scatter with only Markers** chart from the menu.

Unfortunately, the plot this creates is often statistically useless. To make the plot useful, we need to change the display
► With the chart selected click on the **Gridlines** button in the Layout tab to cause the Chart Tools tab to appear.

► Within Primary Horizontal Gridlines, select **None.** This will remove the gridlines from the scatterplot.

► To change the axis scaling, click on the numbers of each axis of the chart, and click on the **Format Selection** button in the Layout tab.

► Select the **Fixed** option instead of the Auto option, and type a value more suited for the scatterplot. You can use the

pop-up dialog window as a straightedge to approximate the appropriate values.

Excel 2007 automatically places the leftmost of the two columns you select on the *x*-axis, and the rightmost one on the *y*-axis. If that's not what you'd prefer for your plot, you'll want to switch them.
To switch the X- and Y-variables
► Click the chart to access the **Chart Tools** tabs.

► Click on the **Select Data** button in the Design tab.

► In the pop-up window's Legend Entries box, click on **Edit.**

► Highlight and delete everything in the Series X Values line, and select new data from the spreadsheet. (Note that selecting the column would inadvertently select the title of the column, which would not work well here.)

► Do the same with the Series Y Values line.

► Press **OK**, then press **OK** again.

Chapter 8. What's My Line?

STATCRUNCH

To compute a regression

1. Click on **Stat**.
2. Choose **Regression » Simple Linear**.
3. Choose X and Y variable names from the list of columns.
4. Click on **Next** (twice) to **Plot the fitted line** on the scatterplot.
5. Click on **Calculate** to see the regression analysis.
6. Click on **Next** to see the scatterplot.

COMMENTS

Remember to check the scatterplot to be sure a linear model is appropriate.

Note that before you **Calculate** clicking on **Next** also allows you to:

► enter an X-value for which you want to find the predicted Y-value;
► save all the fitted values;
► save the residuals;
► ask for a residuals plot.

TI-NSPIRE

To plot and find the equation of the regression line, first create a scatterplot. Using named lists, press ▲ several times so that the first list is highlighted. Then press ⑤ ▶ so that the second list is highlighted. Press ⑩, ③ for Data, and ④ for Quick Graph. Then press ⑩, ③ for Actions, ⑤ for Regression, and ② for Show Linear.

To find the equation of the regression line on a full page, press ⑩, ① for Calculator, ⑩, ⑥ for Statistics, ① for Stat Calculations, and ④ for Linear Regression. Complete the dialog boxes.

To see the plot on a full page, press ⑩, and then ⑤ for Data & Statistics. Move the cursor to "Click to add variable," and then press ⑧ and select the list name. Repeat for the other axis. Then press ⑩, ③ for Actions, ⑤ for Regression, and ② for Show Linear.

TI-89

To find the equation of the regression line (and add the line to a scatterplot), choose **LinReg (a+bx)** from the **Calc Regressions** menu and tell it the list names and a function to store the equation. To make a residuals plot, define a **PLOT** as a scatterplot. Specify your explanatory datalist as Xlist. For Ylist, find the list name **resid** from VAR-LINK by arrowing to the **STATVARS** portion, then press ②, **(r)** and locate the list. Press [ENTER] to finish the plot definition and [F5] to display the plot.

COMMENTS

Each time you execute a **LinReg** command, the calculator automatically computes the residuals and stores them in a data list named RESID.

EXCEL

Make a scatterplot of the data. With the scatterplot front-most, select **Add Trendline...** from the **Chart** menu. Click the **Options** tab and select **Display Equation on Chart**. Click **OK**.

COMMENTS

The computer section for Chapter 7 shows how to make a scatterplot. We don't repeat those steps here.

EXCEL 2007

► Click on a blank cell in the spreadsheet.
► Go to the **Formulas** tab in the Ribbon and click **More Functions → Statistical**.
► Choose the **CORREL** function from the drop-down menu of functions.
► In the dialog that pops up, enter the range of one of the variables in the space provided.
► Enter the range of the other variable in the space provided.
► Click **OK**.

COMMENTS

The correlation is computed in the selected cell. Correlations computed this way will update if any of the data values are changed.

Before you interpret a correlation coefficient, always make a scatterplot to check for nonlinearity and outliers. If the variables are not linearly related, the correlation coefficient cannot be interpreted.

Chapter 9. What's My Curve?

STATCRUNCH

StatCrunch does not fit curved models.

TI-NSPIRE

To plot and find the equation of a curved model, first create a scatterplot. Using named lists, press ▲ several times so that the first list is highlighted. Then press ⑨ ▸ so that the second list is highlighted. Press ⓜ, ③ for Data, and ④ for Quick Graph. Then press ⓜ, ④ for Analyze, ⑤ for Regression, and either ⑦ for ShowPower or ⑧ for Show Exponential.

To find the equation of the model on a full page, press ⓐ, ① for Calculator, ⓜ, ⑥ for Statistics, ① for Stat Calculations. Select the model and complete the dialog boxes.

TI-89

To find the equation of a curved model (and add the curve to a scatterplot), chooose **ExpReg** or **PwrReg** from the **Calc Regressions** menu and tell it the list names and a function to store the equation.

COMMENTS

Each time you execute a regression command, the calculator automatically computes the residuals and stores them in a data list named RESID. Use this list name in **STATVARS** to create a residuals plot.

EXCEL

To fit a curved model
▶ Using the **Chart Wizard,** create the **XY (Scatter)** plot.
▶ In the **Chart** menu, choose **Add Trendline.**
▶ Select a **Power** or **Exponential** model, then click **OK.**

To display the regression equation
▶ Select the curve by clicking on it.
▶ In the Chart toolbar, open the Format Trendline dialog box (first icon).
▶ In the **Options** window, select **Display equation on chart,** then click **OK.**

Chapter 12. Using Randomness

STATCRUNCH

To generate a list of random numbers
1. Click on **Data.**
2. Choose **Simulate data » Uniform.** Enter the number of rows and columns of random numbers you want. (Often you'll specify the desired number of random values as **Rows** and just 1 **Column.**)
3. Enter the interval of possible values for the random numbers ($a \leq x < b$).
4. Click on **Simulate.** The random numbers will appear in the data table.

COMMENTS

To get random *integers* from 0 to 99, set **a** = 0 and **b** = 100, then simply ignore the decimal places in the numbers generated, OR:
1. Under **Data** choose **Compute expression.**
2. As Y, choose the **Column** of random numbers.
3. Choose **Floor(Y)** from the functions.
4. Click on **Set Expression.**
5. Click on **Compute.**

TI-NSPIRE

To generate random integers, press ⓐ, ① for Calculator, ⓜ, ⑤ for Probability, ④ for Random, and ② for Integer. Then type the range for the random integers, such as randInt(1,6).

To create a list of random integers, type the length of the list as the third value, such as randInt(1,6,10).

TI-89

To generate random numbers, move the cursor to highlight the name of a blank list. Use **5:RandInt** from the ☐F4 (Calc) Probability menu. This command will produce any number of random integers in the specified range.

COMMENTS

Some examples:
RandInt(0,10) randomly chooses a 0 or a 1. This is an effective simulation of 10 coin tosses.
RandInt(1,6,2) randomly returns two integers between 1 and 6. This is a good way to simulate rolling two dice.
RandInt(0,56,3) produces three random integers between 0 and 56, a nice way to simulate the chapter's dorm room lottery.

EXCEL

The **RAND** function generates a random value between 0 and 1. You can multiply to scale it up to any range you like and use the INT function to turn the result into an integer.

COMMENTS

Published tests of Excel's random-number generation have declared it to be inadequate. However, for simple simulations, it should be OK. Don't trust it for important large simulations.

Chapter 16. Probability Models

STATCRUNCH

To calculate binomial probabilities

1. Click on **Stat**.
2. Choose **Calculators » Binomial**.
3. Enter the parameters, **n** and **p**.

4. Choose a specific outcome (=) or a lower tail (<= or <) or upper tail (>= or >) sum.
5. Enter the number of successes **x**.
6. Click on **Compute**.

TI-NSPIRE

To compute the mean and standard deviation for a random variable, enter the values in one named list and the probabilities in another. Then press ⓐ, ① for Calculator, ⓜ, ⑥ for Statistics, 1 for Stat Calculations, and ① for One-Variable Statistics. Enter 2 for the prompt for the number of lists, ⓣ to OK, ⓔ, and complete the dialog box.

To compute binomial probabilities, press ⓜ, ⑤ for Probability, and ⑤ for Distributions. Select the menu item. Pdf is for the probability of exactly x successes; Cdf will display the totlal probabilities of 0 to x successes. Complete the dialog box.

TI-89

To calculate the mean and standard deviation of a random variable, enter the probability model in two lists:

▶ In one list (say, list1) enter the x-values of the variable.

▶ In a second list (say, list2) enter the associated probabilities $P(X = x)$.

▶ From the **STAT CALC** (☐F4) menu select **1-VarStats.** Use VAR-LINK to enter the list name list1 in the List box and list2 in the Freq box.

COMMENTS

You can enter the probabilities as fractions; the calculator will change them to decimals for you.
Notice that the calculator knows enough to compute only the standard deviation σ, but mistakenly uses $\bar{x}$ when it should say μ. Make sure you don't make that mistake!

To calculate binomial probabilities, use these commands under the ☐F5 (Distributions) menu:

▶ A: **Binomial Pdf** asks for n, p, and x, reporting the probability of exactly x successes.

▶ B: **Binomial Cdf** asks for n, p, and the lower and upper values of interest, reporting the total probability of those numbers of successes.

EXCEL

To find a binomial probability:
Binomdist(x, n, prob, cumulative)

COMMENTS

For the total probability of from 0 to x successes, set cumulative = *true*.
For the probability of exactly x successes set cumulative = *false*. This function fails when x or n is large.

Chapter 17. Confidence Intervals for a Proportion

STATCRUNCH

To create a confidence interval for a proportion using summaries

1. Click on **Stat**.
2. Choose **Proportions » One sample » with summary**.
3. Enter the **Number of successes** (x) and **Number of observations** (n).
4. Click on **Next**.
5. Indicate **Confidence Interval** (Standard-Wald), then enter the **Level** of confidence.
6. Click on **Calculate**.

To create a confidence interval for a proportion using data

1. Click on **Stat**.
2. Choose **Proportions » One sample » with data**.
3. Choose the variable **Column** listing the **Outcomes**.
4. Enter the outcome to be considered a **Success**.
5. Click on **Next**.
6. Indicate **Confidence Interval**, then enter the **Level** of confidence.
7. Click on **Calculate**.

TI-NSPIRE

To compute a confidence interval for a population proportion, press ⌂, ① for Calculator, (menu), ⑥ for Statistics, ⑥ for Confidence Intervals, and ⑤ for 1-Prop z-interval. Complete the

dialog box. Be sure to enter the number of successes, x, as a whole number, and the C level as a decimal, such as .99.

TI-89

To calculate a confidence interval for a population proportion:

▶ Go to the **Ints** menu (2nd F2) and select **5:1-PropZInt**.

▶ Enter the number of successes observed and the sample size.

▶ Specify a confidence level.

▶ Calculate the interval.

COMMENTS

Beware: When you enter the value of x, you need the count, not the percentage. The count must be a whole number. If the number of successes are given as a percentage, you must first multiply np and round the result.

EXCEL

Excel does not find confidence intervals for proportions.

Chapter 18. Surprised? Testing Hypotheses About Proportions

STATCRUNCH

To test a hypothesis for a proportion using summaries

1. Click on **Stat**.
2. Choose **Proportions » One sample » with summary**.
3. Enter the **Number of successes** (x) and **Number of observations** (n).
4. Click on **Next**.
5. Indicate **Hypothesis Test**, then enter the hypothesized **Null proportion** and choose the **Alternative** hypothesis.
6. Click on **Calculate**.

To test a hypothesis for a proportion using data

1. Click on **Stat**.
2. Choose **Proportions » One sample » with data**.
3. Choose the variable **Column** listing the **Outcomes**.
4. Enter the outcome to be considered a **Success**.
5. Click on **Next**.
6. Indicate **Hypothesis Test**, then enter the hypothesized **Null proportion** and choose the **Alternative** hypothesis.
7. Click on **Calculate**.

TI-NSPIRE

To compute a hypothesis test for a population proportion, press ⌂, ① for Calculator, (menu), ⑥ for Statistics, ⑦ for Stat Tests,

and ⑤ for 1-Prop z-test. Complete the dialog box. Be sure to enter the number of successes, x, as a whole number.

TI-89

To do the mechanics of a hypothesis test for a proportion,

▶ Select **5:1-PropZTest** from the **STAT TESTS** [2nd][F1] menu.

▶ Specify the hypothesized proportion.

▶ Enter the observed value of *x*.

▶ Specify the sample size.

▶ Indicate what kind of test you want: one-tail lower tail, two-tail, or one-tail upper tail.

▶ Specify whether to calculate the result or draw the result (a normal curve with *p*-value area shaded).

COMMENTS

Beware: When you enter the value of *x*, you need the *count,* not the percentage. The count must be a whole number. If the number of successes is given as a percent, you must first multiply *np* and round the result to obtain *x*.

EXCEL

Excel does not do hypothesis tests for proportions.

Chapter 19. Inferences About a Mean

STATCRUNCH

To do inference for a mean

1. Click on **Stat**.

2. Choose **T Statistics » One sample » with data** or **with summary**.

3. If you are using data, choose the variable **Column**.

    OR

    If you are using summaries, enter the **Sample mean, Sample std dev**, and **Sample size**.

4. Click on **Next**.

5. Indicate **Hypothesis Test**, then enter the hypothesized **Null mean** and choose the **Alternative** hypothesis. Indicate **Confidence Interval**, then enter the **Level** of confidence.

6. Click on **Calculate**.

To do inference for the mean of paired differences

1. Click on **Stat**.

2. Choose **T Statistics » Paired**.

3. Choose the **Column** for each variable.

4. Check **Save differences** so you can look at a histogram to be sure the Nearly Normal condition is satisfied.

5. Click on **Next**.

6. Indicate **Hypothesis Test**, then enter the hypothesized **Null mean diff**erence (usually 0), and choose the **Alternative** hypothesis.

    OR

    Indicate **Confidence Interval**, then enter the **Level** of confidence.

7. Click on **Calculate**.

TI-NSPIRE

To compute a confidence interval for a population mean, press (⌂), ① for Calculator, (menu), ⑥ for Statistics, ⑥ for Confidence Intervals, and ② for *t*-interval. Select between Data and Stats, (tab) to OK, and press (enter). Complete the dialog box. Be sure to enter the number of successes, *x*, as a whole number, and the C level as a decimal, such as .99.

To compute a hypothesis test for a population mean, press (⌂), ① for Calculator, (menu), ⑥ for Statistics, ⑦ for Stat Tests, and ② for *t*-test. Select between Data and Stats, (tab) to OK, and (enter). Complete the dialog box.

For inference on a matched pair design, compute a third list of differences such as *diff = time2-time1*. Then construct the confidence interval or conduct the hypothesis test in the same way as 1-sample procedures, using the list of differences.

TI-89

Finding a confidence interval for a mean
In the **STAT Ints** menu, choose **2:TInterval**. Specify whether you are using data stored in a list or whether you will enter the mean, standard deviation, and sample size. You must also specify the desired level of confidence.

Testing a hypothesis for a mean
In the **STAT Tests** menu, choose **2:T-Test**. You must specify whether you are using data stored in a list or whether you will enter the mean, standard deviation, and size of your sample. You must also specify the hypothesized model mean and whether the test is to be two-tail, lower-tail, or upper-tail. Select whether the test is to be simply computed or whether to display the distribution curve and highlight the area corresponding to the P-value of the test.

For inference in paired differences
If the data are stored in two lists, say, list1 and list2, create a list of the differences: Move the cursor to the name of an empty list, and then use VAR-LINK to enter the command list1-list2. Press ENTER to perform the subtraction.

Since inference for paired differences uses one-sample *t*-procedures, select **2:T-Test** or **2:TInterval** from the **STAT Tests** or **Ints** menu. Specify as your data the list of differences you just created, and apply the procedure.

EXCEL

Excel does not do *t*-procedures for one mean.

For inference about paired differences:
In Excel 2003 and earlier, select **Data Analysis** from the **Tools** menu.
In Excel 2007, select **Data Analysis** from the **Analysis** Group on the **Data** Tab.
From the **Data Analysis** menu, choose **t-test: paired two-sample for Means.** Fill in the cell ranges for the two groups, the hypothesized difference, and the alpha level.

Chapter 20. Comparing Proportions or Means

STATCRUNCH

To do inference for the difference between two proportions

1. Click on **Stat**.
2. Choose **Proportions » Two sample » with data** or **with summary**.
3. If you are using data, for each group, choose the variable **Column** listing the **Outcomes**, and enter the outcome to be considered a **Success**.

   OR

   If you are using summaries, enter the **Number of successes** (x) and **Number of observations** (n) in each group.
4. Click on **Next**.
5. Indicate **Hypothesis Test**, then enter the hypothesized **Null prop**ortion **diff**erence (usually 0), and choose the **Alternative** hypothesis.

   OR

   Indicate **Confidence Interval**, then enter the **Level** of confidence.
6. Click on **Calculate**.

To do inference for the difference between two means

1. Click on **Stat**.
2. Choose **T Statistics » Two sample » with data** or **with summary**.
3. If you are using data, choose the variable **Column** for each group.

   OR

   If you are using summaries, enter the **Sample mean**, **Standard deviation**, and sample **Size** for each group.
4. De-select **Pool variances**.
5. Click on **Next**.
6. Indicate **Hypothesis Test**, then enter the hypothesized **Null mean diff**erence (usually 0), and choose the **Alternative** hypothesis.

   OR

   Indicate **Confidence Interval**, then enter the **Level** of confidence.
7. Click on **Calculate**.

TI-NSPIRE

To compute a confidence interval for the difference between two population proportions, press ⌂, ① for Calculator, (menu), ⑥ for Statistics, ⑥ for Confidence Intervals, and ⑥ for 2-Prop *z*-interval. Complete the dialog box. Be sure to enter each number of successes as a whole number, and the C level as a decimal, such as .99.
To compute a hypothesis test for the difference between two population proportions, press ⌂, ① for Calculator, (menu), ⑥ for Statistics, ⑦ for Stat Tests, and ⑥ for 2-Prop *z*-test. Complete the dialog box. Be sure to enter each number of successes as a whole number.

To compute a confidence interval for the difference between two population means, press ⌂, ① for Calculator, (menu), ⑥ for Statistics, ⑥ for Confidence Intervals, and ④ for 2-Sample *t*-interval. Select between Data and Stats, (tab) to OK, and ⏎. Complete the dialog box. Be sure to enter the C level as a decimal, such as .99.

To compute a hypothesis test for the difference between two population means, press ⌂, ① for Calculator, (menu), ⑥ for Statistics, ⑦ for Stat Tests, and ④ for 2-Sample *t*-test. Select between Data and Stats, (tab) to OK, and ⏎. Complete the dialog box.

TI-89

To calculate a confidence interval for the difference between two population proportions
▶ Select **6:2-PropZInt** from the **STAT Ints** menu.
▶ Enter the observed counts and the sample sizes for both samples.
▶ Specify a confidence level.
▶ Calculate the interval.

To do the mechanics of a hypothesis test for equality of population proportions
▶ Select **6:2-PropZTest** from the **STAT Tests** menu.
▶ Enter the observed counts and sample sizes.
▶ Indicate what kind of test you want: one-tail upper tail, lower tail, or two-tail.
▶ Specify whether results should simply be calculated or displayed with the area corresponding to the P-value of the test shaded.

To calculate a confidence interval for the difference between two population means:
In the **STAT Ints** menu, choose **4:2-SampTInt.** You must specify whether you are using data stored in two lists or whether you will enter the means, standard deviations, and sizes of both samples. You must also indicate whether to pool the variances (when in doubt, say no) and specify the desired level of confidence.

To test a hypothesis of equality of population means:
In the **STAT TESTS** menu, choose **4:2-SampTTest.** You must specify whether you are using data stored in two lists or whether you will enter the means, standard deviations, and sizes of both samples. You must also indicate whether to pool the variances (when in doubt, say no) and specify whether the test is to be two-tail, lower-tail, or upper-tail.

EXCEL

From the Data Tab, Analysis Group, choose **Data Analysis.** Alternatively (if the Data Analysis Tool Pack is not installed), in the Formulas Tab, choose More functions > Statistical > TTEST, and specify Type = 3 in the resulting dialog.
Fill in the cell ranges for the two groups, the hypothesized difference, and the alpha level.

COMMENTS

Excel expects the two groups to be in separate cell ranges. Notice that, contrary to Excel's wording, we do not need to assume that the variances are *not* equal; we simply choose not to assume that they *are* equal.
Excel does not do inference for the difference in two proportions.

Chapter 21. Comparing Counts

STATCRUNCH

To perform a Goodness-of-Fit test

1. Enter the observed counts in one column of a data table, and the expected counts in another.
2. Click on **Stat**.
3. Choose **Goodness-of-Fit » Chi-Square test**.
4. Choose the **Observed Column** and the **Expected Column**.
5. Click on **Calculate**.

COMMENTS

These Chi-square tests may also be performed using the actual data table instead of summary counts. See the StatCrunch **Help page** for details.

To perform a 2-way table test

1. Create a table (without totals):
   • Name the first column as one variable, enter the categories underneath.
   • Name the adjacent columns as the categories of the other variable, entering the observed counts underneath.
2. Click on **Stat**.
3. Choose **Tables » Contingency » with summary**.
4. Choose the **Columns** holding counts.
5. Choose the **Row labels column**.
6. Enter the **Column variable** name.
7. Click on **Next** to ask for **expected counts**.
8. Click on **Calculate**.

TI-NSPIRE

To conduct a χ^2 goodness-of-fit test, enter the observed and the expected values into two named lists. Then press ⌂, ① for Calculator, (menu), ⑥ for Statistics, ⑦ for Stat Tests, and ⑦ for χ^2 GOF. Complete the dialog box.

To conduct a χ^2 test for a 2-way table, first enter the data into a matrix. Press ⌃ ⌅ and select the matrix icon. Enter the dimensions and (tab) to OK, and ⏎. Then type the data into the matrix. Then press ▸ to exit the matrix, press ⌃ (var) and a matrix name such as *ma* to store the matrix. To complete the test, press ⌂, ① for Calculator, (menu), ⑥ for Statistics, ⑦ for Stat Tests, and ⑧ for χ^2 2-way Test. Complete the dialog box.

TI-89

To test goodness-of-fit, enter the observed counts in a list and the expected counts in another list. Expected counts can be entered as n*p, and the calculator will compute them for you. From the **STAT Tests** menu, select **7:Chi2 GOF.** Enter the list names using VAR-LINK and the degrees of freedom, $k - 1$, where k is the number of categories. Select whether to simply calculate or display the result with the area corresponding to the P-value highlighted.

To test a 2-way table, you need to enter the data as a matrix. From the home screen, press APPS and select **6:Data/Matrix Editor,** then select **3:New.** Specify type as Matrix and name the matrix in the **Variable** box. Specify the number of rows and columns. Type the entries, pressing ENTER after each. Press 2nd ESC to leave the editor.

To do the test, choose **8:Chi2 2-way** from the **STAT Tests** menu.

EXCEL

CHITEST(actual_range, expected_range) computes a chi-square value for 2-way tables. Both ranges are of the form UpperleftCell: LowerRightCell, specifying two rectangular tables that must hold counts (although Excel will not check for integer values). The two tables must be of the same size and shape.

COMMENTS

Excel does not perform a goodness-of-fit test.

Answers

Here are the "answers" to the exercises for the chapters and the unit reviews. As we said in Chapter 1, the answers are outlines of the complete solution. Your solution should follow the model of the Step-By-Step examples, where appropriate. You should explain the context, show your reasoning and calculations, and draw conclusions. For some problems, what you decide to include in an argument may differ somewhat from the answers here. But, of course, the numerical part of your answer should match the numbers in the answers shown.

Chapter 2

1. Categorical

3. Quantitative

5. *Population*—All bears
   Sample—54 captured bears

7. *Population*—All men
   Sample—25,892 men aged 30–87

9. *Who*—Coffee drinkers at a Newcastle University
        coffee station
   What—Amount of money contributed

11. *Who*—25,892 men aged 30 to 87
    What—Fitness level and cause of death

13. *Who*—2500 cars
    What—Distance from car to bicycle
    Variable—Helmet
        Type—Categorical (yes/no)
    Variable—Distance
        Type—Quantitative
        Units—Inches

15. *Variable*—Type of meat
        Type—Categorical
    Variable—Number of calories
        Type—Quantitative
        Units—Calories
    Variable—Serving size
        Type—Quantitative
        Units—Ounces

17. *Who*—882 births
    What—Mother's age, length of pregnancy, type of
        birth, level of prenatal care, birth weight of
        baby, sex of baby, and baby's health problems
    Population—All newborn babies.

Variable—Mother's age
    Type—Quantitative
    Units—Not specified; probably years
Variable—Length of pregnancy
    Type—Quantitative
    Units—Weeks
Variable—Birth weight of baby
    Type—Quantitative
    Units—Not specified, probably pounds and ounces
Variable—Type of birth
    Type—Categorical
Variable—Level of prenatal care
    Type—Categorical
Variable—Sex
    Type—Categorical
Variable—Baby's health problems
    Type—Categorical

19. *Who*—41 refrigerator models
    What—Brand, cost, size, type, estimated annual
            energy cost, overall rating, and repair history
    Population—All refrigerators currently being sold
    Variable—Brand
        Type—Categorical
    Variable—Cost
        Type—Quantitative
        Units—Not specified (dollars)
    Variable—Size
        Type—Quantitative
        Units—Cubic feet
    Variable—Type
        Type—Categorical
    Variable—Estimated annual energy cost
        Type—Quantitative
        Units—Not specified (dollars)
    Variable—Overall rating
        Type—Categorical (ordinal)

Variable—Percent requiring repair in last 5 years
   Type—Quantitative
   Units—Percent

21. *Who*—Experiment subjects
   What—Treatment (herbal cold remedy or sugar
      solution) and cold severity
   When—Not specified
   Where—Not specified
   Why—To test efficacy of herbal remedy on common
      cold
   How—The scientists set up an experiment.
   Variable—Treatment
     Type—Categorical
   Variable—Cold severity rating
     Type—Quantitative (perhaps ordinal categorical)
     Units—Scale from 0 to 5
   Concerns—The severity of a cold seems subjective and
      difficult to quantify. Scientists may feel
      pressure to report negative findings of
      herbal product.

23. *Who*—Streams
   What—Name of stream, substrate of the stream,
      acidity of the water, temperature, BCI
   When—Field research
   Where—Upstate New York
   Why—To study ecology of streams
   How—Field research
   Variable—Stream name
     Type—Identifier
   Variable—Substrate
     Type—Categorical
   Variable—Acidity of water
     Type—Quantitative
     Units—pH
   Variable—Temperature
     Type—Quantitative
     Units—Degrees Celsius
   Variable—BCI
     Type—Quantitative
     Units—Not specified

25. *Who*—Kentucky Derby races
   What—Date, winner, margin, jockey, net proceed to
      winner, duration, track condition
   When—1875 to 2010
   Where—Churchill Downs, Louisville, Kentucky
   Why—Not specified (To see trends in horse racing?)
   How—Official statistics collected at race
   Variable—Year
     Type—Quantitative
     Units—Day and year
   Variable—Winner
     Type—Identifier
   Variable—Margin
     Type—Quantitative
     Units—Horse lengths
   Variable—Jockey
     Type—Categorical
   Variable—Net proceeds to winner
     Type—Quantitative
     Units—Dollars

Variable—Duration
   Type—Quantitative
   Units—Minutes and seconds
Variable—Track condition
   Type—Categorical

27. Answers will vary.

Chapter 3

1. a) Yes; each is categorized in a single genre.
   b) Thriller/Horror

3. a) Comedy

   b) It is easier to tell from the bar chart; slices of the
     pie chart are too close in size.

5. 1755 students applied for admission to the magnet
   schools program. 53% were accepted, 17% were
   wait-listed, and the other 30% were turned away.

7. a) Yes. We can add because these categories do not
     overlap. (Each person is assigned only one cause
     of death.)
   b) $100 - (27.2 + 23.1 + 6.3 + 5.1 + 4.7) = 33.6\%$

9. Either a bar chart or pie chart with "other" added
   would be appropriate. A bar chart is shown.

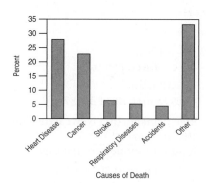

11. a) Yes. These are counts of categorical data and all
     spillages are included in one and only one category.
   b) The bar chart shows that grounding and collision
     are the most frequent causes of oil spills. Very few
     have unknown causes.

13. a) 82.5%   b) 12.9%   c) 11.1%
   d) 13.4%   e) 85.7%

15. Yes, there appears to be an association between brand
   preferences and region of the country. Newport was
   far more popular in the South than in the West, where
   Camel was cited nearly 3 times as often as in the
   South. Nearly twice as many smokers in the West as
   in the South indicated that they had no usual brand
   (12.9% to 6.7%).

17. There's no title, the percentages total only 92%, and
   the three-dimensional display distorts the sizes of the
   regions.

19. a) 59.05%
    b) There are 212 American cars, of which 107 or 107/212 = 50.47% were owned by students.
    c) There are 195 students, of whom 107 or 107/195 = 54.87% owned American cars.
    d) 54.3% of drivers are students and 45.7% are staff.
    e) 50.5% of people driving American cars are students, and 49.5% are staff.

21. a)

	Cars		
		All cars	**Percentage**
Origin	American	212	59.1%
	European	45	12.5%
	Asian	102	28.4%
	Totals	**359**	**100.0%**

    b)

	Driver		
		Student	**Percentage**
Origin	American	107	54.87%
	European	33	16.92%
	Asian	55	28.21%
	Totals	**195**	**100.00%**

    c)

	Driver		
		Staff	**Percentage**
Origin	American	105	64.02%
	European	12	7.32%
	Asian	47	28.66%
	Totals	**164**	**100.00%**

    d) No, the distributions look slightly different. A higher percentage of staff drive American cars, 64% compared to about 55% among students. Bar chars or pie charts could be used to compare.

23. a) 22.7%   b) 52.9%   c) 39.7%
    d) 21.9% of mothers of twins received intensive prenatal care, and another 55.4% adequate care, but 22.7% had inadequate prenatal care.
    e) Of mothers who had induced or cesarean preterm twin births, 23.7% had intensive prenatal care, and for another 60.5% care was adequate, but 15.8% had inadequate prenatal care.

25. a) 27.3% of twin births were preterm cesarean or induced, 25.5% were preterm without procedures, and 47.1% were full term.
    b) Among those women who received inadequate prenatal care, 19% of twin births were preterm cesarean or induced, 21% were preterm without procedures, and 60% were full term.
    c) Yes. It appears that women receiving inadequate prenatal care are more likely to have full-term births.

    d) No. Women with difficult pregnancies may require both intensive prenatal care and have preterm births.

27. No, there's no evidence that Prozac is effective. The relapse rates were nearly identical: 28.6% among the people treated with Prozac, compared to 27.3% among those who took the placebo.

29. 1755 students applied for admission to the magnet schools program: 53% were accepted, 17% were wait-listed, and the other 30% were turned away. While the overall acceptance rate was 53%, 93.8% of blacks and Hispanics were accepted, compared to only 37.7% of Asians and 35.5% of whites. Overall, 29.5% of applicants were black or Hispanic, but only 6% of those turned away were. Asians accounted for 16.6% of all applicants, but 25.4% of those turned away. Whites were 54% of the applicants and 68.5% of those who were turned away. It appears that the admissions decisions were not independent of the applicant's ethnicity.

31. Answers will vary.

33. Answers will vary.

Chapter 4

1. a) Unimodel (near 0) and skewed to the right. Many seniors will have 0 or 1 speeding tickets. Some may have several, and a few may have more than that.
   b) Probably unimodel and slightly skewed to the right. It is easier to score 15 strokes over the mean than 15 strokes under the mean.

3. a) Probably unimodel and symmetric. Weights may be equally likely to be over or under the average.
   b) Probably bimodel. Men's and women's distributions may have different modes. It may also be skewed to the right, since it is possible to have very long hair, but hair length can't be negative.

5. a) Bimodal. Looks like two groups. Modes are near 6% and 46%. No real outliers.
   b) Looks like two groups of cereals, a low-sugar and a high-sugar group.

7. Acreages of vineyards are skewed to the right with at least one high outlier. Most of the vineyards are less than 90 acres with a few over 240 acres. The mode is between 0 and 30 acres.

9. a) The median; the skewed distribution will pull the mean to the right, making it too large to represent a typical vineyard.
   b) The IQR; the outlier will make the standard deviation very large, making it appear there's more variation in vineyard size.

11. The mean and standard deviation because the distribution is unimodal and symmetric.

13. The distribution of Dallas pizza prices is unimodal and roughly symmetric, centering around $2.60 per slice. During most weeks the cost was between $2.40 and $2.80, never getting cheaper than $2.20 and only rarely going above $3.00.

15. a) The mean is closest to $2.60 because that's the balancing point of the histogram.
    b) The standard deviation is closest to $0.15 since that's a typical distance from the mean. There are no prices as far as $0.50 or $1.00 from the mean.

17.
```
Stem | Leaf
   7 | 5
   6 | 115699
   5 | 002345669
   4 | 13445566777
   3 | 01116777899
   2 | 12379
```
Total Points (7|5 means 75)

19. a) 45 points   b) 37 points and 55 points

21. The distribution of total points scored in the Super Bowl is roughly symmetric. Teams typically score a total of about 45 points, with half the games totaling between 37 and 55 points. In only one-fourth of the games have the teams scored fewer than 27 points, and they once totaled 75.

23. Histogram bins are too wide to be useful.

25. a) About 100 minutes
    b) Yes, only 4 of these movies run at least that long.
    c) The mean would be higher. The distribution is skewed high.

27. a) i. The middle 50% of movies ran between 97 and 119 minutes.
        ii. On average, movie lengths varied from the mean run time by 19.6 minutes.
    b) We should be cautious in using the standard deviation because the distribution of run times is skewed to the right.

29. a) Because the distribution is skewed to the right, we expect the mean to be larger.
    b) Bimodal and skewed to the right. Center mode near 8 days. Another mode at 1 day (may represent patients who didn't survive). Most of the patients stay between 1 and 15 days. There are some extremely high values above 25 days.
    c) The median and IQR, because the distribution is strongly skewed.

31. Mean = 2.38, median = 3

33. a) Mean = 8.0, SD = 4.2
    b) Increasing each of the old values by 80 changes the center but not the spread.

35. a) Mean = 15.0, SD = 3.6
    b) The SD will be larger because 11 and 19 are farther from the mean than 14 and 16 were.

37. The publication is using the median; the watchdog group is using the mean, pulled higher by the several very expensive movies in the long right tail.

39. a)
```
Stem | Leaf
  24 | 56
  24 |
  23 | 68
  23 | 23
  22 | 677789
  22 | 1234
```
22|1 = $2.21/gallon

b) The distribution of gas prices is unimodal and skewed to the right (upward), centered around $2.27, with most stations charging between $2.26 and $2.33 per gallon. The lowest and highest prices were $2.21 and $2.46.
c) There are two high prices separated from the other gas stations by a gap.
d) All data: mean = $2.303, SD = $0.076; with two high prices removed, mean = $2.281, SD = $0.051. Omitting the two high prices lowers the mean price about 2 cents per gallon and also decreases the standard deviation by about 2 cents per gallon.

41. a) The median will probably be unaffected. The mean will be larger.
    b) The range and standard deviation will increase; the IRQ will be unaffected.

43. a) Mean $525, median $450
    b) 2 employees earn more than the mean.
    c) The median because of the outlier.
    d) The IQR will be least sensitive to the outlier of $1200, so it would be the best to report.

45. a) Since these data are strongly skewed to the right, the median and IQR are the best statistics to report.
    b) The mean will be larger than the median because the data are skewed to the right.
    c) The median is 4 million. Q1 = 1 million, Q3 = 6 million, Min = 1 million; Max = 3.4 million.
    d) The distribution of populations of the states and Washington, DC, is unimodal and skewed to the right. The median population is 4 million. One state is an outlier, with a population of 34 million.

47. a) This is not a histogram. The horizontal axis should split the number of home runs hit in each year into bins. The vertical axis should show the number of years in each bin.
    b)

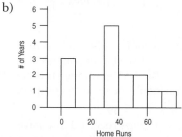

49. The distribution of pH levels from Allegheny County rain and snow samples is skewed to the right and possibly bimodal. One large cluster of readings centers around pH 4.4, but there's a smaller cluster at a much higher pH of 5.6 and a few middle readings that don't seem to belong to either group. Overall the median pH was 4.54, with most measurements below 4.8. No observations were below pH 4.1 or above pH 5.8.

```
Stem | Leaf
  57 | 8
  56 | 27
  55 | 1
  54 |
  53 |
  52 | 9
  51 |
  50 | 8
  49 |
  48 | 2
  47 | 3
  46 | 034
  45 | 267
  44 | 015
  43 | 0199
  42 | 669
  41 | 22
```
41|2 = 4.12 pH

51. The distribution of 8th grade math achievement scores is skewed to the left; it may be bimodal. The typical state average is around 239. The middle 50% of states scored between 233 and 242. Alabama, Mississippi, and New Mexico scores were much lower than other states' scores.

53. Answers will vary.

55. Answers will vary.

Chapter 5

1. The class A is 1, class B is 2, and class C is 3.

3. a) Slightly skewed to the right with two high outliers at 36 and 48. Most victims are between the ages of 16 and 24.
   b) The slight increase between ages 22 and 24 is apparent in the histogram but not in the boxplot. It may be a second mode.
   c) The median would be the most appropriate measure of center because of the slight skew and the extreme outliers.
   d) The IQR would be the most appropriate measure of spread because of the slight skew and the extreme outliers.

5. a) Both girls have a median score of about 17 points per game, but Scyrine is much more consistent. Her IQR is about 2 points, while Alexandra's is over 10.
   b) If the coach wants a consistent performer, she should take Scyrine. She'll almost certainly deliver somewhere between 15 and 20 points. But if she wants to take a chance and needs a "big game," she should take Alexandra. Alex scores over 24 points about a quarter of the time. (On the other hand, she scores under 11 points as often.)

7. No, all data are within the fences.

9. a)

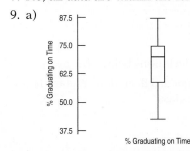

b) Probably slightly left skewed. The mean is slightly below the median, and the 25th percentile is farther from the median than the 75th percentile.
   c) The 48 universities graduate, on average, about 68% of freshmen "on time," with percents ranging from 43% to 87%. The middle 50% of these universities graduate between 59% and 75% of their freshmen in 4 years.

11. a) Day 16 (but any estimate near 20 is okay).
    b) Around day 50

13. a) Min and max each increase $3, range remains the same
    b) Median and quartiles increase $3, IQR remains the same
    c) Mean increases $3, standard deviation remains the same

15. Each summary statistic increases by 10%.

17. a) 72 oz., 40 oz.   b) 4.5 lb, 2.5 lb

19. a) Prices appear to be both higher on average and more variable in Baltimore than in the other three cities. Prices in Chicago may be slightly higher than in Dallas and Denver, but the difference is very small.
    b) There are outliers on the low end in Baltimore and Chicago and one high outlier in Dallas, but these do not affect the overall conclusions reached in part a.

21. Both fuel economy and its spread decrease from 4 to 6 to 8 cylinders (not enough data to compare 5-cylinder cars). The lower 75% of mpgs for the 8-cylinder cars corresponds roughly to the bottom 25% of mpgs for the 6-cylinder cars. All the 8-cylinder cars get less mileage than all the 4-cylinder cars, and their fuel economy is consistently low.

23. a) They should be put on the same scale, from 0 to 20 days.
    b) Lengths of men's stays appear to vary more than for women. Men have a mode at 1 day and then taper off from there. Women have a mode near 5 days, with a sharp drop afterward.
    c) A possible reason is childbirth.

25. a) About 59%   b) Bimodal
    c) Some cereals are very sugary; others are healthier low-sugar brands.
    d) Yes
    e) Although the ranges appear to be comparable for both groups (about 28%), the IQR is larger for the adult cereals, indicating that there's more variability in the sugar content of the middle 50% of adult cereals.

27. a)

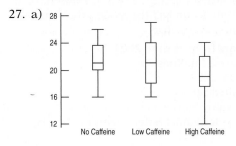

b) The participants scored about the same with no caffeine and low caffeine. The medians for both were 21 points, with slightly more variation for the low-caffeine group. The high-caffeine group generally scored lower than the other two groups with all measures of the 5-number summary lower: min, lower quartile, median, upper quartile, and max.

29. a) The distribution is unimodal and skewed to the right. The mode is near 100, and values range from 95 to 140.
   b) The mean will be larger than the median, since the distribution is right skewed.
   c) Create a boxplot with quartiles at 97 and 105.5, and median at 100. The IQR is 8.5, so the upper fence is at $(1.5 \times 8.5) + 105.5 = 118.25$. There are several outliers to the right. There are no outliers to the left because the minimum at 95 lies well within the left fence at $97 - (1.5 \times 8.5) = 84.25$.

31. a) Skewed to the right; mean is higher than median.
   b) $350 and $950.
   c) Minimum $350. Mean $750. Median $550. Range $1200. IQR $600. Q1 $400. SD $400.
   d) Minimum $330. Mean $770. Median $550. Range $1320. IQR $660. Q1 $385. SD $440.

33. a) The median and IQR, because the means are much larger than the median and the SDs are much larger than the IQR, indicating either right skewness and/or outliers.
   b) Since the median rainfall for seeded clouds is more than 4 times that for unseeded clouds, it appears that seeding clouds may be effective.

35. a)

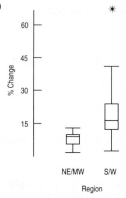

b) Growth rates in NE/MW states are tightly clustered near 5%. S/W states are more variable, and bimodal with modes near 14 and 22. The S/W states have an outlier as well. Around all the modes, the distributions are fairly symmetric.

37. a) *Who*—Years from 1994 to 2003
   What—Bicycle fatalities
   When—1994–2003
   Where—United States
   Why—To study bicycle helmet safety
   How—Bicycle Helmet Safety Institute Report

b)
Stem	Leaf
6	1
6	68
7	2
7	5569
8	12

6/1 = 610 − 619 fatalities

c)

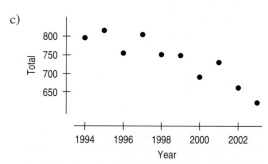

d) The stem-and-leaf display shows the distribution is skewed to the left. It also provides some idea about the center and spread of the annual fatalities.
e) The number of bicycle fatalities has tended to decrease over the 10-year period.
f) In the 10-year period from 1994 to 2003, reported bicycle fatalities decreased fairly steadily from about 800 per year to around 620 a year.

39. Lowest score = 910. Mean = 1230. SD = 120. Q3 = 1350. Median = 1270. IQR = 240.

41. a) About 36 mph
   b) Q₁ about 35 mph and Q₃ about 37 mph
   c) The range appears to be about 7 mph, from about 31 to 38 mph. The IQR is about 2 mph.
   d) We can't know exactly, but the boxplot may look something like this:

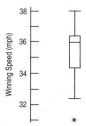

e) The median winning speed has been about 36 mph, with a max of about 38 and a min of about 31 mph. Half have run between about 35 and 37 mph, for an IQR of 2 mph.

43. Answers will vary.

45. Answers will vary.

Chapter 6

1. About 1.81 standard deviations below the mean.

3. Your score was 2.2 standard deviations higher than the mean score in the class.

5. 65

7. 1026 lbs

9. 1000 ($z = 1.81$) is more unusual than 1250 ($z = 1.17$).

11. In January, a high of 55 is not quite 2 standard deviations above the mean, whereas in July a high of 55 is more than 2 standard deviations lower than the mean. So it's less likely to happen in July.

13. Any weight more than 2 standard deviations below the mean, or less than $1152 - 2(84) = 984$ pounds, is unusually low. We expect to see a steer below $1152 - 3(84) = 900$ pounds only rarely.

15.

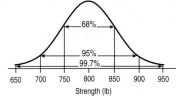

17. a) 650–950 lbs   b) 97.5%
    c) Chances are 84% that the strength is at least 750 pounds. That means 16% of the time it will be below. Seems like much too high a percentage.

19. In the extreme case, I would want a very small probability of failure, maybe less than 1 out of a million. For this I'd need to be sure that 800 was 6 SDs higher than my limit. So I might not use the rivets if I needed them to have shear strength more than 500 pounds.

21. The z-scores, which account for the difference in the distributions of the two tests, are 1.5 and 0 for Derrick and 0.5 and 2 for Julie. Derrick's total is 1.5, which is less than Julie's 2.5.

23. College professors can have between 0 and maybe 40 (or possibly 50) years' experience. A standard deviation of 1/2 year is impossible, because many professors would be 10 or 20 SDs away from the mean, whatever it is. An SD of 16 years would mean that 2 SDs on either side of the mean is plus or minus 32, for a range of 64 years. That's too high. So, the SD must be 6 years.

25. a)

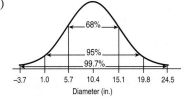

b) Between 1.0 and 19.8 inches
c) 2.5%   d) 34%   e) 16%

27. Since the histogram is not unimodal and symmetric, it is not wise to have faith in numbers from the Normal model.

29. a) 12.2%   b) 71.6%   c) 23.3%

31. a) 1259.7 lb   b) 1081.3 lb
    c) 1108 lb to 1196 lb

33. a) 1130.7 lb   b) 1347.4 lb

35. a) 11.1%   b) (35.9, 40.5) inches
    c) 40.5 inches

37. a) 2.5%
    b) 2.5% of the receivers should gain less than −333 yards, but that's impossible, so the model doesn't fit well.
    c) Data are strongly skewed to the right, not symmetric.

39. a) Megan   b) Anna

41. a) 16%   b) 3.8%
    c) Because the Normal model doesn't fit well.
    d) Distribution is skewed to the right.

43. 113.3 lbs

45. a)

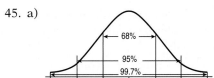

b) 30.85%   c) 17.00%
d) 32 points   e) 212.9 points

Chapter 7

1. a) Weight in ounces: explanatory; Weight in grams: response. (Could be other way around.) To predict the weight in grams based on ounces. Scatterplot: positive, straight, strong (perfectly linear relationship).
   b) Circumference: explanatory. Weight: response. To predict the weight based on the circumference. Scatterplot: positive, linear, moderately strong.
   c) Shoe size: explanatory; GPA: response. To try to predict GPA from shoe size. Scatterplot: no direction, no form, very weak.
   d) Miles driven: explanatory; Gallons remaining: response. To predict the gallons remaining in the tank based on the miles driven since filling up. Scatterplot: negative, straight, moderate.

3. a) Altitude: explanatory; Temperature: response. (Other way around possible as well.) To predict the temperature based on the altitude. Scatterplot: negative, possibly straight, weak to moderate.
   b) Ice cream cone sales: explanatory. Air-conditioner sales: response—although the other direction would work as well. To predict one from the other. Scatterplot: positive, straight, moderate.
   c) Age: explanatory; Grip strength: response. To predict the grip strength based on age. Scatterplot: curved down, moderate. Very young and elderly would have grip strength less than that of adults.
   d) Reaction time: explanatory; Blood alcohol level: response. To predict blood alcohol level from reaction time test. (Other way around is possible.) Scatterplot: positive, nonlinear, moderately strong.

5. a) None   b) 3 and 4   c) 2, 3, and 4
   d) 1 and 2   e) 3 and possibly 1

7. 1; the relationship is curved.

9. a) 0.006   b) 0.777   c) −0.923   d) −0.487

11. There may be an association, but not a correlation unless the variables are quantitative. There could be a correlation between average number of hours of TV watched per week per person and number of crimes committed per year. Even if there is a relationship, it doesn't mean one causes the other.

13. a) Yes. It shows a linear form and no outliers.
    b) Positive and high, near 0.9

15. There is a strong, positive, linear association between drop and speed; the greater the coaster's initial drop, the higher the top speed.

17. a) Assuming the relation is linear, a correlation of −0.772 shows a strong relation in a negative direction.
    b) Continent is a categorical variable. Correlation does not apply.

19. a) The relationship is positive, moderately strong, and curved. As the water's hardness increases from 0 to 250, the pH generally increases, but then stays relatively constant across hardness levels above 250 grains.
    b) The scatterplot is not linear; correlation is not appropriate.

21. The correlation coefficient won't change, because it's based on z-scores. The z-scores of the prediction errors are the same whether they are expressed in nautical miles or miles.

23. a) Actually, yes, taller children will tend to have higher reading scores, but this doesn't imply causation.
    b) Older children are generally both taller and are better readers. Age is the lurking variable.

25. a) No. We don't know this from the correlation alone. There may be a nonlinear relationship or outliers.
    b) No. We can't tell from the correlation what the form of the relationship is.
    c) No. We don't know from the correlation coefficient.
    d) Yes, the correlation doesn't depend on the units used to measure the variables.

27. a) The association is positive, moderately strong, and roughly straight, with several states whose HCI seems high for their median income and one state whose HCI appears low given its median income.
    b) No. We can only say that higher median incomes are associated with higher housing costs, but we don't know why. There may be other economic variables at work.

29. a) The correlation would still be 0.65.
    b) The correlation wouldn't change.
    c) DC would be a moderate outlier whose HCI is high for its median income. It would lower the correlation slightly.

31. a)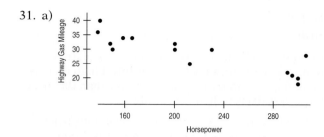

    b) Negative, linear, strong.
    c) −0.869
    d) There is a strong linear relation in a negative direction between horsepower and highway gas mileage. Lower fuel efficiency is associated with higher horsepower.

33. a)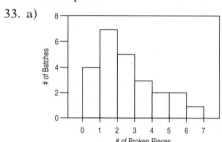

    b) Unimodal, skewed to the right. The skew.
    c) The positive, somewhat linear relation between batch number and broken pieces.

35.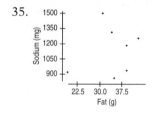

    (Plot could have explanatory and predictor variables swapped.) Correlation is 0.199. There does not appear to be a relation between sodium and fat content in burgers, especially without the low-fat, low-sodium item. The correlation of 0.199 shows a weak relationship with the outlier included.

37. a) Yes, the scatterplot appears to be somewhat linear.
    b) As the number of runs increases, the attendance also increases.
    c) There is a positive association, but it does not *prove* that more fans will come if the number of runs increases. Association does not indicate causality.

39. A scatterplot shows a weak linear pattern with no outliers. The correlation between *Drop* and *Duration* is 0.35, indicating that rides on coasters with greater initial drops generally last somewhat longer, but the association is weak.

41. The correlation may be near 0. We expect nighttime temperatures to be low in January, increase through spring and into the summer months, then decrease again in the fall and winter. The relationship is not linear.

43. a)

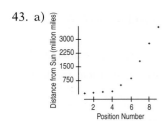

The relation between position and distance is nonlinear, with a positive direction. There is very little scatter from the trend.

b) The relation is not linear.

Chapter 8

1. 281 milligrams

3. The potassium content is actually lower than the model predicts for a cereal with that much fiber.

5. The model predicts that cereals will have approximately 27 more milligrams of potassium for every additional gram of fiber.

7. a) The relationship is not straight.
   b) It will be curved downward.

9. a) Model is appropriate.
   b) Model is not appropriate. Relationship is nonlinear.
   c) Model may not be appropriate. Spread is changing.

11. a) *Price* (in thousands of dollars) is y and *Size* (in square feet) is x.
    b) Slope is thousands of $ per square foot.
    c) Positive. Larger homes should cost more.

13. a) Price should be 0.845 SDs above the mean in price.
    b) Price should be 1.690 SDs below the mean in price.

15. a) *Price* increases by about $0.061 × 1000, or $61.00, per additional sq ft.
    b) 230.82 thousand, or $230,820.
    c) $115,020; $6000 is the residual.

17.

	$\bar{x}$	s_x	$\bar{y}$	s_y	r	$\hat{y} = a + bx$
a)	10	2	20	3	0.5	$\hat{y} = 12.5 + 0.75x$
b)	2	0.06	7.2	1.2	−0.4	$\hat{y} = 23.2 − 8x$
c)	12	6	**152**	**30**	−0.8	$\hat{y} = 200 − 4x$
d)	2.5	1.2	**25**	100	**0.6**	$\hat{y} = −100 + 50x$

19. Perhaps high blood pressure causes high body fat, high body fat causes high blood pressure, or both could be caused by a lurking variable such as a genetic or lifestyle issue.

21. a) Probably. The scatterplot is reasonably straight and there are no outliers.
    b) Nicotine should be 1.92 SDs below average.
    c) Tar should be 0.96 SDs above average.

23. a) $\widehat{Nicotine} = 0.15403 + 0.065052\ Tar$
    b) 0.414 mg
    c) Predicted nicotine content increases by 0.065 mg of nicotine per additional milligram of tar.
    d) We'd expect a cigarette with no tar to have 0.154 mg of nicotine.
    e) 0.1094 mg

25. a) Yes. The scatterplot shows a linear pattern with no outliers.
    b) $\widehat{High\ Jump} = −20.76 + 0.331 × Long\ Jump$.
    c) As the long jump increases by 1 inch, the high jump increases by 0.331 inches, on average, according to the model.
    d) 95.09 inches.

27. a)

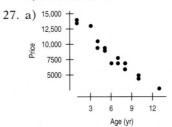

b) $\widehat{Price} = 14,286 − 959 × Years$.
c) Yes.
d) Every extra year of age decreases average value by $959.
e) The average new Corolla costs a predicted $14,286.
f) Negative residual. Its price is below the predicted value for its age.
g) −$1195
h) No. After age 14, the model predicts negative prices. The relationship is no longer linear.

29. a) $\%\ Body\ Fat = −27.4 + 0.25 × Weight$.
    b) Residuals look randomly scattered around 0, so conditions are satisfied.
    c) $\%\ Body\ Fat$ increases, on average, by 0.25 percent per pound of *Weight*.

31. a)

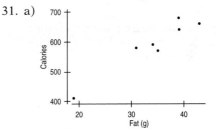

b) $\widehat{Calories} = 211.0 + 11.06 × Fat$.
c)

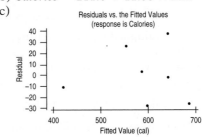

Residuals show no clear pattern, so the model seems appropriate.

d) Could say a fat-free burger still has 211.0 calories, but this is extrapolation (no data close to 0).
e) Every gram of fat adds 11.06 calories, on average.
f) 553.7 calories.

33. a) The regression was for predicting calories from fat, not the other way around.
    b) $\widehat{Fat} = −15.0 + 0.083 × Calories$. Predict 34.8 grams of fat.

35. a) The model isn't cause and effect; it describes birds in general, not the growth of an individual bird.
    b) Predictions based on a regression line are estimates of average values of y for a given x. The actual wingspan will vary around the prediction.

37. a) Probably not. Your score is better than about 97.5% of people, assuming scores follow the Normal model. Your next score is likely to be closer to the mean.
    b) The friend should probably retake the test. His score is better than only about 16% of people. His score is likely to be closer to the mean.

39. a) *Cost* decreases by $2.13 per degree of average daily *Temp*. So warmer temperatures indicate lower costs.
    b) For an avg. monthly temperature of 0°F, the cost is predicted to be $133.
    c) Too high; the residuals (observed − predicted) around 32°F are negative, showing that the model overestimates the costs.
    d) $111.70   e) About $105.70
    f) No, the residuals show a definite curved pattern. The data are probably not linear.
    g) No, there would be no difference. The relationship does not depend on the units.

41. a) Yes; the relationship looks fairly straight. Possibly some outliers (higher-than-expected math scores).
    b) $\widehat{Math} = 217.7 + 0.662 \times Verbal$.
    c) Every point of verbal score adds 0.662 points to the predicted average math score.
    d) 548.5 points   e) 53.0 points

43. a) 0.685   b) $\widehat{Verbal} = 162.1 + 0.71 \times Math$.
    c) The observed verbal score is higher than predicted from the math score
    d) 517.1 points.

Chapter 9

1. a) Linear; y increases at by 150 each time x increases by 2.
   b) Exponential; y doubles each time x increases by 2.

3. a) Exponential; y is multiplied by 2/3 each time x increases by 10.
   b) Linear; y decreases by 15 each time x increases by 10.

5. a) 48   b) 6   c) 27   d) 3

7. a) 500   b) 6% per day   c) 752 bacteria

9. a) 80   b) 3.4% per week   c) 13.24 grams

11. Model b shows curvature.

13. No. Residuals show a curved pattern.

15. Residuals are more randomly spread around 0, with some low outliers.

17. a) The plot shows a wavy pattern, likely part of an annual cycle that repeats each year.
    b) No. Cyclic patterns like this cannot be easily modeled.

19. a) $\widehat{Coins} = 104.95(0.48^{Toss})$; residuals look reasonably random.
    b) 0.48 represents a 52% death rate, approximately 50%.

21. a) Residuals have a curved shape, so linear model is not appropriate.
    b) $\widehat{Distance} = 0.852\,Speed^{1.43}$
    c) 262.5 feet   d) 370.6 feet
    e) Okay for 55 mph, but 70 mph is an extrapolation.

23. a) $\widehat{Distance} = 17.55(1.87^{position})$
    b) Pluto's residual is huge, over 1000 million miles, giving support to the argument that Pluto doesn't behave like a planet.

25. No. The predicted *Distance* of Eris is 9177 million miles, far beyond the actual average distance of 6300 million miles.

27. a) No. The relationship is curved.
    b) $\widehat{Brightness} = 31391\,Distance^{-2.015}$. There are some larger residuals at short distances, but otherwise the residuals are quite small and look random.
    c) 93 cp
    d) The exponent near −2 suggests that this is an example of the inverse square law.

29. $\widehat{Time} = 20.362\,Diameter^{-1.96}$

31. a) The plot looks fairly straight. (It is okay to see a bend in the plot; there's one there.)

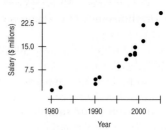

    b) $\widehat{Salary} = -3.18 + 0.965\,Year$

The residuals plot shows a strong bend.
    c) $\widehat{Salary} = 1.2(1.135)^{Year}$

33. a) The scatterplot is curved.
    b) $\widehat{Cost} = 473\,Chips^{-0.502}$

35. The pattern increases and then decreases, not easily modeled using the methods of this chapter.

37. a) $\widehat{Weight} = 0.5987 - 0.000256\,Oranges$, but the residuals plot is curved.
    b) The power model $\widehat{Weight} = 1.222\,Oranges^{-0.159}$ also has curved residuals, and the residuals for the linear model are very small.

Chapter 10

1. a) No. It would be nearly impossible to get exactly 500 males and 500 females from every country by random chance.
   b) A stratified sample, stratified by whether the respondent is male or female.

3. a) Voluntary response.
   b) We have no confidence at all in estimates from such studies.

5. a) Biased toward yes because of "pollute." "Should companies be responsible for any costs of environmental cleanup?"
   b) Biased toward no because of "old enough to vote and serve in the military." "Do you think the drinking age should be lowered from 21?"

7. Bias. Only people watching the news will respond, and their preference may differ from that of other voters. The sampling method may systematically produce samples that don't represent the population of interest.

9. a) Population—All U.S. adults.
   b) Parameter—Proportion who have used and benefited from alternative medicine.
   c) Sampling Frame—All Consumers Union subscribers.
   d) Sample—Those who responded.
   e) Method—Questionnaire to all (nonrandom).
   f) Bias—Nonresponse. Those who respond may have strong feelings one way or another.

11. a) Population—Adults.
    b) Parameter—Proportion who think drinking and driving is a serious problem.
    c) Sampling Frame—Bar patrons.
    d) Sample—Every 10th person leaving the bar.
    e) Method—Systematic sampling (may be random).
    f) Bias—Those interviewed had just left a bar. They may think drinking and driving is less of a problem than do other adults.

13. a) Population—Soil around a former waste dump.
    b) Parameter—Concentrations of toxic chemicals.
    c) Sampling Frame—Accessible soil around the dump.
    d) Sample—16 soil samples.
    e) Method—Not clear.
    f) Bias—Don't know if soil samples were randomly chosen. If not, may be biased toward more or less polluted soil.

15. a) Population—Snack food bags.
    b) Parameter—Weight of bags, proportion passing inspection.
    c) Sampling Frame—All bags produced each day.
    d) Sample—Bags in 10 randomly selected cases, 1 bag from each case for inspection.
    e) Method—Multistage random sampling.
    f) Bias—Should be unbiased.

17. a) Voluntary response. Only those who see the ad, have Internet access, *and* feel strongly enough will respond.
    b) Cluster sampling. One school may not be typical of all.
    c) Attempted census. Will have nonresponse bias.
    d) Stratified sampling with follow-up. Should be unbiased.

19. a) This is a multistage design, with a cluster sample at the first stage and a simple random sample for each cluster.
    b) If any of the three churches you pick at random is not representative of all churches, then you'll increase sampling error by the choice of that church.

21. a) This is a systematic sample.
    b) The sampling frame is patrons willing to wait for the roller coaster on that day at that time. It should be representative of the people in line, but not of all people at the amusement park.
    c) It is likely to be representative of those waiting for the roller coaster. Indeed, it may do quite well if those at the front of the line respond differently (after their long wait) than those at the back of the line.

23. a) Answers will definitely differ. Question 1 will probably get many "No" answers, while Question 2 will get many "Yes" answers. This is response bias.
    b) "Do you think standardized tests are appropriate for deciding whether a student should be promoted to the next grade?" (Other answers will vary.)

25. a) Depends on the Yellow Pages listings used. If from regular (line) listings, this is fair if all doctors are listed. If from ads, probably not, as those doctors may not be typical.
    b) Not appropriate. This cluster sample will probably contain listings for only one or two business types.

27. a) Assign numbers 001 to 120 to each order. Use random numbers to select 10 transactions to examine.
    b) Sample proportionately within each type. (Do a stratified random sample.)

29. a) The population of interest is all adults in the United States aged 18 and older.
    b) The sampling frame is U.S. adults with telephones.
    c) Some members of the population (e.g., many college students) don't have landline phones, which could create a bias.

31. a) Not everyone has an equal chance. Misses people with unlisted numbers, or without landline phones, or at work.
    b) Generate random numbers and call at random times.
    c) Under the original plan, those families in which one person stays home are more likely to be included. Under the second plan, many more are included. People without landline phones are still excluded.
    d) It improves the chance of selected households being included.
    e) This takes care of unlisted phone numbers. Time of day may be an issue. People without landline phones are still excluded.

33. a) Answers will vary.
    b) Parameter is your own arm length; population is all possible measurements of it.
    c) Population is now the arm lengths of you and your friends. The average estimates the mean of these lengths.
    d) Probably not. Friends are likely to be of the same age and not very diverse or representative of the larger population.

Chapter 11

1. a) No. There are no manipulated factors. Observational study.
   b) There may be lurking variables that are associated with both parental income and performance on the SAT.

3. a) This is a retrospective observational study.
   b) That's appropriate because MS is a relatively rare disease.
   c) The subjects were U.S. military personnel, some of whom had developed MS.
   d) The variables were the vitamin D blood levels and whether or not the subject developed MS.

5. a) This was a randomized, placebo-controlled experiment.
   b) 351 women aged 45 to 55 who reported at least two hot flashes a day.
   c) The treatments were black cohosh, a multiherb supplement, a multiherb supplement plus advice, estrogen, and a placebo. The response was the women's symptoms (presumably frequency of hot flashes).
   d) The difference was larger than we'd expect just by random chance.

7. a) Experiment.
   b) Bipolar disorder patients.
   c) Omega-3 fats from fish oil, two levels.
   d) 2 treatments.
   e) Improvement (fewer symptoms?).
   f) Design not specified.
   g) Blind (due to placebo), unknown if double-blind.
   h) Individuals with bipolar disease improve with high-dose omega-3 fats from fish oil.

9. a) Observational study.
   b) Prospective.
   c) Men and women with moderately high blood pressure and normal blood pressure, unknown selection process.
   d) Memory and reaction time.
   e) As there is no random assignment, there is no way to know that high blood pressure *caused* subjects to do worse on memory and reaction-time tests. A lurking variable may also be the cause.

11. a) Observational study.
    b) Retrospective.
    c) Women in Finland, unknown selection process with data from church records.
    d) Women's lifespans.
    e) As there is no random assignment, there is no way to know that having sons or daughters shortens or lengthens the lifespan of mothers.

13. a) Observational study.
    b) Prospective.
    c) People with or without depression, unknown selection process.
    d) Frequency of crying in response to sad situations.
    e) There is no apparent difference in crying response (to sad movies) for depressed and nondepressed groups.

15. a) Experiment.
    b) Postmenopausal women.
    c) Alcohol—2 levels; blocking variable—estrogen supplements (2 levels).
    d) 1 factor (alcohol) at 2 levels = 2 treatments.
    e) Increase in estrogen levels.
    f) Blocked.
    g) Not blind.
    h) Indicates that alcohol consumption *for those taking estrogen supplements* may increase estrogen levels.

17. a) Experiment.
    b) Athletes with hamstring injuries.
    c) 1 factor: type of exercise program (2 levels).
    d) 2 treatments.
    e) Time to return to sports.
    f) Completely randomized.
    g) No blinding—subjects must know what kind of exercise they do.
    h) Can determine which of the two exercise programs is more effective.

19. They need to compare omega-3 results to something. Perhaps bipolarity is seasonal and would have improved during the experiment anyway.

21. a) Subjects' responses might be related to many other factors (diet, exercise, genetics, etc.). Randomization should equalize the two groups with respect to unknown factors.
    b) More subjects would minimize the impact of individual variability in the responses, but the experiment would become more costly and time consuming.

23. People who engage in regular exercise might differ from others with respect to bipolar disorder, and that additional variability could obscure the effectiveness of this treatment.

25. The improvement was greater than one could expect to happen by random chance alone.

27. a) First, they are using athletes who have a vested interest in the success of the shoe by virtue of their sponsorship. They should choose other athletes. Second, they should randomize the order of the runs, not run all the races with their shoes second. They should blind the athletes by disguising the shoes if possible, so they don't know which is which. The timers shouldn't know which athletes are running with which shoes, either. Finally, they should replicate several times, since times will vary under both shoe conditions.
    b) Because of the problems in part a, the results they obtain may favor their shoes. In addition, the results obtained for Olympic athletes may not be generalized to other runners.

29. Randomly assign half the reading teachers in the district to use each method. Students should be randomly assigned to teachers as well. Make sure to block both by school and grade (or control grade by using only one grade). Construct an appropriate reading test to be used at the end of the year, and compare scores.

31. a) Allowing athletes to self-select treatments could confound the results. Other issues such as severity

of injury, diet, age, etc., could also affect time to heal; randomization should equalize the treatment groups with respect to any such variables.

b) A control group could have revealed whether either exercise program was better (or worse) than just letting the injury heal.

c) Doctors who evaluated the athletes to approve their return to sports should not know which treatment the subject had.

d) It's hard to tell. The difference of 15 days seems large, but the standard deviations indicate that there was a great deal of variability in the times.

33. a) Arrange the 20 containers in 20 separate locations. Randomly decide which 10 containers should be filled with water.

b) Guessing, the dowser should be correct about 50% of the time. A record of 60% (12 out of 20) does not appear to be significantly different.

c) Answers may vary. You would need to see a high level of success—say, at least 80%, that is, 16 to 20 correct.

35. a) Observational, prospective study.

b) The supposed relation between health and wine consumption might be explained by the confounding variables of income and education.

c) None of these. While the variables have a relation, there is no causality indicated for the relation.

37. a) They mean that the difference is higher than they would expect from normal sampling variability.

b) An observational study.

c) No. Perhaps the differences are attributable to some confounding variable (e.g., people are more likely to engage in riskier behaviors on the weekend) rather than the day of admission.

d) Perhaps people have more serious accidents and traumas on weekends and are thus more likely to die as a result.

39. a) Answers may vary. Randomly assign the eight patients to either the current medication or the new medication. Have nurses assess the degree of shingles involvement for the patient. Ask patients to rate their pain levels. Administer the medications for a prescribed time. Have nurses reassess the degree of shingles involvement. Ask patients to rate their pain levels post-medication. Compare the improvement levels.

b) Assuming that the ointments look alike, it would be possible to blind the experiment for the subject and for the administrator of the treatment.

c) A block design with factors for gender and for ointment would be appropriate. Subjects would be randomly assigned to each treatment group in the blocked design.

Chapter 12

1. Yes. You cannot predict the outcome beforehand.

3. A machine pops up numbered balls. If it were truly random, the outcome could not be predicted and the outcomes would be equally likely. It is random only if the balls generate numbers in equal frequencies.

5. Use two-digit numbers 00–99; let $00-02 = $ defect, $03-99 = $ no defect

7. a) 45, 10   b) 17, 22

9. Choose students 15(Harris), 11(Burbank), 12(Colongeli), and 01(Campbell).

11. Number the boys from 1 to 9. Use the calculator to generate two random digits 1–9, and choose those boys. Skip repeated digits. Then number the girls from 1 to 11. Generate two random numbers 1–11 (skipping repeats again), and choose those two girls.

13. a) The outcomes are not equally likely; for example, tossing 5 heads does not have the same probability as tossing 0 or 9 heads, but the simulation assumes they are equally likely.

b) The even-odd assignment assumes that the player is equally likely to score or miss the shot. In reality, the likelihood of making the shot depends on the player's skill.

c) The likelihood for the first ace in the hand is not the same as for the second or third or fourth. But with this simulation, the likelihood is the same for each. (And it allows you to get 5 aces, which could get you in trouble in a real poker game!)

15. The conclusion should indicate that the simulation *suggests* that the average length of the line would be 3.2 people. Future results might not match the simulated results exactly.

17. Answers will vary, but average answer will be about 36%.

19. a) One possible approach: Use random digits 0–9. Let $0-7 = $ right and $8-9 = $ wrong.

b) Generate 6 random digits 0–9 and classify each as right (7 or less) or wrong (8, 9).

c) Answers will vary, but the average answer will be about 26%.

21. a) Answers will vary, but you should win about 10% of the time.

b) You should win at the same rate with any number.

23. Answers will vary, but you should win about 10% of the time.

25. Answers will vary, but average answer will be about 3 children.

27. a) Use 2-digit random numbers 00–99. Let $00-71 = $ hit, $72-99 = $ miss.

b) Look at the first random number. If it's a miss, the player gets 0 points. If it's a hit, the player gets one point and we look at the next random number. If it's a hit, the player gets a second point.

c) Answers will vary, but the average answer will be about 1.24 points.

29. Answers will vary, but average answer will be about 7.5 rolls.

31. Do the simulation in two steps. First simulate the payoffs. Then count until $500 is reached. Answers will vary, but average should be near 10.2 customers.

33. No, it will happen about 40% of the time.

35. Three women will be selected about 7.8% of the time.

Chapter 13

1. a) S = {heart, club, diamond, spade}; equally likely.
   b) S = {1,2,3,4,5,6}; not equally likely.
   c) S = {HH,HT,TH,TT}, equally likely.
   d) S = {H,TH,TTH,TTT}, not equally likely.

3. In this context "truly random" should mean that every number is equally likely to occur.

5. This estimate is based on the long run (so far) experience (data) for similar patients.

7. There is no "Law of Averages." She would be wrong to think that they are "due" for a harsh winter.

9. There is no "Law of Averages." If at-bats are independent, his chance for a hit does not change based on recent successes or failures.

11. There is no "Law of Averages." If crashes are independent, it makes no difference. If crashes were due to problems with the aircraft, another crash may be more likely; however, increased maintenance vigilance may *lessen* the chance of another crash.

13. a) $_{12}P_4 = 11,880$ b) $_{12}C_4 = 495$

15. $_{12}C_6 = 924$

17. a)

X	1	2	3	4	5	6
1	1	2	3	4	5	6
2	2	4	6	8	10	12
3	3	6	9	12	15	18
4	4	8	12	16	20	24
5	5	10	15	20	25	30
6	6	12	18	24	30	36

   b) $\frac{1}{4}$

19. a) $_{52}C_5 = 2,598,960$ b) $_{13}C_5 = 1287$
   c) $\frac{4(1287)}{2,598,960} \approx 0.002$

21. a) There is some chance you would have to pay out much more than the $300.
   b) Many customers pay for insurance. The small risk for any one customer is spread among all.

23. a) Representing mother-father:

   O-O   O-A   O-B   O-AB
   A-O   A-A   A-B   A-AB
   B-O   B-A   B-B   B-AB
   AB-O AB-A AB-B AB-AB

   b) Blood types aren't equally likely.

25. a) S = {0,1,2,3}
   b) These aren't equally likely; there's only one birth order that's all boys, but several orders that have only 1 or 2 boys.
   c) Using {BBB,BBG,BGB,GBB,BGG,GBG,GGB, GGG}, P(2 boys) = $\frac{3}{8}$.
   d) That 50% of children are boys, and in a family the sexes are independent.

27. a) S = {1,2,3,4,5,6}

big	1	2	3	4	5	6
1	1	2	3	4	5	6
2	2	2	3	4	5	6
3	3	3	3	4	5	6
4	4	4	4	4	5	6
5	5	5	5	5	5	6
6	6	6	6	6	6	6

   b) These aren't equally likely; it's much rarer to have a 1 as the larger roll.
   c) Based on the table, P(5) = $\frac{1}{4}$.

29. $\frac{_4C_3}{_{12}C_3} \approx 0.018$

31. a) $\frac{_{18}C_5}{_{26}C_5} \approx 0.13$ b) Use a stratified sample.
   c) $_{18}C_3 \cdot _8C_2 = 22,848$ d) $\frac{22,848}{_{26}C_5} \approx 0.347$

33. $\frac{_4C_3 \cdot _{48}C_2 + _4C_4 \cdot _{48}C_1}{_{52}C_5} \approx 0.0018$ (fewer than one hand in 500)

35. $\frac{_{17}C_2 \cdot _{13}C_2 \cdot _4C_2}{_{34}C_6} \approx 0.047$

37. $\frac{_{17}C_6 + _{13}C_6}{_{34}C_6} \approx 0.01$

39. a) $23 \cdot 23 \cdot 23 \cdot 10 \cdot 10 = 1,216,700$
   b) $\frac{6 \cdot 6 \cdot 6 \cdot 10 \cdot 10}{1,216,700} \approx 0.018$

41. a) $\frac{_6C_3 \cdot _{43}C_3}{_{49}C_6} \approx 0.0177$ (about 1 in every 57 tickets)
   b) $\frac{_6C_3 \cdot _{43}C_3 + _6C_4 \cdot _{43}C_2 + _6C_5 \cdot _{43}C_1 + _6C_6}{_{49}C_6} \approx 0.0186$
   (about 1 in every 54 tickets)

43. The probability that all 4 people hired would be among the 10 non-Republicans $\frac{_{10}C_4}{_{22}C_4} \approx 0.029$. In other words, there's only about 1 chance in 35 that this outcome occurred purely by chance, probably an indication of favoritism.

45. a) $\frac{6!}{2!} = 360$ b) $\frac{10!}{3!} = 604,800$
   c) $\frac{11!}{2!2!} = 9,979,200$ d) $\frac{7!}{3!2!} = 420$
   e) $\frac{10!}{3!3!2!} = 50,400$ f) $\frac{11!}{2!2!2!2!} = 2,494,800$

Chapter 14

1. a) Legitimate.   b) Legitimate.
   c) Not legitimate (sum more than 1).
   d) Legitimate.
   e) Not legitimate (can't have negatives or values more than 1).

3. a) 0.04   b) 0.5

5. a) 0.22   b) 0.28
   c) We assume responses are independent.
   d) Yes, because people are being selected at random.

7. a) 0.72   b) 0.89   c) 0.28

9. a) 0.5184   b) 0.0784   c) 0.4816

11. a) Repair needs for the two cars must be independent.
    b) Maybe not. An owner may treat the two cars similarly, taking good (or poor) care of both. This may decrease (or increase) the likelihood that each needs to be repaired.

13. A family may own both a car and an SUV. The events are not disjoint, so the Addition Rule does not apply.

15. When cars are traveling close together, their speeds are not independent, so the Multiplication Rule does not apply.

17. a) 0.30   b) 0.30   c) 0.90   d) 0.0

19. a) 0.027   b) 0.128   c) 0.512   d) 0.271

21. Disjoint (can't be both red and orange).

23. Independent (unless you're drawing from a small bag).

25. a) 0.0046   b) 0.125   c) 0.296   d) 0.421   e) 0.995

27. a) 0.027   b) 0.063   c) 0.973   d) 0.014

29. a) 0.024   b) 0.250   c) 0.543

31. 0.078

33. a) He has multiplied the two probabilities.
    b) He assumes that being accepted at the colleges are independent events.
    c) No. Colleges use similar criteria for acceptance, so the decisions are not independent.

35. a) 0.4712   b) 0.7112
    c) $(1 - 0.76) + 0.76(1 - 0.38)$ *or* $1 - (0.76)(0.38)$

37. a) No car would have both defects.
    b) If a car has one of the defects, it's no more or less likely to also have the other one.
    c) No. If the two defects are disjoint, then knowing a car has the trunk latch problem means it can't have the brake light problem. But if the events were independent, the chances a car had the brake problem would be the same regardless of whether or not it had the trunk problem.

39. a) For any day with a valid three-digit date, the chance is 0.001, or 1 in 1000. For many dates in October through December, the probability is 0. (No three digits will make 10/15, for example.)
    b) There are 65 days when the chance to match is 0. (Oct. 10-31, Nov. 10-30, and Dec. 10-31.) The chance for no matches on the remaining 300 days is 0.741.
    c) 0.259   d) 0.049

Chapter 15

1. 0.68

3. b) i) 0.32   ii) 0.04

5. 0.266

7. No, 26.6% of homes with garages have pools; 21% of homes overall have pools.

9. No, 17% of homes have both.

11. a) 0.50   b) 1.00   c) 0.077   d) 0.333

13. 0.059

15. 0.21

17. a) 0.11   b) 0.27   c) 0.48

19. a) 0.407   b) 0.344

21. No, the probability that the luggage arrives on time depends on whether the flight is on time. The probability is 95% if the flight is on time and only 65% if not.

23. 0.695

25. a) 0.31   b) 0.48   c) 0.31

27. a) 32%   b) 0.135
    c) No, 7% of juniors have taken both.
    d) No, the probability that a junior has taken a computer course is 0.23. The probability that a junior has taken a computer course *given* he or she has taken a Statistics course is 0.135.

29. Yes, P(Ace) = 4/52. P(Ace | any suit) = 1/13.

31. No, only 32% of all men have high cholesterol, but 40.7% of those with high blood pressure do.

33. a) 95.6%
    b) Probably. 95.4% of people with cell phones had landlines, and 95.6% of all people did.

35. No. Only 34% of men were Democrats, but over 41% of all voters were.

37. a) 0.145   b) 0.118   c) 0.414   d) 0.217

39. a) 0.318   b) 0.955   c) 0.071   d) 0.009

41. a) No, the probability of missing work for day-shift employees is 0.01. It is 0.02 for night-shift employees. The probability depends on whether they work day or night shift.
    b) 1.4%

43. a) 0.2025   b) 0.6965   c) 0.2404   d) 0.0402

45. a) 0.011   b) 0.222   c) 0.054   d) 0.337   e) 0.436

47. a) 0.17
    b) No; 13% of the chickens had both contaminants.
    c) No; P(C | S) = 0.87 ≠ P(C). If a chicken is contaminated with salmonella, it's more likely also to have campylobacter.

49. a) 0.20   b) 0.272

51. 0.016

53. a) 0.8377   b) Over 0.999

Chapter 16

1. a) 19   b) 4.2

3. a) 7   b) 1.89

5. a) 1.7   b) 0.9

7. a)

X = defects	3	2	1	0
P(X)	0.07	0.11	0.21	061

   b) $\mu = 0.64$ c) $\sigma = 0.93$

9. a) No. More than two outcomes are possible.
   b) Yes, assuming the people are unrelated to each other.
   c) No. The chance of a heart changes as cards are dealt so the trials are not independent.
   d) No, we're sampling without replacement, so the probability of success changes (and 500 is more than 10% of 3000).
   e) If packages in a case are independent of each other, yes.

11. a) 0.31   b) 0.36   c) 0.17   d) 0.84

13. a) $E(X) = 3.5$ b) $SD(X) = 1.025$

15. a) 0.817   b) 0.167   c) 0.015
    d) 0.0008   e) barely less than 1

17. $\mu = 20, \sigma = 4.43$

19. a)

Amount won	$0	$5	$10	$30
P(Amount won)	$\frac{26}{52}$	$\frac{13}{52}$	$\frac{12}{52}$	$\frac{1}{52}$

   b) $4.13   c) $4 or less (answers may vary)

21. $5.44

23. a) $50   b) $100

25. a)

Number good	0	1	2
P(Number of good)	0.067	0.467	0.467

   b) 1.40   c) 0.61

27. Departures from the same airport during a 2-hour interval may not be independent. All could be delayed by weather, for example.

29. a) 0.058   b) 0.296   c) 0.745
    d) 0.031   e) 0.983

31. a) 0.118   b) 0.324   c) 0.744   d) 0.580

33. a) $\mu = 56, \sigma = 4.10$
    b) Yes, $np = 56 \geq 10$, $nq = 24 \geq 10$, serves are independent.
    c) Normal, approx.: 0.014; Binomial, exact: 0.016

35. a) $\mu = 96, \sigma = 4.38$
    b) Normal, approx.: 0.819; Binomial, exact: 0.848

37. a) $\mu = 18, \sigma = 4.11$
    b) Assuming apples fall and become blemished independently of each other, Binom(300, 0.06) is appropriate. Since $np \geq 10$ and $nq \geq 10$, $N(18, 4.11)$ is also appropriate.
    c) Yes, 50 is 7.8 SDs above the mean.

39. The mean number of sales should be 24 with SD 4.60. Ten sales is more than 3.0 SDs below the mean. He was probably misled.

41. a)

Children	1	2	3
P(Children)	0.5	0.25	0.25

   b) 1.75 children   c) 0.83 children

43. a) No. The probability of winning the second depends on the outcome of the first.
    b) 0.42   c) 0.08
    d)

Games won	0	1	2
P(Games won)	0.42	0.50	0.08

    e) $\mu = 0.66, \sigma = 0.62$

45. a) Use single random digits. Let 0, 1 = LeBron. Examine random digits in groups of five, counting the number of 0's and 1's.
    c) Results will vary.
    d)

x	0	1	2	3	4	5
P(x)	0.33	0.41	0.20	0.05	0.01	0.0

47. Normal, approx.: 0.053; Binomial, exact: 0.061

49. a) We assume serves are independent; both $np = 56$ and $nq = 24$ are at least 10.
    b) $\mu = 56, \sigma = 4.1$

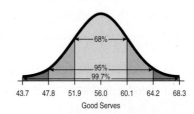

    c) In a match with 80 serves, approximately 68% of the time she will have between 51.9 and 60.1 good serves, approximately 95% of the time she will have between 47.8 and 64.2 good serves, and approximately 99.7% of the time she will have between 43.7 and 68.3 good serves.

51. a) We assume the performance of chips is independent; 1000 is less than 10% of all computer chips and both $np = 20$ and $nq = 980$ are at least 10.
    b) $\mu = 20, \sigma = 4.43$
    c) On 68% of the days, the company can expect to produce between 15.57 and 24.43 defective chips. They should have between 11.14 and 28.86 on 95% of the days, and between 6.71 and 33.29 defective chips on 99.7% of the days.

53. $\mu = 20, \sigma = 4$. I'd want *at least* 32 (3 SDs above the mean). (Answers will vary.)

55. Yes. We'd expect him to make 22 shots, with a standard deviation of 3.15 shots. 32 shots is more than 3 standard deviations above the expected value, an unusually high rate of success.

Chapter 17

1. a) 1/6   b) 0.037

3. a) Yes. Rolls are independent; $np = 100(\frac{1}{6}) = 16.67$, $nq = 100(\frac{5}{6}) = 83.33$, both at least 10.

b) $N(0.167, 0.037)$

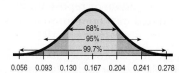

0.056   0.093   0.130   0.167   0.204   0.241   0.278

c) If we roll a pair of fair dice 100 times, 68% of the time we'd expect to get between 13% and 20.4% 7s, 95% if the time between 9.3% and 24.1% 7s, and between 5.6% and 27.8% 7s 99.7% of the time.

5. a) 0.42   b) 0.055

7. a) Yes. Assume the donors are a representative sample; $80 < 10\%$ of all Americans; $np = 80(0.42) = 33.6$, $nq = 80(0.58) = 46.4$, both at least 10.

b)

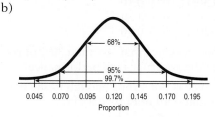

0.045   0.070   0.095   0.120   0.145   0.170   0.195
Proportion

c) At 68% of these drives the Red Cross can expect between 36.5% and 47.5% of donors to be Type A, between 31% and 53% at 95% of the drives, and between 25.5% and 58.5% at 99.7% of blood drives where there are 80 donors.

9. She believes the true proportion is within 4% of her estimate, with some (probably 95%) degree of confidence.

11. a) Not correct. This implies certainty.
    b) Not correct. Different samples will give different results. Many fewer than 95% will have 88% on-time orders.
    c) Not correct. The interval is about the population proportion, not the sample proportion in different samples.
    d) Not correct. In this sample, we *know* 88% arrived on time.
    e) Not correct. The interval is about the parameter, not the days.

13. On the basis of this sample, we are 95% confident that the proportion of Japanese cars is between 29.9% and 47.0%.

15. It's likely that between 44% and 64% of all Scrabble hands contain the letter "A."

17. a) Population—all cars; sample—those actually stopped at the checkpoint; p—proportion of all cars with safety problems; $\hat{p}$—proportion actually seen with safety problems (10.4%); if sample (a cluster sample) is representative, then the methods of this chapter will apply.
    b) Population—general public; sample—those who logged onto the website; p—population proportion of those who favor prayer in school; $\hat{p}$—proportion of those who voted in the poll who favored prayer in school (81.1%); can't use methods of this chapter—sample is biased and nonrandom.

c) Population—parents at the school; sample—those who returned the questionnaire; p—proportion of all parents who favor uniforms; $\hat{p}$—proportion of respondents who favor uniforms (60%); should not use with methods of this chapter, since not SRS (nonresponse may bias results).

d) Population—students at the college; sample—the 1632 students who entered that year; p—proportion of all students who will graduate on time; $\hat{p}$—proportion of that year's students who graduate on time (85.0%); can use methods of this chapter if that year's students (a cluster sample) are viewed as a representative sample of all possible students at the school.

19. a) 0.23   b) 0.01

21. a) Yes; it's a random sample of less than 10% of all students with 417 successes and 1398 failures, both at least 10.
    b) ±0.02
    c) I'm 95% confident that between 21% and 25% of all high school students are current smokers.

23. a) 0.685   b) 0.025

25. a) We assume these cars are a representative sample of the traffic here; $355 < 10\%$ of all cars on this highway; there were 243 successes and 112 failures, both at least 10.
    b) ±0.05
    c) We can be 95% confident that between 63.5% and 73.5% of the cars on this highway exceed the speed limit by more than 5 miles per hour.

27. a) Symmetric, because probability of heads and tails is equal.
    b) 0.5   c) 0.125   d) $np = 8 < 10$

29. a) About 68% should have proportions between 0.4 and 0.6, about 95% between 0.3 and 0.7, and about 99.7% between 0.2 and 0.8.
    b) $np = 12.5, nq = 12.5$; both are ≥ 10.
    c)

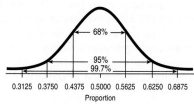

0.3125   0.3750   0.4375   0.5000   0.5625   0.6250   0.6875
Proportion

$np = nq = 32$; both are ≥ 10.
    d) Becomes narrower (less spread around 0.5).

31. Probably nothing. Those who bothered to fill out the survey may be a biased sample.

33. a) Assume the *Consumer Reports* sample is representative of all chickens; $525 < 10\%$ of all broilers for sale, and the 436 successes and 89 failures are both at least 10.
    b) (0.798, 0.863)
    c) We're 95% confident that between 80% and 86% of all broiler chicken sold in U.S. food stores is infected with *Campylobacter*.

35. 3195
37. a) (0.0465, 0.0491). The assumptions and conditions for constructing a confidence interval are satisfied.
    b) The confidence interval gives the set of plausible values (with 95% confidence). Since 0.05 is outside the interval, that seems to be a bit too optimistic.
39. a) Assume these are typical hens; 280 < 10% of all hens; the 221 successes and 59 failures are both at least 10.
    b) 0.79 ± 0.05
    c) We're 95% confident that the daily egg production rate among hens fed this supplement would be between 74% and 84%.
    d) 1120
41. a) False   b) True   c) True   d) False
43. All the histograms are centered near 0.05. As n gets larger, the histograms approach the Normal shape, and the variability in the sample proportions decreases.
45. a)

n	Observed Mean	Theoretical Mean	Observed st. dev.	Theoretical st. dev.
20	0.0497	0.05	0.0479	0.0487
50	0.0516	0.05	0.0309	0.0308
100	0.0497	0.05	0.0215	0.0218
200	0.0501	0.05	0.0152	0.0154

    b) They are all quite close to what we expect from the theory.
    c) The histogram is unimodal and symmetric for $n = 200$.
    d) The Success/Failure Condition says that np and nq should both be at least 10, which is not satisfied until $n = 200$ for $p = 0.05$. The theory predicted my choice.
47. Answers will vary. Using $\mu + 3\sigma$ for "very sure," the restaurant should have 89 nonsmoking seats. Assumes customers at any time are independent of each other, a random sample, and represent less than 10% of all potential customers. $np = 72$, $nq = 48$, so Normal model is reasonable ($\mu = 0.60$, $\sigma = 0.045$).
49. a) 0.025
    b) We're 90% confident that this poll's estimate is within $\pm 2.5\%$ of the true proportion of people who are baseball fans.
    c) Larger. To be more certain, we must be less precise.
    d) 0.039
    e) Less confidence.
    f) No evidence of change; given the margin of error, 0.37 is a plausible value for 2007 as well.
51. a) (12.1%, 19.1%)
    b) We are 98% confident, based on this sample, that the proportion of all auto accidents that involve teenage drivers is between 12.1% and 19.1%.
    c) About 98% of all random samples will produce confidence intervals that contain the true population proportion.
    d) Contradicts. The interval is completely below 20%.

53. a) (15.5%, 26.3%)   b) 612
    c) Sample may not be random or representative. Deer that are legally hunted may not represent all sexes and ages.
55. a) Wider. The sample size is probably about one-fourth of the sample size for all adults, so we'd expect the confidence interval to be about twice as wide.
    b) Smaller. The second poll used a slightly larger sample size.

Chapter 18

1. a) $H_0: p = 0.30$; $H_A: p < 0.30$
   b) $H_0: p = 0.50$; $H_A: p \neq 0.50$
   c) $H_0: p = 0.20$; $H_A: p > 0.20$
3. There is strong evidence that less than 30% of voters disapprove of the governor's job performance.
5. No. The evidence is not strong enough to say the coin is unfair, but that does not prove it's fair.
7. There's not enough evidence to say that the tape helps people quit smoking.
9. a) One-tailed. Let p be the percentage of teenagers who prefer the new formulation. $H_0: p = 0.5$ vs. $H_A: p > 0.5$
   b) One-tailed. Let p be the percentage of people who intend to vote for the override. $H_0: p = 2/3$ vs. $H_A: p > 2/3$.
   c) Two-tailed. Let p be the percentage of days that the market goes up. $H_0: p = 0.5$ vs. $H_A: p \neq 0.5$
11. Any apparent difference seen in the sample could be just sampling error rather than an indication that teens like the new formula better.
13. The company missed the fact that students actually do prefer the new formula.
15. It would be very unusual to get a sample showing this much support for the budget override if it actually won't get the 2/3 majority needed to pass.
17. The newspaper concluded the budget override will pass, but in fact not enough voters support it.
19. Type I: The student thinks the market doesn't go up and down with equal probability, but actually it does. Type II: The student concludes that the market is equally likely to go up or down, but actually it's not.
21. 22 is more than 10% of the population of 150; $(0.20)(22) < 10$.
23. $H_0: p = 0.103$; $H_A: p > 0.103$. Not random, but assume last year is representative of current and (near) future years at this school; 1782(0.103) and 1782(0.897) are both at least 10. $z = 2.06$; These data provide evidence that the dropout rate has increased.
25. $H_0: p = 0.03$ $H_A: p \neq 0.03$. One mother having twins will not affect another, so observations are independent; not an SRS; sample is less than 10% of births to

very young mothers. $(0.03)(469) = 14.07 \geq 10$; $(0.97)(469) \geq 10$. $z = -1.91$; fail to reject H_0. These data show little evidence that the rate of twins born to teenage girls at this hospital is less than the national rate of 3%. It is not clear whether this can be generalized to all teenagers.

27. $H_0: p = 0.40$; $H_A: p < 0.40$. Data are for all executives in this company and may not be able to be generalized to all companies; $(0.40)(43) \geq 10$; $(0.60)(43) \geq 10$. $z = -1.31$; P-value $= 0.0955$. Fail to reject H_0. These data do not show that the selection process for executive positions is biased against the women in the company.

29. $H_0: p = 0.90$; $H_A: p < 0.90$. Assume the 122 people were chosen at random; $122 < 10\%$ of all people who have lost luggage; $122(0.90) = 109.8$ and $122(0.10) = 12.2$ are both at least 10. $\hat{p} = 0.844$; $z = -2.05$; P-value $= 0.0201$. Because the P-value is so low, we reject H_0. There is strong evidence that the actual rate at which passengers with lost luggage are reunited with it within 24 hours is less than the 90% claimed by the airline.

31. a) The null is that the level of home ownership remains the same. The alternative is that it rises.
    b) The city concludes that home ownership is on the rise, but in fact the tax breaks don't help.
    c) The city abandons the tax breaks, but they were helping.
    d) A Type I error causes the city to forego tax revenue, while a Type II error withdraws help from those who might have otherwise been able to buy a home.

33. a) It is decided that the shop is not meeting standards when it is.
    b) The shop is certified as meeting standards when it is not.
    c) Type I   d) Type II

35. a) One-tailed. The company wouldn't be sued if "too many" minorities were hired.
    b) Deciding the company is discriminating when it is not.
    c) Deciding the company is not discriminating when it is.

37. A P-value this low is strong evidence that the company's hiring practices discriminate against minority applicants.

39. No, we can say only that there is a 27% chance of seeing the observed effectiveness just from natural sampling variation. There is no *evidence* that the new formula is more effective, but we can't conclude that they are equally effective.

41. Statement d is correct.

43. If there is no difference in effectiveness, the chance of seeing an observed difference this large or larger by natural sampling variation is 4.7%.

45. $\alpha = 0.10$: Yes. The P-value is less than 0.05, so it's less than 0.10. But to reject H_0 at $\alpha = 0.01$, the P-value must be below 0.01, which isn't necessarily the case.

47. $H_0: p = 0.25$; $H_A: p > 0.25$. SRS; sample is less than 10% of all potential subscribers; $(0.25)(500) \geq 10$; $(0.75)(500) \geq 10$. $z = 1.24$; P-value $= 0.1076$. The P-value is high, so do not reject H_0. These data do not show that more than 25% of current readers would subscribe; the company should not go ahead with the WebZine on the basis of these data.

49. a) $H_0: p = 0.63$, $H_A: p > 0.63$
    b) The sample is representative. $240 < 10\%$ of all law school applicants. We expect $240(0.63) = 151.2$ to be admitted and $240(0.37) = 88.8$ not to be, both at least 10. $z = 1.58$; P-value $= 0.057$
    c) Although the evidence is weak, there is some indication that the program may be successful. Candidates should decide whether they can afford the time and expense.

51. a) Yes; assuming this sample to be a typical group of people, $P = 0.0008$. This cancer rate is very unusual.
    b) No, this group of people may be atypical for reasons that have nothing to do with the radiation.

53. a) No. There's a 25% chance of losing twice in a row. That's not unusual.
    b) 0.125
    c) No, we expect that to happen 1 time in 8.
    d) Maybe 5? The chance of 5 losses in a row is only 1 in 32, which seems unusual.

55. a) Type II error   b) Type I error
    c) By making it easier to get the loan, the bank has reduced the alpha level.
    d) The risk of a Type I error is decreased and the risk of a Type II error is increased.

Chapter 19

1. a) 10 ounces   b) 0.115 ounces

3. a) 10 ounces   b) 0.04 ounces

5. a) The confidence interval is for the population mean, not the individual cows in the study.
   b) The confidence interval is not for individual cows.
   c) We *know* the average gain in this study was 56 pounds!
   d) The average weight gain of all cows does not vary. It's what we're trying to estimate.
   e) No. There is not a 95% chance for another sample to have an average weight gain between 45 and 67 pounds. There is a 95% chance that another sample will have its average weight gain within two standard errors of the true mean.

7. a) No. A confidence interval is not about individuals in the population.
   b) No. It's not about individuals in the sample, either.
   c) No. We know the mean cost for students in the sample was $1196.
   d) No. A confidence interval is not about other sample means.
   e) Yes. A confidence interval estimates a population parameter.

9. a) Based on this sample, we can say, with 95% confidence, that the mean pulse rate of adults is between 70.9 and 74.5 beats per minute.
   b) 1.8 beats per minute

11. H_0: $\mu = 340$; H_A: $\mu > 340$

13. H_0: $\mu = 5.20$; H_A: $\mu \neq 5.20$

15. H_0: $\mu = 48.2$; H_A: $\mu < 48.2$

17. If in fact the mean cholesterol of pizza eaters does not indicate a health risk, then only 7 of every 100 samples would have mean cholesterol levels as high (or higher) as observed in this sample.

19. We'd expect to see this level of unpopped kernels in about 19% of our tests even if the new bag design really is no better than the old one.

21. a) Paired; tests the same students before and after using the study guide.
   b) Not paired; the boys and the girls are two unrelated groups.
   c) Paired; tests the same cars on the same track, once dry and once wet.

23. a) Bimodality should be more apparent, with sample mean and SD closer to the population values.
   b) Closer to Normal, same mean, much smaller SD.

25.
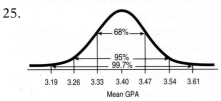

27. a) Given no time trend, the monthly on-time departure rates should be independent. Though not a random sample, these months should be representative, and they're fewer than 10% of all months. The histogram looks unimodal, but slightly left-skewed; not a concern with this large sample.
   b) $80.45 < \mu(\text{OT Departure\%}) < 81.92$
   c) We can be 95% confident that the interval from 80.45% to 81.92% holds the true mean monthly percentage of on-time flight departures.

29. a) Yes. Randomly selected group; less than 10% of the population; the histogram is not unimodal and symmetric, but it is not highly skewed and there are no outliers, so with a sample size of 52, the CLT says $\bar{y}$ is approximately Normal.
   b) (98.1, 98.5) degrees F
   c) We are 95% confident, based on the data, that the average body temperature for an adult is between 98.1°F and 98.5°F.

31. a) H_0: $\mu = 98.6$ b) Two-tail; H_A: $\mu \neq 98.6$
   c) Yes. Randomly selected group; less than 10% of the population; the histogram is not unimodal and symmetric, but it is not highly skewed and there are no outliers, so with a sample size of 52, the CLT says $\bar{y}$ is approximately Normal.
   d) $t = -3.33$; P-value $= 0.0016$

e) With such a low P-value, reject H_0. These data provide strong evidence that the true mean human body temperature is not 98.6°F. It appears to be lower.

33. a) Random sample; the Nearly Normal Condition seems reasonable; the histogram is roughly unimodal and symmetric with no outliers.
   b) (1187.9, 1288.4) chips
   c) Based on this sample, the mean number of chips in an 18-ounce bag is between 1187.9 and 1288.4, with 95% confidence.

35. a) H_0: $\mu = 1000$ b) One-tail; H_A: $\mu > 1000$
   c) Random sample; the Nearly Normal Condition seems reasonable; the histogram is roughly unimodal and symmetric with no outliers.
   d) $t = 10.1$; P-value is barely greater than 0.
   e) With such a low P-value, reject H_0. There's very strong evidence that the bags of cookies average more than 1000 chips.

37. a) H_0: $\mu = 23.3$; H_A: $\mu > 23.3$
   b) We have a random sample of the population. Population may not be normally distributed, as it would be easier to have a few much older men at their first marriage than some very young men. However, with a sample size of 40, $\bar{y}$ should be approximately Normal.
   c) 0.1447
   d) If the average age at first marriage is still 23.3 years, there is a 14.5% chance of getting a sample mean of 24.2 years or older simply from natural sampling variation.
   e) We lack evidence that the average age at first marriage has increased from the mean of 23.3 years.

39. We conclude that there is evidence (P-value 0.0212) that the mean number of cars found on the M25 motorway on Friday the 13th is less than on the previous Friday.

41. a) H_0: $\mu_d = 0$; H_A: $\mu_d > 0$
   b) The price quotes are paired; they were for a random sample of fewer than 10% of the agent's customers; the histogram of differences looks approximately Normal.
   c) $t = 0.83$; P-value $= 0.215$
   d) With a high P-value of 0.215, we don't reject the null hypothesis. These data don't provide evidence that online premiums are lower, on average.

43. a) The histogram is unimodal and slightly skewed to the right, centered at 36 inches with a standard deviation near 4 inches.
   b) All the histograms are centered near 36 inches. As n gets larger, the histograms approach the Normal shape and the variability in the sample means decreases. The histograms are fairly normal by the time the sample reaches size 5.

45. Shape becomes closer to Normal; center does not change; spread becomes narrower.

47. a) Probably a representative sample; the Nearly Normal Condition seems reasonable. The histogram is nearly uniform, with no outliers or skewness.

b) $\bar{y} = 28.78, s = 0.40$ c) $(28.36, 29.21)$ grams

d) Based on this sample, we are 95% confident the average weight of the content of Ruffles bags is between 28.36 and 29.21 grams.

e) The company is erring on the safe side, as it appears that, on average, it is putting in slightly more chips than stated.

49. a) Narrower. A smaller margin of error, so less confident.

b) Advantage: more chance of including the true value. Disadvantage: wider interval.

c) Narrower; due to the larger sample, the SE will be smaller.

51. a) Upper-tail. We want to show it will hold 500 pounds (or more) easily.

b) They will decide the stands are safe when they're not.

c) They will decide the stands are unsafe when they are in fact safe.

53. a) Paired sample test. Data are before/after for the same workers; workers randomly selected; assume fewer than 10% of all this company's workers; boxplot of differences shows them to be symmetric, with no outliers.

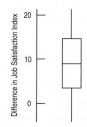

b) $H_0: \mu_D = 0$ vs. $H_A: \mu_D > 0$. $t = 3.60$, P-value $= 0.0029$. Because $P < 0.01$, reject H_0. These data show evidence that average job satisfaction has increased after implementation of the program.

c) Type I

55.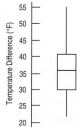

Data are paired for each city; cities are independent of each other; boxplot shows the temperature differences are reasonably symmetric, with no outliers. This is probably not a random sample, so we might be wary of inferring that this difference applies to all European cities. Based on these data, we are 90% confident that the average temperature in European cities in July is between 32.3°F and 41.3°F higher than in January.

Chapter 20

1. It's very unlikely that samples would show an observed difference this large if in fact there is no real difference in the proportions of boys and girls who have used online social networks.

3. The ads may be working. If there had been no real change in name recognition, there'd be only about a 3% chance the percentage of voters who heard of this candidate would be at least this much higher in a different sample.

5. 0.035

7. We are 95% confident, based on these data, that the proportion of pets with a malignant lymphoma in homes where herbicides are used is between 0.36 and 0.49 higher than the proportion of pets with lymphoma in homes where no herbicides are used.

9. The high P-value means that we lack evidence of a difference in mean calorie content.

11. At 90% confidence, plausible values of $\mu_{Meat} - \mu_{Beef}$ are all negative, so the mean fat content is probably higher for beef hot dogs, by between 1.4 and 6.5 g on average.

13. a) False. The confidence interval is about means, not about individual hot dogs.

b) False. The confidence interval is about means, not about individual hot dogs.

c) True.

d) False. CI's based on other samples will also try to estimate the true difference in population means; there's no reason to expect other samples to conform to this result.

e) True.

15. Based on this sample, we are 95% confident that students who learn Math using the CPMP method will score, on average, between 5.57 and 11.43 points better on a test solving applied Algebra problems with a calculator than students who learn by traditional methods.

17. On average, students who learn with the CPMP method do significantly worse on Algebra tests that do not allow them to use calculators than students who learn by traditional methods.

19. 0.25 brands

21. a) Yes. Random sample; less than 10% of the population; samples are independent; more than 10 successes and failures in each sample.

b) $(0.055, 0.140)$

c) We are 95% confident, based on these samples, that the proportion of American women age 65 and older who suffer from arthritis is between 5.5% and 14.0% more than the proportion of American men of the same age who suffer from arthritis.

23. a) Yes, subjects were randomly divided into independent groups, and more than 10 successes and failures were observed in each group.

b) $(4.7\%, 8.9\%)$

c) Yes, we're 95% confident that the rate of infection is 5–9 percentage points lower.

25. a) $(-0.065, 0.221)$
    b) We are 95% confident that the proportion of teens whose parents disapprove of smoking who will eventually smoke is between 22.1% less and 6.5% more than for teens with parents who are lenient about smoking.

27. a) $H_0: p_1 - p_2 = 0$; $H_A: p_1 - p_2 \neq 0$.
    b) Although not a random sample, these people may be representative of and are certainly fewer than 10% of all pregnant women; the two groups are independent; the 94, 3038, 20, and 586 successes and failures are all at least 10.
    c) $z = -0.39$, P-value = 0.6951.
    d) With a P-value this high, we fail to reject H_0. There is no evidence of racial differences in the likelihood of multiple births, based on these data.

29. a) $H_0: p_M - p_F = 0$ $H_A: p_M - p_F \neq 0$
    b) It's a random sample, fewer than 10% of each gender of voter; the two groups are independent; the 246, 227, 235, and 287 successes and failures are all at least 10.
    c) $z = 2.2$; P-value = 0.03
    d) With $z > 2$ we reject H_0; there's evidence of a gender gap in support for this candidate.

31. a) $(1.36, 4.64)$
    b) No; 5 minutes is beyond the high end of the interval.

33.

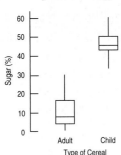

Random sample—questionable, but probably representative, independent samples, less than 10% of all cereals; boxplot shows no outliers—not exactly symmetric, but these are reasonable sample sizes. Based on these samples, with 95% confidence, children's cereals average between 32.49% and 40.80% more sugar content than adults' cereals.

35. a) $H_0: \mu_S - \mu_R = 0$ $H_A: \mu_S - \mu_R > 0$
    b) $t = 9.70$; P-value < 0.0001
    c) We reject the null hypothesis of equal mean distances. There is strong evidence that the mean distance traveled will be greater for golf balls hit off Stinger tees.

37. a) $H_0: \mu_{big} - \mu_{small} = 0$ vs. $H_A: \mu_{big} - \mu_{small} \neq 0$.
    b) With 34.3 df, $t = 2.104$ and P-value = 0.0428.
    c) The low P-value leads us to reject the null hypothesis. There is evidence of a difference in the average amount of ice cream that people scoop when given a bigger bowl.

39. a) $H_0: \mu_2 - \mu_5 = 0$ $H_A: \mu_2 - \mu_5 \neq 0$
    b) Independent Groups Assumption: The runners are different women, so the groups are independent.

The Randomization Condition is satisfied since the runners are selected at random for these heats. Nearly Normal Condition: programs reasonably unimodal and symmetric.
    c) $t = -1.14$, with P = 0.2837.
    d) The P-value is so large that we fail to reject the null hypothesis of equal means and conclude that there is no evidence of a difference in the mean times for runners in unseeded heats.

41. a) Stratified   b) 6% higher among males   c) 4%
    d)

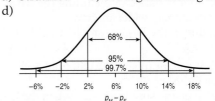

    e) Yes; a poll result showing little difference is only 1–2 standard deviations below the expected outcome.

43. a) No; subjects weren't assigned to treatment groups. It's an observational study.
    b) $H_0: p_1 - p_2 = 0$; $H_A: p_1 - p_2 \neq 0$. $z = 3.56$, P-value = 0.0004. With a P-value this low, we reject H_0. There is a significant difference in the clinic's effectiveness. Younger mothers have a higher birth rate than older mothers. Note that the Success/Failure Condition is not met.
    c) We are 95% confident, based on these data, that the proportion of successful live births at the clinic is between 10.0% and 27.8% higher for mothers under 38 than in those 38 and older. However, the Success/Failure Condition is not met for the older women, since # Successes < 10. We should be cautious in trusting this confidence interval.

45. a) Prospective study
    b) $H_0: p_1 - p_2 = 0$; $H_A: p_1 - p_2 \neq 0$ where p_1 is the proportion of students whose parents disapproved of smoking who became smokers and p_2 is the proportion of students whose parents are lenient about smoking who became smokers.
    c) Yes. We assume the students were randomly selected; they are less than 10% of the population; samples are independent; at least 10 successes and failures in each sample.
    d) $z = -1.17$, P-value = 0.2422. These samples do not show evidence that parental attitudes influence teens' decisions to smoke.
    e) If there is no difference in the proportions, there is about a 24% chance of seeing the observed difference or larger by natural sampling variation.
    f) Type II

47. a) $H_0: \mu_M - \mu_R = 0$ vs. $H_A: \mu_M - \mu_R > 0$. $t = -0.70$, df = 45.88, P-value = 0.7563. Because the P-value is so large, we do not reject H_0. These data provide no evidence that listening to Mozart while studying is better than listening to rap.
    b) With 90% confidence, the average difference in score is between 0.189 and 5.351 objects more for

those who listen to no music while studying, based on these samples.

49. $H_0: \mu_S - \mu_N = 0$ vs. $H_A: \mu_S - \mu_N \neq 0$. $t = -6.08$, df = 213.99, P-value = 5.5×10^{-9}. Because the P-value is low, we reject H_0. These data suggest that ad recall between shows with sexual and neutral content is different; those who saw shows with neutral content had higher average recall. We can be 95% confident that people remember an average of between 0.99 and 1.93 more brands that were advertised.

51.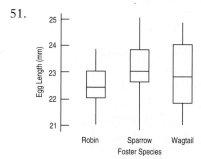

$H_0: \mu_S - \mu_R = 0$ vs. $H_A: \mu_S - \mu_R \neq 0$.

$t = 1.641$, df = 21.60, P-value = 0.115.

Since $P > 0.05$, fail to reject H_0. There is no evidence of a difference in mean cuckoo egg length between robin and sparrow foster parents.

Chapter 21

1. a) H_0: The die is fair; all faces are equally likely to appear.
   b) H_A: The faces are not equally likely to appear.

3. Each face should appear 10 times.

5. We have counts for each face; the rolls are independent and random; we expect at least 5 (10, actually) of each.

7. a) 5 b) 2.5
   c) $0.1 + 0.9 + 0.1 + 2.5 + 0.4 + 1.6 = 5.6$

9. a) There's a 35% chance that a fair die would produce results at least this strange.
   b) There's no evidence this die isn't fair.

11. H_0: Breastfeeding success is independent of having an epidural.
    H_A: There's an association between breastfeeding success and having an epidural.

13. a) 1 b) 159.34
    c) Breastfeeding behavior should be independent for these babies. They are fewer than 10% of all babies; we assume they are representative. We have counts, and all the expected counts are at least 5.

15. a) 5.90
    b) The P-value is very low, so reject the null. There's evidence of an association between having an epidural and subsequent success in breastfeeding.

17. It appears that babies whose mothers had epidurals during childbirth are much less likely to be breastfeeding 6 months later.

19. a) H_0: All numbers 0–9 are equally likely to appear.
       H_A: Some numbers are more likely than others.
    b) 65.4
    c) These are counts; the draws should be independent; these weeks should be representative of all draws, and are fewer than 10% of all possible Pick-3 lotteries. All expected values are at least 5.
    d) 9 df e) $\chi^2 = 6.46$ f) Not significant
    g) It's reasonable to expect this much variability among the numbers drawn; there's no evidence that the digits are not equally likely.

21. a) H_0: Cans of this nut mix contain 5 kinds of nuts in the proportions stated by the company. H_A: The proportions of the 5 kinds of nuts are different from what the company says.
    b) 101.9 Brazil nuts, 203.8 cashews, 203.8 almonds, 101.9 hazelnuts, 407.6 peanuts
    c) These are counts of nuts; this can is independent of the others and contains a representative sample of less than 10% of all nuts produced by this company; all expected values are at least 5.
    d) 4 df e) $\chi^2 = 16.16$
    f) Statistically significant
    g) There's strong evidence that the proportions stated by the company are not correct; it appears these mixes contain more than 40% peanuts.

23. a) 40.2% b) 8.1% c) 62.2% d) 285.48
    e) H_0: Survival was independent of status on the ship.
       H_A: Survival depended on the status.
    f) 3
    g) We reject the null hypothesis. Survival depended on status. We can see that first-class passengers were more likely to survive than passengers of any other class.

25. First-class passengers were most likely to survive, while third-class passengers and crew were underrepresented among the survivors.

27. a) Experiment—actively imposed treatments (different drinks)
    b) H_0: The rate of urinary tract infection is the same for all three groups. H_A: The rate of urinary tract infection is different among the groups.
    c) Count data; random assignment to treatments; all expected frequencies larger than 5.
    d) 2 e) $\chi^2 = 7.776$, P-value = 0.020.
    f) With a P-value this low, we reject H_0. These data provide reasonably strong evidence that there is a difference in urinary tract infection rates between cranberry juice drinkers, lactobacillus drinkers, and the control group. It appears those who drank cranberry juice were less likely to develop urinary tract infections; those who drank lactobacillus were more likely to have infections.

29. a) H_0: *Political Affiliation* is independent of *Sex*.
       H_A: There is a relationship between *Political Affiliation* and *Sex*.
    b) Counted data; probably a random sample, but can't extend results to other states; all expected frequencies greater than 5.
    c) 2 df d) $\chi^2 = 4.851$, df = 2, P-value = 0.0884.

e) Because of the high P-value, we do not reject H_0. These data do not provide evidence of a relationship between *Political Affiliation* and *Sex*.

31. a) H_0: *Political Affiliation* is independent of *Region*. H_A: There is a relationship between *Political Affiliation* and *Region*.
   b) We have counts. The individuals are a random sample of fewer than 10% of all Montana voters; all expected counts are at least 5.
   c) $\chi^2 = 13.849$, df = 4, P-value = 0.0078.
   d) With a P-value this low, we reject H_0. *Political Affiliation* and *Region* are related. It appears that those in the West are more likely to be Democrat than Republican; those in the Northeast are more likely to be Republican than Democrat.

33. H_0: Traits occur in the predicted 9:3:3:1 ratio. H_A: The actual ratio is not as predicted. We have counts. We assume these 100 flies are representative of all fruit flies. They're independent and fewer than 10% of all flies. Expected counts are all at least 5. $\chi^2 = 5.671$, df = 3, P-value = 0.1288. With a P-value this high, we fail to reject H_0. Yes, these data are consistent with those predicted by genetic theory.

35. $\chi^2 = 14.058$, df = 1, P-value = 0.0002. With a P-value this low, we reject H_0. There is evidence of racial steering. Blacks are much less likely to rent in Section A than Section B.

37. $\chi^2 = 5.89$, df = 3, P = 0.117. Because the P-value is >0.05, these data show no evidence of an association between the mother's age group and the outcome of the pregnancy.

39. a) H_0: The grade distribution is the same for both professors. H_A: The grade distributions are different.
   b)

	Prof. Alpha	Prof. Beta
A	6.667	5.333
B	12.778	10.222
C	12.222	9.778
D	6.111	4.889
E	2.222	1.778

   Three cells have expected frequencies less than 5.
   c)

	Prof. Alpha	Prof. Beta
A	6.667	5.333
B	12.778	10.222
C	12.222	9.778
Below C	8.333	6.667

   All expected frequencies are now larger than 5.
   d) Decreased from 4 to 3.
   e) $\chi^2 = 9.306$, P-value = 0.0255. Because the P-value is so low, we reject H_0. The grade distributions for the two professors are different. Prof. Alpha gives fewer A's and more grades below C than Prof. Beta.

Photo Acknowledgments

Page 244 Michael Pettigrew/iStockphoto **Page 245 (Dilbert)** Dilbert copyright © 2010 by Scott Adams/Distributed by United Feature Syndicate **Page 245 (dogs)** Digital Vision **Page 246** Photodisc **Page 248** XKCD.com **Page 251** Galinka/Shutterstock **Page 252** Wizard of Id copyright © 2010 by Parker and Hart/Distributed by Creators Syndicate **Page 253** Vuk Nenezic/Shutterstock **Page 255 (dog food dish)** Jostein Hauge/iStockphoto **Page 255 (medical researcher)** AbleStock/Thinkstock **Page 256** L. Brian Stauffer/University of Illinois News Bureau **Page 257** Courtesy Beth Anderson **Page 265** PhotoDisc **Page 266** Dilbert copyright © 2010 by Scott Adams/Distributed by United Feature Syndicate **Page 267** Nano/iStockphotos **Page 268 (high school band)** Oleg Prikhodko/iStockphoto **Page 268 (math problem)** Andresr/Shutterstock **Page 269** Anastasios Kandris/Shutterstock **Page 269 (LeBron James)** Matthew Healey/UPI/Newscom **Page 269 (Serena Williams)** Reuters/Corbis **Page 269 (David Beckham)** Getty Sports **Page 271** Malerapaso/iStockphoto **Page 272** Jim McIsaac/Getty Images **Page 273** Henrik Larsson/iStockphotos **Page 274** Andrew Rich/iStockphotos **Page 277** Ana de Sousa/Shutterstock **Page 278** PhotoDisc **Page 284** Jessica Bethke/Shutterstock **Page 287 (Keno)** PhotoEdit **Page 287 (lotto balls)** Creative11/Dreamstime **Page 287 (roulette)** PhotoDisc Red **Page 288** Skyline/Shutterstock **Page 289** Olga Koronevska/iStockphotos **Page 291 (log in)** Hermann Liesenfeld/iStockphoto **Page 291 (track)** Shawn Pecor/Shutterstock **Page 292 (exclamation point)** SerhioGrey/Shutterstock **Page 292 (iPod at the beach)** Plustwentyseven/Thinkstock **Page 293** Feng Yu/iStockphoto **Page 294 (county legislature)** Zuma Press/Newscom **Page 294 (dorm room)** James Woodson/Thinkstock **Page 295** Brent Holland/iStockphoto **Page 299** Jessica Bethke/Shutterstock **Page 305** David H. Lewis/Istockphoto **Page 306** University of St. Andrews MacTutor History of Mathematics Archive **Page 307** Jessica Bethke/Shutterstock **Page 308** ChrisHepburn/iStockphoto **Page 309** Ilja Mašík/Shutterstock **Page 312** Yuri Arcurs/Shutterstock **Page 313** Leland Bobbe/Photonica/Getty Images **Page 314** David H. Lewis/iStockphoto **Page 318** CNN/Getty Images **Page 319** Corbis Royalty Free **Page 320 (money)** US Dept of Treasury **Page 320 (college cafeteria food)** Otokimus/Dreamstime **Page 321** Lisa F. Young/iStockphoto **Page 323** Pearson **Page 325** Joshua Hodge Photography/iStockphoto **Page 327** Randy Miramontez/Alamy **Page 328** Exactostock/SuperStock **Page 330** Huguette Roe/Shutterstock **Page 331** Bee-nana/Dreamstime **Page 332** Evgeny Murtola/Shutterstock **Page 335** Stock Vector/iStockphotos **Page 336** Corbis Royalty Free **Page 342** Digital Vision **Page 345** Sean Locke/iStockphoto **Page 346** Nuno Silva/iStockphoto **Page 347** Erich Schlegel/Dallas Morning News/Newscom **Page 348** Mikkel William Nielsen/iStockphoto **Page 349 (LeBron James)** Matthew Healey/UPI/Newscom **Page 349 (Daniel Bernoulli)** St. Andrews University History of Mathematics Archive **Page 349 (Calvin and Hobbes)** Calvin and Hobbes copyright © 2010 Bill Watterson/Distributed by Universal Uclick **Page 350** Tatiana Popova/iStockphoto **Page 351** Ene/iStockphoto **Page 352** Roman Okopny/iStockphoto **Page 353** Vladislav Mitic/iStockphoto **Page 354** Abimelec Olan/iStockphoto **Page 355 (phone)** Mark Stay/iStockphoto **Page 355 (blood bank with patients)** PhotoDisc **Page 357 (annoying phone call)** Marilyn Nieves/iStockphoto **Page 357 (coin tossing)** Bryan Myhr/iStockphoto **Page 358** Henrik Jonsson/iStockphoto **Page 360** Twitter/Newscom **Page 361** Digital Vision **Page 370** Dejan Lazervic/Shutterstock **Page 374 (coin flip)** Ronstik/Shutterstock **Page 374 (giving blood saves lives)** Valerie Loiseleux/iStockphoto **Page 375** PunchStock **Page 377** Joshua Wanyama/Dreamstime **Page 378** Chaoss/Dreamstime **Page 379** PRNewsFoto/Newsweek/AP Images **Page 381** Goldenangel/Shutterstock **Page 382** Garfield copyright © 2010 by Jim Davis/Distributed by Universal Uclick **Page 383** Erlend Kvalsvik/iStockphoto **Page 384 (crowd of people)** Lawrence Sawyer/iStockphoto **Page 384 (polling)** ZF/Shutterstock **Page 386** Yuri Arcurs/Shutterstock **Page 387** Dejan Lazervic/Shutterstock **Page 396** Adam Woolfitt/Corbis **Page 398** XKCD.com **Page 399 (teen taking drivers test)** Lisa F. Young/Shutterstock **Page 399 (allergies)** Trent Chambers/iStockphotos **Page 400** Lisafx/Dreamstime **Page 401** Clint Scholz/iStockphoto **Page 402 (cartoon sneezing)** Lorelyn Medina/Shutterstock **Page 402 (teen with driver's license)** Lisafx/Dreamstime **Page 403 (allergy relief)** Aldo Murillo/iStockphoto **Page 403 (scoreboard)** Mark Rose/iStockphoto **Page 406 (DMV sign)** Rachel Epstein/PhotoEdit **Page 406 (placebo)** Ugurhan Betin/iStockphoto **Page 407** Idreamphoto/Shutterstock **Page 410** Cmsphoto/Newscom

Page 411 (football injury) George Holland/Cal Sport Media/Newscom **Page 411 (metal casting)** Adam Woolfitt/Corbis **Page 419** Digital Vision/Getty Images **Page 422** Robert Crow/Shutterstock **Page 423 (pregnant women)** Blend Images/SuperStock **Page 423 (crowded elevator)** Keith Brofsky/Photodisc/Thinkstock **Page 425** University of York Department of Mathematics **Page 427** John Angerson/Alamy **Page 428** Rudy Umans/Shutterstock **Page 431** Prism68/Shutterstock **Page 433** AVTG/iStockphotos **Page 435 (strong battery)** Yaroslav Manyuk/iStockphoto **Page 435 (speed skaters)** Yuri Kadobnov/AFP/Getty Images **Page 436** Courtesy of authors **Page 438** Stockbyte/Thinkstock **Page 443** Public Domain **Page 444** Bonnie jacobs/iStockphoto **Page 445** Digital Vision/Getty Images **Page 457** Digital Vision/Getty Images **Page 458** MGM/The Kobal Collection **Page 459** All Canada Photos/SuperStock **Page 460** Christopher Futcher/iStockphoto **Page 461 (social networking diagram)** CurvaBezier/iStockphoto **Page 461 (teen using seatbelt)** Jupiter Images/Getty Images/Thinkstock **Page 464 (PBS logo)** KRT Kids Elements/Newscom **Page 464 (man sleeping)** PhotoDisc/Getty Images **Page 467** Joerg Habermeier/iStockphoto **Page 468** Jason Lugo/iStockphoto **Page 469** Julija Sapic/iStockphoto **Page 470** Juanmonino/iStockphoto **Page 472** Jamie Grill/Corbis **Page 475** Jennifer Gilroy/Shutterstock **Page 477** Aldo Murillo/iStockphoto **Page 479** Daniel Villeneuve/iStockphoto **Page 480** Photo collage by Beth Anderson **Page 481** Sean Locke/iStockphoto **Page 482** Digital Vision/Getty Images **Page 493** Royalty Free/Corbis **Page 495** Linda Kloosterhof/iStockphoto **Page 496 (baseball player)** Richard Paul Kane/Shutterstock **Page 496 (earlobes)** Malerapaso/iStockphoto; MorePixels/iStockphoto **Page 498** Baloncici/iStockphoto **Page 503** Richard Becker/FLPA/Photolibrary New York **Page 504** Imagestate Media Partners Limited/Impact Photos/Alamy **Page 506** PhotoDisc/Getty Images **Page 508** Rob Byron/Shutterstock **Page 509** Royalty Free/Corbis **Images of students appearing in some elements of the design** Patrick Hermans/Shutterstock, Anton Albert/Shutterstock, Lev Olkha/Shutterstock, Szefei/Shutterstock **Icon art** Pedro Tavares/Shutterstock, SkillUp/Shutterstock

Index

Note: Page numbers in **boldface** indicate chapter-level topics; page numbers in *italics* indicate definitions; FE indicates For Example references.

Index of TI Tips

Tables

Row					TABLE OF RANDOM DIGITS					
1	96299	07196	98642	20639	23185	56282	69929	14125	38872	94168
2	71622	35940	81807	59225	18192	08710	80777	84395	69563	86280
3	03272	41230	81739	74797	70406	18564	69273	72532	78340	36699
4	46376	58596	14365	63685	56555	42974	72944	96463	63533	24152
5	47352	42853	42903	97504	56655	70355	88606	61406	38757	70657
6	20064	04266	74017	79319	70170	96572	08523	56025	89077	57678
7	73184	95907	05179	51002	83374	52297	07769	99792	78365	93487
8	72753	36216	07230	35793	71907	65571	66784	25548	91861	15725
9	03939	30763	06138	80062	02537	23561	93136	61260	77935	93159
10	75998	37203	07959	38264	78120	77525	86481	54986	33042	70648
11	94435	97441	90998	25104	49761	14967	70724	67030	53887	81293
12	04362	40989	69167	38894	00172	02999	97377	33305	60782	29810
13	89059	43528	10547	40115	82234	86902	04121	83889	76208	31076
14	87736	04666	75145	49175	76754	07884	92564	80793	22573	67902
15	76488	88899	15860	07370	13431	84041	69202	18912	83173	11983
16	36460	53772	66634	25045	79007	78518	73580	14191	50353	32064
17	13205	69237	21820	20952	16635	58867	97650	82983	64865	93298
18	51242	12215	90739	36812	00436	31609	80333	96606	30430	31803
19	67819	00354	91439	91073	49258	15992	41277	75111	67496	68430
20	09875	08990	27656	15871	23637	00952	97818	64234	50199	05715
21	18192	95308	72975	01191	29958	09275	89141	19558	50524	32041
22	02763	33701	66188	50226	35813	72951	11638	01876	93664	37001
23	13349	46328	01856	29935	80563	03742	49470	67749	08578	21956
24	69238	92878	80067	80807	45096	22936	64325	19265	37755	69794
25	92207	63527	59398	29818	24789	94309	88380	57000	50171	17891
26	66679	99100	37072	30593	29665	84286	44458	60180	81451	58273
27	31087	42430	60322	34765	15757	53300	97392	98035	05228	68970
28	84432	04916	52949	78533	31666	62350	20584	56367	19701	60584
29	72042	12287	21081	48426	44321	58765	41760	43304	13399	02043
30	94534	73559	82135	70260	87936	85162	11937	18263	54138	69564
31	63971	97198	40974	45301	60177	35604	21580	68107	25184	42810
32	11227	58474	17272	37619	69517	62964	67962	34510	12607	52255
33	28541	02029	08068	96656	17795	21484	57722	76511	27849	61738
34	11282	43632	49531	78981	81980	08530	08629	32279	29478	50228
35	42907	15137	21918	13248	39129	49559	94540	24070	88151	36782
36	47119	76651	21732	32364	58545	50277	57558	30390	18771	72703
37	11232	99884	05087	76839	65142	19994	91397	29350	83852	04905
38	64725	06719	86262	53356	57999	50193	79936	97230	52073	94467
39	77007	26962	55466	12521	48125	12280	54985	26239	76044	54398
40	18375	19310	59796	89832	59417	18553	17238	05474	33259	50595

TABLE Z	Second decimal place in z										
	0.09	0.08	0.07	0.06	0.05	0.04	0.03	0.02	0.01	0.00	z
	0.0001	0.0001	0.0001	0.0001	0.0001	0.0001	0.0001	0.0001	0.0001	0.0001	−3.8
	0.0001	0.0001	0.0001	0.0001	0.0001	0.0001	0.0001	0.0001	0.0001	0.0001	−3.7
	0.0001	0.0001	0.0001	0.0001	0.0001	0.0001	0.0001	0.0001	0.0002	0.0002	−3.6
	0.0002	0.0002	0.0002	0.0002	0.0002	0.0002	0.0002	0.0002	0.0002	0.0002	−3.5
	0.0002	0.0003	0.0003	0.0003	0.0003	0.0003	0.0003	0.0003	0.0003	0.0003	−3.4
	0.0003	0.0004	0.0004	0.0004	0.0004	0.0004	0.0004	0.0005	0.0005	0.0005	−3.3
	0.0005	0.0005	0.0005	0.0006	0.0006	0.0006	0.0006	0.0006	0.0007	0.0007	−3.2
	0.0007	0.0007	0.0008	0.0008	0.0008	0.0008	0.0009	0.0009	0.0009	0.0010	−3.1
	0.0010	0.0010	0.0011	0.0011	0.0011	0.0012	0.0012	0.0013	0.0013	0.0013	−3.0
	0.0014	0.0014	0.0015	0.0015	0.0016	0.0016	0.0017	0.0018	0.0018	0.0019	−2.9
	0.0019	0.0020	0.0021	0.0021	0.0022	0.0023	0.0023	0.0024	0.0025	0.0026	−2.8
	0.0026	0.0027	0.0028	0.0029	0.0030	0.0031	0.0032	0.0033	0.0034	0.0035	−2.7
	0.0036	0.0037	0.0038	0.0039	0.0040	0.0041	0.0043	0.0044	0.0045	0.0047	−2.6
	0.0048	0.0049	0.0051	0.0052	0.0054	0.0055	0.0057	0.0059	0.0060	0.0062	−2.5
	0.0064	0.0066	0.0068	0.0069	0.0071	0.0073	0.0075	0.0078	0.0080	0.0082	−2.4
	0.0084	0.0087	0.0089	0.0091	0.0094	0.0096	0.0099	0.0102	0.0104	0.0107	−2.3
	0.0110	0.0113	0.0116	0.0119	0.0122	0.0125	0.0129	0.0132	0.0136	0.0139	−2.2
	0.0143	0.0146	0.0150	0.0154	0.0158	0.0162	0.0166	0.0170	0.0174	0.0179	−2.1
	0.0183	0.0188	0.0192	0.0197	0.0202	0.0207	0.0212	0.0217	0.0222	0.0228	−2.0
	0.0233	0.0239	0.0244	0.0250	0.0256	0.0262	0.0268	0.0274	0.0281	0.0287	−1.9
	0.0294	0.0301	0.0307	0.0314	0.0322	0.0329	0.0336	0.0344	0.0351	0.0359	−1.8
	0.0367	0.0375	0.0384	0.0392	0.0401	0.0409	0.0418	0.0427	0.0436	0.0446	−1.7
	0.0455	0.0465	0.0475	0.0485	0.0495	0.0505	0.0516	0.0526	0.0537	0.0548	−1.6
	0.0559	0.0571	0.0582	0.0594	0.0606	0.0618	0.0630	0.0643	0.0655	0.0668	−1.5
	0.0681	0.0694	0.0708	0.0721	0.0735	0.0749	0.0764	0.0778	0.0793	0.0808	−1.4
	0.0823	0.0838	0.0853	0.0869	0.0885	0.0901	0.0918	0.0934	0.0951	0.0968	−1.3
	0.0985	0.1003	0.1020	0.1038	0.1056	0.1075	0.1093	0.1112	0.1131	0.1151	−1.2
	0.1170	0.1190	0.1210	0.1230	0.1251	0.1271	0.1292	0.1314	0.1335	0.1357	−1.1
	0.1379	0.1401	0.1423	0.1446	0.1469	0.1492	0.1515	0.1539	0.1562	0.1587	−1.0
	0.1611	0.1635	0.1660	0.1685	0.1711	0.1736	0.1762	0.1788	0.1814	0.1841	−0.9
	0.1867	0.1894	0.1922	0.1949	0.1977	0.2005	0.2033	0.2061	0.2090	0.2119	−0.8
	0.2148	0.2177	0.2206	0.2236	0.2266	0.2296	0.2327	0.2358	0.2389	0.2420	−0.7
	0.2451	0.2483	0.2514	0.2546	0.2578	0.2611	0.2643	0.2676	0.2709	0.2743	−0.6
	0.2776	0.2810	0.2843	0.2877	0.2912	0.2946	0.2981	0.3015	0.3050	0.3085	−0.5
	0.3121	0.3156	0.3192	0.3228	0.3264	0.3300	0.3336	0.3372	0.3409	0.3446	−0.4
	0.3483	0.3520	0.3557	0.3594	0.3632	0.3669	0.3707	0.3745	0.3783	0.3821	−0.3
	0.3859	0.3897	0.3936	0.3974	0.4013	0.4052	0.4090	0.4129	0.4168	0.4207	−0.2
	0.4247	0.4286	0.4325	0.4364	0.4404	0.4443	0.4483	0.4522	0.4562	0.4602	−0.1
	0.4641	0.4681	0.4721	0.4761	0.4801	0.4840	0.4880	0.4920	0.4960	0.5000	−0.0

Areas under the standard normal curve

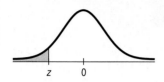

z   0

For $z \leq -3.90$, the areas are 0.0000 to four decimal places.

	Second decimal place in z									
TABLE Z (cont.)										
Areas under the standard normal curve										
z	*0.00*	*0.01*	*0.02*	*0.03*	*0.04*	*0.05*	*0.06*	*0.07*	*0.08*	*0.09*
0.0	0.5000	0.5040	0.5080	0.5120	0.5160	0.5199	0.5239	0.5279	0.5319	0.5359
0.1	0.5398	0.5438	0.5478	0.5517	0.5557	0.5596	0.5636	0.5675	0.5714	0.5753
0.2	0.5793	0.5832	0.5871	0.5910	0.5948	0.5987	0.6026	0.6064	0.6103	0.6141
0.3	0.6179	0.6217	0.6255	0.6293	0.6331	0.6368	0.6406	0.6443	0.6480	0.6517
0.4	0.6554	0.6591	0.6628	0.6664	0.6700	0.6736	0.6772	0.6808	0.6844	0.6879
0.5	0.6915	0.6950	0.6985	0.7019	0.7054	0.7088	0.7123	0.7157	0.7190	0.7224
0.6	0.7257	0.7291	0.7324	0.7357	0.7389	0.7422	0.7454	0.7486	0.7517	0.7549
0.7	0.7580	0.7611	0.7642	0.7673	0.7704	0.7734	0.7764	0.7794	0.7823	0.7852
0.8	0.7881	0.7910	0.7939	0.7967	0.7995	0.8023	0.8051	0.8078	0.8106	0.8133
0.9	0.8159	0.8186	0.8212	0.8238	0.8264	0.8289	0.8315	0.8340	0.8365	0.8389
1.0	0.8413	0.8438	0.8461	0.8485	0.8508	0.8531	0.8554	0.8577	0.8599	0.8621
1.1	0.8643	0.8665	0.8686	0.8708	0.8729	0.8749	0.8770	0.8790	0.8810	0.8830
1.2	0.8849	0.8869	0.8888	0.8907	0.8925	0.8944	0.8962	0.8980	0.8997	0.9015
1.3	0.9032	0.9049	0.9066	0.9082	0.9099	0.9115	0.9131	0.9147	0.9162	0.9177
1.4	0.9192	0.9207	0.9222	0.9236	0.9251	0.9265	0.9279	0.9292	0.9306	0.9319
1.5	0.9332	0.9345	0.9357	0.9370	0.9382	0.9394	0.9406	0.9418	0.9429	0.9441
1.6	0.9452	0.9463	0.9474	0.9484	0.9495	0.9505	0.9515	0.9525	0.9535	0.9545
1.7	0.9554	0.9564	0.9573	0.9582	0.9591	0.9599	0.9608	0.9616	0.9625	0.9633
1.8	0.9641	0.9649	0.9656	0.9664	0.9671	0.9678	0.9686	0.9693	0.9699	0.9706
1.9	0.9713	0.9719	0.9726	0.9732	0.9738	0.9744	0.9750	0.9756	0.9761	0.9767
2.0	0.9772	0.9778	0.9783	0.9788	0.9793	0.9798	0.9803	0.9808	0.9812	0.9817
2.1	0.9821	0.9826	0.9830	0.9834	0.9838	0.9842	0.9846	0.9850	0.9854	0.9857
2.2	0.9861	0.9864	0.9868	0.9871	0.9875	0.9878	0.9881	0.9884	0.9887	0.9890
2.3	0.9893	0.9896	0.9898	0.9901	0.9904	0.9906	0.9909	0.9911	0.9913	0.9916
2.4	0.9918	0.9920	0.9922	0.9925	0.9927	0.9929	0.9931	0.9932	0.9934	0.9936
2.5	0.9938	0.9940	0.9941	0.9943	0.9945	0.9946	0.9948	0.9949	0.9951	0.9952
2.6	0.9953	0.9955	0.9956	0.9957	0.9959	0.9960	0.9961	0.9962	0.9963	0.9964
2.7	0.9965	0.9966	0.9967	0.9968	0.9969	0.9970	0.9971	0.9972	0.9973	0.9974
2.8	0.9974	0.9975	0.9976	0.9977	0.9977	0.9978	0.9979	0.9979	0.9980	0.9981
2.9	0.9981	0.9982	0.9982	0.9983	0.9984	0.9984	0.9985	0.9985	0.9986	0.9986
3.0	0.9987	0.9987	0.9987	0.9988	0.9988	0.9989	0.9989	0.9989	0.9990	0.9990
3.1	0.9990	0.9991	0.9991	0.9991	0.9992	0.9992	0.9992	0.9992	0.9993	0.9993
3.2	0.9993	0.9993	0.9994	0.9994	0.9994	0.9994	0.9994	0.9995	0.9995	0.9995
3.3	0.9995	0.9995	0.9995	0.9996	0.9996	0.9996	0.9996	0.9996	0.9996	0.9997
3.4	0.9997	0.9997	0.9997	0.9997	0.9997	0.9997	0.9997	0.9997	0.9997	0.9998
3.5	0.9998	0.9998	0.9998	0.9998	0.9998	0.9998	0.9998	0.9998	0.9998	0.9998
3.6	0.9998	0.9998	0.9999	0.9999	0.9999	0.9999	0.9999	0.9999	0.9999	0.9999
3.7	0.9999	0.9999	0.9999	0.9999	0.9999	0.9999	0.9999	0.9999	0.9999	0.9999
3.8	0.9999	0.9999	0.9999	0.9999	0.9999	0.9999	0.9999	0.9999	0.9999	0.9999

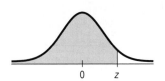

For *z* ≥ 3.90, the areas are 1.0000 to four decimal places.

		Two tail probability	0.20	0.10	0.05	0.02	0.01		
		One tail probability	0.10	0.05	0.025	0.01	0.005		
		df						df	
TABLE T		1	3.078	6.314	12.706	31.821	63.657	1	
Values of t_α		2	1.886	2.920	4.303	6.965	9.925	2	
		3	1.638	2.353	3.182	4.541	5.841	3	
		4	1.533	2.132	2.776	3.747	4.604	4	
		5	1.476	2.015	2.571	3.365	4.032	5	
		6	1.440	1.943	2.447	3.143	3.707	6	
		7	1.415	1.895	2.365	2.998	3.499	7	
		8	1.397	1.860	2.306	2.896	3.355	8	
		9	1.383	1.833	2.262	2.821	3.250	9	
		10	1.372	1.812	2.228	2.764	3.169	10	
		11	1.363	1.796	2.201	2.718	3.106	11	
		12	1.356	1.782	2.179	2.681	3.055	12	
		13	1.350	1.771	2.160	2.650	3.012	13	
		14	1.345	1.761	2.145	2.624	2.977	14	
		15	1.341	1.753	2.131	2.602	2.947	15	
		16	1.337	1.746	2.120	2.583	2.921	16	
		17	1.333	1.740	2.110	2.567	2.898	17	
		18	1.330	1.734	2.101	2.552	2.878	18	
		19	1.328	1.729	2.093	2.539	2.861	19	
		20	1.325	1.725	2.086	2.528	2.845	20	
		21	1.323	1.721	2.080	2.518	2.831	21	
		22	1.321	1.717	2.074	2.508	2.819	22	
		23	1.319	1.714	2.069	2.500	2.807	23	
		24	1.318	1.711	2.064	2.492	2.797	24	
		25	1.316	1.708	2.060	2.485	2.787	25	
		26	1.315	1.706	2.056	2.479	2.779	26	
		27	1.314	1.703	2.052	2.473	2.771	27	
		28	1.313	1.701	2.048	2.467	2.763	28	
		29	1.311	1.699	2.045	2.462	2.756	29	
		30	1.310	1.697	2.042	2.457	2.750	30	
		32	1.309	1.694	2.037	2.449	2.738	32	
		35	1.306	1.690	2.030	2.438	2.725	35	
		40	1.303	1.684	2.021	2.423	2.704	40	
		45	1.301	1.679	2.014	2.412	2.690	45	
		50	1.299	1.676	2.009	2.403	2.678	50	
		60	1.296	1.671	2.000	2.390	2.660	60	
		75	1.293	1.665	1.992	2.377	2.643	75	
		100	1.290	1.660	1.984	2.364	2.626	100	
		120	1.289	1.658	1.980	2.358	2.617	120	
		140	1.288	1.656	1.977	2.353	2.611	140	
		180	1.286	1.653	1.973	2.347	2.603	180	
		250	1.285	1.651	1.969	2.341	2.596	250	
		400	1.284	1.649	1.966	2.336	2.588	400	
		1000	1.282	1.646	1.962	2.330	2.581	1000	
		∞	1.282	1.645	1.960	2.326	2.576	∞	
	Confidence levels		80%	90%	95%	98%	99%		

Two tails

$-t_{\alpha/2}$ 0 $t_{\alpha/2}$

One tail

0   t_α

Right tail probability		0.10	0.05	0.025	0.01	0.005
TABLE χ Values of χ_α^2	df					
	1	2.706	3.841	5.024	6.635	7.879
	2	4.605	5.991	7.378	9.210	10.597
	3	6.251	7.815	9.348	11.345	12.838
	4	7.779	9.488	11.143	13.277	14.860
	5	9.236	11.070	12.833	15.086	16.750
	6	10.645	12.592	14.449	16.812	18.548
	7	12.017	14.067	16.013	18.475	20.278
	8	13.362	15.507	17.535	20.090	21.955
	9	14.684	16.919	19.023	21.666	23.589
	10	15.987	18.307	20.483	23.209	25.188
	11	17.275	19.675	21.920	24.725	26.757
	12	18.549	21.026	23.337	26.217	28.300
	13	19.812	22.362	24.736	27.688	29.819
	14	21.064	23.685	26.119	29.141	31.319
	15	22.307	24.996	27.488	30.578	32.801
	16	23.542	26.296	28.845	32.000	34.267
	17	24.769	27.587	30.191	33.409	35.718
	18	25.989	28.869	31.526	34.805	37.156
	19	27.204	30.143	32.852	36.191	38.582
	20	28.412	31.410	34.170	37.566	39.997
	21	29.615	32.671	35.479	38.932	41.401
	22	30.813	33.924	36.781	40.290	42.796
	23	32.007	35.172	38.076	41.638	44.181
	24	33.196	36.415	39.364	42.980	45.559
	25	34.382	37.653	40.647	44.314	46.928
	26	35.563	38.885	41.923	45.642	48.290
	27	36.741	40.113	43.195	46.963	49.645
	28	37.916	41.337	44.461	48.278	50.994
	29	39.087	42.557	45.722	59.588	52.336
	30	40.256	43.773	46.979	50.892	53.672
	40	51.805	55.759	59.342	63.691	66.767
	50	63.167	67.505	71.420	76.154	79.490
	60	74.397	79.082	83.298	88.381	91.955
	70	85.527	90.531	95.023	100.424	104.213
	80	96.578	101.879	106.628	112.328	116.320
	90	107.565	113.145	118.135	124.115	128.296
	100	118.499	124.343	129.563	135.811	140.177